T0377100

Differential Equations

A Problem Solving Approach Based on MATLAB®

Differential Equations

A Problem Solving Approach Based on MATLAB®

P. Mohana Shankar

Drexel University

Department of Electrical & Computer Engineering

Philadelphia, Pennsylvania, USA

CRC Press is an imprint of the
Taylor & Francis Group, an **informa** business

A SCIENCE PUBLISHERS BOOK

MATLAB® and Simulink® are trademarks of The MathWorks, Inc. and are used with permission. The MathWorks does not warrant the accuracy of the text or exercises in this book. This book's use or discussion of MATLAB® and Simulink® software or related products does not constitute endorsement or sponsorship by The MathWorks of a particular pedagogical approach or particular use of the MATLAB® and Simulink® software.

CRC Press
Taylor & Francis Group
6000 Broken Sound Parkway NW, Suite 300
Boca Raton, FL 33487-2742

Printed on acid-free paper
Version Date: 20171219

International Standard Book Number-13: 978-1-1385-0160-7 (Hardback)

Library of Congress Cataloging-in-Publication Data

Names: Shankar, P. M., author.
Title: Differential equations : a problem solving approach based on MATLAB / P. Mohana Shankar, Drexel University, Department of Electrical & Computer Engineering, Philadelphia, Pennsylvania, USA.
Description: Boca Raton, FL : CRC Press, Taylor & Francis Group, 2018. | "A science publishers book." | Includes bibliographical references and index.
Identifiers: LCCN 2017057524 | ISBN 9781138501607 (hardback : alk. paper)
Subjects: LCSH: Differential equations--Numerical solutions--Data processing. | MATLAB.
Classification: LCC QA371.5.D37 S45 2018 | DDC 515/.350285536--dc23
LC record available at https://lccn.loc.gov/2017057524

Visit the Taylor & Francis Web site at
http://www.taylorandfrancis.com

and the CRC Press Web site at
http://www.crcpress.com

Dedicated to my parents, Padmanabharao and Kanakabai, who were school teachers

Preface

This book grew out of instructional material and notes the author created while teaching courses in Linear Algebra and Differential Equations to undergraduate engineering students. Starting from simple demos prepared for instructing students, all aspects associated with the first order differential equations, linear second and higher order differential equations with constant coefficients and coupled first order differential equations were incorporated into several modules for obtaining solutions to any exercise. Taking the pulse of the students to determine what they felt was essential for their understanding, these modules were constantly updated and put in to a final from where they display solutions to differential equations with a simple input in textual form. The solutions are displayed as figures such that they provide theoretical aspects, explanations, step-by-step generation of solutions, solutions using multiple approaches, including verification through numerical techniques. The examples in the book capture this holistic approach to problem solving. MATLAB® (version 2016a) with its symbolic toolbox provided the appropriate software for the creation of the solutions in this format.

The book provides theoretical background containing a large number of examples, far exceeding what is seen in typical textbooks and a substantial number of exercises. The goal of the book has been the creation of materials that provide the theory, ample examples to cover the diverse theoretical concepts in each chapter, and multiple ways of solving problems providing the pedagogy in terms of verification of solutions, etc. The solutions manual is available separately.

This book has been a family project with the support and encouragement from my wife Raja and daughter Raji. Their assistance in the preparation of materials was crucial to the timely completion of the project. I want to thank my editor Mr. Vijay Primlani for his support during this endeavor. The support provided by the staff at CRC press is also greatly appreciated.

Contents

Dedication v

Preface vii

1. Introduction 1
2. First Order Differential Equations 5
3. Linear Second Order Differential Equations with Constant Coefficients 81
4. Linear Higher Order Differential Equations with Constant Coefficients 181
5. First Order Coupled Differential Equations with Constant Coefficients 236

Appendices 383

Suggested Readings 445

Index 447

CHAPTER 1

Introduction

In the sciences, engineering, social sciences, business and other areas that impact our lives, changes take place continuously. This dynamic behavior can be easily demonstrated, studied and critically analyzed through modeling carried out by examining how changes take place in time. This time dependent or temporal behavior leads us to the notion of derivatives and differential equations. Therefore, a study of differential equations, their solutions, and interpretation of the solutions is essential for the proper understanding of systems operating in all walks of life.

Consider a phenomenon where the observable quantity is identified by y. If changes are taking place in time, it is obvious that **y** will be a function of time **t**, expressed as y(t). If we take the first order derivative of **y** (differentiation once with respect to the independent variable **t**) and equate it to another function $f(y,t)$, we obtain a first order differential equation expressed as

$$\frac{dy}{dt} = f(y,t). \tag{1.1}$$

It should be noted that f(y,t) may be a constant, a function of time, a function of **y** (implicit dependence on **t**) or a function of **y** and **t**. For example, we may have

$$\frac{dy}{dt} = f(y,t) = y + y^2 t + t^2 + 2. \tag{1.2}$$

If $f(y,t)$ only depends on y, the first order differential equation is an **autonomous** one as in the case of

$$\frac{dy}{dt} = f(y,t) = h(y). \tag{1.3}$$

In eqn. (1.3) h(y) is a linear or nonlinear function of y only.

It is possible to take the derivative of eqn. (1.1) once more. In this case, we get a **second** order differential equation. We may also take higher order derivatives. The order of the differential equation is determined by the highest order derivative existing in the equation. In addition to differential equations in a single variable of the first or higher orders, it is also possible that there may be two or more observables such as $y_1(t)$, $y_2(t)$,.., in which case, we may have interconnected or coupled differential equations.

In this book, we study all these differential equations starting from a single first order differential equation, followed by second order and higher order differential equations and culminating in coupled first order differential equations. If the differential equations only contain terms in y, $\frac{dy}{dt}$, $\frac{dy}{dt}$,…, or $\frac{d^n y}{dt^n}$, the differential equations are **linear**. An example of a **4th order linear differential** equation is

$$a(t)\frac{d^4 y}{dt^4}+b(t)\frac{d^2 y}{dt^2}+c(t)y=h(t). \tag{1.4}$$

If a(t), b(t) and c(t) in eqn. (1.4) are constants (no dependence on t), the differential equation is identified as a **linear differential equation with constant coefficients.** An example of a second order linear differential equation with constant coefficients (A, B, and, C) is

$$A\frac{d^2 y}{dt^2}+B\frac{dy}{dt}+Cy=h(t). \tag{1.5}$$

Additionally, if h(t)=0, eqn. (1.5) will be a **homogeneous** differential equation while a **non-homogeneous** differential equation will have h(t)≠0. Note that h(t) may be number (not equal to zero) in a non-homogeneous differential equation.

Chapter 2 is devoted to the study of first order differential equations with special emphasis on autonomous systems to develop an appreciation and understanding of the equilibrium conditions of associated systems. The chapter starts with the method of integrating factors and the method of separable functions for obtaining the solution of first order differential equations. A description of D-fields is provided to offer a pictorial description of the behavior of the system. All aspects necessary to understand an autonomous system are presented and well supported by an extensive set of examples. The examples are annotated to provide insights into the form of the solution, type of equilibrium as well as verification of the solution through additional approaches including use of dsolve(.) in MATLAB®, along with plots if and when necessary.

The second order differential equations are studied in Chapter 3. The differential equations are limited to linear ones with constant coefficients. Starting with the homogeneous differential equations, the concept of the characteristic equation or polynomial is introduced and the evolution of the solution from the roots of the characteristic equation is described. Solutions are also obtained using Laplace transforms and the state of the system is examined using the phase portraits to ensure that the analytical solution and their properties match the conclusions that can be drawn from phase portraits. An additional approach for obtaining the solution based on converting the second order equation into a pair of coupled first order differential equations is also offered as yet another level of verification of the solutions along with numerical solution obtained using Runge-Kutta methods. Finally, examples are provided which encompass every aspect of the solution with necessary equations, annotations, explanations and verification. Non-homogeneous differential equations are studied next in Chapter 3 by offering two methods for obtaining the particular solutions, namely the method of undetermined coefficients and the method of variation of parameters. The particular solutions obtained using both methods are compared offering a comprehensive view of the solutions to the non-homogeneous differential equations. The Laplace transform based approach and Runge-Kutta method are used to provide verification. Numerous examples are provided to cover diverse forcing functions and annotations in figures that clearly illustrate steps in the generation of the particular solutions, e.g., the need to scale the initial guess by t or t^2 as the case may be. The examples are fully annotated including steps taken to evaluate unknown constants when the initial conditions are given along with the background theoretical aspects.

Higher order differential equations are studied in Chapter 4. The solution to the homogeneous differential equation is obtained using the roots of the characteristic polynomial and issues with the diversity of relationships among the multiple roots are discussed. Based on this, the characteristic polynomial is used only for differential equations up to the 4th order. The Laplace transform method is used for all orders. The particular solution is obtained using the method of variation of parameters (up to the 4th order). The Runge-Kutta method is used as a means to verify the results. Once again, the examples are annotated with theoretical aspects, relationships among the roots (if they exist), formulation of the solution, determination of the unknown constants (initial conditions given), and, plots of the theoretical and numerical solutions.

Chapter 5 is devoted to coupled first order systems. A pair of coupled equations is studied first. The use of eigenvectors and eigenvalues in obtaining the solution to the homogeneous set is presented. The notion of defective matrices and the need for generalized eigenvectors is described. Another method based on the conversion of a pair of first order equations into a single second order equation is also presented along with the method based on Laplace transforms, as additional means for verification. The Runge-Kutta method is used as an extra step of verification. Phase portraits are used to justify conclusions about the stability of the system based on the solution obtained analytically. The concept of the fundamental matrix is invoked to obtain the particular solution when forcing functions are present. The fundamental matrix is used to obtain solutions when multiple coupled differential equations are present. Taking note of the existence of defective matrices, the analysis is limited to coefficient matrices of size [4 x 4] for the use of eigenvalues and eigenvectors. For larger matrices, only the Laplace transform based method (which is used throughout regardless of the size of the coefficient matrix) and the Runge-Kutta method are used for verification. The examples are annotated by displaying eigenvalues and eigenvectors, identifying defective matrices (algebraic and geometric multiplicities), degenerate eigenvalues, generalized eigenvectors, etc.

Appendices provide in depth coverage of topics such as numerical methods for solving differential equations (Euler's and Runge-Kutta methods), theory of Laplace transforms and applications to differential equations and phase plane analysis. In all cases, theoretical aspects are enunciated with appropriate examples to demonstrate the relevance to the issues presented in the main body of the book. The final appendix (Appendix D) is devoted to concepts of linear algebra that are absolutely essential for gaining a better understanding of the methods employed to find solutions to differential equations. These include properties of matrices, solution of a set of equations using row reduced echelon forms, Cramer's rule if applicable, matrix inversion if permitted and most importantly, the topic of eigenvalues and eigenvectors of square matrices. An expanded view of defective matrices and generalized eigenvectors is presented to illustrate how they play an important role in obtaining solutions to a set of first order coupled differential equations.

A common thread throughout this book is the use of MATLAB (version 2016a) for the creation of the displays associated with every example. In every chapter, MATLAB scripts for solving example problems, creating phase plots, obtaining Laplace transforms, etc. are given for the benefit of the reader. Through the use of LaTex conversion and extensive use of symbolic toolbox, the displays of the examples and solutions provide equations, explanations, sketches, plots, etc. reproducing the class room work by an instructor. MATLAB is used not merely as a means to verify the solution. It is also used to create a step-by-step solution matching the theory in every respect and thus providing a pedagogical element which is a unique feature of this book. Even though it is not possible to assure the absolute accuracies, every effort has been made to ensure that the MATLAB scripts (written in version 2016a) and the results provided are correct.

As indicated in the preface, the book is the culmination of the author's participation in teaching courses in Linear Algebra and Differential Equations to engineering students. The books used during that period of instruction and a number of journal publications by the author that came out of the preliminary work are listed below.

Bibliography

Books

Brannan, J. R. and W. E. Boyce. Differential Equations: An Introduction to Modern Methods and Applications, 2nd Edition, John Wiley, NJ. 2011.

Farlow, J., J. E. Hall, J. M. McDill and B. H. West. Differential Equations and Linear Algebra, 2nd Edition, Pearson, NY, 2007.

Lay, D. C. Linear Algebra and Its Applications, 5th Edition, Pearson, NY, 2015.

Nagle, R. K., E. B. Saff and A. D. Snider. Fundamentals of Differential Equations, 8th Edition, Person, NY 2012.

Publications

Shankar, P. M. A MATLAB workbook on the pedagogy of generalized eigenvectors. Computer Applications in Engineering Education, Vol. 25, No. 3, pp. 411–419, March 2017. doi:10.1002/cae.21808

Shankar, P. M. Pedagogy of Cramer's Rule and beyond: A MATLAB workbook. Mathematics & Computer Education, Fall 2016, Vol. 50 Issue 3, pp. 207–215, December 2016.

Shankar, P. M. Pedagogy of autonomous differential equations and equilibria using a MATLAB workbook. Mathematics & Computer Education. Winter 2016, Vol. 50 Issue 1, pp. 57–72.

Shankar, P. M. Pedagogy of second order homogeneous differential equations: A holistic approach using a MATLAB workbook. Computer Applications in Engineering Education, Vol. 24, No. 1, January 2016, pp. 114–121.

Shankar, P. M. Pedagogy of solutions to a set of linear equations using a MATLAB workbook. Computer Applications in Engineering Education, Vol. 25, No. 3, pp. 345–351, March 2017. doi:10.1002/cae.21803

CHAPTER 2

First Order Differential Equations

2.1 Introduction 5
2.2 D-field Plots 6
2.3 Methods of Solving First Order Differential Equations 9
2.3.1 Method of integrating factors 9
2.3.2 Separable differential equations 17
2.4 Additional Examples on D-field Plots 26
2.5 Autonomous Differential Equations 31
2.5.1 A general example 35
2.5.2 An example using MATLAB® and the script 44
2.5.3 Additional examples 46
2.6 Summary 77
2.7 Exercises 77

2.1 Introduction

Most physical and chemical phenomena involve changes taking place with respect to one or more parameters. This means that in its simplest form, systems responsible for these phenomena can be modeled in terms of an independent parameter (or variable) and a dependent parameter (or variable). Since changes take place over time, the easiest way to represent the system is through an expression describing the variation of a parameter (temperature, pressure, concentration of a chemical, force, pressure, etc.) in time. Variations are often represented in terms of derivatives resulting in first and higher order differential equations to model these systems.

Differential equations of first order are characterized by the existence of the derivative of the first order in an equation containing an independent variable (t) and a dependent variable (y) such as

$$\frac{dy}{dt} = f(t, y). \tag{2.1}$$

In eqn. (2.1), f(.) is generally taken to be a function of two variables, even though f(.) may be a function of y or t alone. It is also possible that f(.) is a constant. If f(t,y) does not depend on t, the differential equation in eqn. (2.1) is identified as an **autonomous** differential equation. If f(t,y) only contains terms that are linear in y, eqn. (2.1) becomes a first order **linear** differential equation. If f(t,y) can be separable and written as the product of two functions h(t) and g(y), a function of t and a function of y, the differential equation is a **separable** one.

2.2 D-field Plots

Considerable insight into the systems described through eqn. (2.1) can be gained by examining the behavior of the differential equation through the use of plots known as directional field plots or D-field plots. They rely on the fact that eqn. (2.1) also is the slope of y, describing how y varies with time t. Consider a simple differential equation

$$\frac{dy}{dt} = 6 - y. \tag{2.2}$$

If a value of y=6 is chosen, the slope is 0. For y=0, the slope is 6. If there is a way to indicate the values of the slope at various values of y by short arrows on a plot of t vs. y, the pattern created will provide information on how the solution varies with time and what is likely to happen when t goes up or down. The plot so generated is called the directional field or D-field plot. The plot can be generated with MATLAB using the quiver(.) command which draws small arrows. The simple code is given below and the resulting display is shown in Figure 2.1.

```
clear;clc
y=0:10; % pick values of y
t=0:.4:8; % pick values of t
[tt,yt]=meshgrid(t,y); % create a 2-d array of samples of t and y
dy=6-yt; % create dy/dt
dt=ones(size(dy)); % create corresponding time coordinates
figure, quiver(tt,yt,dt,dy,1.,'b') % quiver displays the directional arrows
xlabel('time'),ylabel('y(t)'), axis tight
title('dy/dt=6-y')
hold on
plot(t,6*ones(length(t)),'-r','linewidth',2)
CC=[-5:5];% values of C
% now plot the solution
tt=t;
clear t
syms y(t) C
ys=dsolve(diff(y,t)==6-y); % get the symbolic solution
vars=symvar(ys);% find the symbolic variables
ys=subs(ys,vars(1),C); % replace the unknown constant with C
yss=MATLABFunction(ys); % create in-line function for plotting the solution
for k=1:length(CC)
  C=CC(k);
ysS=yss(C,tt);
plot(tt,ysS)
end;
```

A solution to eqn. (2.2) can easily be found by rewriting it as

$$\frac{dy}{(6-y)} = dt \tag{2.3}$$

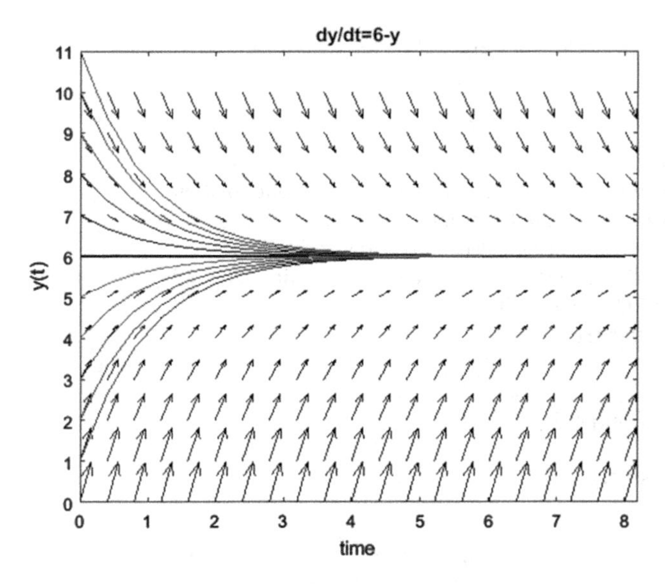

Figure 2.1 D-field of dy/dt=6-y.

The solution is obtained by integrating both sides of eqn. (2.3) as

$$y(t) = 6 + Ce^{-t}. \tag{2.4}$$

Regardless of the value of the unknown constant C, it can be seen that as t→∞, the solution approaches a stable value of 6. This can be observed from the plot. If a second step is undertaken to plot the solution for several values of C [for example, ranging from 0 to 5] and displayed on the same D-field plot, the solutions (shown by continuous lines) will follow the field directions arrows as shown in Figure 2.1. If an analytical solution is unavailable, numerical techniques (Appendix A) can be used to superimpose the samples of solutions as shown.

While eqn. (2.2) was autonomous, a slightly different case is that of a non-autonomous linear differential equation

$$\frac{dy}{dt} = y + 5 - 3t. \tag{2.5}$$

Proceeding as before and taking an additional note of the existence of t by choosing values of t along with values of y, the D-field can be plotted. The MATLAB script appears below and results are shown in Figure 2.2.

```
clear;clc;close all
y=-5:.5:5; % pick values of y
t=0:.25:5; % pick values of t
[tt,yt]=meshgrid(t,y); % create a 2-d array of samples of t and y
dy=yt+5-3*tt; % create dy/dt
dt=ones(size(dy)); % create corresponding time coordinates
figure,
quiver(tt,yt,dt,dy,1.5,'b') % quiver displays the directional arrows
```

```
xlabel('time'),ylabel('y(t)'), axis tight
hold on
t1=t;
clear t
syms y(t) C
ys=dsolve(diff(y,t)==y+5-3*t); % obtain the symbolic solution
vars=symvar(ys); % find the symbolic variables
ys=subs(ys,vars(1),C); % replace the constant with C
ys1=MATLABFunction(ys);% create in-line function for plotting the solution
CC=[-2:.25:2];% values of C
for k=1:length(CC)
  C=CC(k);
yss=ys1(C,t1);
plot(t1,yss),ylim([-5,5])
end;
title('dy/dt=y+5-3*t')
```

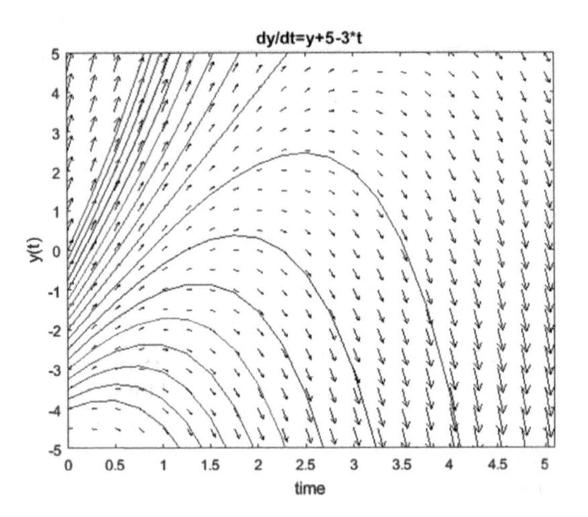

Figure 2.2 D-field of dy/dt=5+y-3t.

Unlike the example in eqn. (2.2), the contours seem to diverge indicating that the system represented in eqn. (2.5) is unstable. This can be established by getting the solution of eqn. (2.5) as (methods of obtaining solutions are discussed in the next section)

$$y(t) = 3t - 2 + Ce^{t}. \tag{2.6}$$

The theoretical plots are superimposed (continuous lines). The benefits of D-field patterns are clearly evident in terms of the directional arrows providing information on whether the system is likely to attain a stable value once the system starts functioning and how the behavior is likely to change with time, without actually knowing the solution.

Additional examples of D-field plots are given in Section 2.4 following the discussion of methods for finding solutions of first order differential equations.

2.3 Methods of Solving First Order Differential Equations

While D-field patterns provide useful information, it is necessary to have access to solutions of the differential equations. Solutions may be obtained analytically and if analytical solutions do not exist, solutions may be obtained numerically. Numerical techniques are discussed in Appendix A and the two different methods of solving first order differential equations analytically are given next. These are based on the method of integrating factors and the concept of separable functions.

2.3.1 Method of integrating factors

First order linear differential equations can be solved using the concept of integrating factors. Consider a first order linear differential equation with a dependent variable y(t) and independent variable t given by

$$a(t)Dy + p(t)y = g(t). \tag{2.7}$$

In eqn. (2.7), a(t), p(t) and g(t) are either constants or functions of time and

$$Dy = \frac{dy}{dt}. \tag{2.8}$$

Equation (2.7) is rewritten as

$$Dy + \frac{p(t)}{a(t)}y = \frac{g(t)}{a(t)}. \tag{2.9}$$

It is understood from eqn. (2.9) that

$$a(t) \neq 0. \tag{2.10}$$

Rewriting eqn. (2.9) as

$$Dy + q(t)y(t) = h(t). \tag{2.11}$$

The integrating factor $\mu(t)$ is a function of time that satisfies the following relationship

$$\frac{d}{dt}\left[\mu(t)y(t)\right] = \mu(t)\left[Dy + q(t)y(t)\right] \tag{2.12}$$

Multiplying eqn. (2.11) by $\mu(t)$ leads to

$$\mu(t)Dy + \mu(t)q(t)y(t) = \mu(t)h(t). \tag{2.13}$$

Comparing eqns. (2.12) and (2.13), the second term on the left hand side of eqn. (2.13) yields

$$\frac{d}{dt}\mu(t) = \mu(t)q(t). \tag{2.14}$$

Equation (2.14) can be solved for $\mu(t)$ by writing

$$\frac{d\mu(t)}{\mu(t)} = q(t)dt. \tag{2.15}$$

Integrating both sides leads to the solution for the integrating factor $\mu(t)$ as

$$\mu(t) = e^{\int q(t)dt} = e^{\int \frac{p(t)}{a(t)}dt} \tag{2.16}$$

Equation (2.13) can now be rewritten as

$$\frac{d}{dt}\left[\mu(t)y(t)\right] = \mu(t)h(t). \tag{2.17}$$

Integrating both sides of eqn. (2.17) results in

$$\mu(t)y(t) = \int_{t_0}^{t} \mu(z)h(z)dz + C \tag{2.18}$$

Note that t_0 is arbitrary (usually taken as 0) and C is an unknown constant. Rewriting eqn. (2.18), leads to the solution to the differential equation in eqn. (2.7) as

$$y(t) = \frac{1}{\mu(t)}\left[\int_{t_0}^{t} \mu(z)h(z)dz + C\right] = \frac{1}{\mu(t)}\left[\int_{t_0}^{t} \mu(z)\frac{g(z)}{a(z)}dz + C\right]. \tag{2.19}$$

If the initial conditions are given, the unknown constant C can be determined and an exact solution can be obtained.

MATLAB is a perfect vehicle to solve these problems through the use of the Symbolic Toolbox. A few examples (including one where no analytical solution exists appear below). All these were done in MATLAB to incorporate appropriate theory and results as well as explanations if necessary such as the case where no analytical solutions exist. The results are verified using dsolve(.) in MATLAB. It is noted that the constants created with the integrating factor based approach will be different from the unknown constants displayed in the solution obtained using dsolve(.). If the initial condition $y(t_0)=y_0$ with $t_0=0$ or the boundary condition with $y(t_0)=y_0$, $t_0 \neq 0$, the unknown coefficients can be evaluated and an exact solution can be obtained. If the analytical solution to the differential equation is available, the implicit solution in eqn. (2.19) can be plotted by choosing a few values of C.

If the example falls in a special category such as the availability of the exact or particular solution, unavailability of analytical solution, etc. these are indicated in the displays created in MATLAB.

Example # 2.1

1st order differential equation: Solution using Integrating Factor

Input Differential Equation ⇒ $\mathrm{Dy} = 2\,y - \frac{t}{2} + 4$

$$\text{The Differential Equation} \Rightarrow a(t)Dy + p(t)y = g(t)$$

$$\text{Integrating factor} \Rightarrow \mu(t) = e^{\int \frac{p(t)}{a(t)}dt}$$

$$\text{General Solution} \Rightarrow y(t) = \frac{1}{\mu(t)}\left[\int_{t_0}^{t} \mu(s)\frac{g(s)}{a(s)}ds + C\right]$$

$$a(t) \Rightarrow 1 \quad p(t) \Rightarrow -2 \quad g(t) \Rightarrow 4 - \frac{t}{2}$$

Integrating factor μ(t) ⇒ $\mathrm{e}^{-2\,t}$

General solution (Integrating factor) ⇒ $y(t) = \frac{t}{4} + C\,\mathrm{e}^{2\,t} - \frac{15}{8}$

General Solution (dsolve in Matlab) ⇒ $y(t) = \frac{t}{4} + \frac{B\,\mathrm{e}^{2\,t}}{8} - \frac{15}{8}$

Note that the constants might not MATCH for the case of general solutions!!

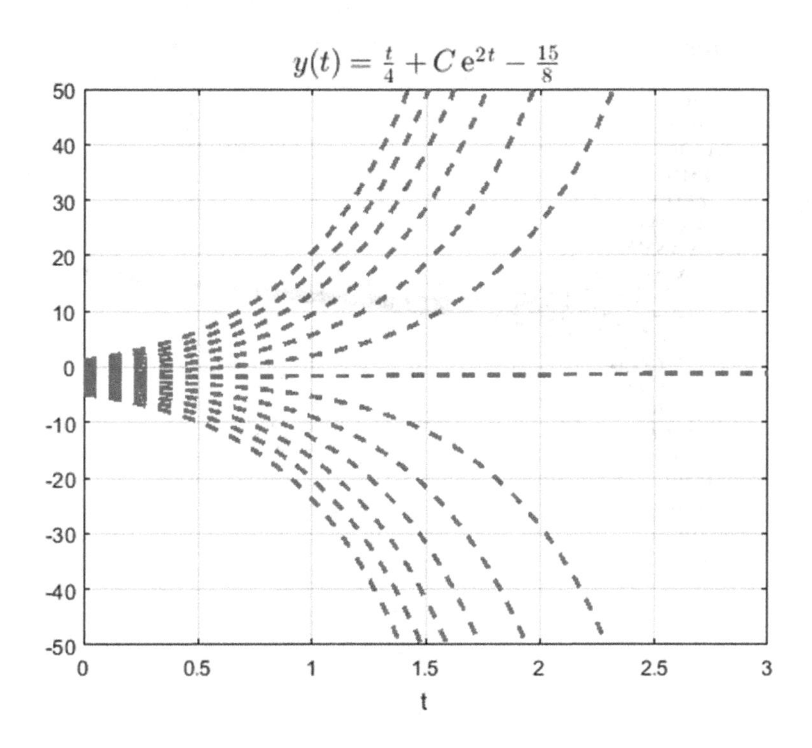

Example # 2.2 Boundary conditions given. An exact solution is available.

1st order differential equation: Solution using Integrating Factor

Input Differential Equation ⇒ $\mathrm{Dy} = 4\,t - \frac{2y}{t}$

The Differential Equation	$\Rightarrow a(t)Dy + p(t)y = g(t)$
Integrating factor	$\Rightarrow \mu(t) = e^{\int \frac{p(t)}{a(t)} dt}$
General Solution	$\Rightarrow y(t) = \frac{1}{\mu(t)}\left[\int_{t_0}^{t} \mu(s)\frac{g(s)}{a(s)}ds + C\right]$

$a(t) \Rightarrow t \qquad p(t) \Rightarrow 2 \qquad g(t) \Rightarrow 4t^2$

Integrating factor μ(t) ⇒ t^2

General solution (Integrating factor) ⇒ $y(t) = \frac{t^4+C}{t^2}$

General Solution (dsolve in Matlab) $y(t) = \frac{t^4+B}{t^2}$

Boundary conditions ⇒ $t_0 = 1$ $y(t_0) = 3$

Apply Boundary Conditions ⇒ $C = 2$ $B = 2$

Particular solution (Integrating factor) ⇒ $y(t) = \frac{t^4+2}{t^2}$

Particular Solution (dsolve in Matlab) $y(t) = \frac{t^4+2}{t^2}$

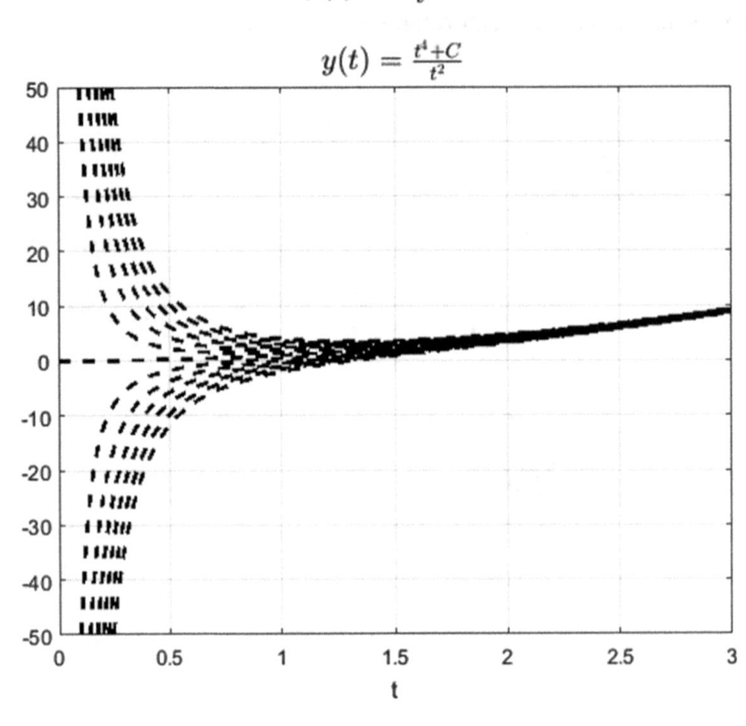

Example # 2.3

1st order differential equation: Solution using Integrating Factor

Input Differential Equation ⇒ $\mathrm{Dy} = t - 3\,y + \mathrm{e}^{-2\,t}$

The Differential Equation	$\Rightarrow a(t)Dy + p(t)y = g(t)$
Integrating factor	$\Rightarrow \mu(t) = e^{\int \frac{p(t)}{a(t)}\,dt}$
General Solution	$\Rightarrow y(t) = \dfrac{1}{\mu(t)}\left[\int_{t_0}^{t} \mu(s)\dfrac{g(s)}{a(s)}ds + C\right]$

$a(t) \Rightarrow 1 \quad p(t) \Rightarrow 3 \quad g(t) \Rightarrow t + \mathrm{e}^{-2\,t}$

Integrating factor μ(t) ⇒ $\mathrm{e}^{3\,t}$

General solution (Integrating factor) ⇒ $y(t) = \frac{t}{3} + \mathrm{e}^{-2\,t} + C\,\mathrm{e}^{-3t} - \frac{1}{9}$

General Solution (dsolve in Matlab) ⇒ $y(t) = \frac{t}{3} + \mathrm{e}^{-2\,t} + B\,\mathrm{e}^{-3t} - \frac{1}{9}$

Note that the constants might not MATCH for the case of general solutions!!

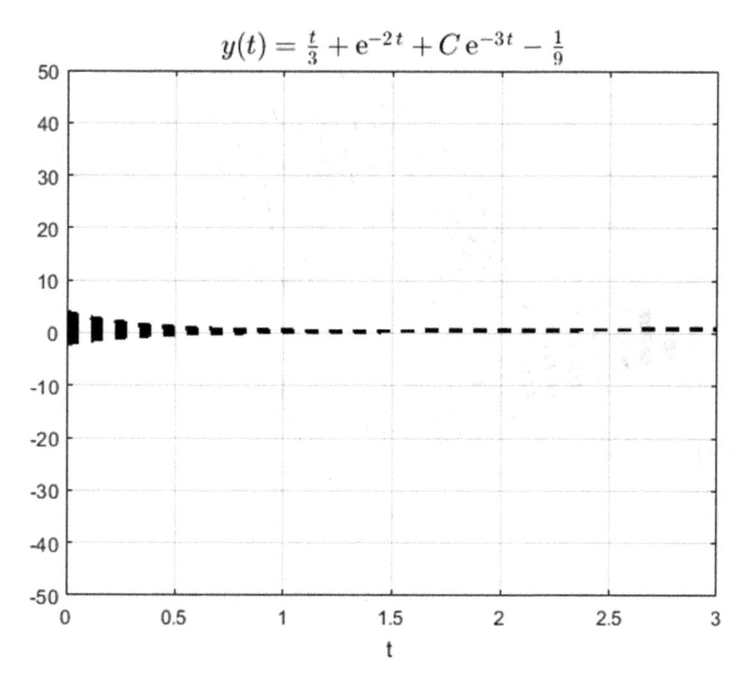

Example # 2.4

1st order differential equation: Solution using Integrating Factor

Input Differential Equation ⇒ $\mathrm{Dy} = 2\,y + t^2\,e^{2\,t}$

$$\text{The Differential Equation} \quad \Rightarrow a(t)Dy + p(t)y = g(t)$$

$$\text{Integrating factor} \quad \Rightarrow \mu(t) = e^{\int \frac{p(t)}{a(t)}dt}$$

$$\text{General Solution} \quad \Rightarrow y(t) = \frac{1}{\mu(t)}\left[\int_{t_0}^{t} \mu(s)\frac{g(s)}{a(s)}ds + C\right]$$

$$a(t) \Rightarrow \; 1 \quad p(t) \Rightarrow \; -2 \qquad g(t) \Rightarrow \; t^2\,e^{2t}$$

Integrating factor μ(t) ⇒ $e^{-2\,t}$

General solution (Integrating factor) ⇒ $y(t) = \frac{e^{2\,t}\,(t^3+3\,C)}{3}$

General Solution (dsolve in Matlab) ⇒ $y(t) = \frac{e^{2\,t}\,(t^3+3\,B)}{3}$

Note that the constants might not MATCH for the case of general solutions!!

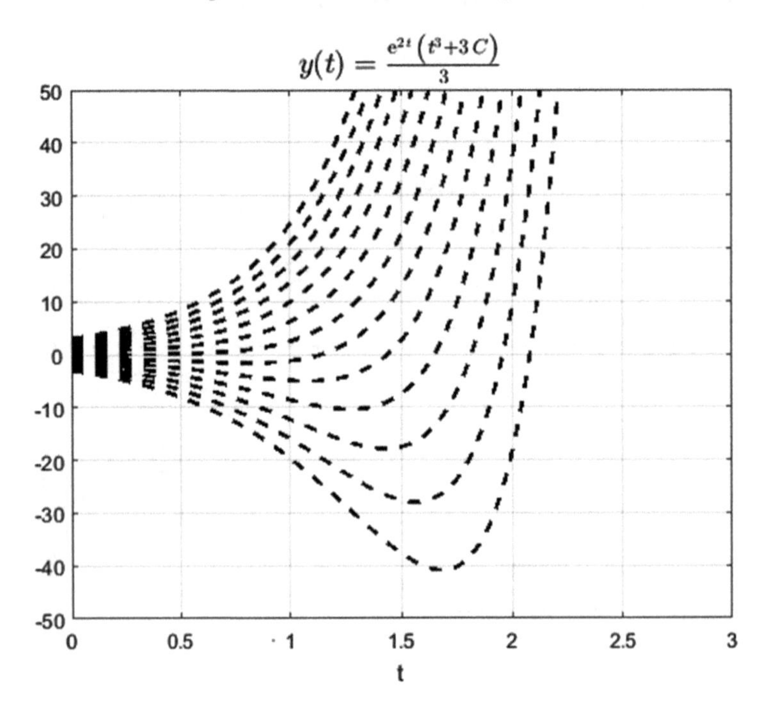

Example # 2.5

1st order differential equation: Solution using Integrating Factor

Input Differential Equation ⇒ $\mathrm{Dy} = \frac{\sin(t)}{t} - \frac{2y}{t}$

$$\text{The Differential Equation} \quad \Rightarrow a(t)Dy + p(t)y = g(t)$$

$$\text{Integrating factor} \quad \Rightarrow \mu(t) = e^{\int \frac{p(t)}{a(t)} dt}$$

$$\text{General Solution} \quad \Rightarrow y(t) = \frac{1}{\mu(t)} \left[\int_{t_0}^{t} \mu(s) \frac{g(s)}{a(s)} ds + C \right]$$

$$a(t) \Rightarrow \quad t \quad p(t) \Rightarrow \quad 2 \qquad g(t) \Rightarrow \quad \sin(t)$$

Integrating factor μ(t) ⇒ t^2

General solution (Integrating factor) ⇒ $y(t) = \frac{C + \sin(t) - t\cos(t)}{t^2}$

General Solution (dsolve in Matlab) ⇒ $y(t) = \frac{B + \sin(t) - t\cos(t)}{t^2}$

Note that the constants might not MATCH for the case of general solutions!!

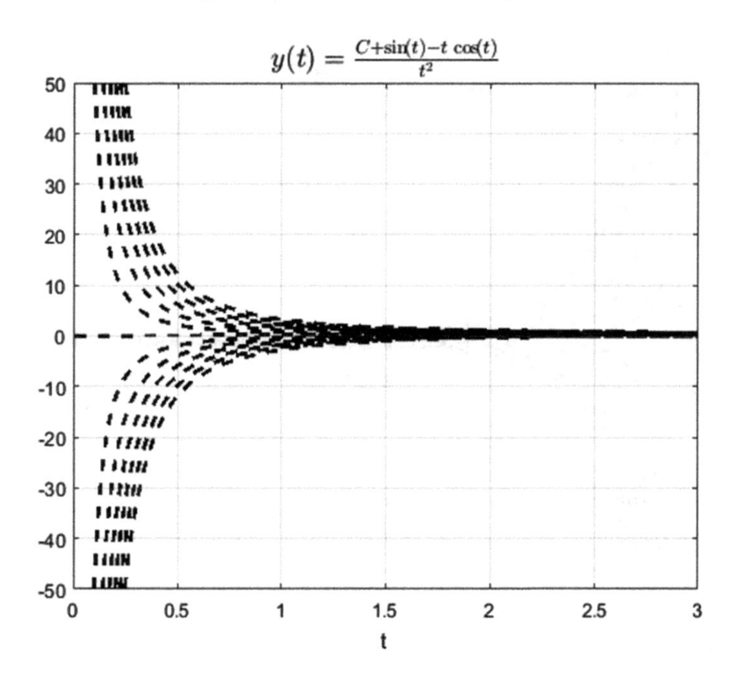

Example # 2.6 Boundary condition available.

1st order differential equation: Solution using Integrating Factor

Input Differential Equation ⇒ $\mathrm{Dy} = \frac{\cos(t)}{t^2} - \frac{2\,y}{t}$

The Differential Equation ⇒ $a(t)Dy + p(t)y = g(t)$

Integrating factor ⇒ $\mu(t) = e^{\int \frac{p(t)}{a(t)} dt}$

General Solution ⇒ $y(t) = \frac{1}{\mu(t)}\left[\int_{t_0}^{t} \mu(s)\frac{g(s)}{a(s)} ds + C\right]$

$a(t) \Rightarrow t^2 \quad p(t) \Rightarrow 2t \quad g(t) \Rightarrow \cos(t)$

Integrating factor μ(t) ⇒ t^2

General solution (Integrating factor) ⇒ $y(t) = \frac{C+\sin(t)}{t^2}$

General Solution (dsolve in Matlab) $y(t) = -\frac{B-\sin(t)}{t^2}$

Boundary conditions ⇒ $t_0 = 2\pi$ $y(t_0) = 0$

Apply Boundary Conditions ⇒ $C = 0$ $B = 0$

Particular solution (Integrating factor) ⇒ $y(t) = \frac{\sin(t)}{t^2}$

Particular Solution (dsolve in Matlab) $y(t) = \frac{\sin(t)}{t^2}$

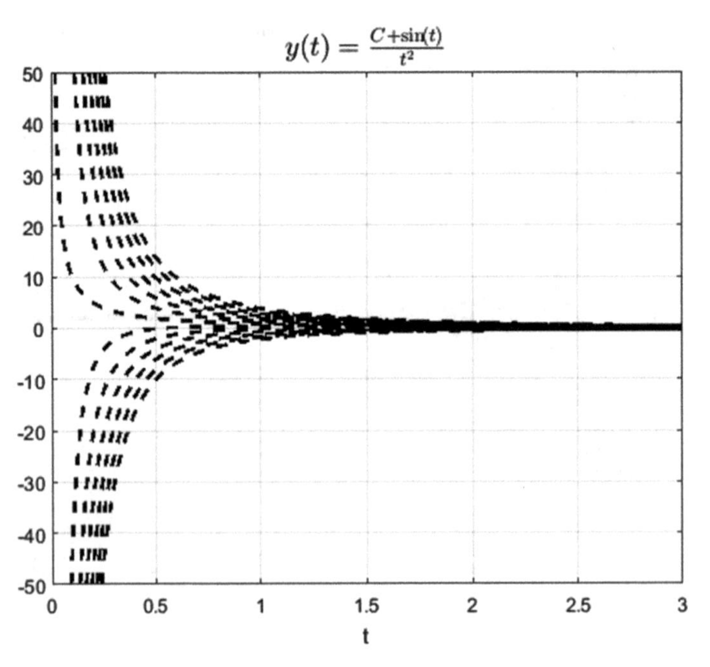

Example # 2.7 In this case, the analytical solution is not available for plotting because it is given in terms of the error function of a complex input. The MATLAB script displays this information.

1st order differential equation: Solution using Integrating Factor

Input Differential Equation ⇒ $Dy = 1 - \frac{ty}{2}$

The Differential Equation	$\Rightarrow a(t)Dy + p(t)y = g(t)$
Integrating factor	$\Rightarrow \mu(t) = e^{\int \frac{p(t)}{a(t)} dt}$
General Solution	$\Rightarrow y(t) = \frac{1}{\mu(t)} \left[\int_{t_0}^{t} \mu(s) \frac{g(s)}{a(s)} ds + C \right]$

$a(t) \Rightarrow 2 \quad p(t) \Rightarrow t \quad g(t) \Rightarrow 2$

Integrating factor $\mu(t)$ ⇒ $e^{\frac{t^2}{4}}$

General solution (Integrating factor) ⇒ $y(t) = e^{-\frac{t^2}{4}} \left(C - \sqrt{\pi}\, \mathrm{erf}\left(\frac{t\,i}{2}\right) i\right)$

General Solution (dsolve in Matlab) ⇒ $y(t) = e^{-\frac{t^2}{4}} \left(B - \sqrt{\pi}\, \mathrm{erf}\left(\frac{t\,i}{2}\right) i\right)$

Simple Analytical Solution Does not Exist; EXIT

Note that the constants might not MATCH for the case of general solutions!!

Note that erf(ix) in the display is

$$erfi(x) = -i\,erf(ix) = \frac{2}{\sqrt{\pi}} \int_0^z e^{t^2} dt \tag{2.20}$$

2.3.2 Separable differential equations

While the method of integrating factors is applicable only to linear first order differential equations, linear and non-linear first order differential equations may be solved using the method of separable differential equations. This method takes advantage of the form of the differential equation when the derivative is expressed as the product of two functions of y (dependent variable) and x (independent variable) as in

$$\frac{dy}{dx} = f(x)h(y). \tag{2.21}$$

Rewriting eqn. (2.21),

$$\frac{dy}{h(y)} = f(x)dx. \tag{2.22}$$

Integrating both sides,

$$\int \frac{dy}{h(y)} = \int f(x)\,dx - C. \tag{2.23}$$

Consider the example of

$$\frac{dy}{dx} = \frac{4x^2}{y^2}. \tag{2.24}$$

The differential equation in eqn. (2.24) can be expressed as

$$\frac{dy}{dx} = f(x)h(y) \tag{2.25}$$

$$\begin{aligned} f(x) &= x^2 \\ h(y) &= \frac{4}{y^2} \end{aligned} \tag{2.26}$$

The integration of the separable equations leads to

$$\int \frac{y^2}{4}dy - \int x^2 dx + C = 0. \tag{2.27}$$

The solution in implicit form will be,

$$\frac{y^3}{12} - \frac{x^3}{3} + C = 0. \tag{2.28}$$

Often it might not be possible to get an explicit solution for y in the form

$$y = G(x, C). \tag{2.29}$$

In other words, it is easy to get an implicit solution of the form in eqn. (2.28) and it may or may not be possible to get a solution in explicit form as in eqn. (2.29) where G(.) is only a function of x and C. The implicit solutions can be plotted by choosing a range of values of the unknown constant C. The plotting can be done using the ezplot(.) command available in the symbolic toolbox in MATLAB. It should be noted that the method based on separable functions will provide an implicit solution, while the use of dsolve(.) in MATLAB leads to explicit solutions and these may involve complex functions. Use of dsolve(.) may also lead to implicit solutions expressed in the form containing phrases ‘RootOf’, suggesting that a simple representation of the solution is not possible.

A few examples are given below demonstrating the various aspects of the approach described above. All these were created in MATLAB and keeping up with the theme of the book, the displays provide the theoretical aspects, steps involved, solution obtained using the approach and the solution (if it exists) using dsolve(.). Note that the unknown constants in the solutions obtained using the method of separable differential equations and dsolve(.) in MATLAB may be different.

Example # 2.8 Explicit solution available using dsolve(.) as seen by the three separate solutions for y, with two of them being complex.

First Order Separable Differential Equation $\Rightarrow \frac{dy}{dx} = f(x)h(y)$

Differential Equation $\quad \frac{dy}{dx} = \frac{4x^2}{y^2}$

Separable Functions $\Rightarrow$

$$f(x) = x^2$$

$$h(y) = \frac{4}{y^2}$$

$$\int f(x)dx \Rightarrow \frac{x^3}{3} \qquad \int \frac{1}{h(y)}dy \Rightarrow \frac{y^3}{12}$$

$$\text{Solution} \Rightarrow \int \frac{1}{h(y)}dy - \int f(x)dx + C = 0$$

Solution using Separable Equations

$$-\frac{x^3}{3} + \frac{y^3}{12} + C = 0$$

Explicit Solution using dsolve(.)

$$y(x) = 3^{\frac{1}{3}} \left(\frac{4x^3}{3} + B\right)^{\frac{1}{3}}$$

$$y(x) = 3^{\frac{1}{3}} \left(-\frac{1}{2} + \frac{\sqrt{3}\,\mathrm{i}}{2}\right) \left(\frac{4x^3}{3} + B\right)^{\frac{1}{3}}$$

$$y(x) = -3^{\frac{1}{3}} \left(\frac{1}{2} + \frac{\sqrt{3}\,\mathrm{i}}{2}\right) \left(\frac{4x^3}{3} + B\right)^{\frac{1}{3}}$$

$$-\frac{x^3}{3} + \frac{y^3}{12} + C = 0$$

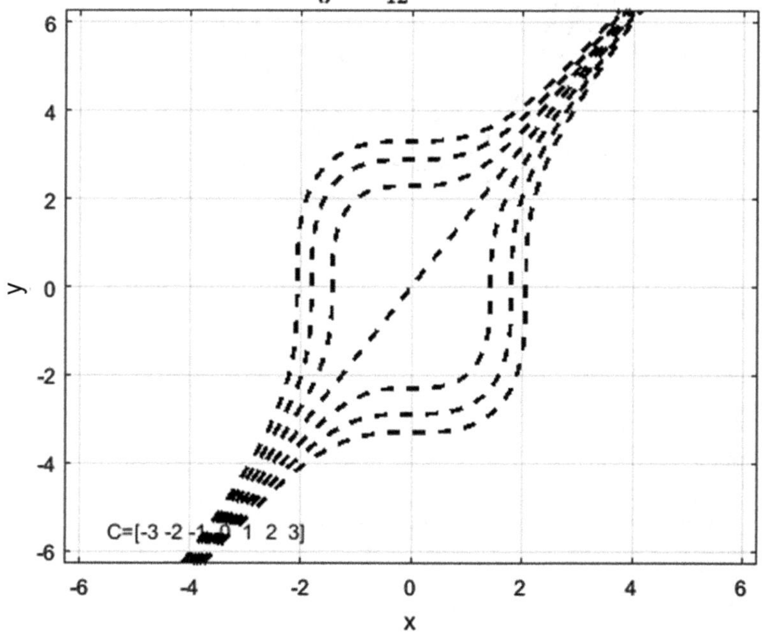

C=[-3 -2 -1 0 1 2 3]

y

x

Example # 2.9 Explicit solution available using dsolve(.).

First Order Separable Differential Equation $\Rightarrow \frac{dy}{dx} = f(x)h(y)$

Differential Equation $\frac{dy}{dx} = y^{\frac{1}{3}}\,(x+1)$

Separable Functions $\Rightarrow$ $f(x) = x+1$, $h(y) = y^{\frac{1}{3}}$

$\int f(x)dx \Rightarrow \frac{x\,(x+2)}{2}$ $\qquad \int \frac{1}{h(y)}dy \Rightarrow \frac{3\,y^{\frac{2}{3}}}{2}$

$$\text{Solution} \Rightarrow \int \frac{1}{h(y)}dy - \int f(x)dx + C = 0$$

Solution using Separable Equations $C - \frac{x\,(x+2)}{2} + \frac{3\,y^{\frac{2}{3}}}{2} = 0$

Explicit Solution using dsolve(.)

$$y(x) = 0$$

$$y(x) = \left(B + \frac{x\,(x+2)}{3}\right)^{\frac{3}{2}}$$

$$C - \frac{x\,(x+2)}{2} + \frac{3\,y^{\frac{2}{3}}}{2} = 0$$

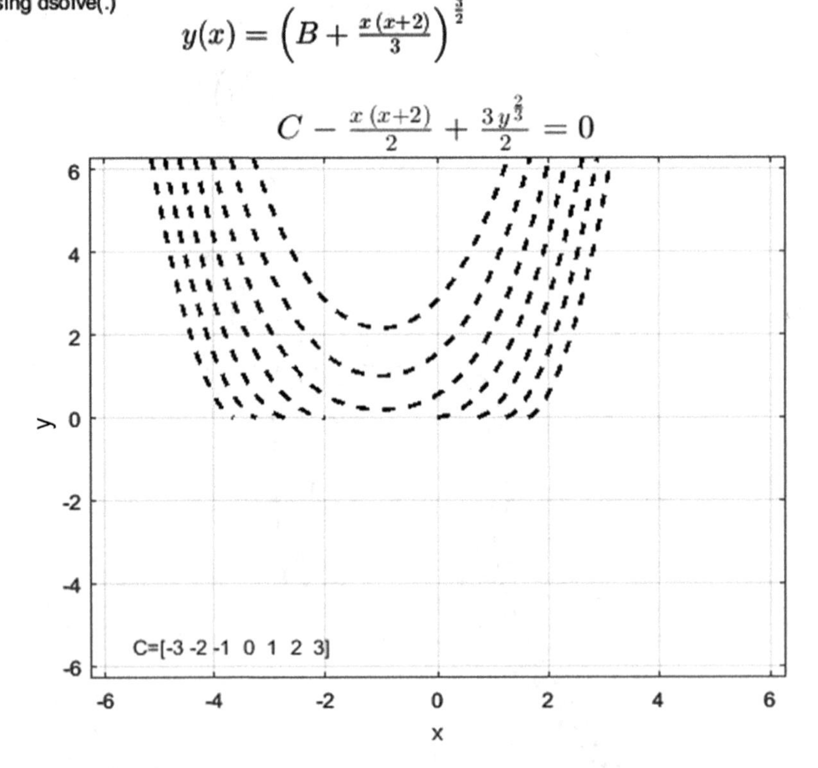

Example # 2.10 Explicit solution available using dsolve(.).

First Order Separable Differential Equation $\Rightarrow \frac{dy}{dx} = f(x)h(y)$

Differential Equation $\frac{dy}{dx} = -\frac{4x-3x^2}{2y-4}$

Separable Functions $\Rightarrow$ $f(x) = 3x^2 - 4x$, $h(y) = \frac{1}{2y-4}$

$$\int f(x)dx \Rightarrow x^2(x-2) \quad \int \frac{1}{h(y)}dy \Rightarrow y(y-4)$$

$$\text{Solution} \Rightarrow \int \frac{1}{h(y)}dy - \int f(x)dx + C = 0$$

Solution using Separable Equations $C - x^2(x-2) + y(y-4) = 0$

Explicit Solution using dsolve(.)

$$y(x) = \sqrt{2}\sqrt{B + \frac{x^2(x-2)}{2} + 2} + 2$$

$$y(x) = 2 - \sqrt{2}\sqrt{B + \frac{x^2(x-2)}{2} + 2}$$

$$C - x^2(x-2) + y(y-4) = 0$$

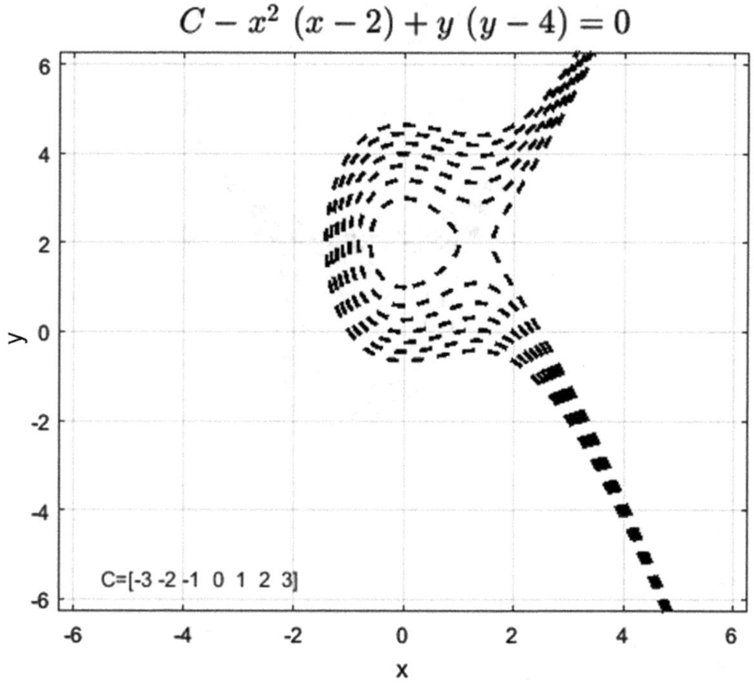

Example # 2.11 Explicit solution (complex) available using dsolve(.).

$$\text{First Order Separable Differential Equation} \Rightarrow \frac{dy}{dx} = f(x)h(y)$$

Differential Equation $\quad \frac{dy}{dx} = \frac{x^3+x}{(y-2)^2}$

Separable Functions $\Rightarrow$ $f(x) = x^3 + x$, $h(y) = \frac{1}{(y-2)^2}$

$$\int f(x)dx \Rightarrow \frac{x^2\left(x^2+2\right)}{4} \qquad \int \frac{1}{h(y)}dy \Rightarrow \frac{(y-2)^3}{3}$$

$$\text{Solution} \Rightarrow \int \frac{1}{h(y)}dy - \int f(x)dx + C = 0$$

Solution using Separable Equations

$$C + \frac{(y-2)^3}{3} - \frac{x^2\left(x^2+2\right)}{4} = 0$$

Explicit Solution using dsolve(.)

$$y(x) = 3^{\frac{1}{3}}\left(B + \frac{x^2\left(x^2+2\right)}{4}\right)^{\frac{1}{3}} + 2$$

$$y(x) = 3^{\frac{1}{3}}\left(B + \frac{x^2\left(x^2+2\right)}{4}\right)^{\frac{1}{3}}\left(-\frac{1}{2} + \frac{\sqrt{3}\mathrm{i}}{2}\right) + 2$$

$$y(x) = 2 - 3^{\frac{1}{3}}\left(B + \frac{x^2\left(x^2+2\right)}{4}\right)^{\frac{1}{3}}\left(\frac{1}{2} + \frac{\sqrt{3}\mathrm{i}}{2}\right)$$

$$C + \frac{(y-2)^3}{3} - \frac{x^2\left(x^2+2\right)}{4} = 0$$

Example # 2.12 Explicit solution unavailable using dsolve(.) as seen by the presence of the phrase ‘RootOf’ in the MATLAB generated solution.

First Order Separable Differential Equation $\Rightarrow \frac{dy}{dx} = f(x)h(y)$

Differential Equation $\quad \frac{dy}{dx} = \frac{4x - x^3}{y^3+8}$

Separable Functions $\Rightarrow$

$$f(x) = 4x - x^3$$

$$h(y) = \frac{1}{y^3+8}$$

$$\int f(x)dx \Rightarrow -\frac{x^2\left(x^2-8\right)}{4} \qquad \int \frac{1}{h(y)}dy \Rightarrow \frac{y\left(y^3+32\right)}{4}$$

$$\text{Solution} \Rightarrow \int \frac{1}{h(y)}dy - \int f(x)dx + C = 0$$

Solution using Separable Equations

$$C + \frac{y\left(y^3+32\right)}{4} + \frac{x^2\left(x^2-8\right)}{4} = 0$$

Explicit Solution using dsolve(.)

$$y(x) = \mathrm{RootOf}(\mathrm{z1}^4 + 32\,\mathrm{z1} - 4\,B + x^2\left(x^2-8\right), \mathrm{z1})$$

Roots cannot be found

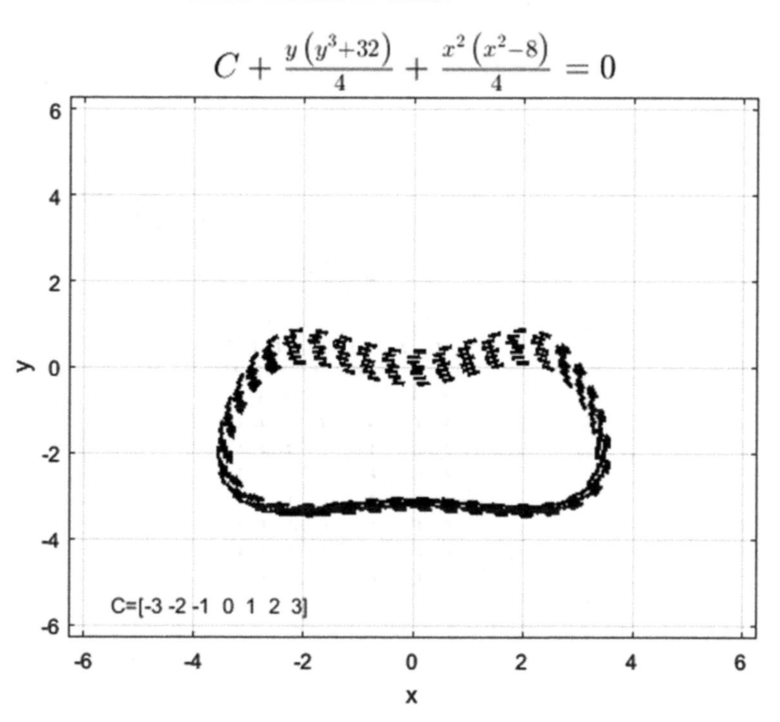

Example # 2.13 Explicit solution available using dsolve(.).

First Order Separable Differential Equation $\Rightarrow \frac{dy}{dx} = f(x)h(y)$

Differential Equation $\quad \frac{dy}{dx} = \mathrm{e}^{-y}\left(x + \mathrm{e}^{-x}\right)$

Separable Functions $\Rightarrow$

$$f(x) = x + \mathrm{e}^{-x}$$

$$h(y) = \mathrm{e}^{-y}$$

$$\int f(x)dx \Rightarrow \frac{x^2}{2} - \mathrm{e}^{-x} \qquad \int \frac{1}{h(y)}dy \Rightarrow \mathrm{e}^{y}$$

$$\text{Solution} \Rightarrow \int \frac{1}{h(y)}dy - \int f(x)dx + C = 0$$

Solution using Separable Equations $\quad C + \mathrm{e}^{-x} + \mathrm{e}^{y} - \frac{x^2}{2} = 0$

Explicit Solution using dsolve(.) $\quad y(x) = \log\left(B - \mathrm{e}^{-x} + \frac{x^2}{2}\right)$

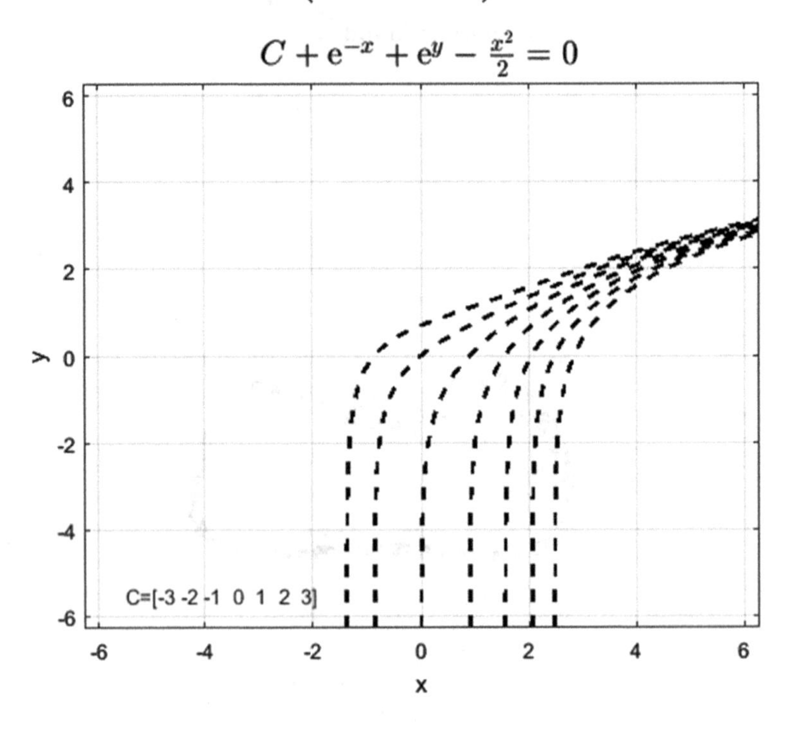

Example # 2.14 Explicit solution unavailable using dsolve(.) as seen by the presence of the phrase 'RootOf' in the MATLAB generated solution.

First Order Separable Differential Equation $\Rightarrow \frac{dy}{dx} = f(x)h(y)$

Differential Equation $\quad \frac{dy}{dx} = -\frac{2\left(x^2+4\right)}{y^2-1}$

Separable Functions $\Rightarrow$
$$f(x) = x^2 + 4$$
$$h(y) = -\frac{2}{y^2-1}$$

$$\int f(x)dx \Rightarrow \frac{x\left(x^2+12\right)}{3} \qquad \int \frac{1}{h(y)}dy \Rightarrow -\frac{y\left(y^2-3\right)}{6}$$

Solution $\Rightarrow \int \frac{1}{h(y)}dy - \int f(x)dx + C = 0$

Solution using Separable Equations
$$C - \frac{x\left(x^2+12\right)}{3} - \frac{y\left(y^2-3\right)}{6} = 0$$

Explicit Solution using dsolve(.)
$$y(x) = \mathrm{RootOf}(\mathrm{z1}^3 - 3\,\mathrm{z1} + 2x\left(x^2+12\right) - 3B, \mathrm{z1})$$

Roots cannot be found

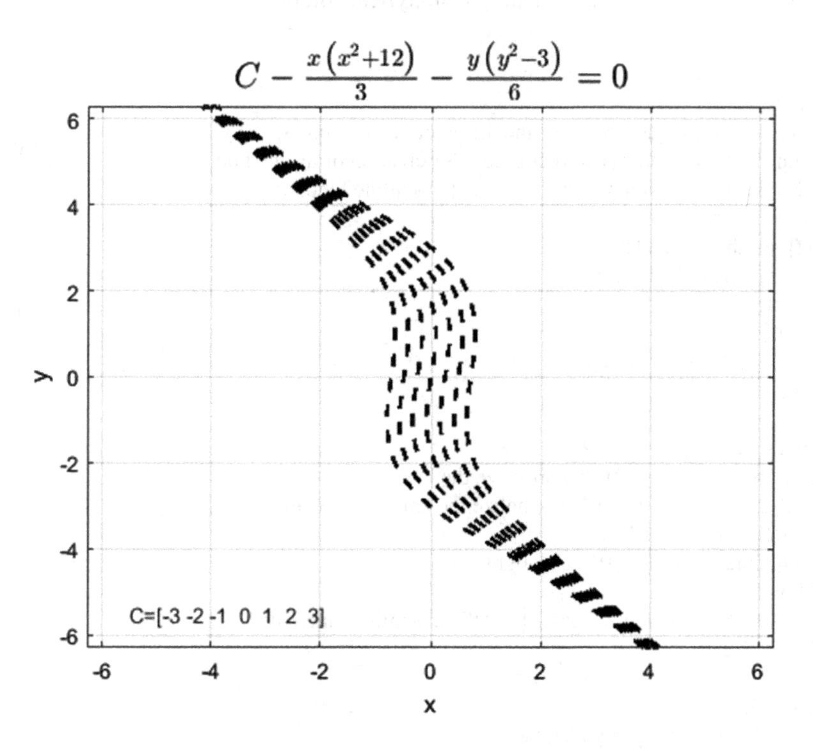

2.4 Additional Examples on D-field Plots

As seen in Section 2.2, D-field plots are useful in understanding the behavior of the system modeled using a first order differential equation. The plots can be generated using numerical integration based on the Runge-Kutta method (Appendix A), implicit solutions generated using the separable function method or explicit solutions obtained using dsolve(.) in MATLAB. The Runge-Kutta method requires the choice of initial conditions (a single line of plot for each initial condition) which makes this approach a little bit more difficult than the plotting the implicit solutions (choice of the unknown constants needed). The explicit solutions also pose problems because the implicit solution might be an equation in the form of y^n ($n>1$). This means that explicit solutions require plots of both solutions when n=2, all three when n=3. This also might pose problems if the explicit solutions are complex. The following examples illustrate the issues related to D-field plots created using numerical and analytical solutions. While example # 2.15 compares the D-field plots generated using numerical and analytical solutions, example # 2.16 compares D-field plots generated using numerical and analytical solutions (implicit and explicit). The results clearly demonstrate the advantage of creating the plots using the implicit solution. It should be noted that the plots created using the quiver(.) command in MATLAB without plots of the solutions can provide all the necessary information on the state of the equilibrium.

Example # 2.15 Numerical and simple analytical solutions of dy/dt=1-y.

```
close all;
syms t y Dy ff C
dy_dt='(1-y)';
ff=['Dy=',char(dy_dt)];% create the differential equation Dy=f(t,y)
V = odeToVectorField(ff);% vector field for creation of in-line Function
F = MATLABFunction(V,'vars', {'t','Y'});% inline function
t1=0:.5:5;y1=-5:.5:5;
[tt,yt]=meshgrid(t1,y1);
 LT=length(t1);LY=length(y1);
 for k=1:LT
   for kk=1:LY
       dy(kk,k)=F(t1(k),y1(kk));
   end;
 end;
dt=ones(size(dy));
dtn=dt./sqrt(dt.^2+dy.^2); % normalize the length of the arrows
dyn=dy./sqrt(dt.^2+dy.^2);% normalize the length of the arrows
quiver(tt,yt,dtn,dyn,1,'b')
xlabel('time'),ylabel('y(t)'), axis tight
hold on
F = MATLABFunction(V,'vars', {'t','Y'});% inline function
 y0=[-20:2:20];
 tspan=[0,5];
 for k=1:length(y0)
 [T,yode]=ode45(F,tspan,y0(k));%
 plot(T,yode,'color','r','linewidth',2)
 end;
```

```
xlim([0,5]),ylim([-5,5])
title('Directional Field of dy/dt=1-y: Numerical solution')
figure
quiver(tt,yt,dtn,dyn,1,'b')
xlabel('time'),ylabel('y(t)'), axis tight, hold on
yy=dsolve(ff); % get the symbolic solution
vr=symvar(yy); % determine the existing symbolic variable
yy=subs(yy,vr(1),C); % replace the constant MATLAB outputs to C
CC=[-20:2:20]; % choose a number of unknown constants
for k=1:length(CC)
  ffr=subs(yy,C,sym(CC(k)));
 p1 = ezplot(ffr);
 set(p1,'Color','red', 'LineWidth', 2)
  hold on
end;
title('Directional Field of dy/dt=1-y: Analytical solution')
ylim([-5,5]),xlim([0,5])
hold off
```

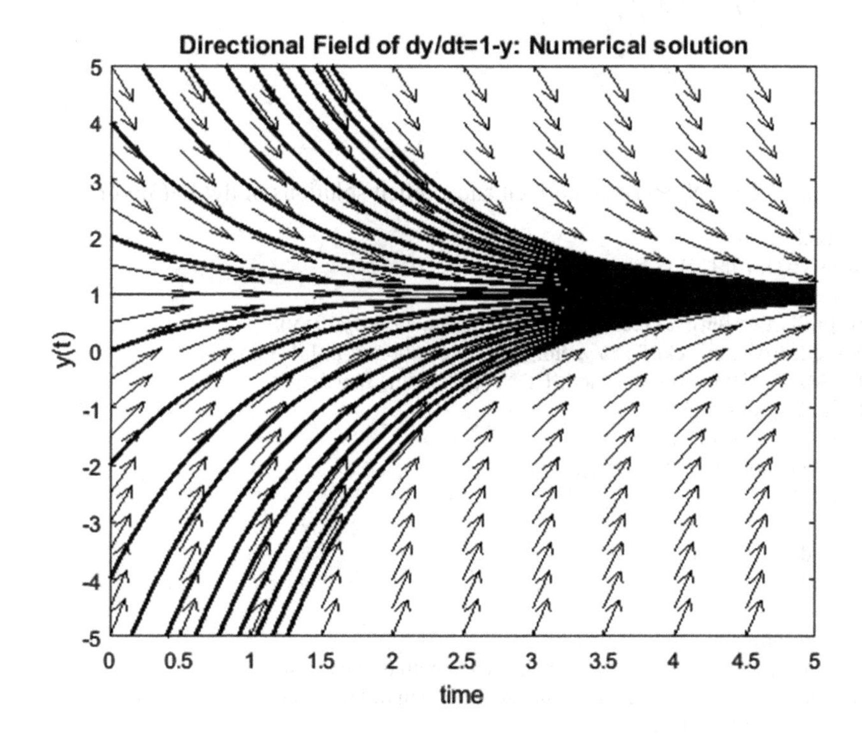

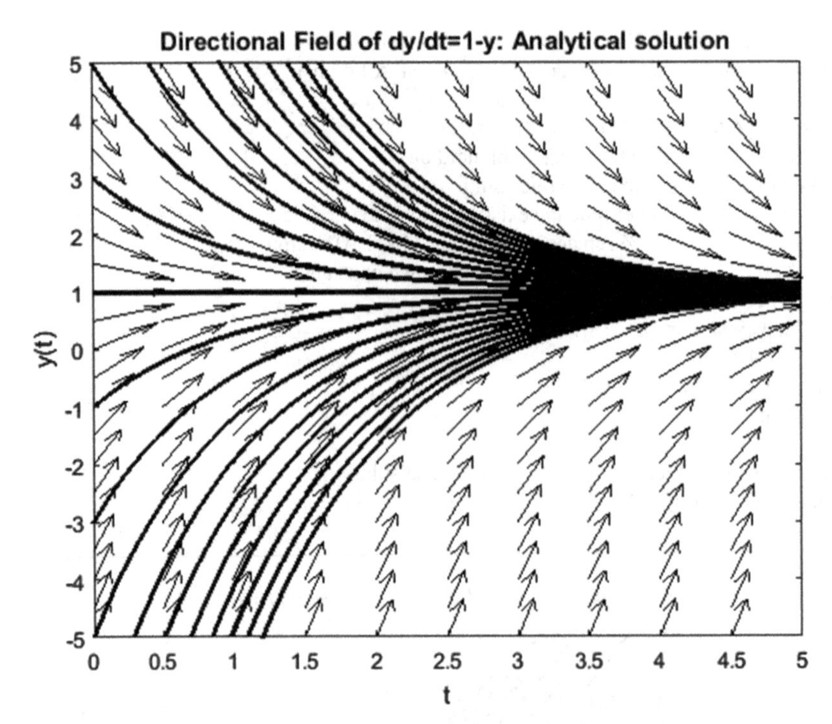

Example # 2.16 Numerical, implicit and explicit solutions of $dy/dt=(3t^2+4t)/(y-1)$.

```
clear;close all
syms y(t) Dy ff C B fx fy
fun=(3*t^2+4*t)/(y-1);
ff=[diff(y,t)==fun];% create the differential equation Dy=f(t,y)
V = odeToVectorField(ff);% vector field for creation of in-line Function
F = MATLABFunction(V,'vars', {'t','Y'});% inline function
t1=1:.25:3;y1=-5:.5:5;
[tt,yt]=meshgrid(t1,y1);
 LT=length(t1);LY=length(y1);
 for k=1:LT
    for kk=1:LY
        dy(kk,k)=F(t1(k),y1(kk));
    end;
 end;
dt=ones(size(dy));
dtn=dt./sqrt(dt.^2+dy.^2); % normalize the length of the arrows
dyn=dy./sqrt(dt.^2+dy.^2);% normalize the length of the arrows
figure,quiver(tt,yt,dtn,dyn,1,'b')
xlabel('time'),ylabel('y(t)'), axis tight
hold on
F = MATLABFunction(V,'vars', {'t','Y'});% inline function
 y0=[-4,-2,0,2,4];
 tspan=[0 3];
 for k=1:length(y0)
 [T,yode]=ode45(F,tspan,y0(k));%
```

```
plot(T,yode,'color','r','linewidth',2)
end;
xlim([1,3]),ylim([-5,5])
tit1='$$ \frac{dy}{dt}=\frac{3t^2+4t}{y-1} $$';
title(tit1,'interpreter','latex','color','b')
text(2.2,1,'[numerical solution]','color','r','fontweight','bold')
%title('dy/dt=(3t^2+4t)/(y-1): Numerical solution')
figure
quiver(tt,yt,dtn,dyn,1,'b')
xlabel('time'),ylabel('y(t)'), axis tight, hold on
yy=dsolve(ff); % get the symbolic solution
vr=symvar(yy); % determine the existing symbolic variable
yy=subs(yy,vr(1),C); % replace the unknown constant to C
% use of dsolve leads to two separate solutions and they need to be
% plotted separately. See the next cell on how to avoid this issue by
% seeking the implicit solution
yy1=yy(1);
yy2=yy(2);
CC=[-20:4:16]; % choose a number of unknown constants
for k=1:length(CC)
  ffr=subs(yy1,C,sym(CC(k)));
 p1 = ezplot(ffr);
 set(p1,'Color','red', 'LineWidth', 2)
  hold on
end;
xlim([1,3])
for k=1:length(CC)
  ffr=subs(yy2,C,sym(CC(k)));
 p1 = ezplot(ffr);
 set(p1,'Color','black','Linestyle','--', 'LineWidth', 2)
  hold on
end;
tit2='$$ \frac{dy}{dt}=\frac{3t^2+4t}{y-1} $$';
title(tit2,'interpreter','latex','color','b')
text(1.8,1,{'[explicit solution # 1 (continuous lines)]';'[explicit solution # 2 (dotted
lines)]'},...
   'color','r','fontweight','bold')

xlim([1,3]),ylim([-5,5])
figure
quiver(tt,yt,dtn,dyn,1,'b')
xlabel('time'),ylabel('y(t)'), axis tight, hold on
% in this case, the method of separable equations is used.
clear y(t)
syms y t % define y and t as symbolic variables
fy=int(y-1,y);
fx=int((3*t*t+4*t),t);
fq=[fy-fx+B==0]; % create an implicit solution B is the unknown constant
CC=[4:2:16]; % choose a number of unknown constants
for k=1:length(CC)
  ffr=subs(fq,B,sym(CC(k)));
```

```
 p1 = ezplot(ffr);
 set(p1,'Color','red', 'LineWidth', 2)
  hold on
end;
tit3='$$ \frac{dy}{dt}=\frac{3t^2+4t}{y-1} $$';
 title(tit3,'interpreter','latex','color','b')
xlim([1,3]),ylim([-5,5])
ylabel('y(t)'),xlabel('time t')
text(2.2,1,'[implicit solution]','color','r','fontweight','bold')
```

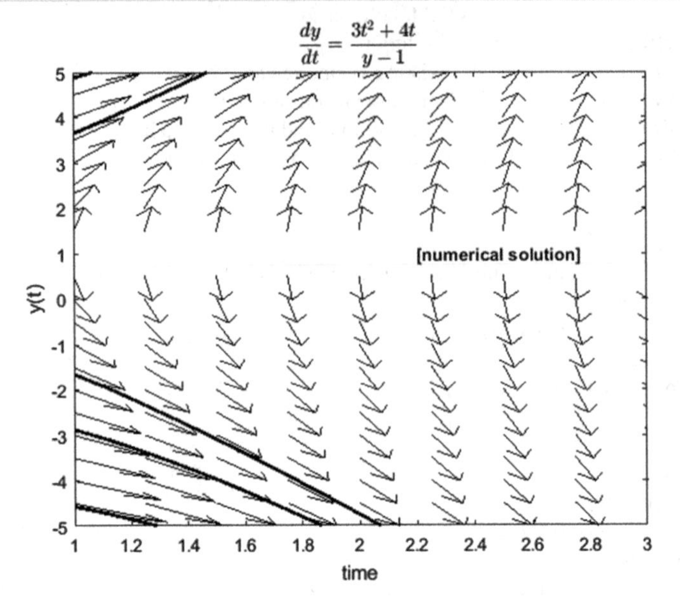

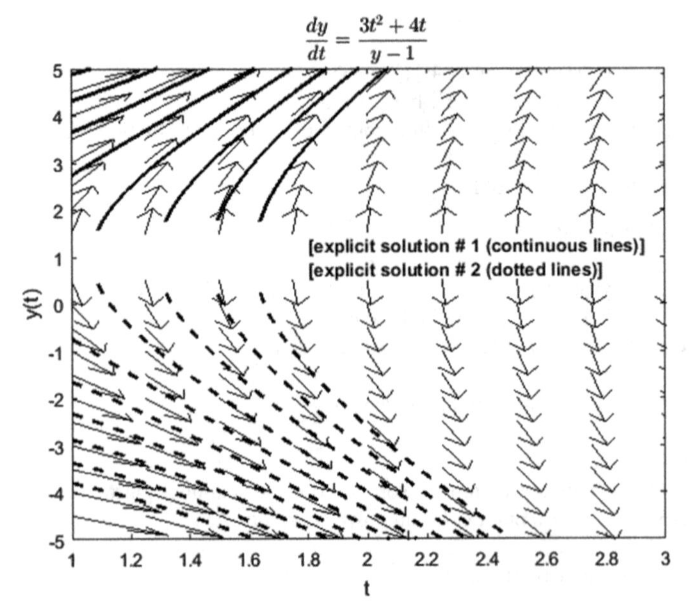

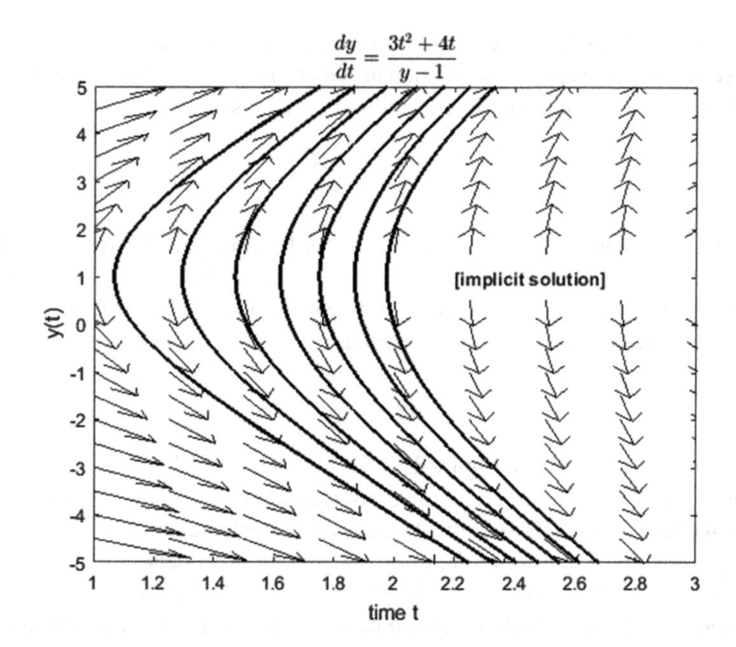

2.5 Autonomous Differential Equations

As mentioned in Section 2.1, a first order differential equation with no explicit dependence on the independent variable is called an autonomous differential equation. A typical autonomous first order differential equation can be expressed as

$$y' = \frac{dy}{dt} = f(y). \tag{2.30}$$

The right hand side of eqn. (2.30) depends only on the dependent variable y and has no explicit dependence on the independent variable t. These differential equations occur in the modeling of growth, decline, decay, or expansion of biological, chemical, physical or other systems that exhibit temporal (time dependent) changes. An autonomous eqn. such as the one in eqn. (2.30) plays an important role in understanding the stability of the system such as whether the system is likely to be stable as time passes or unstable as time passes. An example of the former might be the case of a spread of a disease that slows down with time. An example of the latter may be the unimpeded growth of nuclear radiation in the event of a nuclear disaster. One way to understand this behavior is to examine what happens to the system when equilibrium conditions are achieved. Equilibrium occurs when

$$y' = \frac{dy}{dt} = 0. \tag{2.31}$$

The resulting solutions ($y=y_0$), namely the values of y that satisfy eqn. (2.31) are classified as equilibrium solutions, equilibrium points or critical points of the system represented by the differential equation in eqn. (2.30). These identifications of the solutions to eqn. (2.31) merely reflect the fact that they correspond to absence of variation in y when t varies.

As an example to illustrate the significance of the critical points or equilibrium points, we consider the case of population growth modeled using the Gompertz's law. The first order differential equation describing growth is

$$\frac{dy}{dt} = ry \log_e\left(\frac{K}{y}\right). \tag{2.32}$$

Because the quantity modeled is growth, the function y is always positive. The two constants **r** and **K** are also positive. To understand the stability of the system described through the differential equation in eqn. (2.32), the first step is to examine a plot of the differential equation expressed as

$$f(y) = \frac{dy}{dt}. \tag{2.33}$$

The function f(y) is plotted in Figure 2.15 for K=4 and r=2.

The plot of f(y) in Figure 2.15 shows two critical points, one at y=0 and the other one at y=4 (value of K). These critical points can be obtained by solving f(y)=0 and invoking L'Hospital's rule for limits if necessary. Figure 2.15 also shows the plots of df/dy. The value of df/dy at y=0 is positive while the value of df/dy at y=4 is negative. The reasons for this classification of the two critical points become obvious if one examines the direction fields given in Figure 2.16.

It can be seen that around y=4, the solutions are converging while around y=0 the solutions are diverging, justifying the classification of the critical points as asymptotically stable (y=K=4) and unstable (y=0).

Critical points may also be classified as semi-stable. Consider the system described in terms of the following autonomous differential equation,

$$f(y) = \frac{dy}{dt} = (y-1)^2(y+1) \tag{2.34}$$

Figure 2.17 shows the plot containing f(y) and df/dy. There are two critical points, one at y=–1 and the other one at y=1. Notice that df/dy is positive at y=–1 making the EQM at y =–1 an example of the unstable equilibrium. On the other hand, df/dy is exactly equal to 0 at y=1 and it goes from positive to negative on either side of the equilibrium. This equilibrium point at y=1 is identified as semi-stable and the reasons for this classification are seen in Figure 2.18 containing the directional field. While fields move away from y=–1 indicating unstable conditions, the field shows unique characteristics around y=1. On one side of y=1, field moves towards the equilibrium point while on the other side, they move away from the equilibrium point. This is the characteristic property that separates the semi-stable equilibrium from the stable and unstable equilibrium conditions.

One can now state the rules for classifying the critical points, $y=y_0$, as the following:

If **df/dy at $y=y_0$ is positive**, the equilibrium point y_0 is unstable.
If **df/dy at $y=y_0$ is negative**, the equilibrium point y_0 is asymptotically stable.
If **df/dy at $y=y_0$ is zero and df/dy goes from +ve to –ve or –ve to +ve on either side of $y=y_0$**, the equilibrium point y_0 is semi-stable.

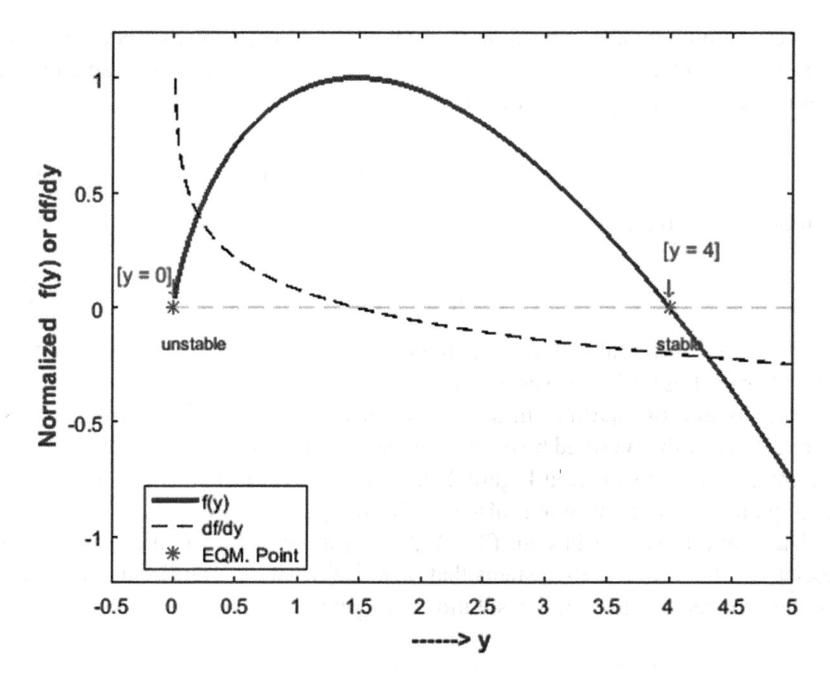

Figure 2.15 Plot of f(y) and df/dy associated with Gompertz's Equation.

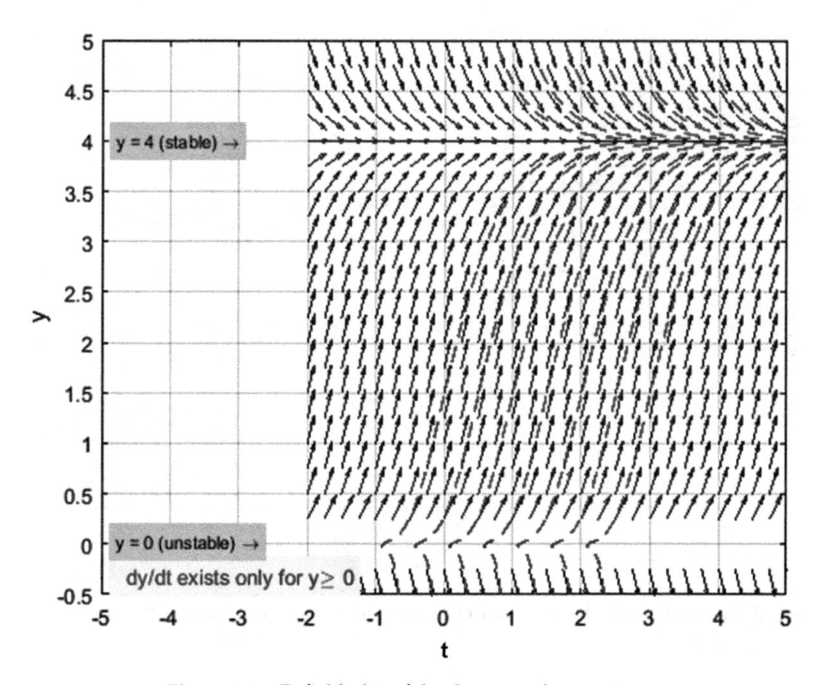

Figure 2.16 D-field plot of the Gompertz's equation.

There is another interesting class of autonomous systems which produce real and complex values for the roots of the equation f(y)=0. Consider the case of an autonomous system described by the differential equation

$$f(y) = \frac{dy}{dt} = (y-1)(y^2+y+1) \tag{2.35}$$

The roots of f(y)=0 are

$$y = 1,\ \frac{1}{2} \pm \frac{\sqrt{3}}{2} j. \tag{2.36}$$

In addition to a real root at y=1, there is a pair of roots that form a complex conjugate pair. Figure 2.19 gives a plot of f(y) and df/dy.

The absence of equilibrium state associated with the complex roots seen in Figure 2.19 is further verified by examining the directional fields associated with the differential equation shown in Figure 2.20. The reasons for the classification of the critical point at y=1 is unstable is also seen from Figures 2.19 and 2.20.

The example shown in eqn. (2.35) also illustrates another interesting aspect associated with autonomous systems that there is no state of equilibrium associated with those systems with complex solutions of f(y)=0.

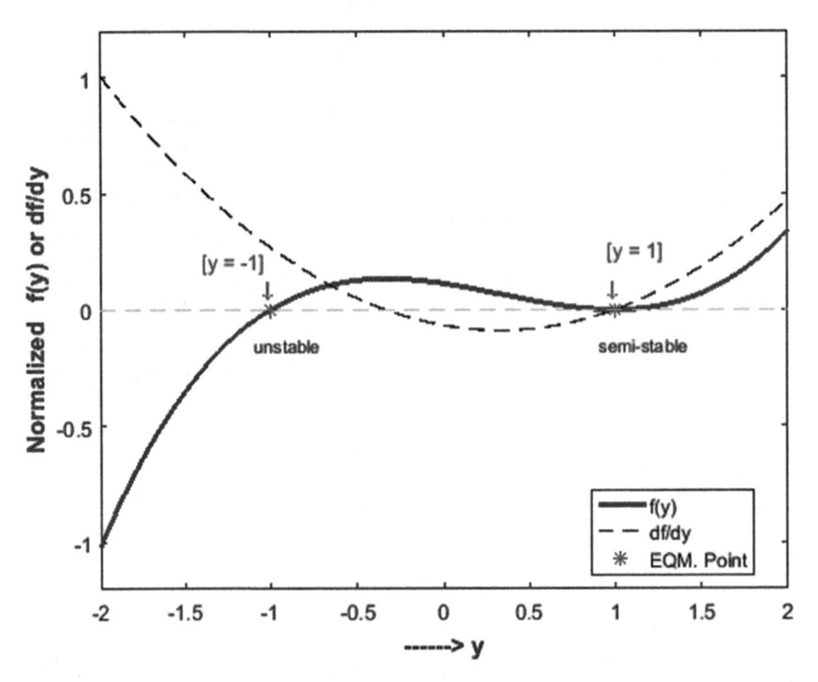

Figure 2.17 Plots of f(y) and df/dy vs. y for the autonomous system described by $f(y)=(1-y)^2(1+y)$.

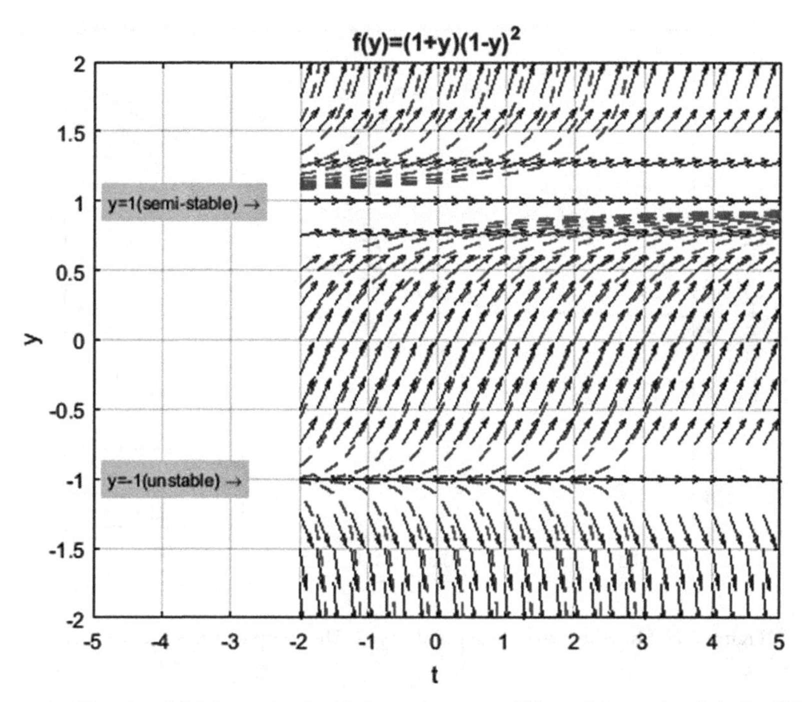

Figure 2.18 Directional field associated with the autonomous differential equation $f(y)=(1-y)^2(1+y)$.

2.5.1 A general example

An example that can illustrate all these form of equilibria can be built on an autonomous first order differential equation such as a quadratic equation,

$$f(y)=\frac{dy}{dt}=ay^2+by+c. \tag{2.37}$$

Depending on the values of a, b, and c, the autonomous system described in eqn. (2.37) will encompass all the three types of critical points as well as absence of equilibrium (complex roots of f(y)). The various cases arising out of the different values of a, b, and c are shown below. For each case, two displays are provided, one containing the plots of f(y) and df/dy with the stability state clearly marked and the other one containing the D-field plot with the stability values shown. The D-field plots can be compared to the plots of f(y) vs. y and df/dy vs. y showing the matching descriptions of the state of equilibrium associated with the autonomous system described by the differential equation in eqn. (2.37).

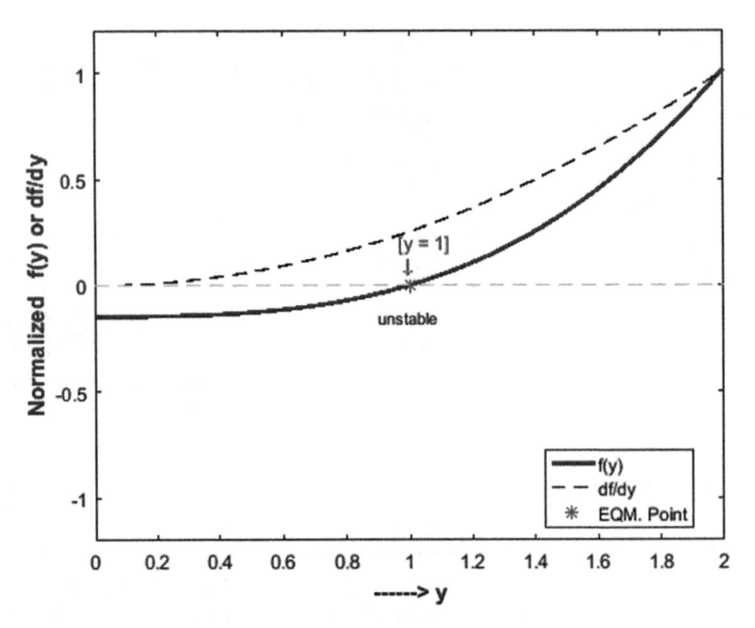

Figure 2.19 Showing the critical point at y=1. The complex roots are not seen.

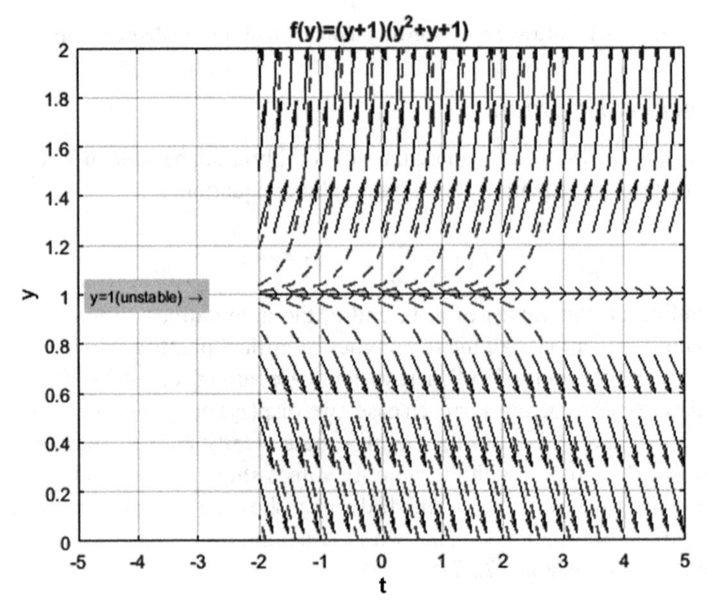

Figure 2.20 Directional field associated with the autonomous differential equation $f(y)=(1+y)(1+y+y^2)$.

Case 1 a=–1, b=–1, c=0: One asymptotically stable (y=0) and one unstable solution (y=–1)

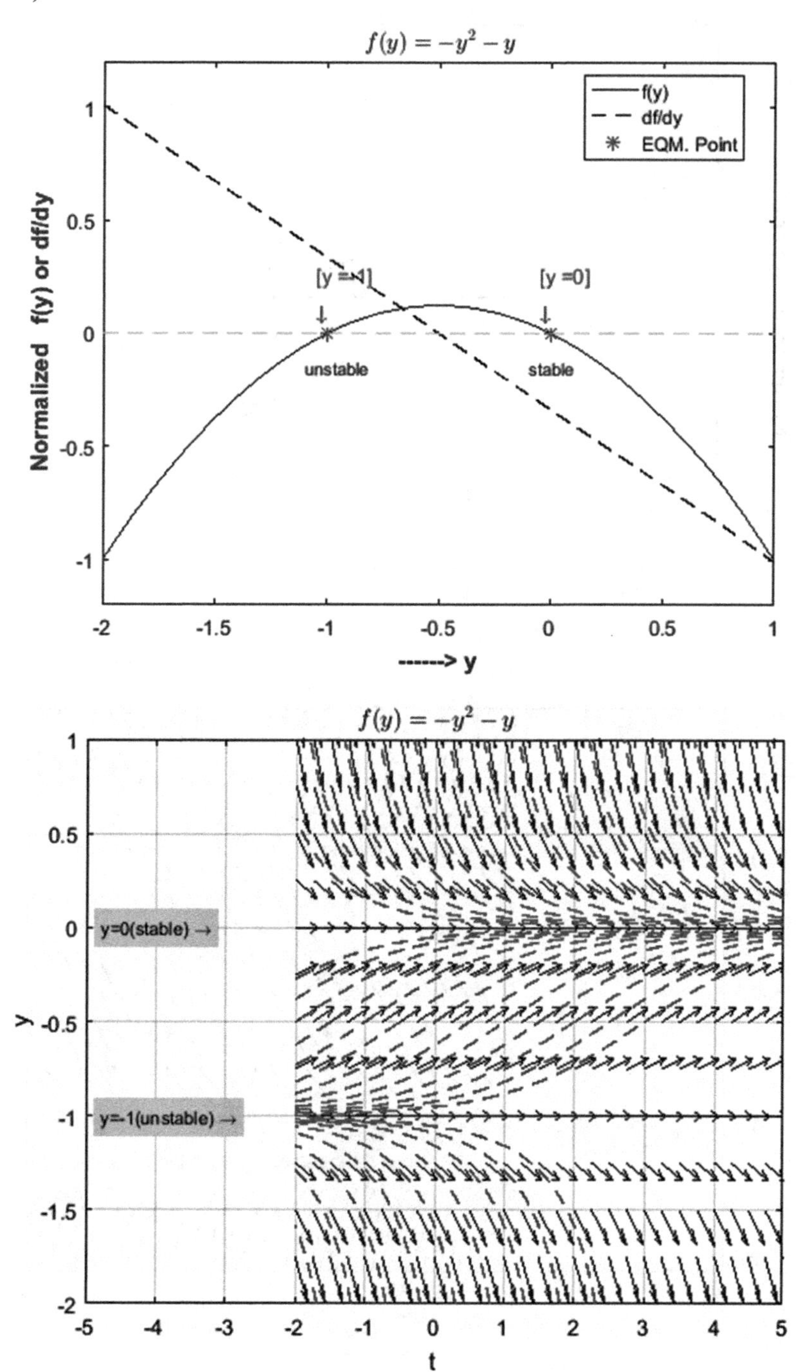

Case 2 a=–1, b=–1, c=1: One asymptotically stable (y=0.62) and one unstable solution (y=–1.62)

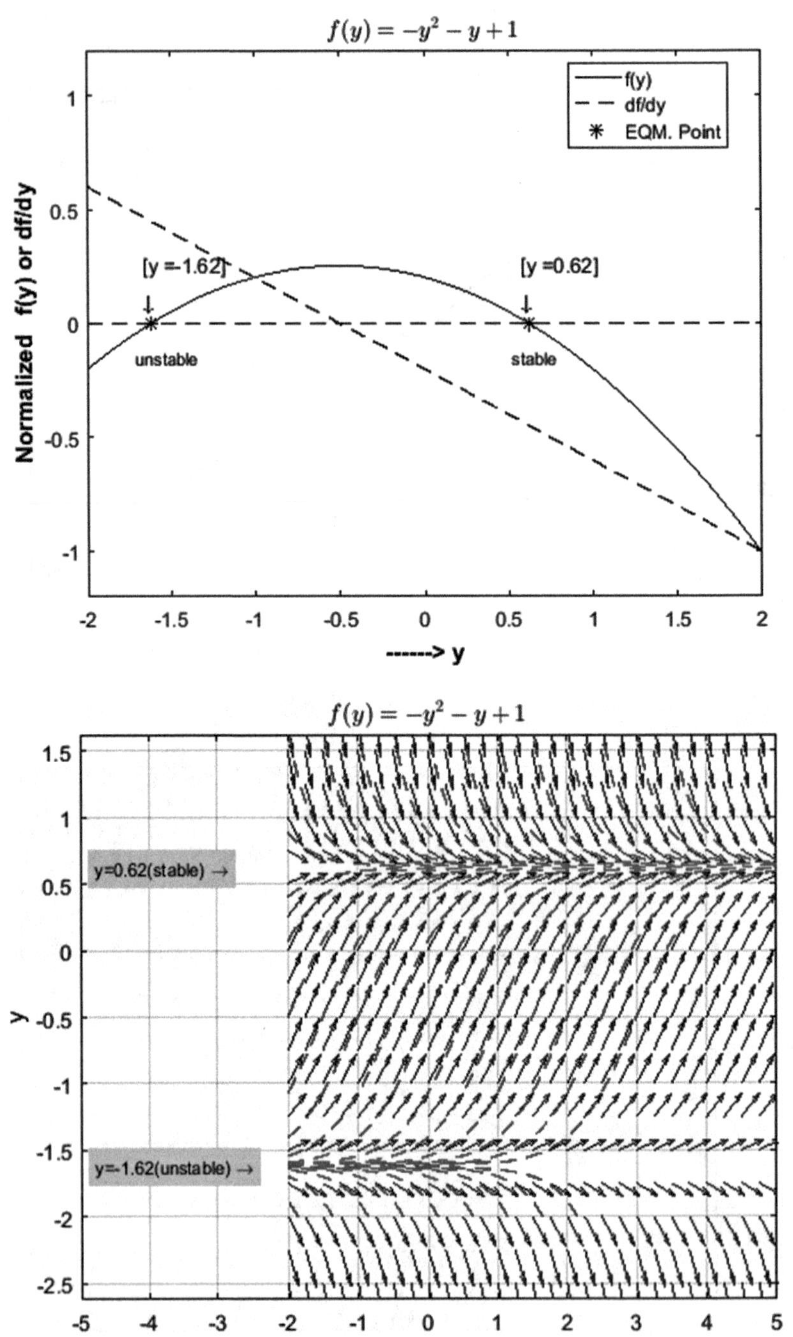

Case 3 a=–1, b=0, c=0: a single semi-stable solution (y=0)

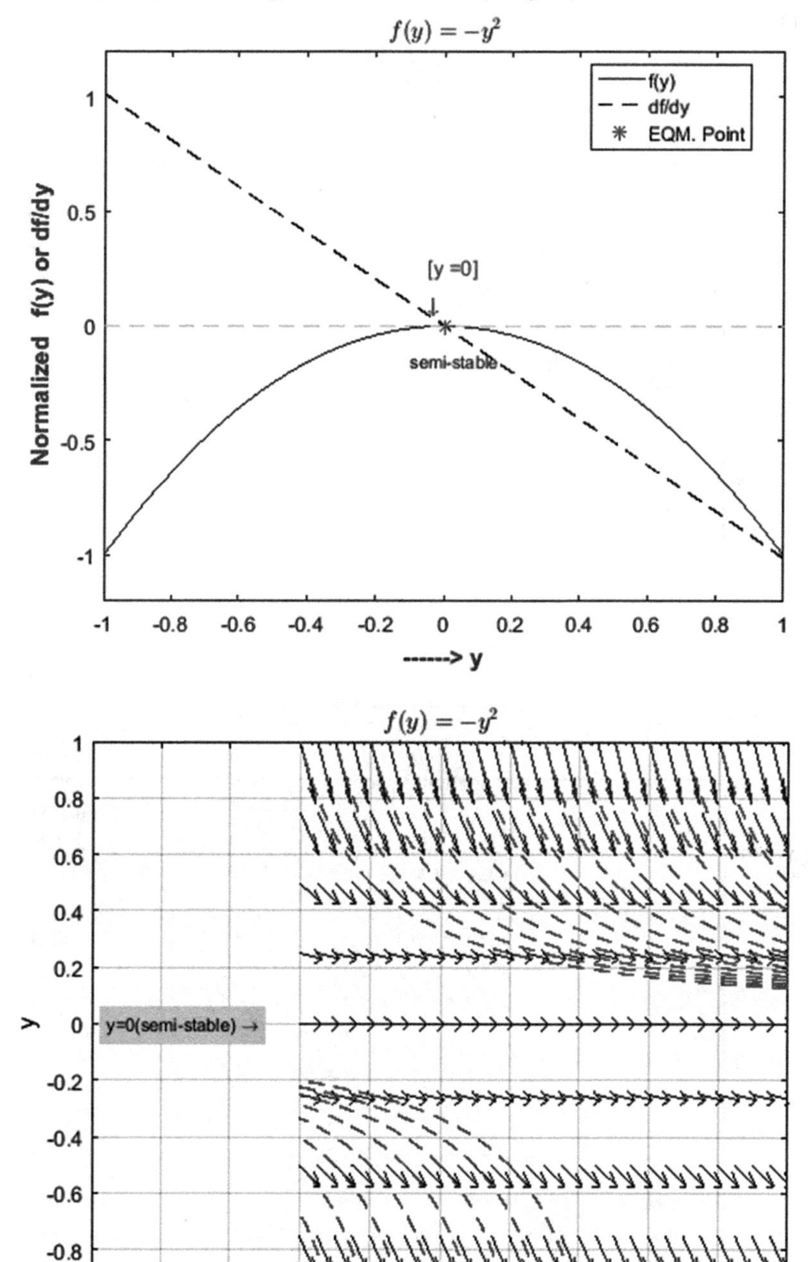

Case 4 a=–1, b=0, c=1: One asymptotically stable (y=1) and one unstable solution (y=–1)

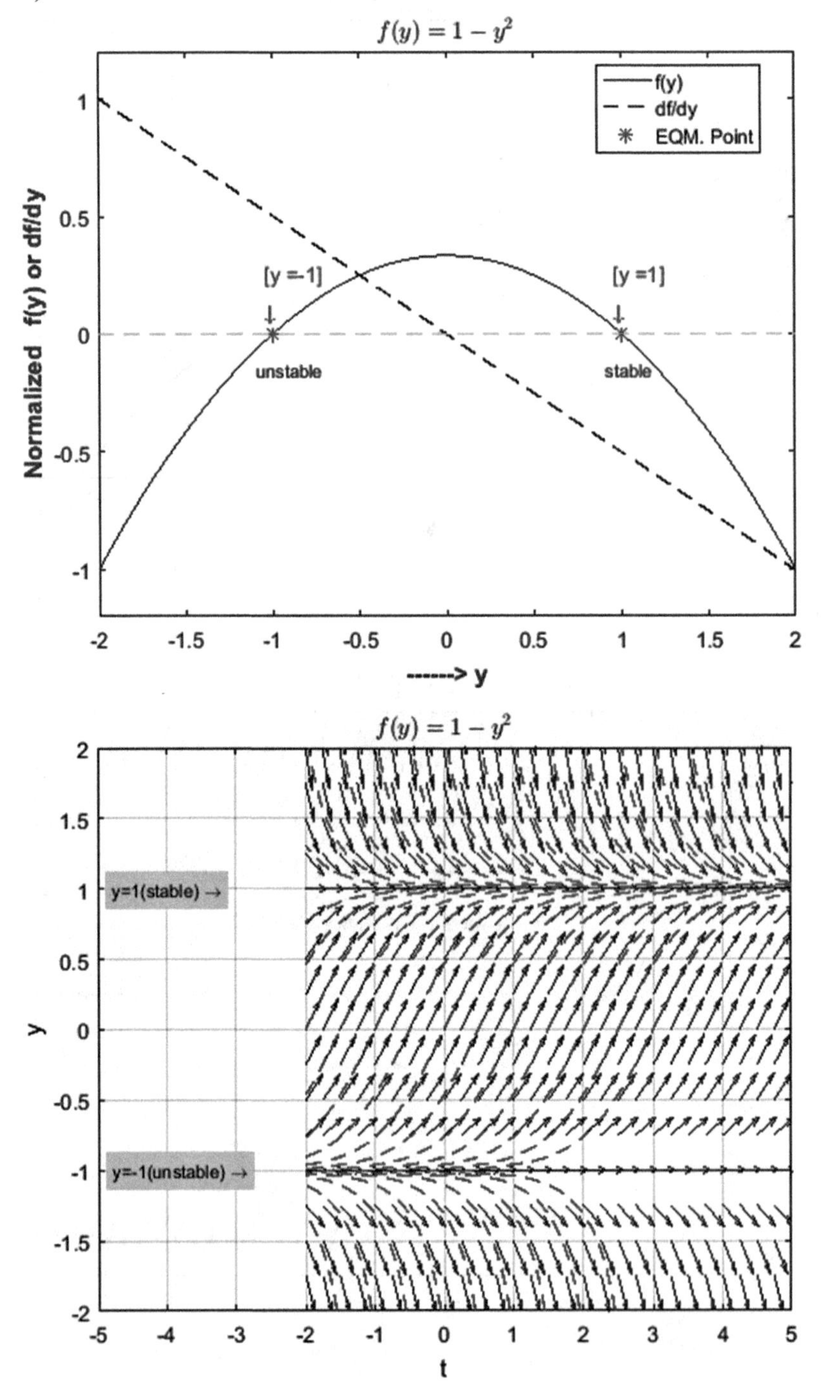

Case 5 a=–1, b=1, c=0: One asymptotically stable (y=1) and one unstable solution (y=0)

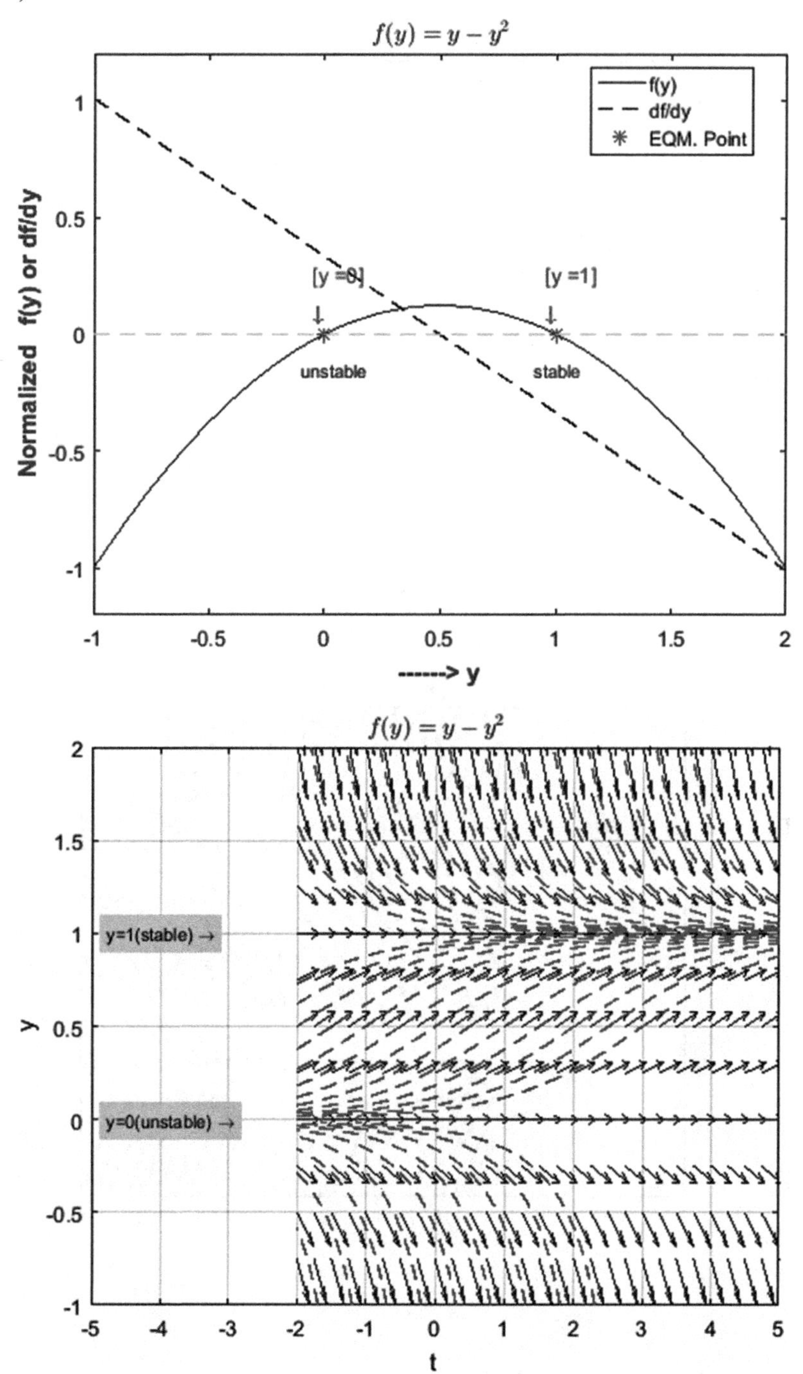

Case 6 a=–1, b=1, c=1: One asymptotically stable (y=1.62) and one unstable solution (y=–0.62)

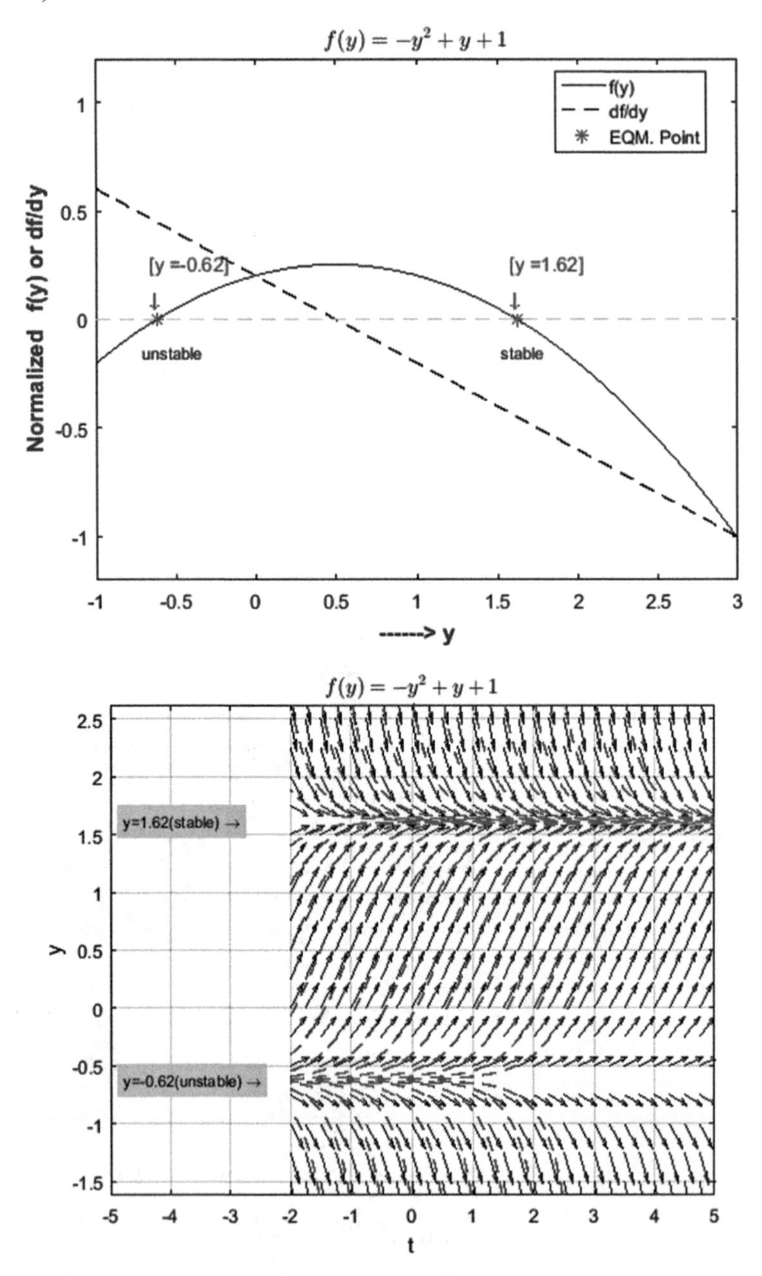

Case 7 a=1, b=–1, c=1: Equilibrium conditions do not exist

$$f(y) = y^2 - y + 1$$

In this case, the two roots form a complex pair

$$\frac{1 \pm i\sqrt{3}}{2}.$$

For this set of values of a, b, and c, no equilibrium conditions exist for the autonomous system.

Case 8 a=1, b=0, c=0: A single semi-stable solution (y=0)

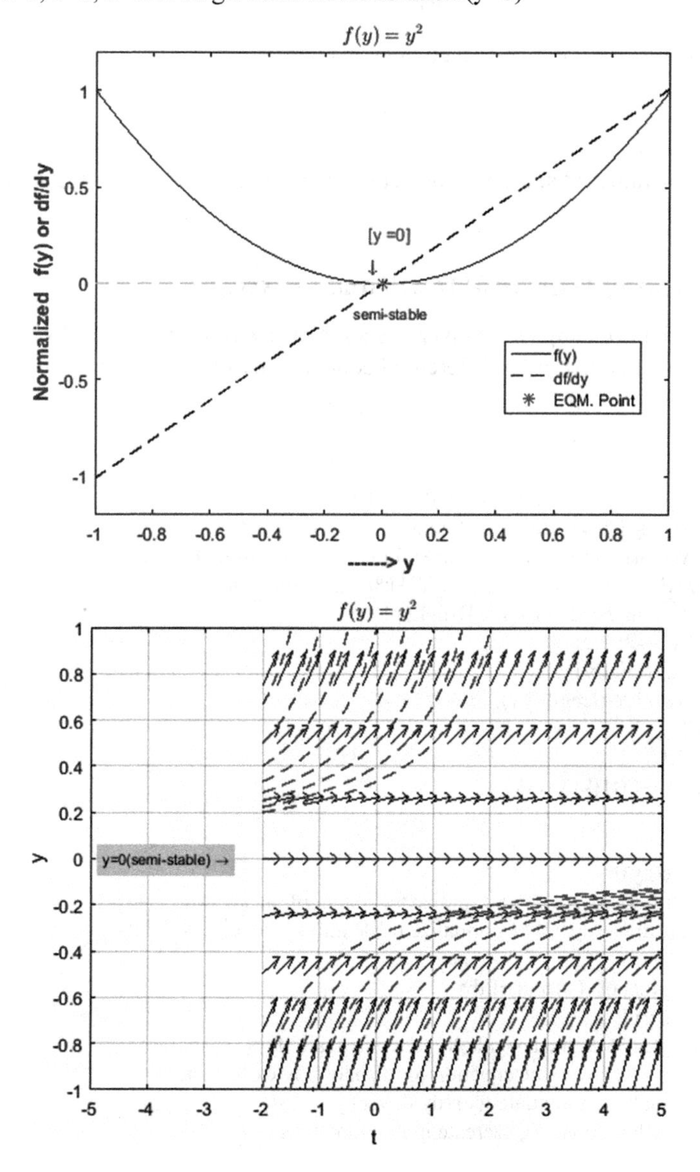

Case 9 a=1, b=0 and c=1: No equilibrium solutions

$$f(y) = 1 + y^2$$

In this case, the two roots form a complex pair

$$0 \pm i\,.$$

For this set of values of a, b, and c, no equilibrium conditions exist for the autonomous system.
Case 10 a=1, b=1, c=1 No equilibrium solutions

$$f(y) = y^2 + y + 1$$

In this case, the two roots form a complex pair

$$\frac{-1 \pm i\sqrt{3}}{2}.$$

For this set of values of a, b, and c, no equilibrium conditions exist for the autonomous system.

2.5.2 An example using MATLAB and the script

Before additional examples are given, a short MATLAB script and results for the case of an autonomous first order differential equation, dy/dt=(1+y)(1–y), are provided.

```
close all; clear
syms y(t)
dy_dt='(1+y)*(1-y)';
fy=eval(dy_dt); % creates fy=(1+y(t))*(1-y(t));
ff=[diff(y,t)==fy];% create the differential equation Dy=f(t,y)
V=odeToVectorField(ff);% vector field for creation of in-line Function
F=MATLABFunction(V,'vars', {'t','Y'});% inline function
% create the data needed for the D-field plot
t1=-2:.2:2;y1=-2:.2:2;
[tt,yt]=meshgrid(t1,y1);
LT=length(t1);LY=length(y1);
for k=1:LT
   for kk=1:LY
      dy(kk,k)=F(t1(k),y1(kk));
   end;
end;
dt=ones(size(dy));
dtn=dt./sqrt(dt.^2+dy.^2); % normalize the length of the arrows
dyn=dy./sqrt(dt.^2+dy.^2);% normalize the length of the arrows
clear y(t)
% generate plots of f(y) and df/dy
syms y t fy f(y) C
fy=eval(dy_dt); % create a symbolic function again: fy=(1+y)*(1-y)
ffy=[f(y)==fy];% create the equation for display (LaTex format
fdy=diff(fy,y); % differentiate f(y)=dy/dt w.r.t y
fin=MATLABFunction(fy); %create in line functions
```

```
fdn=MATLABFunction(fdy); %create in line functions
solF=[int(1/fy,y)-t+C==0]; %implicit solution using separation of equations
% plot everything
yn1=-2:.2:2;
plot(yn1,fin(yn1),'r-',yn1,fdn(yn1),'--k')
legend('f(y)','df/dy')
hold on
plot(yn1,zeros(1,length(yn1)),'color','b','linewidth',1.5)%line through y=0
xlabel('y'),ylabel('f(y) or df/dy')
title(['$' latex(ffy) '$'],'interpreter','latex','color','b')
fs=solve(fy==0,y); % get the solutions of f(y)=0; EQM points
plot(double(fs(1)),0,'*g',double(fs(2)),0,'*g') % mark the EQM points
text(double(fs(1))-.3,-.3,...
   ['EQM Point=',num2str(double(fs(1)))]) % label EQM points
text(double(fs(2))-.3,-.3,...
   ['EQM Point =',num2str(double(fs(2)))])
text(-.8,3.1,'Implicit Solution')
text(-1,2.5,['$' latex(solF) '$'],'interpreter','latex',...
   'color','b','fontweight','bold','fontsize',16) % display solution

figure, quiver(tt,yt,dtn,dyn,.75,'k')
xlim([-2,2])
xlabel('time t'),ylabel('y(t)'), axis tight
title(['$' latex(ffy) '$'],'interpreter','latex','color','b')
```

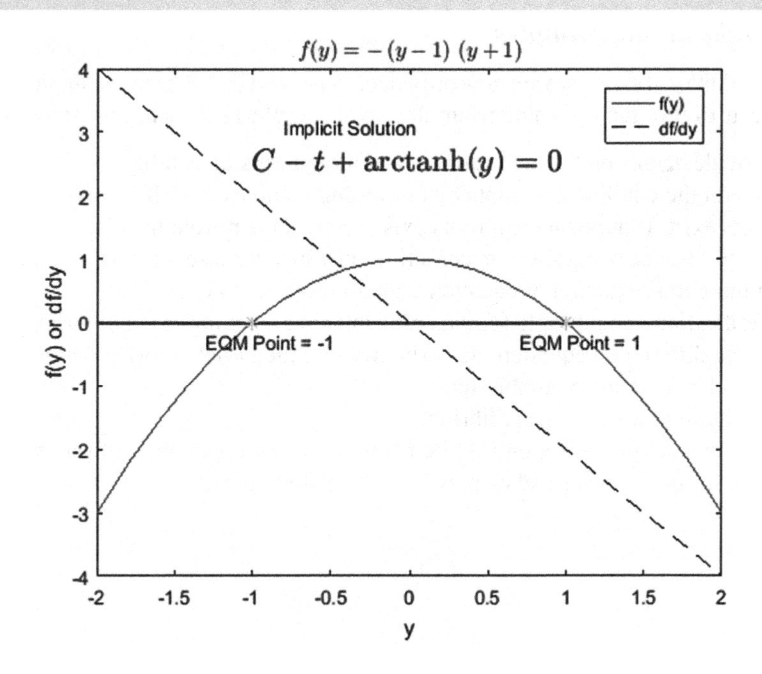

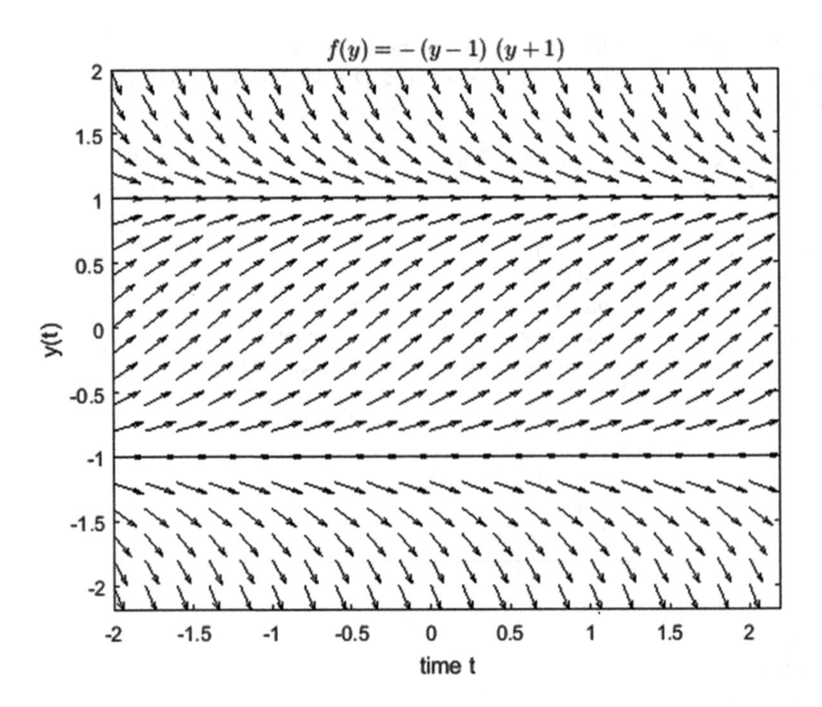

2.5.3 Additional examples

Several additional examples are now provided. These MATLAB based examples cover all aspects of the study of equilibrium displaying results as the analysis proceeds.

1. Provide details on the definitions of the three forms of stability.
2. Obtain the equilibrium points by discarding complex ones if complex and real roots exist. If duplicate real roots exist, remove them from the plots.
3. In the absence of equilibrium conditions in a specific case, state the reason (non-autonomous equation or equation with all roots complex).
4. Verification and multiple formats of displaying the points of equilibrium (plots of the differential equation, its derivative and the D-field plot) and indicate the equilibrium points with the state.
5. Justify the state of the equilibrium.
6. Obtain analytical solution (implicit solution using the method of separation of functions) if it exists and display it on the D-field plot.

7. If analytical solution is unavailable, use numerical techniques to obtain the D-field plot and indicate the absence of the analytical solution.
8. In every example, the first display provides the 'rules' while the last display provides a summary of the analysis.

```
dy_dt='y*(1-y/2)*(1-y/3)'
```

Stability of Autonomous Systems

Autonomous Systems ⇒ $\dfrac{dy}{dt} = f(t, y) \equiv f(y)$

Solve for the roots ⇒ $\dfrac{dy}{dt} = f(y) = 0 \Rightarrow f(y_r) = 0$

Equilibrium Points (y_0) correspond to Real Roots

If Real and Complex Roots exist, discard Complex Ones

Examine behavior of df/dy at and around the Equilibrium Points (y_0)

Rules for determining the stability of Equilibrium Points (y_0)
$df/dy\|_{y=y_0}$ is +ve: unstable
$df/dy\|_{y=y_0}$ is -ve: asymptotically stable
$df/dy\|_{y=y_0}$ = 0 & $df/dy\|_{\|y\|>y_0}$ goes from +/-ve or -/+ve: semi-Stable

Equilibrium Point(s) & Stability

Input f=dy/dt ⇒ $f(y) = y\left(\frac{y}{2} - 1\right)\left(\frac{y}{3} - 1\right)$

Range of roots of [f=0]⇒ [- ∞ + ∞]

Derivative of f w.r.t y $f'(y) = \frac{y^2}{2} - \frac{5y}{3} + 1$

Get roots (y_r) of [f=0]; REAL roots (y_0) are the EQM Points

Roots (y_r) ⇒ [0 2 3]

All roots y_r are Real (y_0); Examine df/dy at y = y_0 for stability

EQM Points y_0 ⇒ [0 2 3]

Examine & justify Stability
EQM at y = 0 is unstable; df/dy is +ve
EQM at y = 2 is asymptotically stable; df/dy is -ve
EQM at y = 3 is unstable; df/dy is +ve

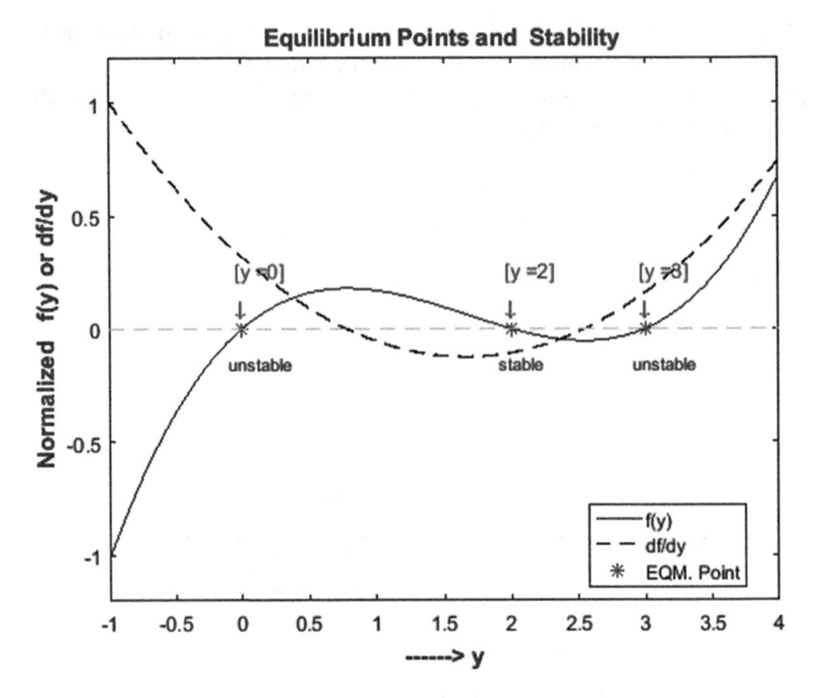
Equilibrium Points and Stability
Normalized f(y) or df/dy
------> y
[y =0]
[y =2]
[y =3]
unstable
stable
unstable
f(y)
df/dy
EQM. Point

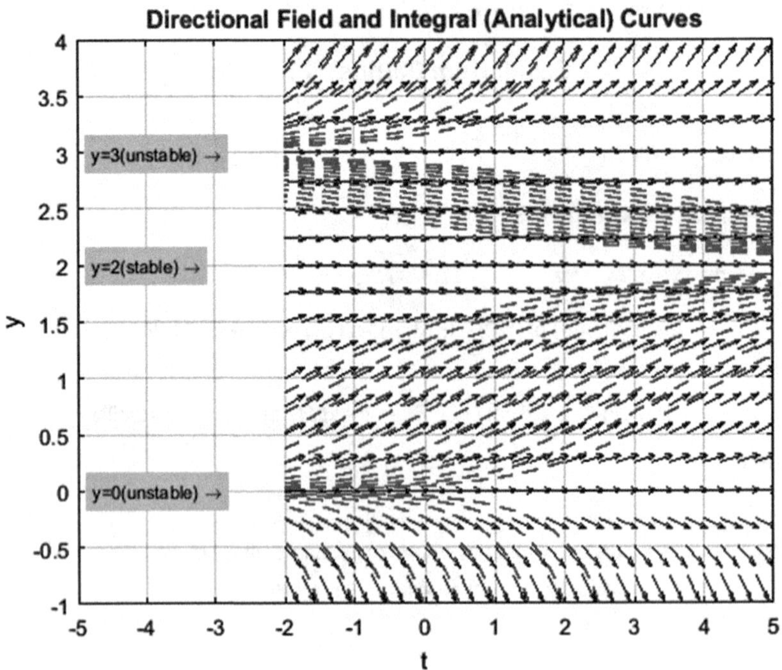
Directional Field and Integral (Analytical) Curves
y=3(unstable) →
y=2(stable) →
y=0(unstable) →
y
t

Summary of Equilibrium Study

Input dy/dt $f(y) = y\left(\frac{y}{2} - 1\right)\left(\frac{y}{3} - 1\right)$

EQM Points y_0 = [0 2 3]

EQM at y = 0 is unstable; df/dy is +ve
EQM at y = 2 is asymptotically stable; df/dy is -ve
EQM at y = 3 is unstable; df/dy is +ve

Implicit Solution of the Autonomous Differential Equation

$$C + 2\log(y - 3) + \log(y) = t + 3\log(y - 2)$$

```
dy_dt='y^2+y+1'
```

Stability of Autonomous Systems

Autonomous Systems ⇒ $\dfrac{dy}{dt} = f(t, y) \equiv f(y)$

Solve for the roots ⇒ $\dfrac{dy}{dt} = f(y) = 0 \Rightarrow f(y_r) = 0$

Equilibrium Points (y_0) correspond to Real Roots

If Real and Complex Roots exist, discard Complex Ones

Examine behavior of df/dy at and around the Equilibrium Points (y_0)

Rules for determining the stability of Equilibrium Points (y_0)
$df/dy|_{y=y_0}$ is +ve: unstable
$df/dy|_{y=y_0}$ is -ve: asymptotically stable
$df/dy|_{y=y_0} = 0$ & $df/dy|_{\lvert y\rvert>y_0}$ goes from +/-ve or -/+ve: semi-Stable

Equilibrium Point(s) & Stability

Input f=dy/dt ⇒ $$f(y) = y^2 + y + 1$$

Range of roots of [f=0]⇒ [- ∞ + ∞]

Derivative of f w.r.t y $$f'(y) = 2y + 1$$

Get roots (y_r) of [f=0]; REAL roots (y_0) are the EQM Points

All roots COMPLEX. No EQM POINTS. Roots given below

-0.5-0.87i
-0.5+0.87i

```
dy_dt='(y-1)*(y^3+1)';
```

Stability of Autonomous Systems

Autonomous Systems ⇒ $$\frac{dy}{dt} = f(t, y) \equiv f(y)$$

Solve for the roots ⇒ $$\frac{dy}{dt} = f(y) = 0 \Rightarrow f(y_r) = 0$$

Equilibrium Points (y_0) correspond to Real Roots

If Real and Complex Roots exist, discard Complex Ones

Examine behavior of df/dy at and around the Equilibrium Points (y_0)

Rules for determining the stability of Equilibrium Points (y_0)
$df/dy\|_{y=y_0}$ is +ve: unstable
$df/dy\|_{y=y_0}$ is -ve: asymptotically stable
$df/dy\|_{y=y_0}$ = 0 & $df/dy\|_{\|y\|>y_0}$ goes from +/-ve or -/+ve: semi-Stable

Equilibrium Point(s) & Stability

Input f=dy/dt ⇒ $$f(y) = (y^3 + 1)\ (y - 1)$$

Range of roots of [f=0]⇒ [- ∞ + ∞]

Derivative of f w.r.t y $$f'(y) = 4\,y^3 - 3\,y^2 + 1$$

Get roots (y_r) of [f=0]; REAL roots (y_0) are the EQM Points

Complex Roots (y_r) [-1+0i 1+0i 0.5+0.87i 0.5-0.87i]

2 of 4 roots REAL (y_0). Examine df/dy at y = y_0 for stability

EQM Points y_0 ⇒ [-1 1], after discarding complex ones

Examine & justify Stability EQM at y = -1 is asymptotically stable; df/dy is -ve

EQM at y = 1 is unstable; df/dy is +ve

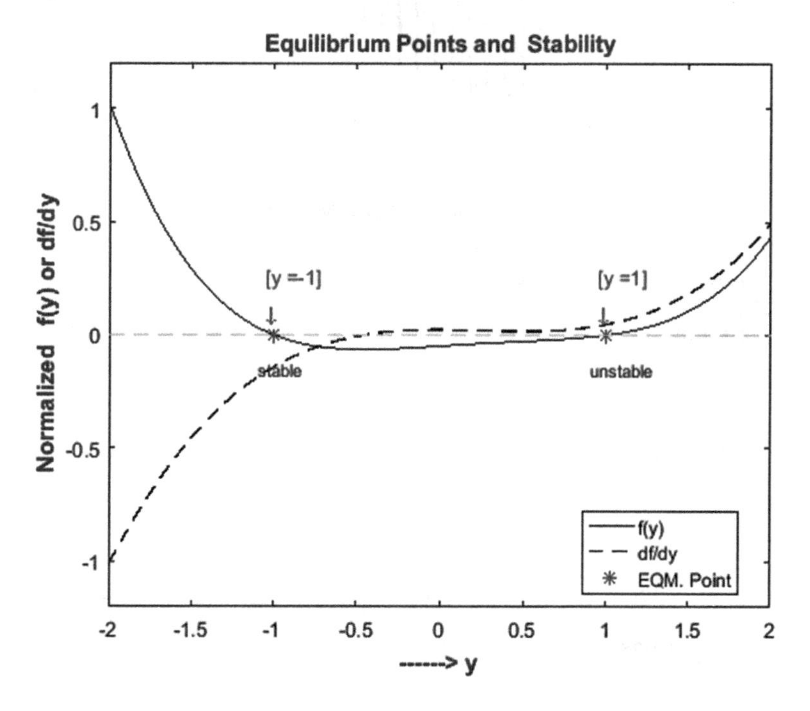

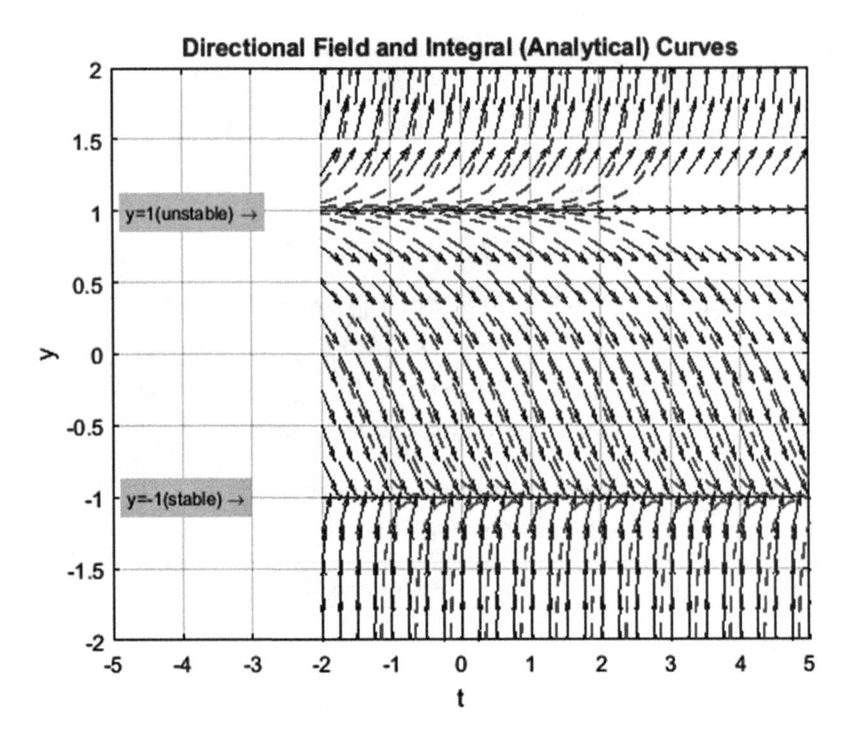

Summary of Equilibrium Study

Input dy/dt $f(y) = (y^3 + 1)\ (y - 1)$

EQM Points y_0 = [-1 1]

EQM at y = -1 is asymptotically stable; df/dy is -ve
EQM at y = 1 is unstable; df/dy is +ve

Implicit Solution of the Autonomous Differential Equation

$$6C + 3\log(y-1) + \log\left(y - \frac{1}{2} - \frac{\sqrt{3}\mathrm{i}}{2}\right)\ (-1 + \sqrt{3}\mathrm{i}) = 6t + \log(y+1) + \log\left(y - \frac{1}{2} + \frac{\sqrt{3}\mathrm{i}}{2}\right)\ (1 + \sqrt{3}\mathrm{i})$$

```
dy_dt='y^2*(y^2-1/25)';
```

Stability of Autonomous Systems

Autonomous Systems ⇒ $\frac{dy}{dt} = f(t,y) \equiv f(y)$

Solve for the roots ⇒ $\frac{dy}{dt} = f(y) = 0 \Rightarrow f(y_r) = 0$

Equilibrium Points (y_0) correspond to Real Roots

If Real and Complex Roots exist, discard Complex Ones

Examine behavior of df/dy at and around the Equilibrium Points (y_0)

Rules for determining the stability of Equilibrium Points (y_0)
$df/dy\|_{y=y_0}$ is +ve: unstable
$df/dy\|_{y=y_0}$ is -ve: asymptotically stable
$df/dy\|_{y=y_0}$ = 0 & $df/dy\|_{\|y\|>y_0}$ goes from +/-ve or -/+ve: semi-Stable

Equilibrium Point(s) & Stability

Input f=dy/dt ⇒ $f(y) = y^2\left(y^2 - \frac{1}{25}\right)$

Range of roots of [f=0]⇒ [- ∞ + ∞]

Derivative of f w.r.t y $f'(y) = 4y^3 - \frac{2y}{25}$

Get roots (y_r) of [f=0]; REAL roots (y_0) are the EQM Points

Roots (y_r) ⇒ [-0.2 0 0 0.2]

All roots y_r are Real (y_0); Examine df/dy at y = y_0 for stability

EQM Points y_0 ⇒ [-0.2 0 0.2] after discrading duplicates

Examine & justify Stability
EQM at y = -0.2 is asymptotically stable; df/dy is -ve
EQM at y = 0 is semi-stable; df/dy=0 & goes from +/-ve <--> -/+ve
EQM at y = 0.2 is unstable; df/dy is +ve

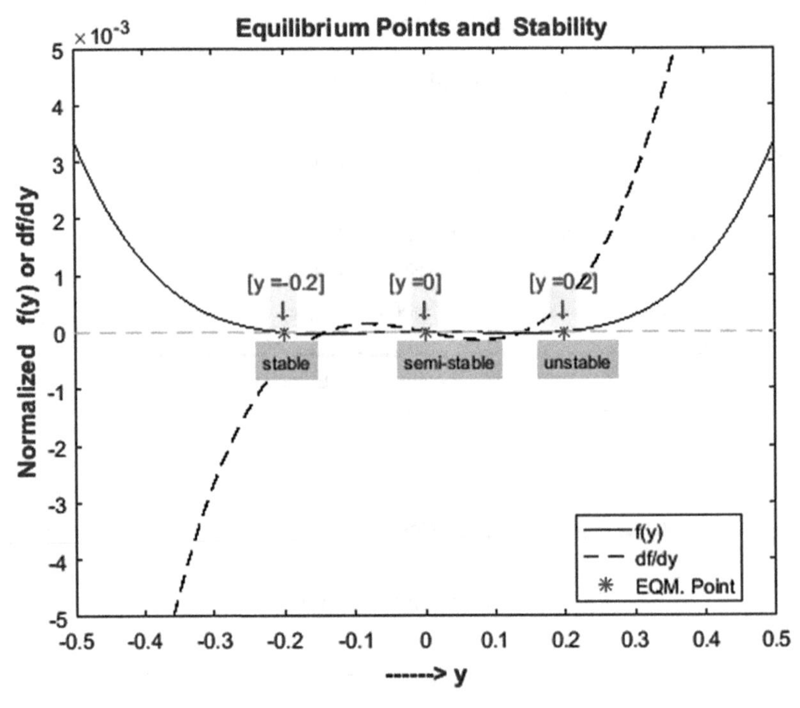
Equilibrium Points and Stability
Normalized f(y) or df/dy
×10^-3
[y =-0.2]
[y =0]
[y =0.2]
stable
semi-stable
unstable
f(y)
df/dy
EQM. Point
------> y

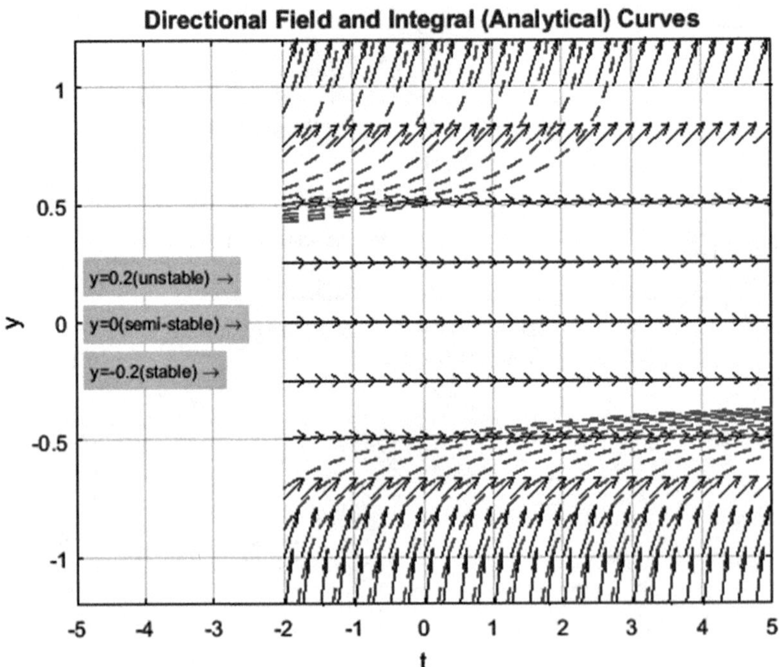
Directional Field and Integral (Analytical) Curves
y=0.2(unstable) →
y=0(semi-stable) →
y=-0.2(stable) →
y
t

Summary of Equilibrium Study

Input dy/dt $f(y) = y^2\ \left(y^2 - \frac{1}{25}\right)$

EQM Points y_0 = [-0.2 0 0.2]

EQM at y = -0.2 is asymptotically stable; df/dy is -ve
EQM at y = 0 is semi-stable; df/dy=0 & goes from +/-ve <--> -/+ve
EQM at y = 0.2 is unstable; df/dy is +ve

Implicit Solution of the Autonomous Differential Equation

$$y\ (t + 125\ \text{arctanh}(5\,y)) = C\,y + 25$$

For the remaining examples, the display of the rules has been suppressed.

```
dy_dt='2*y-sqrt(y)';
```

Equilibrium Point(s) & Stability

Input f=dy/dt ⇒ $f(y) = 2\,y - \sqrt{y}$

Range of roots of [f=0]⇒ Range of the roots of [f=0]--> [0 ∞]

Derivative of f w.r.t y $f'(y) = 2 - \frac{1}{2\,\sqrt{y}}$

Get roots (y_r) of [f=0]; REAL roots (y_0) are the EQM Points

Roots (y_r) ⇒ [0 0.25]

All roots y_r are Real (y_0); Examine df/dy at y = y_0 for stability

EQM Points y_0 ⇒ [0 0.25]

Examine & justify Stability
EQM at y = 0 is asymptotically stable; df/dy is -ve
EQM at y = 0.25 is unstable; df/dy is +ve

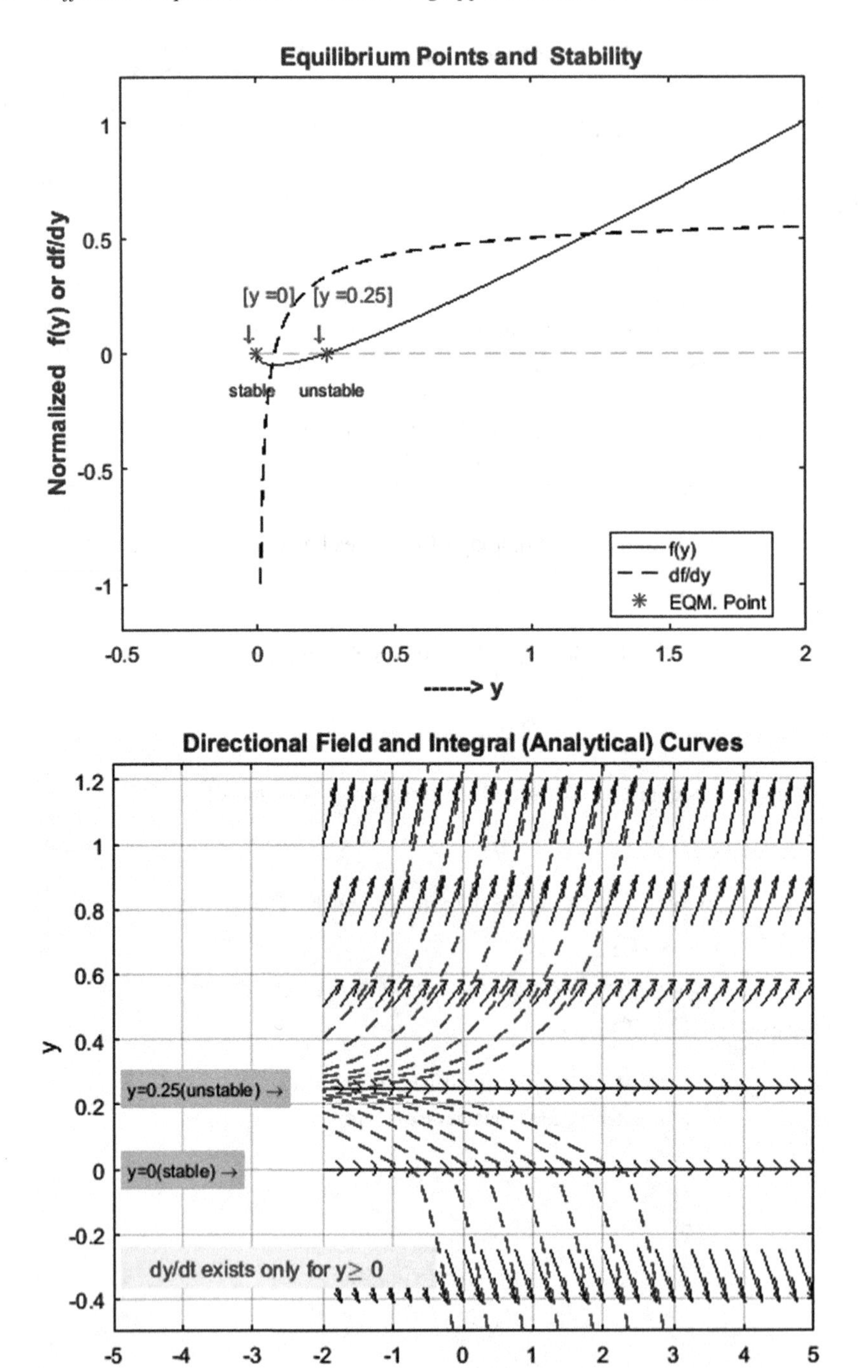
Equilibrium Points and Stability
Normalized f(y) or df/dy
[y =0]
[y =0.25]
stable
unstable
f(y)
df/dy
EQM. Point
------> y
Directional Field and Integral (Analytical) Curves
y
y=0.25(unstable) →
y=0(stable) →
dy/dt exists only for y≥ 0
t

Summary of Equilibrium Study

Input dy/dt $f(y) = 2\,y - \sqrt{y}$

EQM Points y_0 = [0 0.25]

EQM at y = 0 is asymptotically stable; df/dy is -ve
EQM at y = 0.25 is unstable; df/dy is +ve

Implicit Solution of the Autonomous Differential Equation

$$C + \log\left(\sqrt{y} - \tfrac{1}{2}\right) = t$$

```
dy_dt='(1+y)*(y-1)^2';
```

Equilibrium Point(s) & Stability

Input f=dy/dt ⇒ $f(y) = (y-1)^2\,(y+1)$

Range of roots of [f=0]⇒ [- ∞ + ∞]

Derivative of f w.r.t y $f'(y) = 3\,y^2 - 2\,y - 1$

Get roots (y_r) of [f=0]; REAL roots (y_0) are the EQM Points

Roots (y_r) ⇒ [-1 1 1]

All roots y_r are Real (y_0); Examine df/dy at y = y_0 for stability

EQM Points y_0 ⇒ [-1 1] after discrading duplicates

Examine & justify Stability
EQM at y = -1 is unstable; df/dy is +ve
EQM at y = 1 is semi-stable; df/dy=0 & goes from +/-ve <--> -/+ve

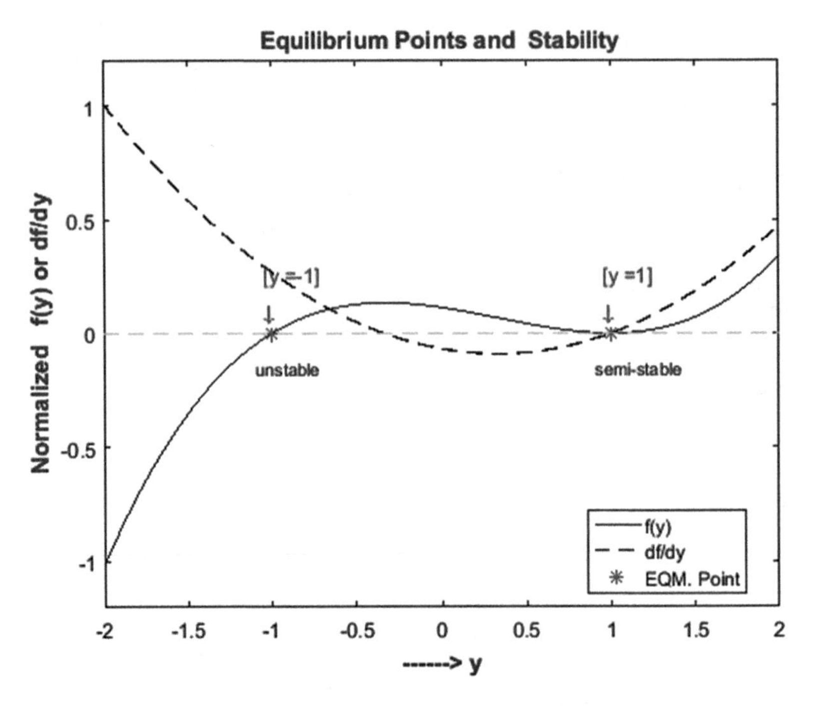
Equilibrium Points and Stability
Normalized f(y) or df/dy
------> y
[y=-1]
[y =1]
unstable
semi-stable
f(y)
df/dy
EQM. Point

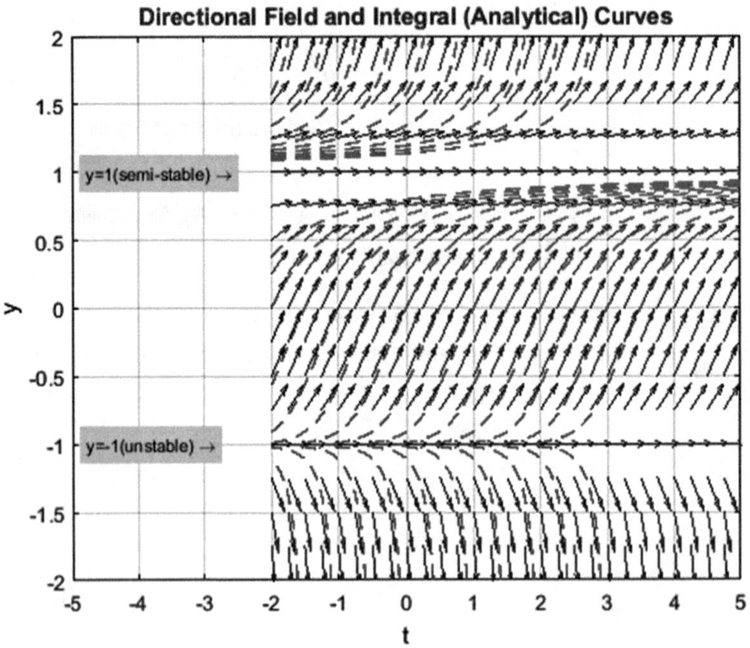
Directional Field and Integral (Analytical) Curves
y=1(semi-stable) →
y=-1(unstable) →
y
t

Summary of Equilibrium Study

Input dy/dt $f(y) = (y-1)^2\ (y+1)$

EQM Points y_0 = [-1 1]

EQM at y = -1 is unstable; df/dy is +ve
EQM at y = 1 is semi-stable; df/dy=0 & goes from +/-ve <--> -/+ve

Implicit Solution of the Autonomous Differential Equation

$$\operatorname{arctanh}(y)\ (y-1) + 2C\ (y-1) = 2t\ (y-1) + 1$$

```
dy_dt='(y-1)*(y^2+y+1)';
```

Equilibrium Point(s) & Stability

Input f=dy/dt ⇒ $f(y) = (y-1)\ (y^2+y+1)$

Range of roots of [f=0]⇒ [- ∞ + ∞]

Derivative of f w.r.t y $f'(y) = 3\,y^2$

Get roots (y_r) of [f=0]; REAL roots (y_0) are the EQM Points

Complex Roots (y_r) [1+0i -0.5+0.87i -0.5-0.87i]

1 of 3 roots REAL (y_0). Examine df/dy at y = y_0 for stability

EQM Points y_0 ⇒ [1], after discarding complex ones

Examine & justify Stability EQM at y = 1 is unstable; df/dy is +ve

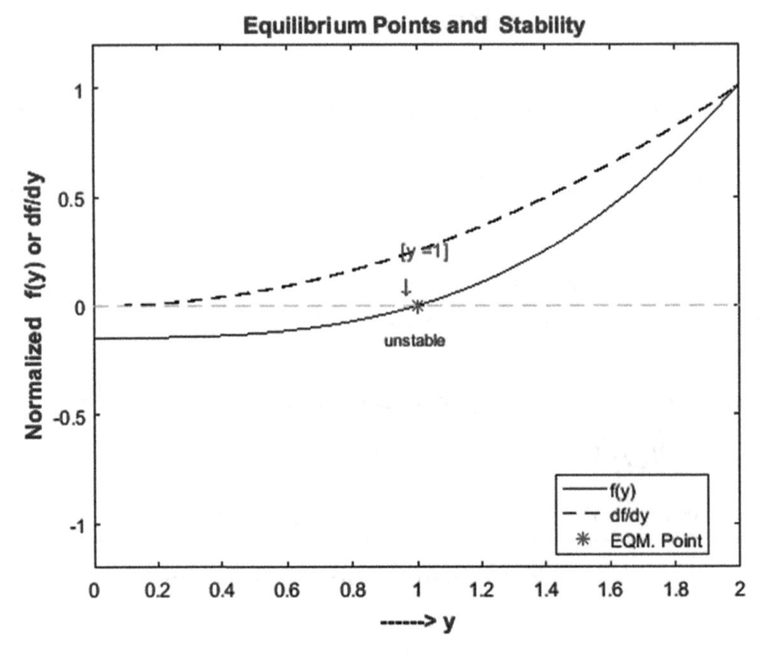
Equilibrium Points and Stability
Normalized f(y) or df/dy
[y =1]
unstable
f(y)
df/dy
EQM. Point
------> y

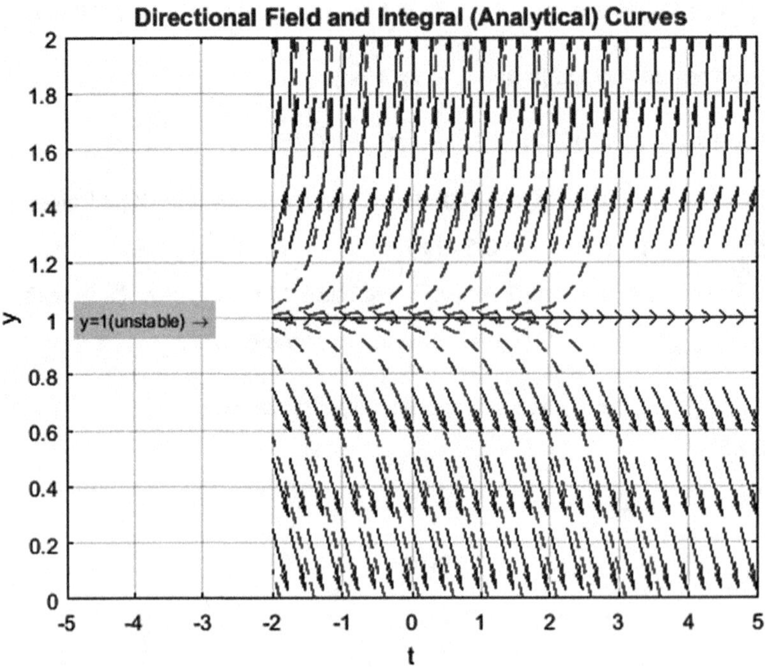
Directional Field and Integral (Analytical) Curves
y=1(unstable) →
y
t

Summary of Equilibrium Study

Input dy/dt $f(y) = (y-1)\,(y^2+y+1)$

EQM Points $y_0 = [\,1\,]$

EQM at y = 1 is unstable; df/dy is +ve

Implicit Solution of the Autonomous Differential Equation

$$3\,C + \log(y-1) + \log\left(y + \tfrac{1}{2} - \tfrac{\sqrt{3}\,\mathrm{i}}{2}\right)\left(-\tfrac{1}{2} + \tfrac{\sqrt{3}\,\mathrm{i}}{2}\right) = 3\,t + \log\left(y + \tfrac{1}{2} + \tfrac{\sqrt{3}\,\mathrm{i}}{2}\right)\left(\tfrac{1}{2} + \tfrac{\sqrt{3}\,\mathrm{i}}{2}\right)$$

```
dy_dt='-2*atan(y)/(1+y^2)';
```

Equilibrium Point(s) & Stability

Input f=dy/dt ⇒ $f(y) = -\frac{2\,\arctan(y)}{y^2+1}$

Range of roots of [f=0]⇒ [- ∞ + ∞]

Derivative of f w.r.t y $f'(y) = \frac{2\,(2\,y\,\arctan(y)-1)}{(y^2+1)^2}$

Get roots (y_r) of [f=0]; REAL roots (y_0) are the EQM Points

Roots (y_r) ⇒ [0]

All roots y_r are Real (y_0); Examine df/dy at y = y_0 for stability

EQM Points y_0 ⇒ [0]

Examine & justify Stability EQM at y = 0 is asymptotically stable; df/dy is -ve

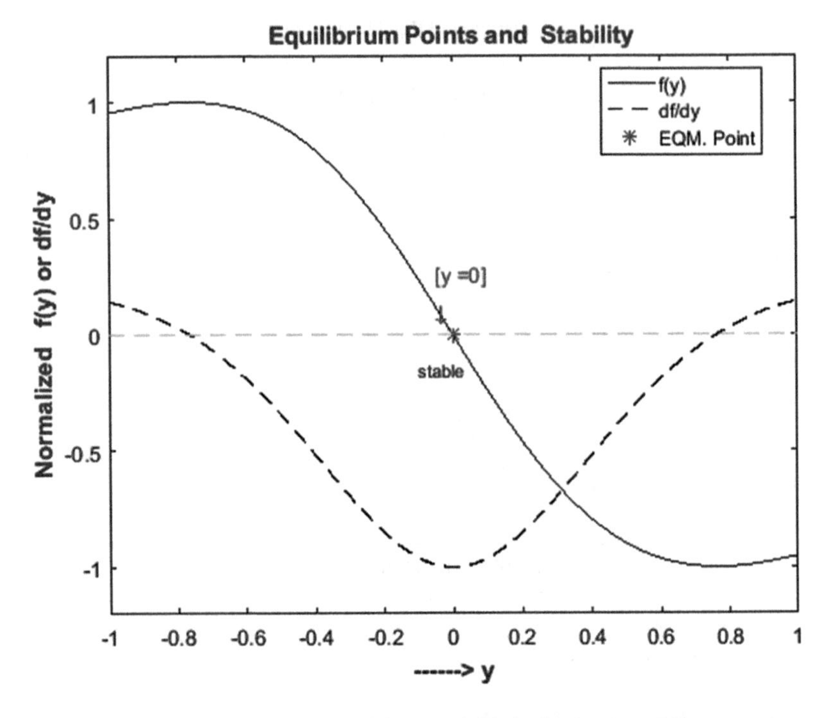
Equilibrium Points and Stability
f(y)
df/dy
EQM. Point
[y =0]
stable
Normalized f(y) or df/dy
------> y

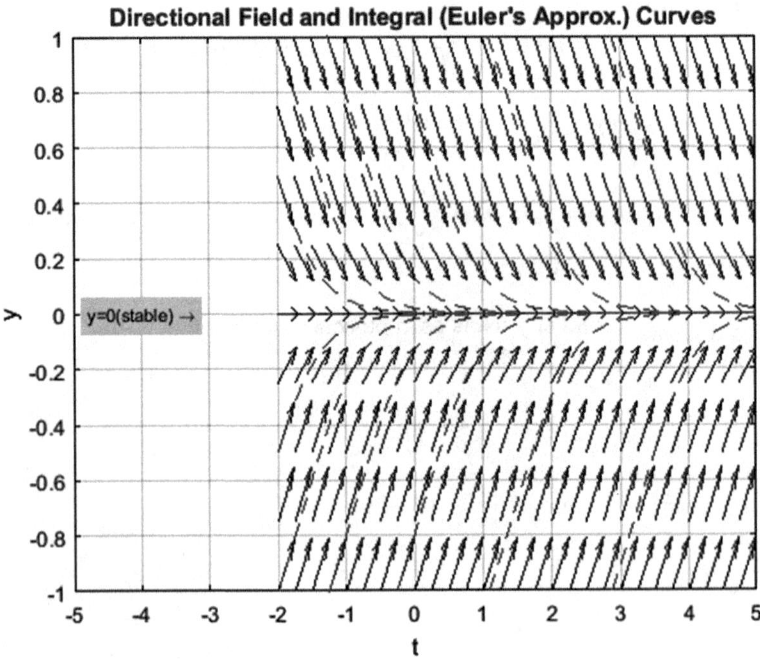
Directional Field and Integral (Euler's Approx.) Curves
y=0(stable) →
y
t

Summary of Equilibrium Study

Input dy/dt $f(y) = -\frac{2\,\mathrm{arctan}(y)}{y^2+1}$

EQM Points $y_0 = [\,0\,]$

EQM at y = 0 is asymptotically stable; df/dy is -ve

No Analytical Solution of the Autonomous Differential Equation

```
dy_dt='2*y-3*sqrt(y)';
```

Equilibrium Point(s) & Stability

Input f=dy/dt ⇒ $f(y) = 2\,y - 3\,\sqrt{y}$

Range of roots of [f=0]⇒ Range of the roots of [f=0]--> [0 ∞]

Derivative of f w.r.t y $f'(y) = 2 - \frac{3}{2\,\sqrt{y}}$

Get roots (y_r) of [f=0]; REAL roots (y_0) are the EQM Points

Roots (y_r) ⇒ [0 2.25]

All roots y_r are Real (y_0); Examine df/dy at y = y_0 for stability

EQM Points y_0 ⇒ [0 2.25]

Examine & justify Stability
EQM at y = 0 is asymptotically stable; df/dy is -ve
EQM at y = 2.25 is unstable; df/dy is +ve

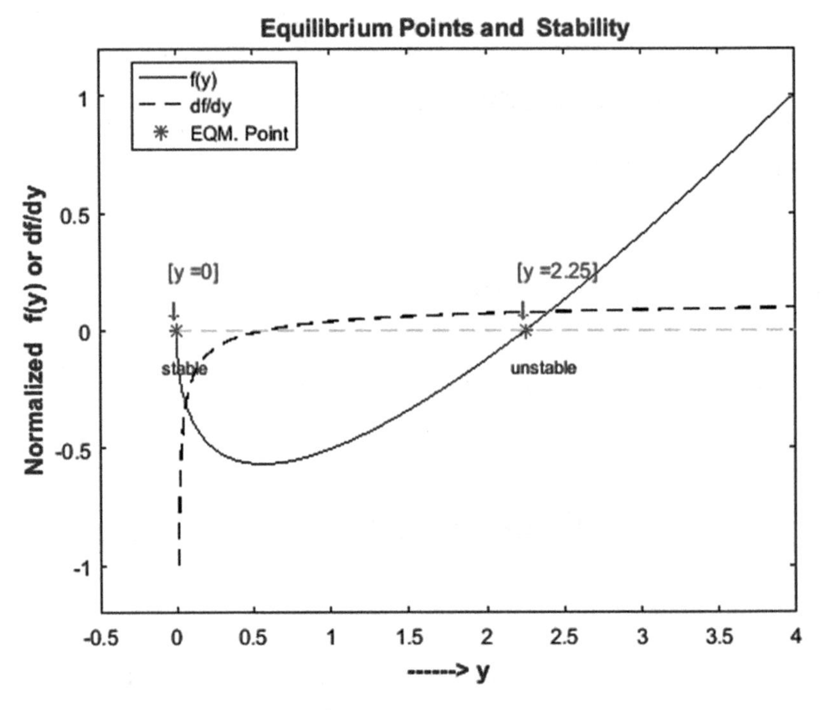
Equilibrium Points and Stability
f(y)
df/dy
EQM. Point
[y =0]
[y =2.25]
stable
unstable
Normalized f(y) or df/dy
------> y

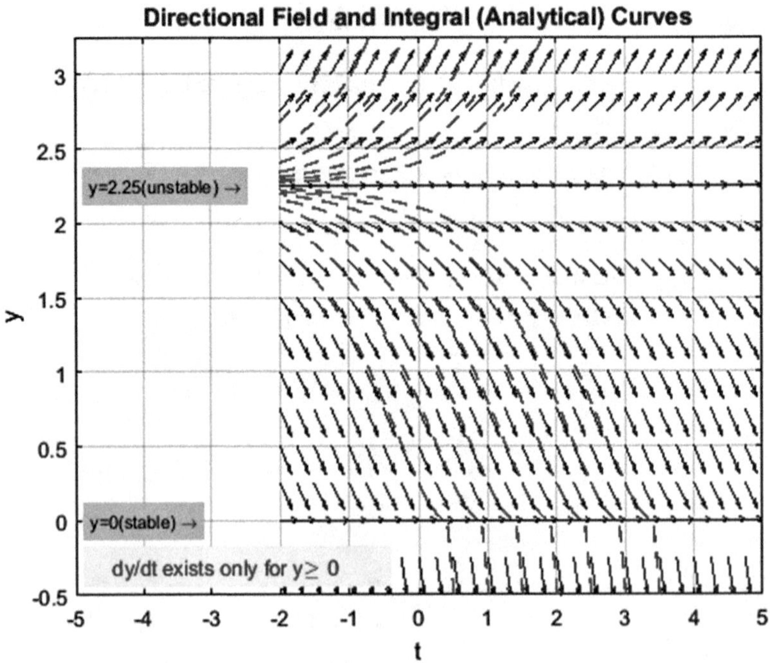
Directional Field and Integral (Analytical) Curves
y=2.25(unstable) →
y=0(stable) →
dy/dt exists only for y≥ 0
y
t

Summary of Equilibrium Study

Input dy/dt $f(y) = 2\,y - 3\,\sqrt{y}$

EQM Points y_0 = [0 2.25]

EQM at y = 0 is asymptotically stable; df/dy is -ve
EQM at y = 2.25 is unstable; df/dy is +ve

Implicit Solution of the Autonomous Differential Equation

$$C + \log\left(\sqrt{y} - \tfrac{3}{2}\right) = t$$

```
dy_dt='2-3*sqrt(y)';
```

Equilibrium Point(s) & Stability

Input f=dy/dt ⇒ $f(y) = 2 - 3\,\sqrt{y}$

Range of roots of [f=0]⇒ Range of the roots of [f=0]--> [0 ∞]

Derivative of f w.r.t y $f'(y) = -\frac{3}{2\,\sqrt{y}}$

Get roots (y_r) of [f=0]; REAL roots (y_0) are the EQM Points

Roots (y_r) ⇒ [0.44]

All roots y_r are Real (y_0); Examine df/dy at y = y_0 for stability

EQM Points y_0 ⇒ [0.44]

Examine & justify Stability EQM at y = 0.44 is asymptotically stable; df/dy is -ve

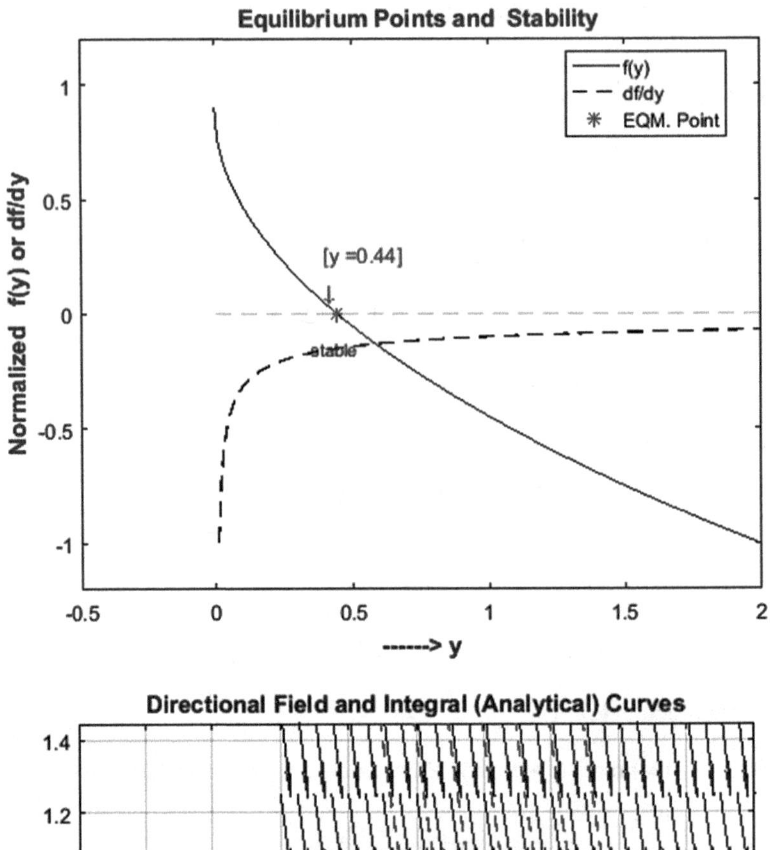
Equilibrium Points and Stability
f(y)
df/dy
EQM. Point
[y =0.44]
stable
Normalized f(y) or df/dy
------> y

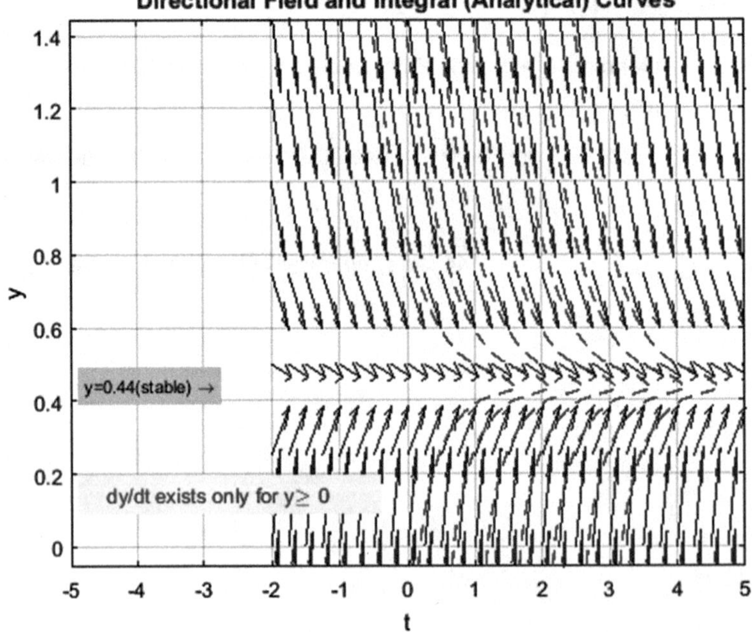
Directional Field and Integral (Analytical) Curves
y=0.44(stable) →
dy/dt exists only for y≥ 0
y
t

Summary of Equilibrium Study

Input dy/dt $f(y) = 2 - 3\sqrt{y}$

EQM Points y_0 = [0.44]

EQM at y = 0.44 is asymptotically stable; df/dy is -ve

Implicit Solution of the Autonomous Differential Equation

$$9\,t + 4\log\left(\sqrt{y} - \tfrac{2}{3}\right) + 6\sqrt{y} = 9\,C$$

```
dy_dt='y+2*y^2';
```

Equilibrium Point(s) & Stability

Input f=dy/dt ⇒ $f(y) = 2\,y^2 + y$

Range of roots of [f=0]⇒ [- ∞ + ∞]

Derivative of f w.r.t y $f'(y) = 4\,y + 1$

Get roots (y_r) of [f=0]; REAL roots (y_0) are the EQM Points

Roots (y_r) ⇒ [-0.5 0]

All roots y_r are Real (y_0); Examine df/dy at y = y_0 for stability

EQM Points y_0 ⇒ [-0.5 0]

Examine & justify Stability
EQM at y = -0.5 is asymptotically stable; df/dy is -ve
EQM at y = 0 is unstable; df/dy is +ve

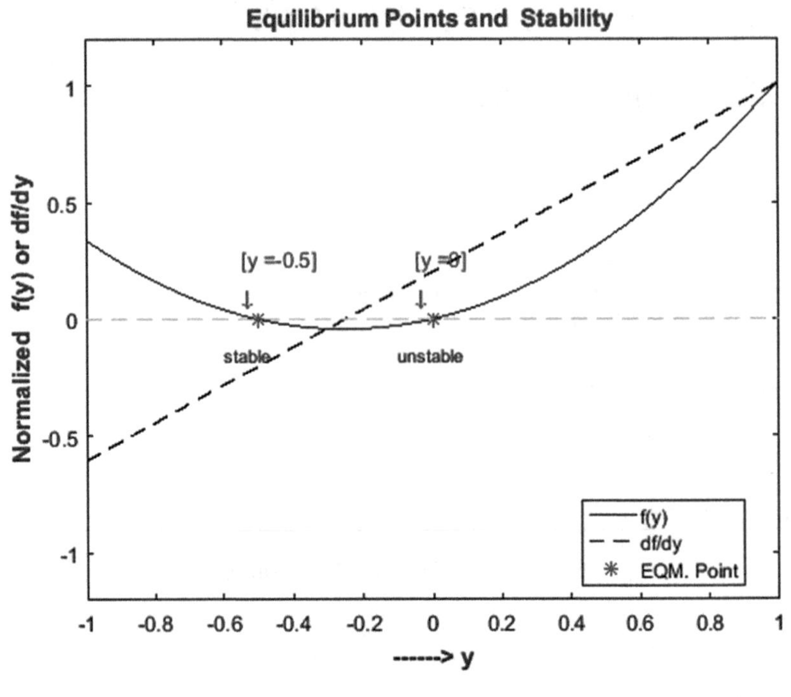
Equilibrium Points and Stability
Normalized f(y) or df/dy
[y =-0.5]
[y =0]
stable
unstable
f(y)
df/dy
EQM. Point
------> y

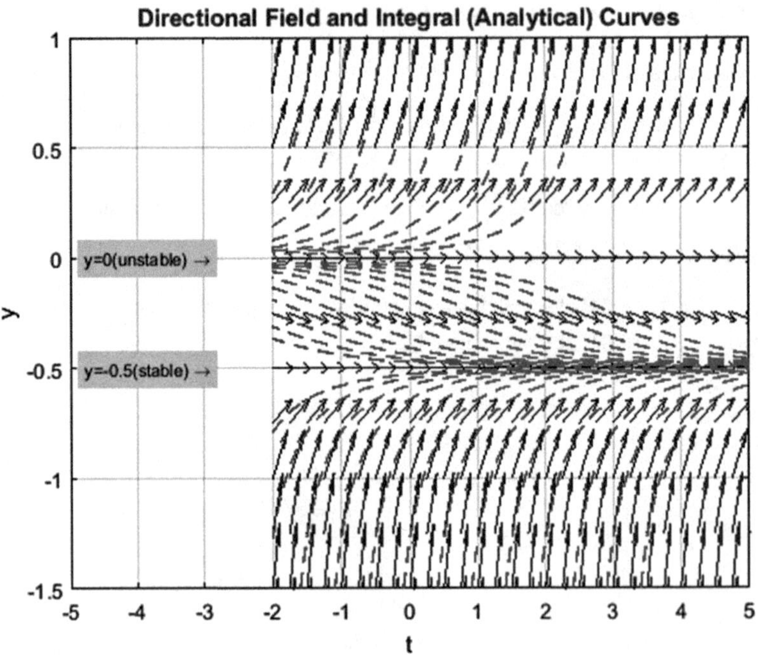
Directional Field and Integral (Analytical) Curves
y=0(unstable) →
y=-0.5(stable) →
y
t

Summary of Equilibrium Study

Input dy/dt $f(y) = 2\,y^2 + y$

EQM Points y_0 = [-0.5 0]

EQM at y = -0.5 is asymptotically stable; df/dy is -ve
EQM at y = 0 is unstable; df/dy is +ve

Implicit Solution of the Autonomous Differential Equation

$$C = t + 2\ \mathrm{arctanh}(4\,y + 1)$$

```
dy_dt='exp(y)-1';
```

Equilibrium Point(s) & Stability

Input f=dy/dt ⇒ $f(y) = e^y - 1$

Range of roots of [f=0]⇒ [$-\infty$ $+\infty$]

Derivative of f w.r.t y $f'(y) = e^y$

Get roots (y_r) of [f=0]; REAL roots (y_0) are the EQM Points

Roots (y_r) ⇒ [0]

All roots y_r are Real (y_0); Examine df/dy at y = y_0 for stability

EQM Points y_0 ⇒ [0]

Examine & justify Stability EQM at y = 0 is unstable; df/dy is +ve

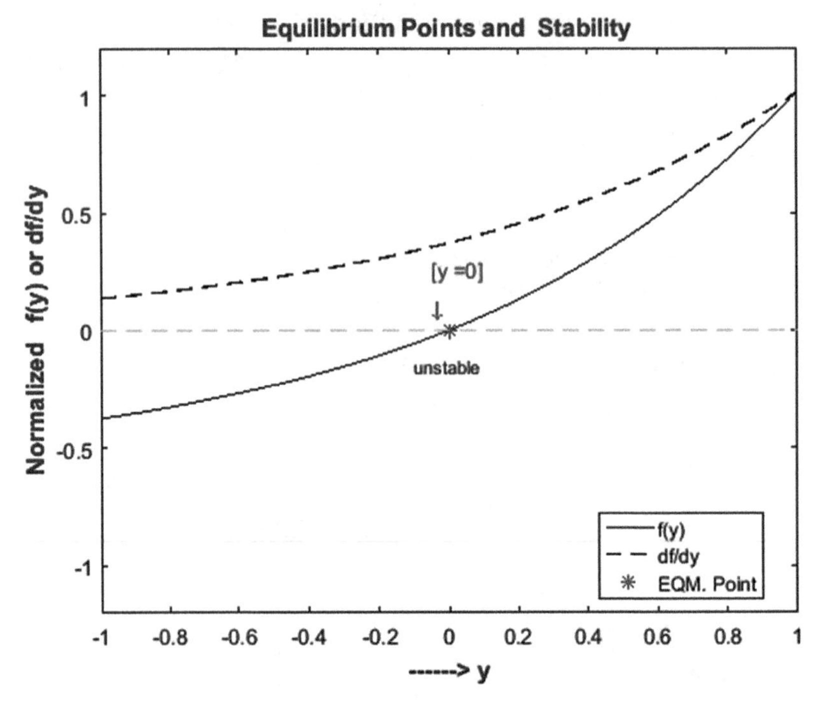
Equilibrium Points and Stability
Normalized f(y) or df/dy
[y =0]
unstable
f(y)
df/dy
EQM. Point
------> y

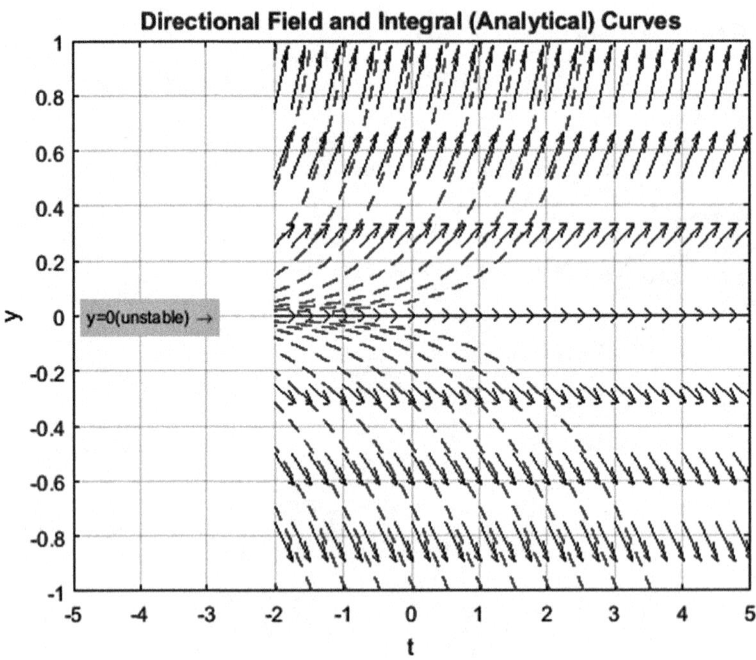
Directional Field and Integral (Analytical) Curves
y
y=0(unstable) →
t

Summary of Equilibrium Study

Input dy/dt $f(y) = e^y - 1$

EQM Points $y_0 = [\,0\,]$

EQM at y = 0 is unstable; df/dy is +ve

Implicit Solution of the Autonomous Differential Equation

$$C + \log(e^y - 1) = t + y$$

```
dy_dt='y*(1-y)^2';
```

Equilibrium Point(s) & Stability

Input f=dy/dt ⇒ $f(y) = y\,(y-1)^2$

Range of roots of [f=0]⇒ [- ∞ + ∞]

Derivative of f w.r.t y $f'(y) = 3\,y^2 - 4\,y + 1$

Get roots (y_r) of [f=0]; REAL roots (y_0) are the EQM Points

Roots (y_r) ⇒ [0 1 1]

All roots y_r are Real (y_0); Examine df/dy at y = y_0 for stability

EQM Points y_0 ⇒ [0 1] after discrading duplicates

Examine & justify Stability
EQM at y = 0 is unstable; df/dy is +ve
EQM at y = 1 is semi-stable; df/dy=0 & goes from +/-ve <--> -/+ve

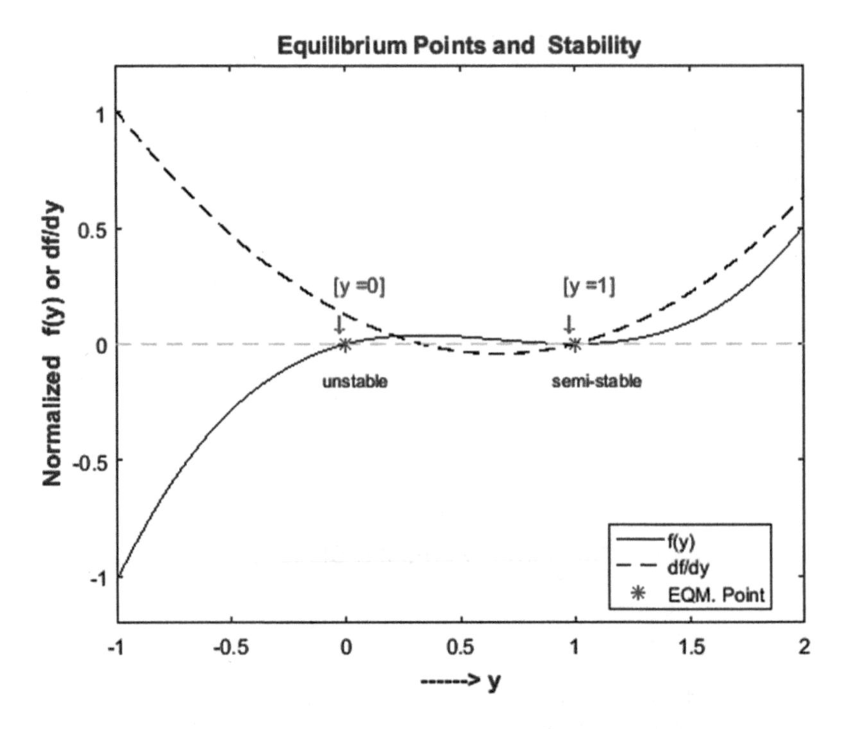
Equilibrium Points and Stability
Normalized f(y) or df/dy
[y =0]
[y =1]
unstable
semi-stable
f(y)
df/dy
EQM. Point
------> y

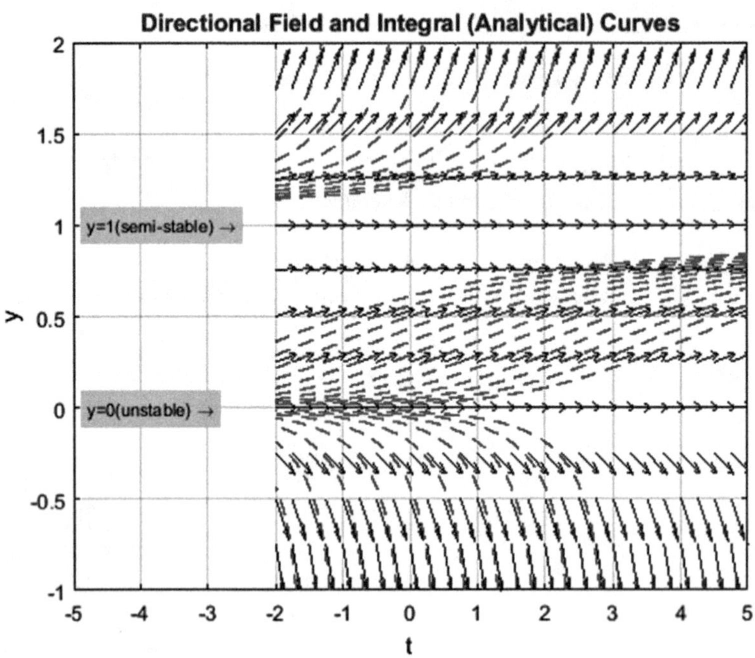
Directional Field and Integral (Analytical) Curves
y=1(semi-stable) →
y=0(unstable) →
y
t

Summary of Equilibrium Study

Input dy/dt $f(y) = y\,(y-1)^2$

EQM Points y_0 = [0 1]

EQM at y = 0 is unstable; df/dy is +ve
EQM at y = 1 is semi-stable; df/dy=0 & goes from +/-ve <--> -/+ve

Implicit Solution of the Autonomous Differential Equation

$$C\;(y-1) = t\;(y-1) + \log\left(\frac{y-1}{y}\right)\;(y-1) + 1$$

```
dy_dt='2*(1-y/3)*y-y';
```

Equilibrium Point(s) & Stability

Input f=dy/dt ⇒ $f(y) = -y - y\,\left(\frac{2y}{3} - 2\right)$

Range of roots of [f=0]⇒ [- ∞ + ∞]

Derivative of f w.r.t y $f'(y) = 1 - \frac{4y}{3}$

Get roots (y_r) of [f=0]; REAL roots (y_0) are the EQM Points

Roots (y_r) ⇒ [0 1.5]

All roots y_r are Real (y_0); Examine df/dy at y = y_0 for stability

EQM Points y_0 ⇒ [0 1.5]

Examine & justify Stability
EQM at y = 0 is unstable; df/dy is +ve
EQM at y = 1.5 is asymptotically stable; df/dy is -ve

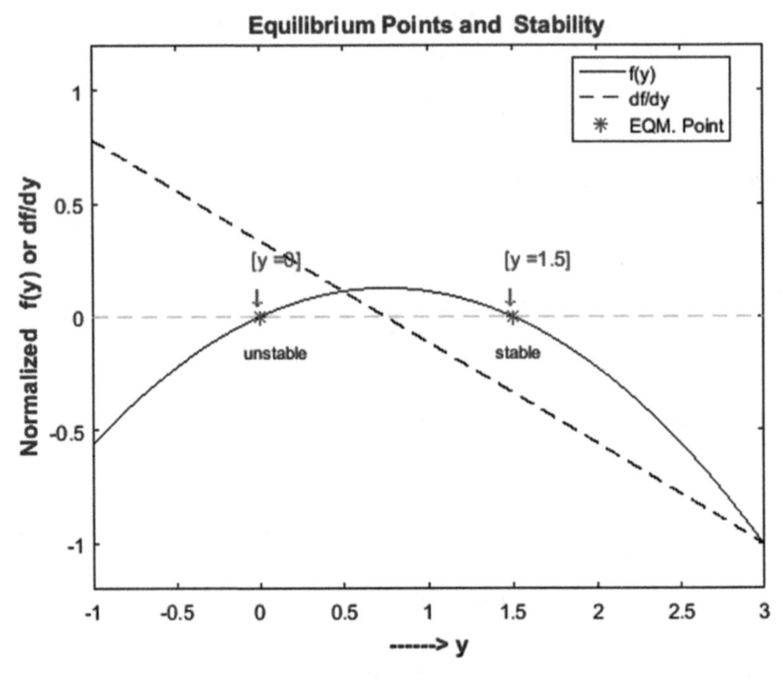

Equilibrium Points and Stability
f(y)
df/dy
EQM. Point
[y =0]
[y =1.5]
unstable
stable
Normalized f(y) or df/dy
------> y

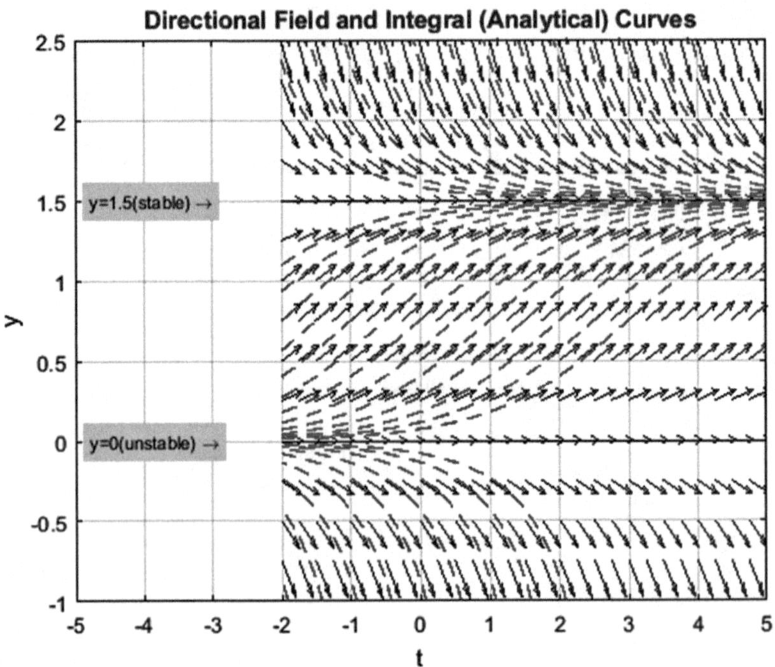

Directional Field and Integral (Analytical) Curves
y=1.5(stable) →
y=0(unstable) →
y
t

Summary of Equilibrium Study

Input dy/dt $f(y) = -y - y\left(\frac{2y}{3} - 2\right)$

EQM Points y_0 = [0 1.5]

EQM at y = 0 is unstable; df/dy is +ve
EQM at y = 1.5 is asymptotically stable; df/dy is -ve

Implicit Solution of the Autonomous Differential Equation

$$C + 2\,\mathrm{arctanh}\left(\frac{4y}{3} - 1\right) = t$$

```
dy_dt='y-abs(y)/y';
```

Equilibrium Point(s) & Stability

Input f=dy/dt ⇒ $f(y) = y - \frac{|y|}{y}$

Range of roots of [f=0]⇒ [- ∞ + ∞]

Derivative of f w.r.t y $f'(y) = 1 - \frac{\mathrm{sign}(y)^2 - 1}{y\,\mathrm{sign}(y)}$

Get roots (y_r) of [f=0]; REAL roots (y_0) are the EQM Points

Roots (y_r) ⇒ [-1 1]

All roots y_r are Real (y_0); Examine df/dy at y = y_0 for stability

EQM Points y_0 ⇒ [-1 1]

Examine & justify Stability
EQM at y = -1 is unstable; df/dy is +ve
EQM at y = 1 is unstable; df/dy is +ve

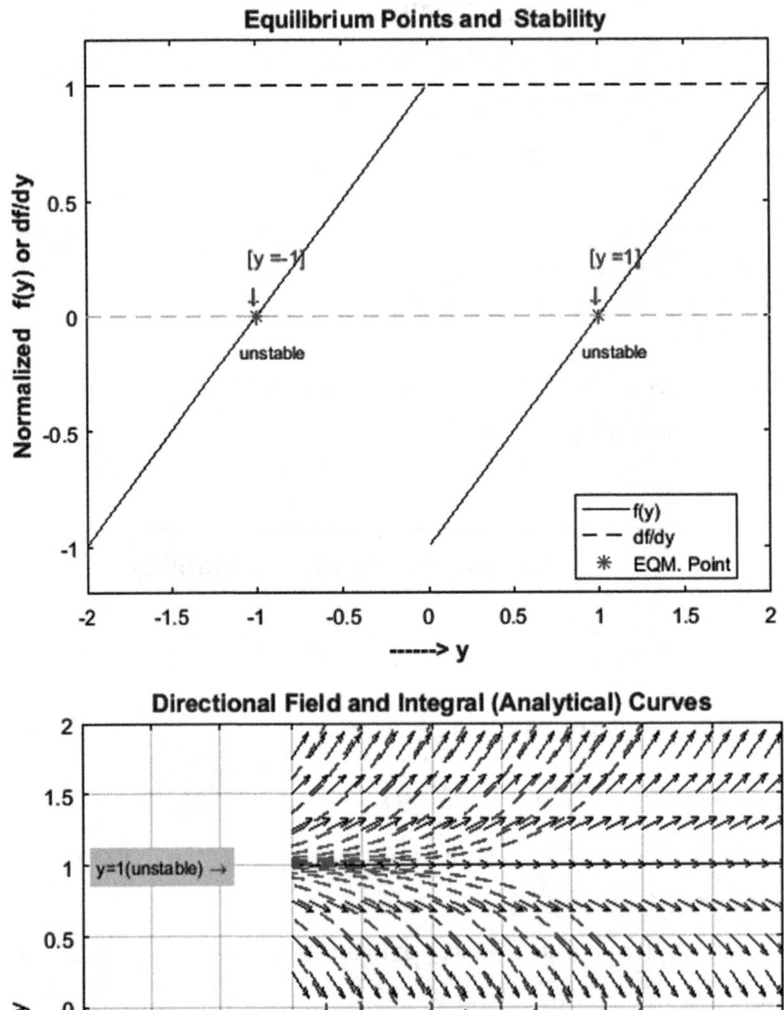

Equilibrium Points and Stability
Normalized f(y) or df/dy
[y =-1]
unstable
[y =1]
unstable
f(y)
df/dy
EQM. Point
------> y

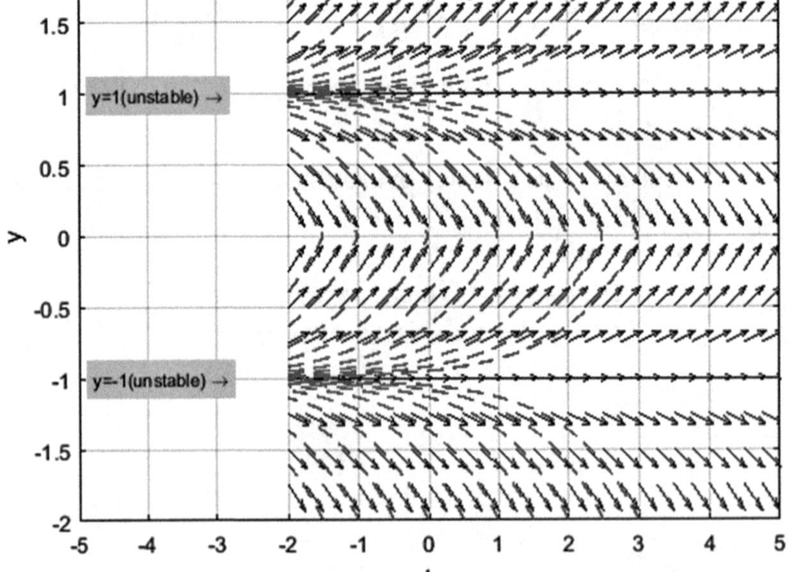

Directional Field and Integral (Analytical) Curves
y=1(unstable) →
y=-1(unstable) →
y
t

Summary of Equilibrium Study

Input dy/dt $f(y) = y - \frac{|y|}{y}$

EQM Points y_0 = [-1 1]

EQM at y = -1 is unstable; df/dy is +ve
EQM at y = 1 is unstable; df/dy is +ve

Implicit Solution of the Autonomous Differential Equation

$$C + \log(\text{sign}(y) - y) = t$$

2.6 Summary

First order differential equations have been discussed in this chapter. Differential equations may be classified as linear, nonlinear, or autonomous. Solutions can be found using the method based on integrating factors or the method on the separation of functions. Solutions may be obtained using the dsolve(.) command in MATLAB for numerical approaches based on Runge-Kutta methods (when initial conditions are available). Even when analytical solutions are available, dsolve(.) and the Runge-Kutta methods can provide means for the verification of the results. Particular attention is paid to obtaining exact solutions using the method of integrating factors when the first order differential equation is linear (depends only on the independent variable y) and use of the separable function approach when the first order differential equation is linear or nonlinear. The D-field patterns which can provide insight into the behavior of the system described by the first order differential equation, are also explored. The analysis of autonomous systems is carried out in detail. The method of separable functions is used to obtain implicit solutions (in the absence of exact solutions, use of Runge-Kutta methods) and multiple ways of understanding the three forms of equilibria associated with these systems are provided.

A number of descriptive and self-contained examples are given providing a broader view of the topics covered through multiple ways of verification of the results.

2.7 Exercises

1. Newton's Law of cooling
   a) A metal rod that has been heated to 180°F is brought to a room with an ambient temperature of 70°F. If y represents the temperature of the rod, the rate at which the rod is cooling can be expressed as

$$\frac{dy}{dt} = -k(y - R)$$

The constant R is the ambient temperature in the room (70°F) and k a constant (units of t^{-1}) that depends on the room and its characteristics. It will determine how fast or how slow the rod cools. Note that y(0)= 180°F. Analyze the equilibrium conditions if k=0.25°F/min.

b) Often cooling does not flow a continuous decrease with time. If the rate of cooling is expressed as (k_1=5)

$$\frac{dy}{dt} = -k\left[y - R - k_1 \sin\left(2\frac{\pi}{10}t\right)\right]$$

2. In a rural area with a population of 3000, the population fluctuates due to births, deaths and influx of a few people who enjoy the scenic areas. The population growth is proportional to its population size at any given time. If 30 people move into the area every year, the overall growth in population is

$$\frac{dy}{dt} = 30 + \frac{y}{k}$$

The population at any given time is y(t) and y(0)=3000. Obtain an expression for the population at any given time. Use k=20;

3. Obtain the integrating factors and solutions for the following first order differential equations. Verify your results in MATLAB. Note that Dy=dy/dt.

$$Dy = t + y - 4$$

$$Dy = y + 4$$

$$Dy = 4t + y, \quad y(0) = -1$$

$$Dy = y + \sin(t) + \cos(t) + 5$$

$$Dy = y + \cosh(t)$$

$$Dy = \frac{\sinh(t) - y}{t}$$

$$Dy = \frac{\sinh(t)}{t} - \frac{y}{t^2}$$

$$Dy = 2y + \frac{1}{1+t^2}$$

4. For the following differential equations, obtain solutions using the method of separation of functions. Verify your results in MATLAB.

$$\frac{dy}{dx} = 4x^2y^2$$

$$\frac{dy}{dx} = (x-1)(y+3)$$

$$\frac{dy}{dx} = \frac{x^3 - x^2}{2y}$$

$$\frac{dy}{dx} = \frac{x^3 + x^{-1}}{(y-2)^2}$$

$$\frac{dy}{dx} = \frac{4x - x^3}{y^3 - 8}$$

$$\frac{dy}{dx} = \frac{2(x^2+4)}{1-y^2}$$

5. For the following autonomous differential equations, identify the state of equilibrium that might exist and verify your conclusions using appropriate plots of the D-fields.

$$\frac{dy}{dt} = \frac{|y|}{y} - y$$

$$\frac{dy}{dt} = y(y-1)(y+1)$$

$$\frac{dy}{dt} = 1 - e^y$$

6. For the following differential equation, examine the state of equilibrium for the set of values (combinations) of b and c

$$\frac{dy}{dt} = y^2 + by + c, \quad b = [-3,-2,0,2,3], c = [-3,-2,2,3]$$

7. Bacterial growth
   The bacterial growth might often be characterized by the fact that the growth factor depends on the square of the bacteria present at any time while the bacteria is growing at a rate proportional to its current strength. This means that

$$\frac{dy}{dt} = k(y^2)y = (r_1 y^2 + r)y$$

8. Effect of antibiotics and elimination of bacteria
   The effect of antibiotics can be modeled as the case where the rate of elimination is proportional to the amount of bacterial as well as time.

$$\frac{dy}{dt} = -ry - kt$$

9. Effect of antibiotics and elimination of bacteria. It is also possible to have the model where elimination is modeled as

$$\frac{dy}{dt} = -kyt^2$$

10. Effect of antibiotics and elimination of bacteria. Another model will be

$$\frac{dy}{dt} = -ky^2 + ry$$

Obtain an expression for the level of bacteria at any given time.

CHAPTER 3

Linear Second Order Differential Equations with Constant Coefficients

3.1 Introduction 81
3.2 Homogeneous Differential Equations 82
3.2.1 Solution from the roots of the characteristic equation 83
3.2.2 Solution using Laplace transforms 84
3.2.3 Solution using Runge-Kutta methods 84
3.2.4 Solution using decomposition into a pair of first order differential equations 85
3.2.5 Interpretation of the solution of the homogeneous differential equation 88
3.2.6 Additional examples 99
3.3 Non-homogeneous Differential Equations: Particular Solutions and Complete Solutions 119
3.3.1 Particular solution 119
3.3.2 Complete solution 121
3.3.3 Additional examples 126
3.4 Summary 177
3.5 Exercises 178

3.1 Introduction

Second order differential equations occur in the study of models such as those that describe the behavior of mechanical, electrical, chemical, transportation systems, etc. For example, the best way to describe the motion of a pendulum, the characteristics of the current or voltage in electric circuits, or the flow of chemicals, etc. is through the use of second order differential equations. If the dependent variable is y and the independent variable is t, a general second order system can be expressed in functional form F(.) as

$$F\left(y, \frac{dy}{dt}, \frac{d^2y}{dt^2}, t\right) = 0. \tag{3.1}$$

Equation (3.1) will be a linear second order differential equation (or system) if it only contains terms in y, y' and y''. A linear second order differential equation is

$$a(t)\frac{d^2}{dt^2}y(t) + b(t)\frac{d}{dt}y(t) + c(t)y(t) = g(t), \quad a(t) \neq 0. \tag{3.2}$$

The function g(t) on the right hand side in eqn. (3.2) is the forcing function. If $a(t)$, **b**(t) and $c(t)$ have no time dependence and are expressed as a, b, and c, respectively, the differential equation in eqn. (3.2) becomes a linear second order differential equation with constant coefficients. The analysis here is limited to such differential equations.

A second order differential equation with constant coefficients in eqn. (3.2) can be expressed using primes as the notation for the derivatives as

$$ay'' + by' + cy = g(t). \tag{3.3}$$

If g(t) is zero, the differential equation in (3.3) is homogeneous and if g(t)≠0, the differential equation is of a non-homogeneous type. It is also implied that y=y(t).

3.2 Homogeneous Differential Equations

A second order homogeneous differential equation with constant coefficients is

$$ay'' + by' + cy = 0, \quad a \neq 0. \tag{3.4}$$

The starting point of the analysis is the assumption that a solution to a linear differential equation with constant coefficients will be of the form e^{rt}, with r being a constant to be determined. By substituting $y = e^{rt}$ in eqn. (3.4), the differential equation in eqn. (3.4) becomes

$$\left[ar^2 + br + c\right]e^{rt} = 0. \tag{3.5}$$

Because e^{rt} cannot be zero, eqn. (3.5) leads to

$$ar^2 + br + c = 0. \tag{3.6}$$

Equation (3.6) is referred to as the characteristic equation or the characteristic polynomial associated with the differential equation in eqn. (3.4). It is also called the auxiliary equation. The roots of the characteristic equation are

$$\begin{aligned} r_1 &= \frac{-b + \sqrt{b^2 - 4ac}}{2a} \\ r_2 &= \frac{-b - \sqrt{b^2 - 4ac}}{2a} \end{aligned}. \tag{3.7}$$

Depending on whether the discriminant $\sqrt{b^2 - 4ac}$ is positive, zero or negative, the roots will have the following characteristics

$$
\begin{aligned}
b^2 - 4ac > 0 &\quad \rightarrow r_1 \neq r_2 \,(real) \\
b^2 - 4ac = 0 &\quad \rightarrow r_1 = r_2 = r \,(real) \\
b^2 - 4ac < 0 &\quad \rightarrow r_{1,2} = r_{\pm} = \alpha \pm j\beta;\ (\alpha, \beta \ real)
\end{aligned}
\quad . \tag{3.8}
$$

3.2.1 Solution from the roots of the characteristic equation

Based on the nature roots of the characteristic equation in eqn. (3.8), the general solution to the differential equation in eqn. (3.4) can be written using the superposition principle (general solution must be the weighted sum of the two possible solutions) as

$$
\begin{aligned}
y_h(t) &= c_1 e^{r_1 t} + c_2 e^{r_2 t} \Rightarrow \quad r_1 \neq r_2 \,(real) \\
y_h(t) &= c_1 e^{rt} + c_2 t e^{rt} \Rightarrow \quad r_1 = r_2 = r \,(real) \\
y_h(t) &= e^{\alpha t}\left[c_1 \cos(\beta t) + c_2 \sin(\beta t)\right] \Rightarrow \quad r_1 = \alpha + j\beta, r_2 = \alpha - j\beta \,(conjugate\ pair)
\end{aligned}
\quad . \tag{3.9}
$$

If the initial conditions y(0) and y'(0) are given, the constants c_1 and c_2 can be evaluated and the exact solution to the differential equation can be obtained. Otherwise, the solution in eqn. (3.9) is identified as the **general** solution to the homogeneous differential equation in eqn. (3.4). The subscript *h* on the left hand side of eqn. (3.9) merely is an indicator of the fact that the solution is associated with a homogeneous differential equation (or a homogeneous system). The solution with the subscript *h* is also identified as the homogeneous solution.

A second order equation that requires a special mention is

$$
y''(t) = 0 . \tag{3.10}
$$

It is clear that the roots of the characteristic equation are both 0's and this case can be handled either from the simple observation of the differential equation or using the general solution in eqn. (3.9) with r=0 resulting in a solution of the form

$$
y(t) = c_1 + c_2 t . \tag{3.11}
$$

Thus, if $b=c=0$, roots of the characteristic equation will be identical and equal to zero and the solution will be a polynomial of t of order 1 (absence of any exponential terms in t).

If on the other hand, $b \neq 0,\ c = 0$ the characteristic equation becomes

$$
ar^2 + br = r(ar + b) = 0 . \tag{3.12}
$$

Equation (3.12) results in one of the roots being a zero.

For the case of $b = 0,\ c > 0,\ a > 0$, the characteristic equation becomes

$$
ar^2 + c = 0 . \tag{3.13}
$$

Equation (3.13) leads to purely imaginary values and the solution will consist of only sine(.) and cosine(.) terms as seen in eqn. (3.9). As stated earlier, for other specific combinations of a, b and c, roots may be complex or real. For all the analysis presented here, $a \neq 0$.

If the initial conditions are given, two additional possibilities exist for obtaining the solution to the differential equation. The first approach is based on the use of Laplace transforms and the second one is based on Runge-Kutta methods resulting in a numerical solution. The Laplace transform based method is discussed first.

3.2.2 Solution using Laplace transforms

If we take the Laplace transform (Appendix B) of the differential equation in eqn. (3.4), we get

$$a\left[s^2Y(s)-sy_0-y_1\right]+b\left[sY(s)-y_0\right]+cY(s)=0 \tag{3.14}$$

In eqn. (3.14),

$$\begin{aligned} y_0 &= y(0) \\ y_1 &= y'(0) \\ Y(s) &= L\left[y(t)\right] \end{aligned} \quad . \tag{3.15}$$

Note that $L[y(t)]$ represents the Laplace transform of y(t). Solving for Y(s), the expression for the Laplace transform of y(t) becomes

$$Y(s)=\frac{(as+b)y_0+ay_1}{as^2+bs+c}. \tag{3.16}$$

Notice the similarity of the denominator in eqn. (3.16) to the characteristic equation in eqn. (3.6). The solution y(t) is obtained by taking the inverse Laplace transform of eqn. (3.16).

3.2.3 Solution using Runge-Kutta methods

As described in **Appendix A**, the Runge-Kutta methods are available to solve a set of first order differential equations. This means that the differential equation in (3.4) needs to be converted into a set of two first order coupled differential equations (specific details of coupled first order differential equations are given in Chapter 5). The first step in that process is to define

$$z(t)=\frac{dy}{dt}. \tag{3.17}$$

Using eqn. (3.17), the homogeneous differential equation in eqn. (3.4) becomes

$$az'+bz+cy=0. \tag{3.18}$$

A set of coupled first order differential equations can be written from eqns. (3.17) and (3.18) as

$$\begin{aligned} \frac{dy}{dt} &= z \\ \frac{dz}{dt} &= -\frac{c}{a}y - \frac{b}{a}z \end{aligned} \tag{3.19}$$

Using the matrix notation, eqn. (3.19) becomes

$$\begin{bmatrix} \frac{dy}{dt} \\ \frac{dz}{dt} \end{bmatrix} = \begin{bmatrix} 0 & 1 \\ -\frac{c}{a} & -\frac{b}{a} \end{bmatrix} \begin{bmatrix} y \\ z \end{bmatrix} \Rightarrow \vec{X}' = A\vec{X} \tag{3.20}$$

In eqn. (3.20) A is the coefficient matrix and $\vec{X}$ is the vector given as

$$\begin{aligned} A &= \begin{bmatrix} 0 & 1 \\ -\frac{c}{a} & -\frac{b}{a} \end{bmatrix} \\ \vec{X} &= \begin{bmatrix} y(t) \\ y'(t) \end{bmatrix} = \begin{bmatrix} y \\ z \end{bmatrix} \\ \vec{X}(0) &= \begin{bmatrix} y(0) \\ y'(0) \end{bmatrix} = \begin{bmatrix} y_0 \\ y_1 \end{bmatrix} \end{aligned} \tag{3.21}$$

Equation (3.21) contains the initial conditions in vectorial form $\vec{X}(0)$.

The approach seen with the decomposition of the second order differential equation into a pair of coupled first order equations in eqn. (3.19) offers another possibility of obtaining the solution for the homogeneous second order differential equation.

3.2.4 Solution using decomposition into a pair of first order differential equations

The discussion of the linear second order homogeneous differential equation will be incomplete without describing another analytical means to solve the eqn. (3.4) using the concept of a pair of coupled differential equations (Chapter 5). The approach used in solving the differential equation in this format is drawn from the Runge-Kutta method in Section 3.2.3.

The second order homogeneous differential equation expressed in matrix form is

$$\begin{bmatrix} \frac{dy}{dt} \\ \frac{dz}{dt} \end{bmatrix} = \begin{bmatrix} \frac{dy}{dt} \\ \frac{d^2y}{dt^2} \end{bmatrix} = \begin{bmatrix} 0 & 1 \\ -\frac{c}{a} & -\frac{b}{a} \end{bmatrix} \begin{bmatrix} y(t) \\ z(t) \end{bmatrix}. \tag{3.22}$$

One can see that the two sets of equations embedded in eqn. (3.22) correspond to y(t) and y'(t) which represent the two coupled variables (Chapter 5). The coefficient matrix A is

$$A = \begin{bmatrix} 0 & 1 \\ -\frac{c}{a} & -\frac{b}{a} \end{bmatrix}. \tag{3.23}$$

The eigenvalues of the matrix are given by (Appendix D)

$$\det\left|A - \lambda I_2\right| = 0. \tag{3.24}$$

Expanding the determinant in eqn. (3.24), the eigenvalues will be solutions of

$$a\lambda^2 + b\lambda + c = 0. \tag{3.25}$$

Therefore, the eigenvalues are identical to the roots of the characteristic equation defined in eqn. (3.6). The corresponding eigenvectors $\vec{v}$ are obtained by solving

$$\left[A - \lambda I_2\right]\vec{v} = 0. \tag{3.26}$$

A solution of eqn. (3.22) resulting in y(t) and y'(t) can be obtained using the concept of eigenvalues and eigenvectors described in **Appendix D**. If the eigenvalues are distinct and real, the solution y(t) and its derivative y'(t) are given by

$$\begin{bmatrix} y(t) \\ y'(t) \end{bmatrix} = c_1 e^{\lambda_1 t} \begin{bmatrix} v_{11} \\ v_{21} \end{bmatrix} + c_2 e^{\lambda_2 t} \begin{bmatrix} v_{12} \\ v_{22} \end{bmatrix}. \tag{3.27}$$

In eqn. (3.27), the eigenvalues are λ_1 and λ_2 and the corresponding eigenvectors are $\vec{v}_1$ and $\vec{v}_2$

$$\vec{v}_1 = \begin{bmatrix} v_{11} \\ v_{21} \end{bmatrix}, \quad \vec{v}_2 = \begin{bmatrix} v_{12} \\ v_{22} \end{bmatrix}. \tag{3.28}$$

Note that the interest in the present case is only in y(t). Thus, the solution for y'(t) can be ignored and the solution for the differential equation becomes

$$y(t) = c_1 v_{11} e^{\lambda_1 t} + c_2 v_{12} e^{\lambda_2 t}. \tag{3.29}$$

The analysis becomes further simpler because A_{11} of the coefficient matrix A in eqn. (3.23) is zero. This leads to

$$\vec{v}_1 = \begin{bmatrix} 1 \\ \lambda_1 \end{bmatrix}, \quad \vec{v}_2 = \begin{bmatrix} 1 \\ \lambda_2 \end{bmatrix}. \tag{3.30}$$

Using the eigenvectors in eqn. (3.30), eqn. (3.29) becomes

$$y(t) = c_1 e^{\lambda_1 t} + c_2 e^{\lambda_2 t}. \tag{3.31}$$

If the eigenvalues are equal (λ), the eigenvectors will be equal resulting in

$$\vec{v}_1 = \begin{bmatrix} 1 \\ \lambda \end{bmatrix}, \quad \vec{v}_2 = \begin{bmatrix} 1 \\ \lambda \end{bmatrix}. \tag{3.32}$$

Using the condition for equal eigenvalues from eqn. (3.8), the coefficient matrix A can be rewritten as

$$A = \begin{bmatrix} 0 & 1 \\ -\frac{b^2}{4a^2} & -\frac{b}{a} \end{bmatrix}. \tag{3.33}$$

The eigenvalue (equal) is

$$\lambda = -\frac{b}{2a}. \tag{3.34}$$

It is clear from eqn. (3.32) that the coefficient matrix A is defective because there is only a single unique eigenvector. Therefore, we need to use the concept of generalized eigenvectors described in Appendix D. For a [2 x 2] matrix, the generalized eigenvectors will be

$$\vec{v}_{g1} = \begin{bmatrix} 1 \\ 0 \end{bmatrix}, \quad \vec{v}_{g2} = \begin{bmatrix} 0 \\ 1 \end{bmatrix}. \tag{3.35}$$

Following the procedure described in Appendix D, the solution set can be written in terms of a basic solutions set

$$\begin{aligned} \vec{x}_1(t) &= e^{\lambda t}\left(\vec{v}_{g1} + t[A - \lambda I_2]\vec{v}_{g1}\right) = e^{\lambda t}\left(\begin{bmatrix} 1 \\ 0 \end{bmatrix} + t[A - \lambda I_2]\begin{bmatrix} 1 \\ 0 \end{bmatrix}\right) = e^{\lambda t}\begin{bmatrix} 1 + t\frac{b}{2a} \\ -t\frac{b^2}{4a^2} \end{bmatrix} \\ \vec{x}_2(t) &= e^{\lambda t}\left(\vec{v}_{g2} + t[A - \lambda I_2]\vec{v}_{g2}\right) = e^{\lambda t}\left(\begin{bmatrix} 0 \\ 1 \end{bmatrix} + t[A - \lambda I_2]\begin{bmatrix} 0 \\ 1 \end{bmatrix}\right) = e^{\lambda t}\begin{bmatrix} t \\ 1 - t\frac{b}{2a} \end{bmatrix} \end{aligned} \tag{3.36}$$

In eqn. (3.36), I_2 is the 2 x 2 identity matrix. The solution y(t) and its derivative can now be written as the weighted sum of the two basic solutions:

$$\begin{bmatrix} y(t) \\ y'(t) \end{bmatrix} = c_1\vec{x}_1(t) + c_2\vec{x}_2(t) = c_1 e^{\lambda t}\begin{bmatrix} 1 + t\frac{b}{2a} \\ -t\frac{c}{a} \end{bmatrix} + c_2 e^{\lambda t}\begin{bmatrix} t \\ 1 - t\frac{b}{2a} \end{bmatrix}. \tag{3.37}$$

The solution y(t) can now be written as

$$y(t) = c_1 e^{\lambda t}\left(1 + t\frac{b}{2a}\right) + c_2 t e^{\lambda t} = (b_1 + b_2 t)e^{\lambda t} \tag{3.38}$$

In eqn. (3.38), b_1 and b_2 are two constants and it is seen that eqn. (3.38) matches the form of the solution for equal roots given in eqn. (3.9).

Consider the case when the eigenvalues form a conjugate pair

$$\lambda_{\pm} = \alpha \pm j\beta. \tag{3.39}$$

In this case, the eigenvectors will be

$$\vec{v}_1 = \begin{bmatrix} 1 \\ \alpha + j\beta \end{bmatrix}, \quad \vec{v}_2 = \begin{bmatrix} 1 \\ \alpha - j\beta \end{bmatrix}. \tag{3.40}$$

Using eqn. (3.27), the solution set becomes

$$\begin{bmatrix} y(t) \\ y'(t) \end{bmatrix} = c_1 e^{\alpha t} \begin{bmatrix} \cos(\beta t) \\ \alpha\cos(\beta t) - \beta\sin(\beta t) \end{bmatrix} + c_2 e^{\alpha t} \begin{bmatrix} \sin(\beta t) \\ \alpha\sin(\beta t) + \alpha\cos(\beta t) \end{bmatrix}. \quad (3.41)$$

Separating the solution y(t) in eqn. (3.41) leads to

$$y(t) = e^{\alpha t}\left[c_1\cos(\beta t) + c_2\sin(\beta t)\right]. \quad (3.42)$$

The solutions obtained using the decomposition match, those obtained directly in eqn. (3.9). While acknowledging the fact that decomposition is the basis for the Runge-Kutta method, there is no specific advantage in using this approach to obtain the analytical solution of a second order differential equation. It must be noted that the approach based on eigenvalues and eigenvectors also requires an extra step when the coefficient matrix is defective.

3.2.5 Interpretation of the solution of the homogeneous differential equation

The state of equilibrium associated with the system described by the homogeneous differential equation can be understood from the behavior of the solution when $t \to \infty$. If the non-zero roots of the characteristic equation are real and positive, the system is unstable. The system is unstable even if one root is positive and other one is negative. The system is also unstable if the roots form a complex conjugate pair and the real part is positive. The system is asymptotically stable if both values are distinct and negative. The system is asymptotically stable when the roots are a conjugate pair with the real part being negative. If the roots are complex and the real part is zero, the system is purely oscillatory. When one of the roots is zero and the other root is negative, the system can be considered to reach a stable state when $t \to \infty$. If both eigenvalues are 0's as with the case of the differential eqn. in eqn. (3.10), the system is unstable.

The state of the stability is seen by observing the trajectory of the plots of y(t) vs. dy/dt w.r.t. the critical point of the system. The critical point is the solution to the matrix equation

$$A\vec{y} = \begin{bmatrix} 0 & 1 \\ -\dfrac{c}{a} & -\dfrac{b}{a} \end{bmatrix}\vec{y} = \begin{bmatrix} 0 \\ 0 \end{bmatrix}. \quad (3.43)$$

It can be easily seen that the critical point will be the origin, [0,0]. If one of the roots of the characteristics is zero, the critical points will lie along the line through the origin.

The state of equilibrium of the system represented by the second order differential equation and its relationship to the roots of the characteristic equation is summarized in Table 3.1.

Table 3.1 Relationship between the roots of the characteristic equation and stability.

Properties of the Roots of the Characteristic Equation	**Stability**
$r_1, r_2 > 0$	Unstable
$r_1, r_2 < 0$	Asymptotically stable
$r_1 < 0 < r_2$	Unstable
$r_1 = r_2 > 0$	Unstable
$r_1 = r_2 < 0$	Asymptotically stable
$r_1 = 0, r_2 < 0$	Stable
$r_1 = 0, r_2 > 0$	Unstable
$r_1 = r_2 = 0$	Unstable
$r_1, r_2 = \alpha \pm j\,\beta, \alpha > 0$	Unstable
$r_1, r_2 = \alpha \pm j\,\beta, \alpha < 0$	Asymptotically stable
$r_1, r_2 = \alpha \pm j\,\beta, \alpha = 0$	Oscillatory

As stated earlier, these conclusions can be visualized and understood by displaying the phase portrait showing how y(t) and y'(t) vary. This plot is obtained by varying the values of c_1 and c_2. The phase portrait as discussed in Appendix C can also be obtained using the quiver(.) command in MATLAB®. The second order differential equation is broken down to two first order differential equations, one in y and the other one in y' as in eqn. (3.22). An example of the phase plot creation along with the MATLAB script is given below.

Example # 3.1 The differential equation $y''+2y'+3=0$. It has the roots, $-1 \pm i\sqrt{2}$.

```
clear;close all;
% y"+2y'+3y=0
a=1;b=2;c=3;
roots([1,2,3])
A=[0,1;-c/a,-b/a];
[x1,x2]=meshgrid(-4:.5:4,-4:.5:4);
DX1=A(1,1)*x1+A(1,2)*x2;
DX2=A(2,1)*x1+A(2,2)*x2;
quiver(x1,x2,DX1,DX2,1),xlabel('y(t)'),ylabel('dy(t)/dt')
hold on
syms y t
y=dsolve('D2y+2*Dy+3*y=0');
dy=diff(y,t);
fy=MATLABFunction(y);
fdy=MATLABFunction(dy);
tt=0:.1:.5;
C=-4:1:4;
L=length(C);
for k1=1:L
```

```
    for k2=1:L
      plot(fy(C(k1),C(k2),tt),fdy(C(k1),C(k2),tt))
    end;
end;
xlim([-4,4]),ylim([-4,4])
-1.0000 + 1.4142i
 -1.0000 - 1.4142i
```

It can be seen that the roots are complex with the real part being negative and leading to a stable equilibrium as seen by the direction of the arrows pointing to the origin. The phase portrait is shown in Figure 3.1.

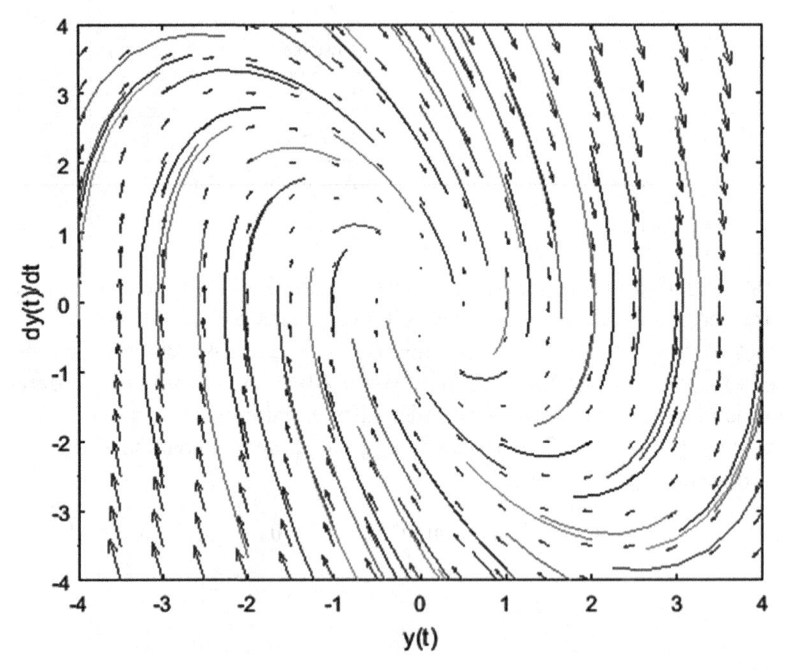

Figure 3.1 Phase portrait associated with the differential equation in Example # 3.1. The directional arrows are moving towards the center, demonstrating that the system is asymptotically stable as suggested in Table 3.1 (complex roots with real part being negative).

All the cases listed in Table 3.1 have been simulated through solving the differential equation ay''+by'+cy=0 by choosing pairs of values of b and c. The selected displays from the output of the MATLAB run are displayed below showing all eleven cases. Three cases deserve special mention. These correspond to cases where either one of the roots is zero or both roots are zeros. In all these three cases, the notion of the critical point is less obvious. In fact, it can be seen that when one of the roots is zero, the solutions move away from a line through the origin, dy/dt=0 (unstable equilibrium) or move towards a line through the origin, dy/dt=0 (stable). When both roots are zero, solutions appear parallel to the line dy/dt=0, moving in opposite directions.

For all the other cases, the critical or equilibrium point is [0,0] and the solutions either move towards it (stable) or move away from it (unstable). The only exception is the case when the roots are purely imaginary in which case, the orbits or the trajectories are circles or ellipses suggesting oscillatory behavior.

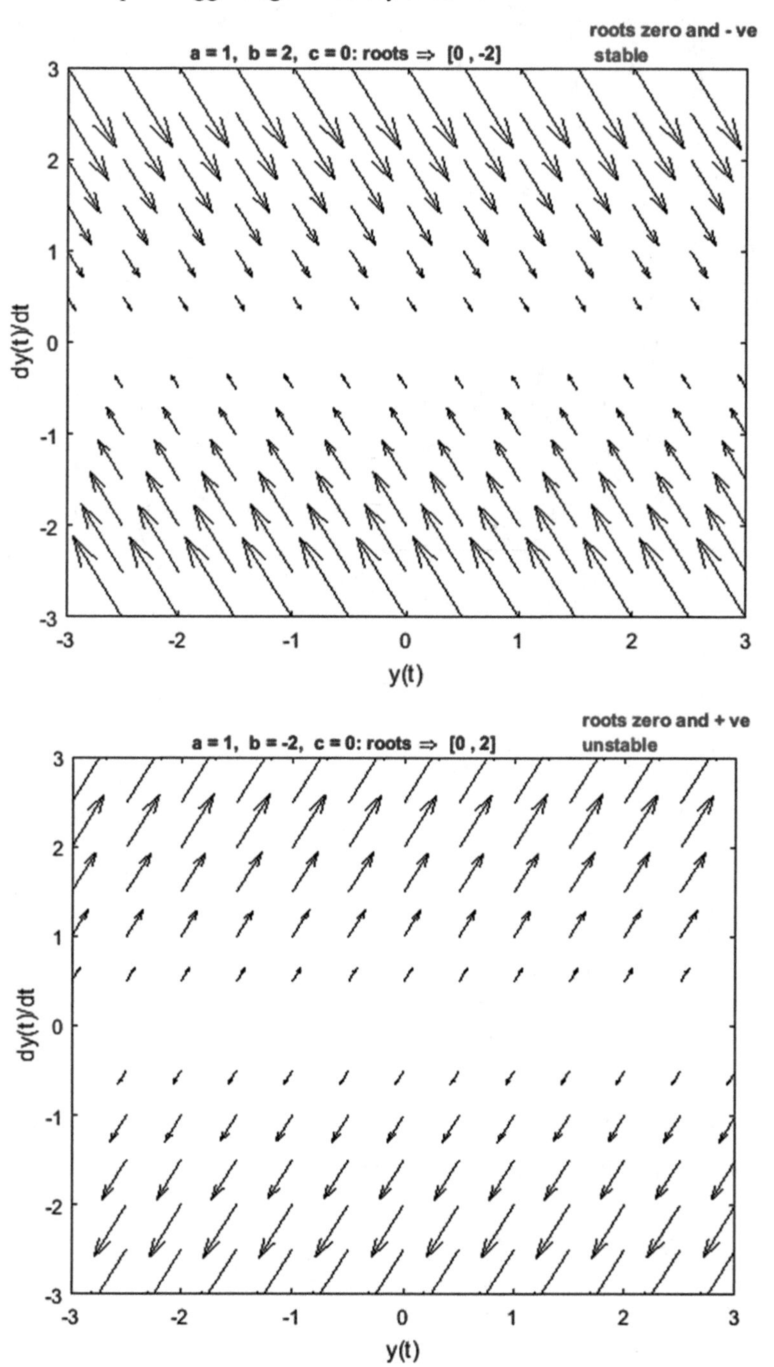

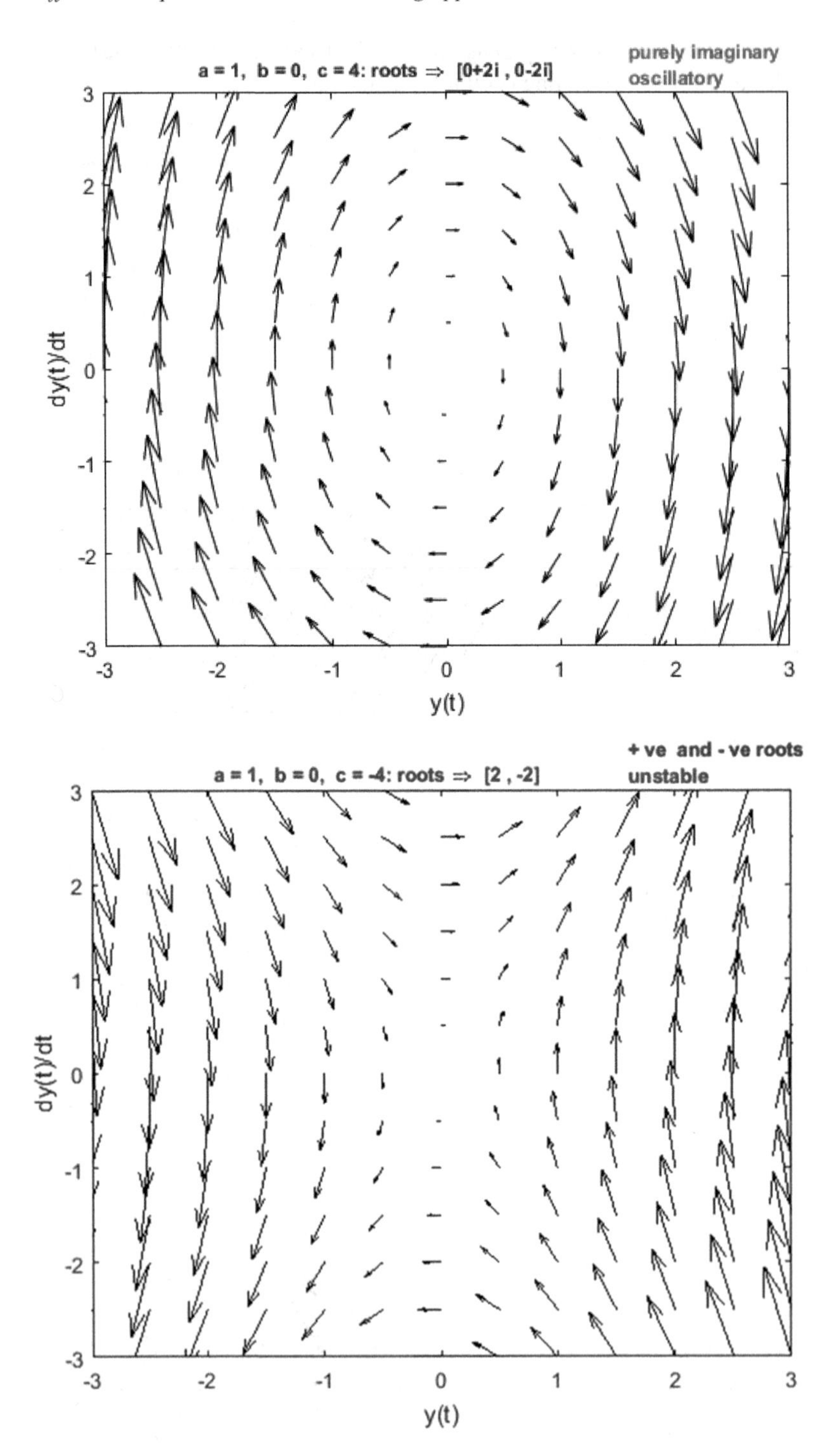
purely imaginary
oscillatory
a = 1, b = 0, c = 4: roots ⇒ [0+2i , 0-2i]
dy(t)/dt
y(t)
+ ve and - ve roots
unstable
a = 1, b = 0, c = -4: roots ⇒ [2 , -2]
dy(t)/dt
y(t)

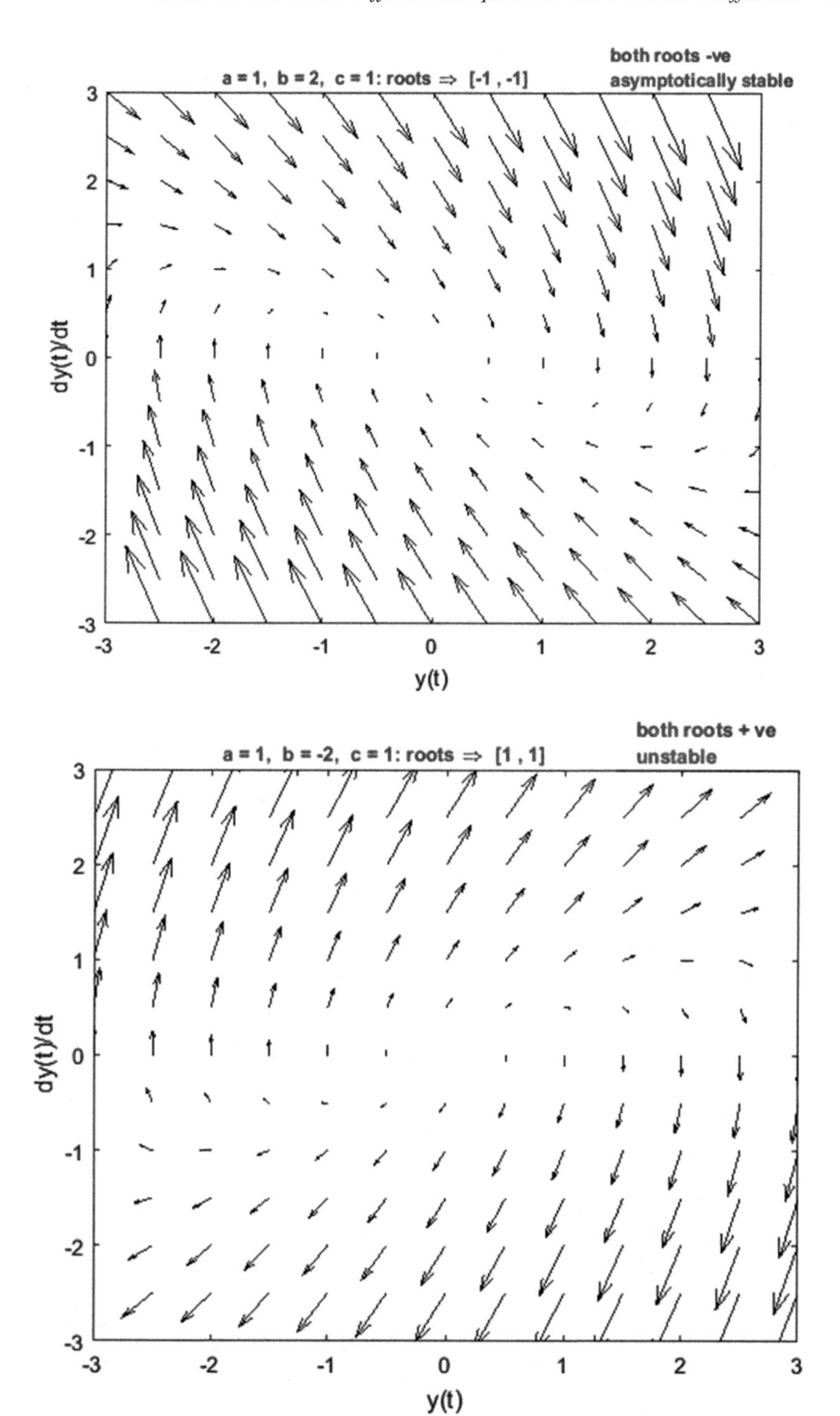
both roots -ve
asymptotically stable
a = 1, b = 2, c = 1: roots ⇒ [-1 , -1]
dy(t)/dt
y(t)
both roots + ve
unstable
a = 1, b = -2, c = 1: roots ⇒ [1 , 1]
dy(t)/dt
y(t)

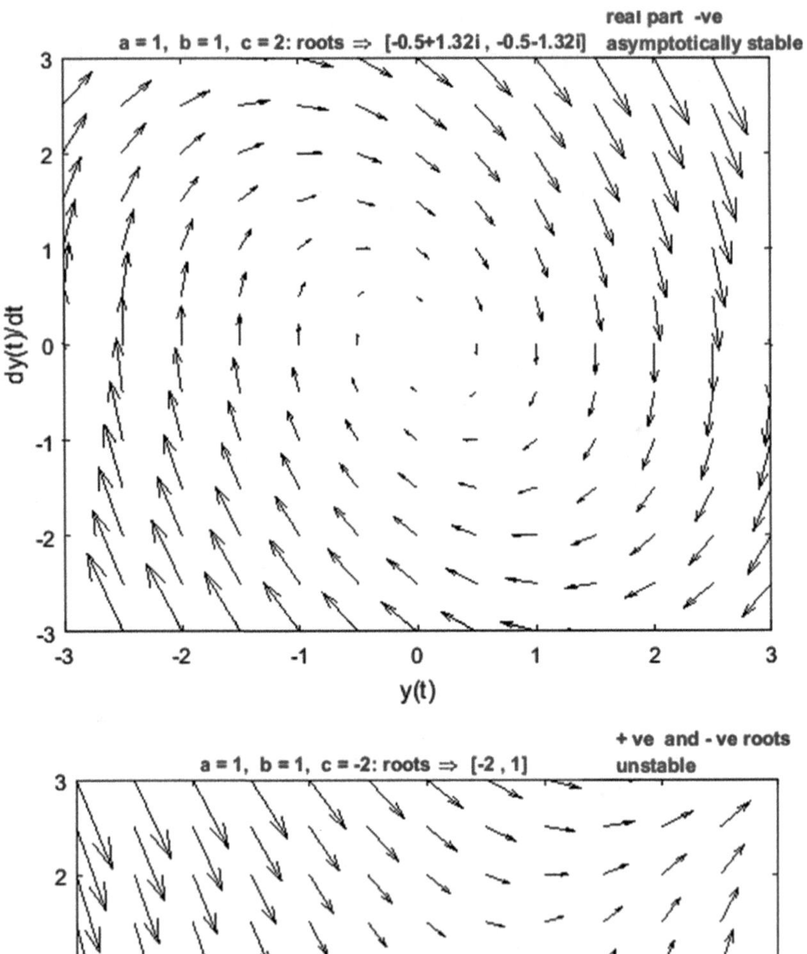
real part -ve
asymptotically stable
a = 1, b = 1, c = 2: roots ⇒ [-0.5+1.32i , -0.5-1.32i]
dy(t)/dt
y(t)
3
2
1
0
-1
-2
-3

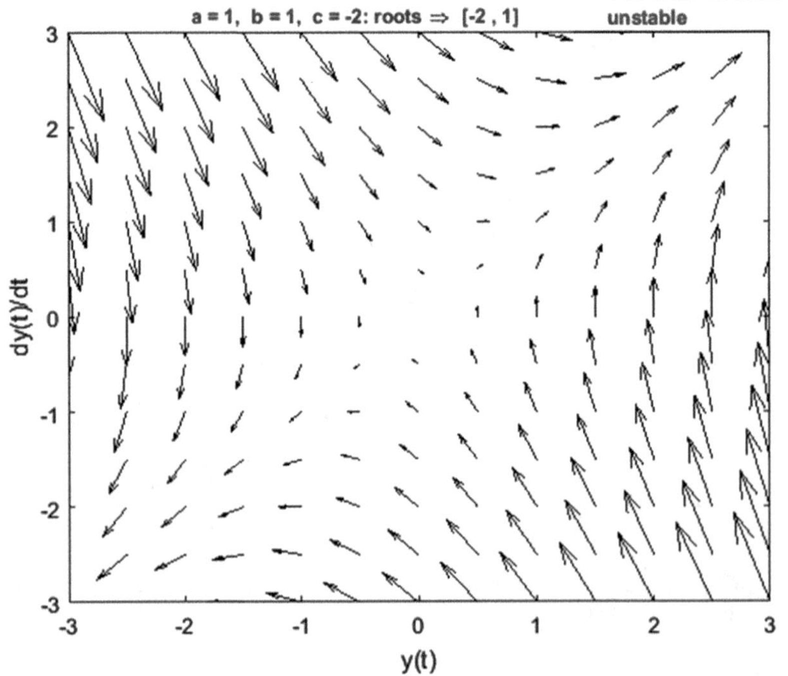
+ve and -ve roots
unstable
a = 1, b = 1, c = -2: roots ⇒ [-2 , 1]
dy(t)/dt
y(t)
3
2
1
0
-1
-2
-3

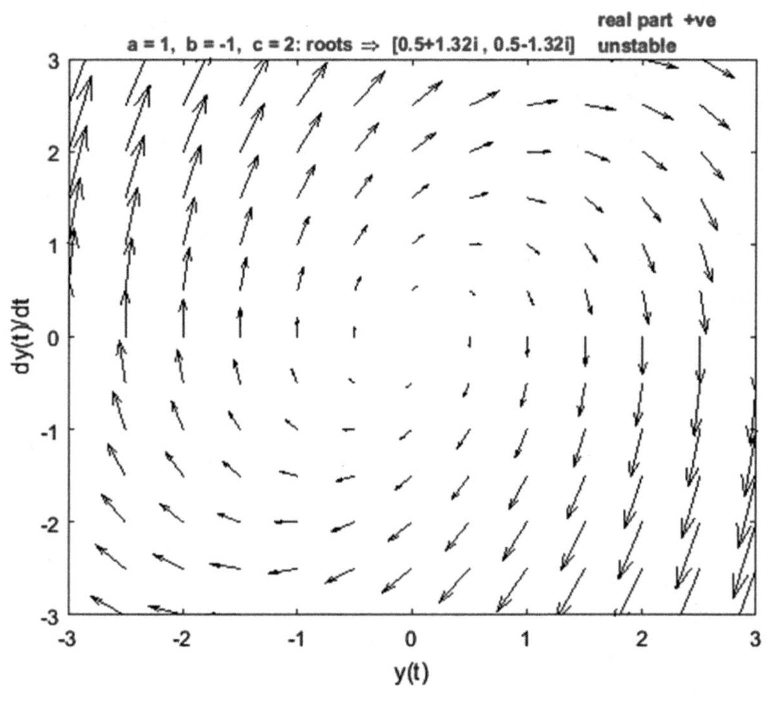
real part +ve
a = 1, b = -1, c = 2: roots ⇒ [0.5+1.32i , 0.5-1.32i]
unstable
3
2
1
0
-1
-2
-3
dy(t)/dt
-3
-2
-1
0
1
2
3
y(t)

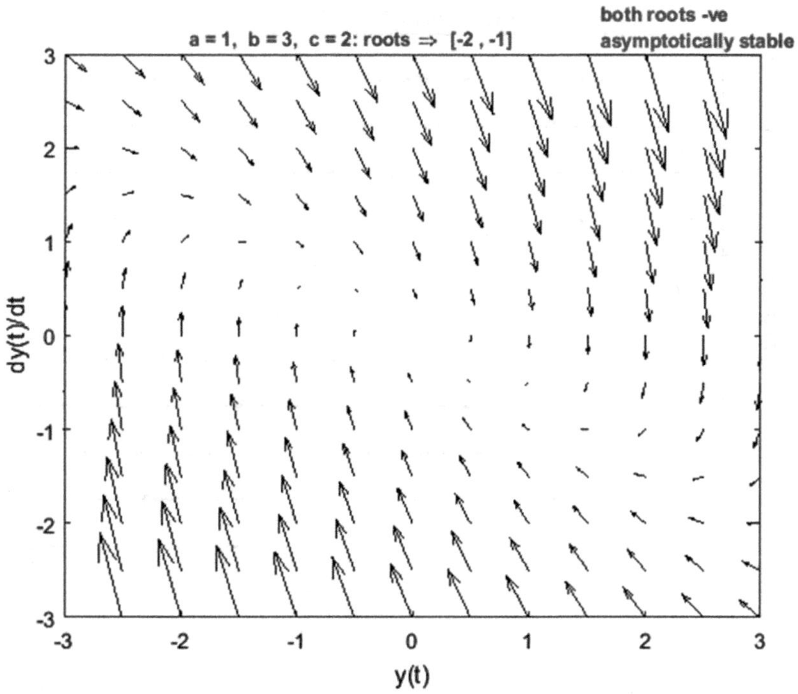
both roots -ve
a = 1, b = 3, c = 2: roots ⇒ [-2 , -1]
asymptotically stable
3
2
1
0
-1
-2
-3
dy(t)/dt
-3
-2
-1
0
1
2
3
y(t)

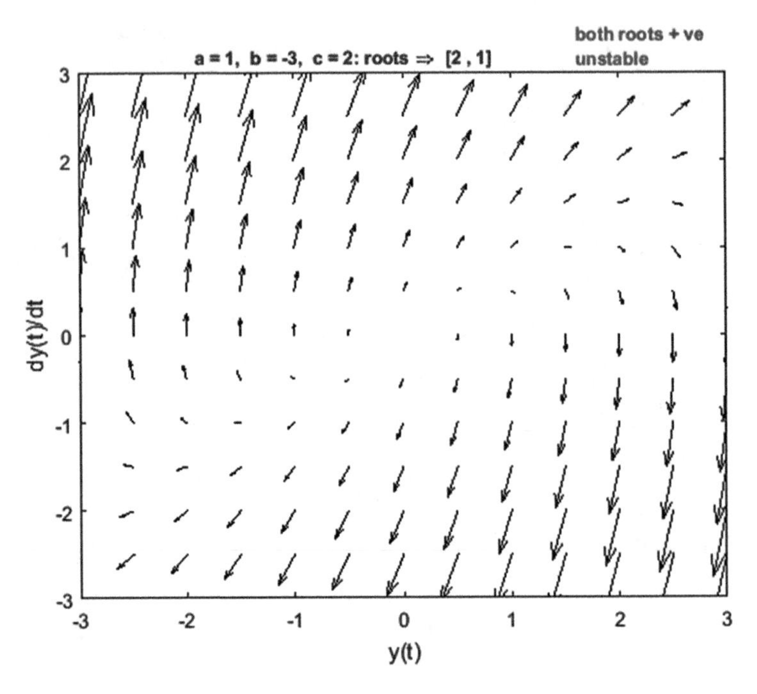
a = 1, b = -3, c = 2: roots ⇒ [2 , 1]
both roots + ve
unstable
dy(t)/dt
y(t)
3
2
1
0
-1
-2
-3
-3
-2
-1
0
1
2
3

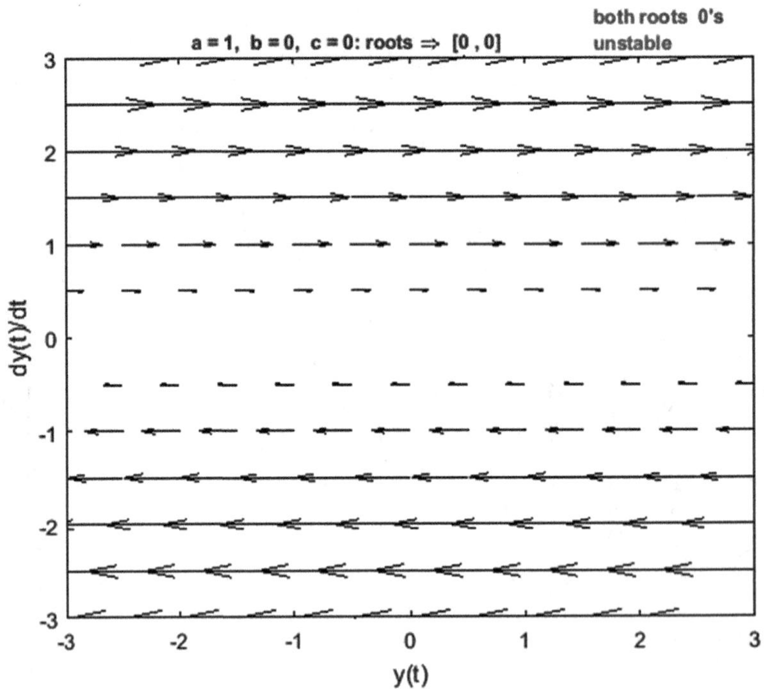
a = 1, b = 0, c = 0: roots ⇒ [0 , 0]
both roots 0's
unstable
dy(t)/dt
y(t)
3
2
1
0
-1
-2
-3
-3
-2
-1
0
1
2
3

Consider a homogeneous linear second order differential equation with constant coefficients (Example # 3.2)

$$y''(t)+5y'(t)+6y(t)=0, \quad [y(0)=1, y'(0)=0]. \tag{3.44}$$

The coefficients a, b, and c respectively are 1, 5 and 6, resulting in the characteristic equation

$$r^2+5r+6=0. \tag{3.45}$$

The roots of the characteristic equation are

$$r_1=-3 \quad r_2=-2. \tag{3.46}$$

The phase plane plot associated with the differential equation in eqn. (3.44) is shown in Figure 3.2. It is clear that both roots are negative and the system is therefore asymptotically stable as seen by the direction of the arrows that move towards the origin, the equilibrium, or the critical point [0,0]. Figure 3.2 also displays the state of equilibrium along with the roots.

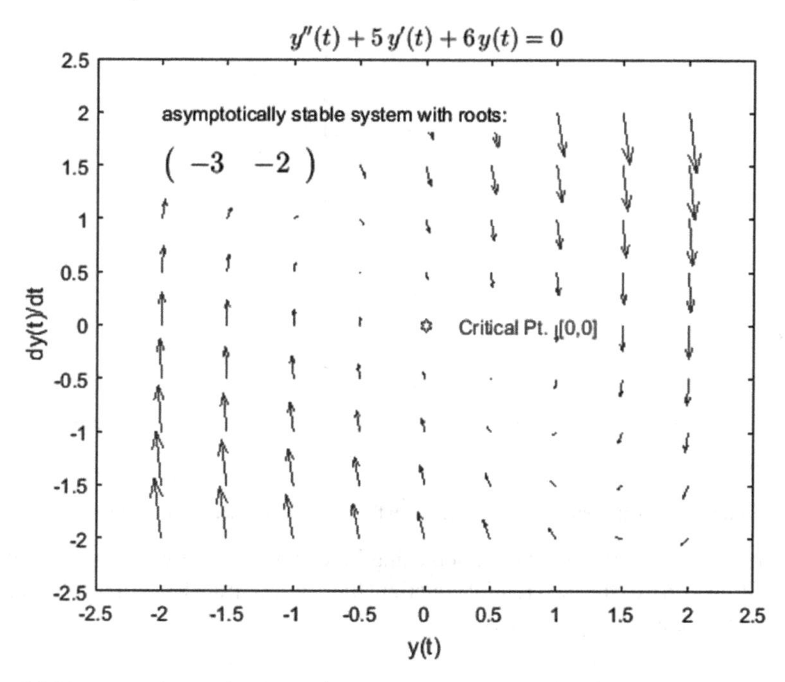

Figure 3.2 Phase portrait associated with the differential equation in Example # 3.2. The asymptotic stability of the system is seen.

The general solution to the homogeneous differential equation is obtained as

$$y(t)=c_1e^{-3t}+c_2e^{-2t}. \tag{3.47}$$

Applying the two initial conditions appearing alongside the differential equation in eqn. (3.44) to eqn. (3.47) and its derivative, two linear equations are obtained as

$$\begin{aligned} c_1 + c_2 &= 1 \\ -3c_1 - 2c_2 &= 0 \end{aligned} \tag{3.48}$$

The values of the constants can be obtained from eqn. (3.48) as

$$\begin{aligned} c_1 &= -2 \\ c_2 &= 3 \end{aligned} \tag{3.49}$$

The solution in eqn. (3.47) can now be written as

$$y(t) = -2e^{-3t} + 3e^{-2t} \tag{3.50}$$

A component plot of t vs. y(t) and a phase plot obtained as y(t) vs. y'(t) are shown in Figure 3.3.

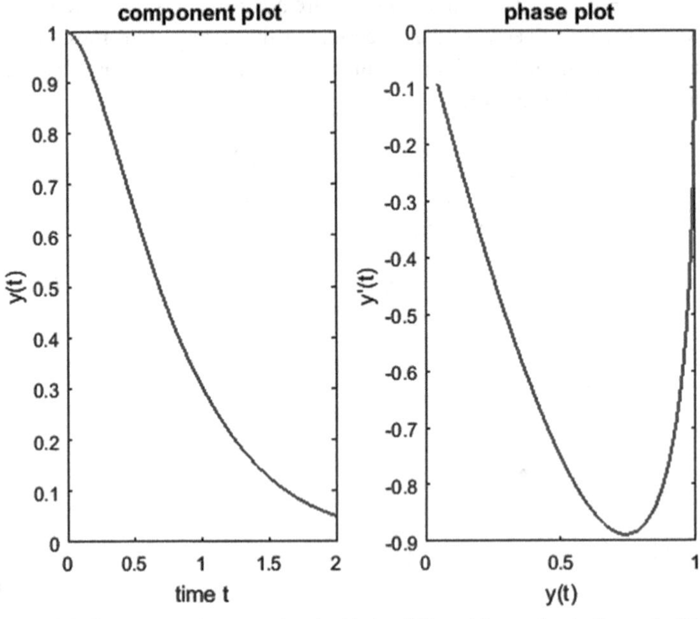

Figure 3.3 Component plots associated with the differential equation in Example # 3.2.

The solution may also be obtained using inverse Laplace transforms. Taking the Laplace transform of the differential equation, the Laplace transform of the differential equation Y(s) can be expressed as

$$Y(s) = \frac{s+5}{(s+2)(s+3)} = \frac{3}{(s+2)} - \frac{2}{(s+3)}. \tag{3.51}$$

Taking the inverse transform, the solution is obtained as

$$y(t) = -2e^{-3t} + 3e^{-2t}. \tag{3.52}$$

Several additional examples are now given providing more details and verification using Runge-Kutta methods.

3.2.6 Additional examples

A number of examples are now provided. In each case, the solution generated in MATLAB provides the theory first, followed by the details of the solution. The input is the set of values of *a*, b, c. If the initial conditions are given, the input set is [*a*,b,c,y(0),y'(0)]. The results consist of the following displays:

1. Theory of Second Order Homogeneous Equations with constant coefficients.
2. Phase plane plot indicating the roots of the characteristic equation and the state of stability.
3. The characteristic equation, its roots, properties of the roots, explanation on the type of equilibrium.
4. If no initial conditions are provided (determined by the number of input values), statement appears indicating the absence.
5. If initial conditions are given, the unknown constants are evaluated (with all the steps shown).
6. Theory of Laplace transform based solution, inverse Laplace transform of y(t) and Laplace transform based solution shown.
7. Comparison of all results including those obtained using dsolve(.) in MATLAB and numerical techniques shown. The stability status is again indicated.
8. Plots of the solution (component plot and phase plot) are also provided.

Example # 3.3

```
abc=[1,-4,0];
```

Linear 2nd Order Differential Equation with Constant Coefficients

Homogeneous Differential Equation $$a\frac{d^2y}{dt^2}+b\frac{dy}{dt}+cy=0,\quad a\neq 0$$

Characteristic Equation $$ar^2+br+c=0$$

Roots of Characteristic Equation or Eigenvalues $$r_{1,2}=r_{\pm}=\frac{-b\pm\sqrt{b^2-4ac}}{2a}$$

Properties of the Roots ⇒
$$b^2-4ac>0,\quad r_1\neq r_2\ (real\ roots)$$
$$b^2-4ac=0,\quad r_1=r_2=r_0\ (real\ root)$$
$$c=0,\quad r_1=0, r_2\neq 0\ or\ r_1\neq 0, r_2=0$$
$$b^2-4ac<0,\quad r_{1,2}=r_{\pm}=\alpha\pm j\beta$$

Stability of the System ⇒
- [$r_1>0$ or $r_2>0$] or [$r_0\geq 0$] or [$\alpha>0$], unstable
- [$r_1<0$ and $r_2<0$] or [$r_0<0$] or [$\alpha<0$], asymptotically stable
- [$r_1=0, r_2<0$] or [$r_1<0, r_2=0$], stable
- [$\alpha=0$], oscillatory

Solution y(t) ⇒
$$c_1e^{r_1t}+c_2e^{r_2t},\quad r_1\neq r_2$$
$$[c_1+c_2t]\,e^{r_0t},\quad r_1=r_2=r_0$$
$$[c_1cos(\beta t)+c_2sin(\beta t)]\,e^{\alpha t},\quad r_{\pm}=\alpha\pm j\beta$$

$$\text{Special Case}\Rightarrow\frac{d^2y}{dt^2}=0,\quad r_1=r_2=0,\quad y(t)=c_1+c_2t$$

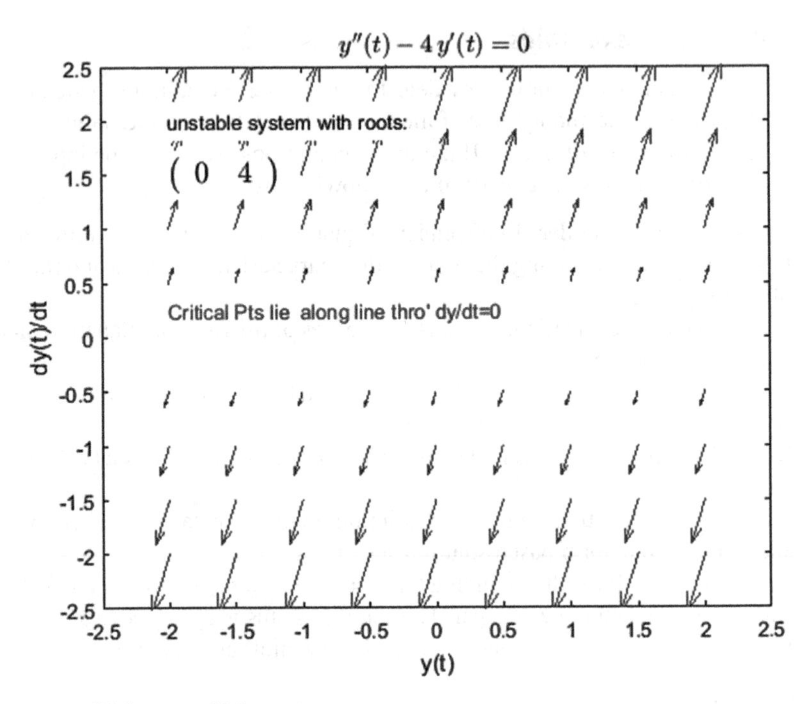

$$y''(t) - 4y'(t) = 0$$

Characteristic Equation $r^2 - 4r = 0$

Roots (r_1 & r_2) $(0 \quad 4)$

Roots are +ve or zero; r_1=0 r_2=r_0; unstable system

------ General Solution using the roots y(t) = $[c_1+c_2e^{r_0t}]$ -----

$$y(t) = c_1 + c_2\,e^{4t}$$

------ General Solution using dsolve(.) obtained directly ---

$$y_s(t) = b_1 + b_2\,e^{4t}$$

No Initial Conditions Given

Example # 3.4

```
abc=[1,0,9,1,-1];
```

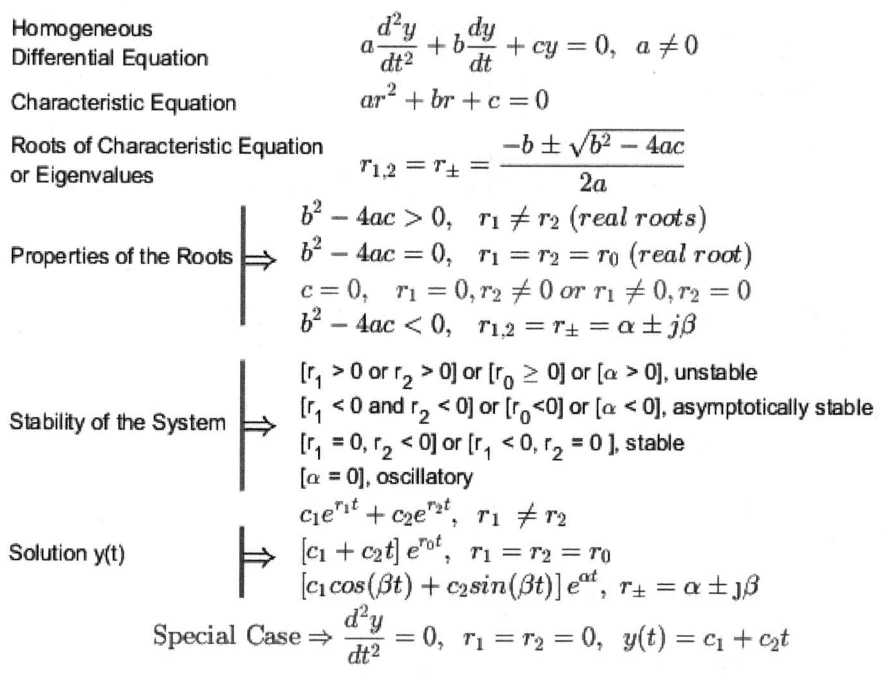

Linear 2nd Order Differential Equation with Constant Coefficients

Homogeneous Differential Equation	$a\frac{d^2y}{dt^2} + b\frac{dy}{dt} + cy = 0, \quad a \neq 0$
Characteristic Equation	$ar^2 + br + c = 0$
Roots of Characteristic Equation or Eigenvalues	$r_{1,2} = r_{\pm} = \frac{-b \pm \sqrt{b^2 - 4ac}}{2a}$
Properties of the Roots ⇒	$b^2 - 4ac > 0, \quad r_1 \neq r_2 \ (real\ roots)$ $b^2 - 4ac = 0, \quad r_1 = r_2 = r_0 \ (real\ root)$ $c = 0, \quad r_1 = 0, r_2 \neq 0 \ or \ r_1 \neq 0, r_2 = 0$ $b^2 - 4ac < 0, \quad r_{1,2} = r_{\pm} = \alpha \pm j\beta$
Stability of the System ⇒	[r_1 > 0 or r_2 > 0] or [$r_0 \geq 0$] or [α > 0], unstable [r_1 < 0 and r_2 < 0] or [r_0<0] or [α < 0], asymptotically stable [r_1 = 0, r_2 < 0] or [r_1 < 0, r_2 = 0], stable [α = 0], oscillatory
Solution y(t) ⇒	$c_1e^{r_1t} + c_2e^{r_2t}, \quad r_1 \neq r_2$ $[c_1 + c_2t]\, e^{r_0t}, \quad r_1 = r_2 = r_0$ $[c_1 cos(\beta t) + c_2 sin(\beta t)]\, e^{\alpha t}, \quad r_{\pm} = \alpha \pm j\beta$

$$\text{Special Case} \Rightarrow \frac{d^2y}{dt^2} = 0, \quad r_1 = r_2 = 0, \quad y(t) = c_1 + c_2t$$

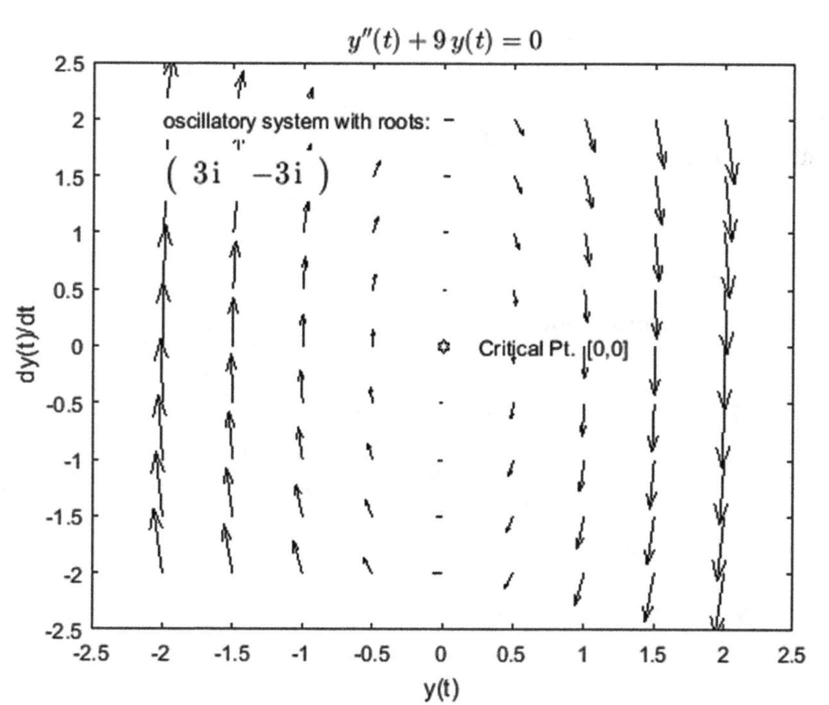

$$y''(t) + 9\,y(t) = 0 \qquad [\,y(0) = 1,\ y'(0) = -1\,]$$

Characteristic Equation $\quad r^2 + 9 = 0$

Roots (r_1 & r_2) $\quad (\ 3i \quad -3i\ )$

Roots imaginary $r_{1,2} = \pm\, j\beta$; oscillatory system

----- General Solution using the roots $y(t) = [\,c_1\cos(\beta t) - c_2\sin(\beta t)\,]$ ----

$$y(t) = c_1\,\cos(3\,t) - c_2\,\sin(3\,t)$$

Differentiate y(t) & apply initial conditions ==> [y(0) = 1, y'(0) = -1]

$$\begin{pmatrix} c_1 = 1 \\ -3\,c_2 = -1 \end{pmatrix} \quad ===> \quad \begin{pmatrix} c_1 = 1 \\ c_2 = \frac{1}{3} \end{pmatrix}$$

----- Solution after substitution (c_1 & c_2) in the general solution -----

$$y(t) = \cos(3\,t) - \frac{\sin(3\,t)}{3}$$

$$y''(t) + 9\,y(t) = 0 \qquad [\,y(0) = 1,\ y'(0) = -1\,]$$

oscillatory system

solution
(Laplace Transform) $\quad y_L(t) = \cos(3\,t) - \frac{\sin(3\,t)}{3}$
details on next page

solution
(dsolve) $\quad y_d(t) = \cos(3\,t) - \frac{\sin(3\,t)}{3}$

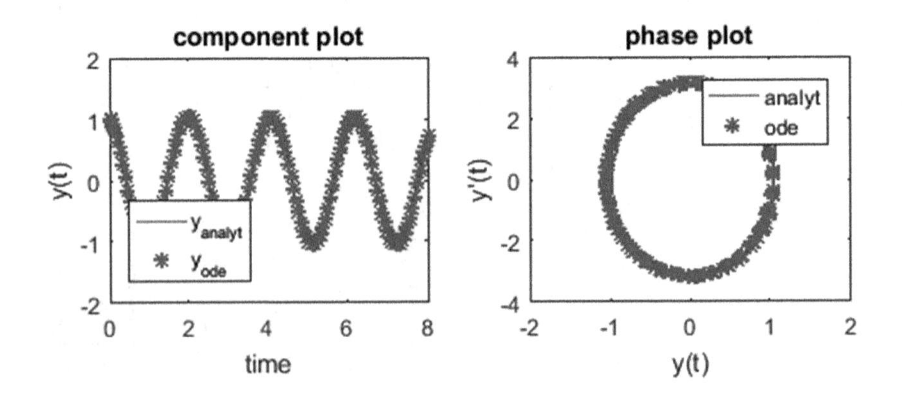

Inverse Laplace based solution of a 2nd order Differential Equation

$$ay''(t) + by'(t) + cy(t) = 0 \quad [y(0) = y_0, y'(0) = y_1] \rightarrow Y(s) = \frac{(as+b)y_0 + ay_1}{as^2 + bs + c}$$

Differential Equation with Initial Conditions

$$y''(t) + 9\,y(t) = 0 \qquad (\,y_0 = 1 \quad y_1 = -1\,)$$

Laplace transform of y(t) using the expression on top

$$Y_L(s) = \frac{s-1}{s^2+9}$$

Laplace transform of y(t) as partial fractions

$$Y_L(s) = \frac{s-1}{s^2+9}$$

Solution using inverse Laplace

$$y_L(t) = \cos(3\,t) - \frac{\sin(3\,t)}{3}$$

oscillatory system

Example # 3.5

```
abc=[1,4,4,1,0];
```

For the remaining examples, the first display consisting of the theory is not provided.

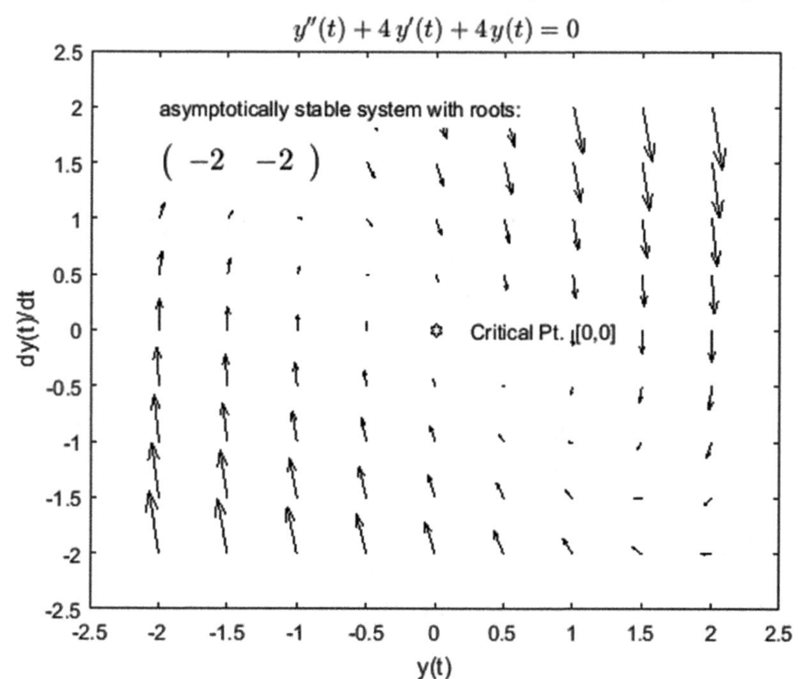

$$y''(t) + 4\,y'(t) + 4\,y(t) = 0 \qquad [\,y(0) = 1,\, y'(0) = 0\,]$$

Characteristic Equation $r^2 + 4r + 4 = 0$

Roots (r_1 & r_2) $\begin{pmatrix} -2 & -2 \end{pmatrix}$

Equal Real roots (r_1, r_2 = r): -ve; asymptotically stable system

------ General Solution using the roots $y(t) = e^{rt}\,[c_1 + c_2 t]$ -----

$$y(t) = e^{-2t}\,(c_2 + c_1\,t)$$

Differentiate y(t) & apply initial conditions ==> [y(0) = 1, y'(0) = 0]

$$\begin{pmatrix} c_2 = 1 \\ c_1 - 2\,c_2 = 0 \end{pmatrix} \quad ===> \quad \begin{pmatrix} c_1 = 2 \\ c_2 = 1 \end{pmatrix}$$

----- Solution after substitution (c_1 & c_2) in the general solution -----

$$y(t) = e^{-2t}\,(2\,t + 1)$$

$$y''(t) + 4\,y'(t) + 4\,y(t) = 0 \qquad [\,y(0) = 1,\, y'(0) = 0\,]$$

asymptotically stable system

solution (Laplace Transform) details on next page $\quad y_L(t) = e^{-2t}\,(2\,t + 1)$

solution (dsolve) $\quad y_d(t) = e^{-2t}\,(2\,t + 1)$

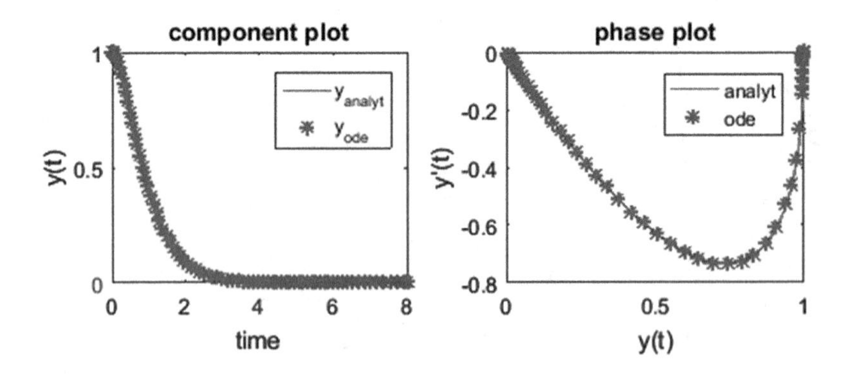

Inverse Laplace based solution of a 2nd order Differential Equation

$$ay''(t) + by'(t) + cy(t) = 0 \quad [y(0) = y_0, y'(0) = y_1] \rightarrow Y(s) = \frac{(as+b)y_0 + ay_1}{as^2 + bs + c}$$

Differential Equation with Initial Conditions

$$y''(t) + 4\,y'(t) + 4\,y(t) = 0 \qquad (\ y_0 = 1 \quad y_1 = 0\ )$$

Laplace transform of y(t) using the expression on top

$$Y_L(s) = \frac{s+4}{(s+2)^2}$$

Laplace transform of y(t) as partial fractions

$$Y_L(s) = \frac{1}{s+2} + \frac{2}{(s+2)^2}$$

Solution using inverse Laplace

$$y_L(t) = \mathrm{e}^{-2t}\ (2\,t + 1)$$

asymptotically stable system

Example # 3.6

```
abc=[1,-4,8,1,0];
```

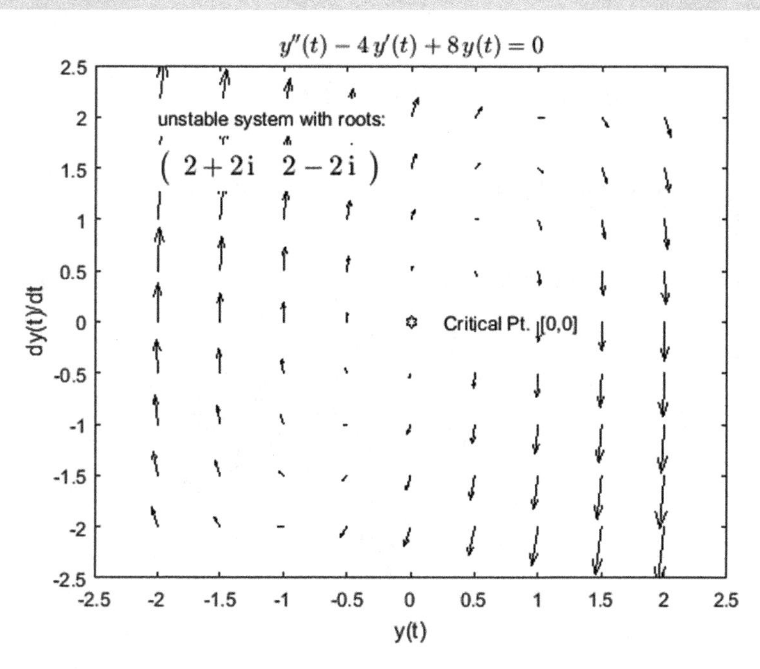

$$y''(t) - 4\,y'(t) + 8\,y(t) = 0 \qquad [\,y(0) = 1,\ y'(0) = 0\,]$$

Characteristic Equation $\quad r^2 - 4r + 8 = 0$

Roots (r_1 & r_2) $\quad \left(\ 2+2\mathrm{i} \quad 2-2\mathrm{i}\ \right)$

Real Parts of roots: + ve; $r_{1,2} = \alpha \pm j\beta$; unstable system

------ General Solution using the roots $y(t) = e^{\alpha t}\,[\,c_1\cos(\beta t) - c_2\sin(\beta t)\,]$ ----

$$y(t) = -\mathrm{e}^{2t}\,\left(c_2\,\sin(2t) - c_1\,\cos(2t)\right)$$

Differentiate y(t) & apply initial conditions ==> $[\,y(0) = 1,\ y'(0) = 0\,]$

$$\begin{pmatrix} c_1 = 1 \\ 2\,c_1 - 2\,c_2 = 0 \end{pmatrix} \quad ===> \quad \begin{pmatrix} c_1 = 1 \\ c_2 = 1 \end{pmatrix}$$

----- Solution after substitution (c_1 & c_2) in the general solution -----

$$y(t) = \mathrm{e}^{2t}\,\left(\cos(2\,t) - \sin(2\,t)\right)$$

$$y''(t) - 4\,y'(t) + 8\,y(t) = 0 \qquad [\,y(0) = 1,\ y'(0) = 0\,]$$

unstable system

solution (Laplace Transform) details on next page $\quad y_L(t) = \sqrt{2}\,\cos\!\left(2\,t + \frac{\pi}{4}\right)\,\mathrm{e}^{2\,t}$

solution (dsolve) $\quad y_d(t) = \sqrt{2}\,\cos\!\left(2\,t + \frac{\pi}{4}\right)\,\mathrm{e}^{2\,t}$

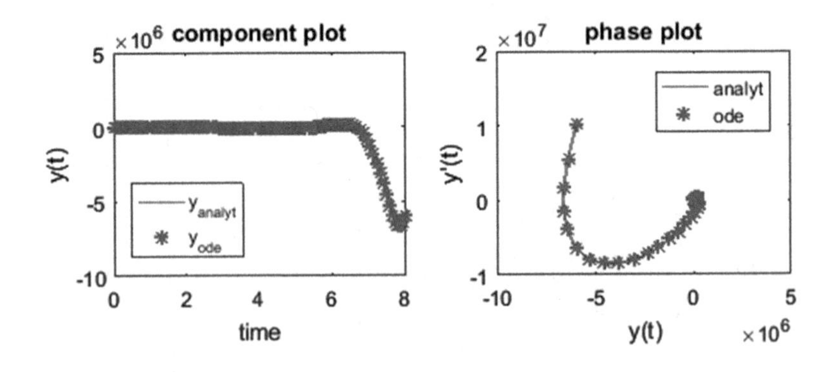

Inverse Laplace based solution of a 2nd order Differential Equation

$$ay''(t) + by'(t) + cy(t) = 0 \quad [y(0) = y_0, y'(0) = y_1] \quad \rightarrow \quad Y(s) = \frac{(as+b)y_0 + ay_1}{as^2 + bs + c}$$

Differential Equation with Initial Conditions

$$y''(t) - 4\,y'(t) + 8\,y(t) = 0 \qquad (\; y_0 = 1 \quad y_1 = 0 \;)$$

Laplace transform of y(t) using the expression on top

$$Y_L(s) = \frac{s-4}{s^2-4\,s+8}$$

Laplace transform of y(t) as partial fractions

$$Y_L(s) = \frac{s-4}{s^2-4\,s+8}$$

Solution using inverse Laplace

$$y_L(t) = \sqrt{2}\,\cos\left(2\,t + \frac{\pi}{4}\right)\,e^{2t}$$

unstable system

Example # 3.7

```
abc=[1,0,-9,-1,1];
```

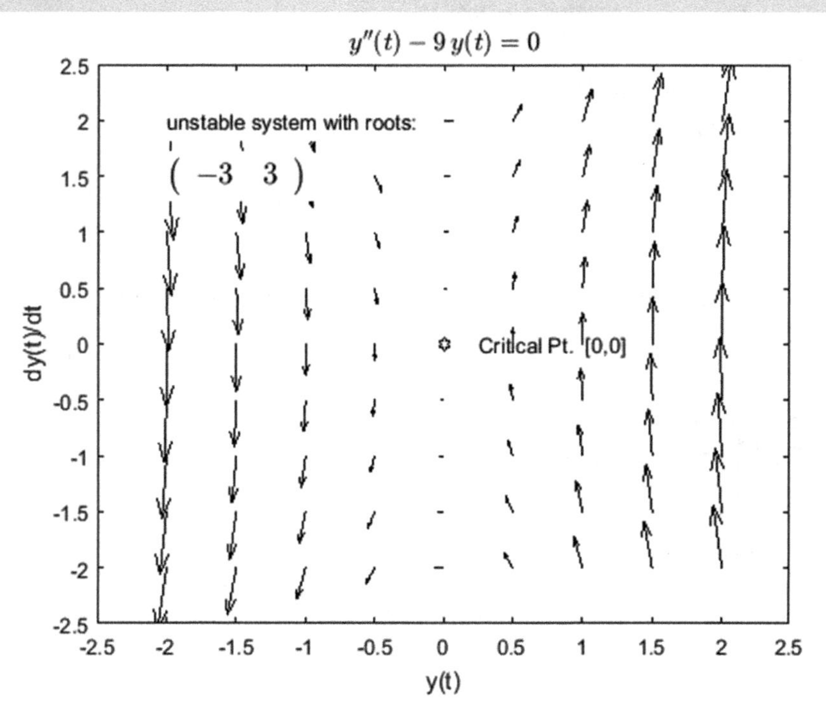

$$y''(t) - 9\,y(t) = 0 \qquad [\,y(0) = -1,\ y'(0) = 1\,]$$

Characteristic Equation $\quad r^2 - 9 = 0$

Roots (r_1 & r_2) $\quad \begin{pmatrix} -3 & 3 \end{pmatrix}$

One/both roots: + ve; unstable system

----- General Solution using the roots $y(t) = c_1 e^{r_1 t} + c_2 e^{r_2 t}$ -----

$$y(t) = c_1\,e^{-3t} + c_2\,e^{3t}$$

Differentiate y(t) & apply initial conditions ==> [y(0) = -1, y'(0) = 1]

$$\begin{pmatrix} c_1 + c_2 = -1 \\ 3\,c_2 - 3\,c_1 = 1 \end{pmatrix} \quad ===> \quad \begin{pmatrix} c_1 = -\frac{2}{3} \\ c_2 = -\frac{1}{3} \end{pmatrix}$$

----- Solution after substitution (c_1 & c_2) in the general solution -----

$$y(t) = -\frac{2\,e^{-3t}}{3} - \frac{e^{3t}}{3}$$

$$y''(t) - 9\,y(t) = 0 \qquad [\,y(0) = -1,\ y'(0) = 1\,]$$

unstable system

solution (Laplace Transform) details on next page $\quad y_L(t) = -\frac{e^{-3t}\left(e^{6t}+2\right)}{3}$

solution (dsolve) $\quad y_d(t) = -\frac{e^{-3t}\left(e^{6t}+2\right)}{3}$

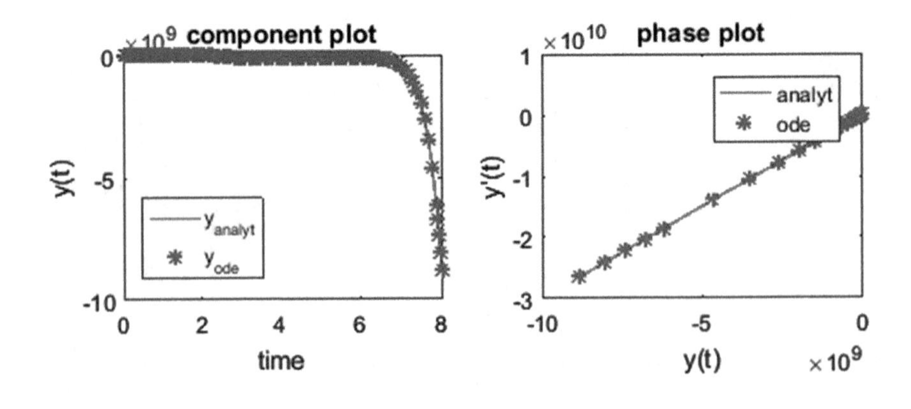

Inverse Laplace based solution of a 2nd order Differential Equation

$$ay''(t) + by'(t) + cy(t) = 0 \quad [y(0) = y_0, y'(0) = y_1] \quad \rightarrow \quad Y(s) = \frac{(as+b)y_0 + ay_1}{as^2 + bs + c}$$

Differential Equation with Initial Conditions

$$y''(t) - 9\,y(t) = 0 \qquad (\; y_0 = -1 \quad y_1 = 1 \;)$$

Laplace transform of y(t) using the expression on top

$$Y_L(s) = -\frac{s-1}{s^2-9}$$

Laplace transform of y(t) as partial fractions

$$Y_L(s) = -\frac{1}{3\,(s-3)} - \frac{2}{3\,(s+3)}$$

Solution using inverse Laplace

$$y_L(t) = -\frac{e^{-3t}\left(e^{6t}+2\right)}{3}$$

unstable system

Example # 3.8

```
abc=[1,6,2,1,0];
```

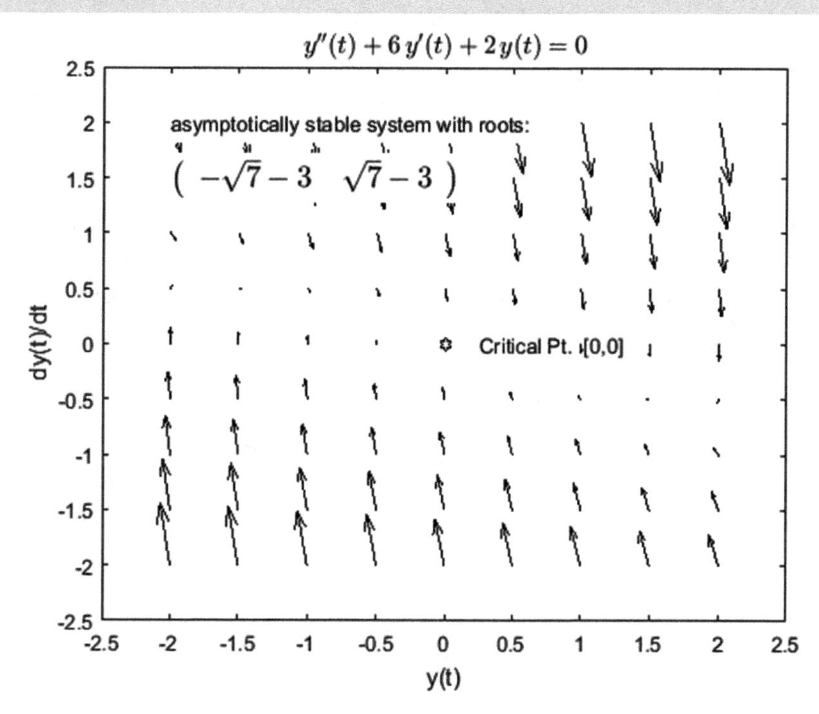

$$y''(t) + 6\,y'(t) + 2\,y(t) = 0 \qquad [\,y(0) = 1,\ y'(0) = 0\,]$$

Characteristic Equation $\quad r^2 + 6r + 2 = 0$

Roots (r_1 & r_2) $\quad \left(\ -\sqrt{7} - 3 \quad \sqrt{7} - 3\ \right)$

Real Parts of roots: - ve; asymptotically stable system

------ General Solution using the roots $y(t) = c_1 e^{r_1 t} + c_2 e^{r_2 t}$ -----

$$y(t) = c_1\, e^{-t\,(\sqrt{7}+3)} + c_2\, e^{t\,(\sqrt{7}-3)}$$

Differentiate y(t) & apply initial conditions ==> [y(0) = 1, y'(0) = 0]

$$\begin{pmatrix} c_1 + c_2 = 1 \\ c_2\,(\sqrt{7} - 3) - c_1\,(\sqrt{7} + 3) = 0 \end{pmatrix} \quad \text{===>} \quad \begin{pmatrix} c_1 = \frac{1}{2} - \frac{3\sqrt{7}}{14} \\ c_2 = \frac{\sqrt{7}(\sqrt{7}+3)}{14} \end{pmatrix}$$

---- Solution after substitution (c_1 & c_2) in the general solution -----

$$y(t) = \frac{\sqrt{7}\, e^{t(\sqrt{7}-3)}\,(\sqrt{7}+3)}{14} - e^{-t(\sqrt{7}+3)}\left(\frac{3\sqrt{7}}{14} - \frac{1}{2}\right)$$

$$y''(t) + 6\,y'(t) + 2\,y(t) = 0 \qquad [\,y(0) = 1,\ y'(0) = 0\,]$$

asymptotically stable system

solution (Laplace Transform) details on next page $\quad y_L(t) = e^{-3t-\sqrt{7}t}\left(\frac{e^{2\sqrt{7}t}}{2} + \frac{3\sqrt{7}\,e^{2\sqrt{7}t}}{14} - \frac{3\sqrt{7}}{14} + \frac{1}{2}\right)$

solution (dsolve) $\quad y_d(t) = e^{t\,(\sqrt{7}-3)}\left(\frac{3\sqrt{7}}{14} + \frac{1}{2}\right) - \frac{e^{-t\,(\sqrt{7}+3)}\,(3\sqrt{7}-7)}{14}$

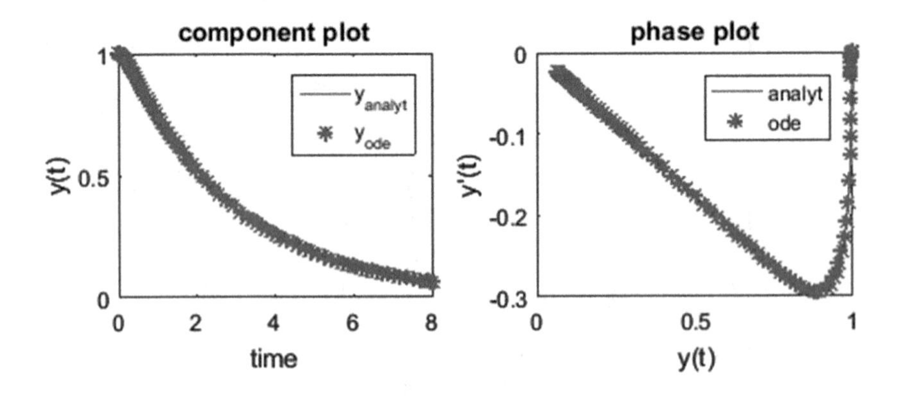

Inverse Laplace based solution of a 2nd order Differential Equation

$$ay''(t) + by'(t) + cy(t) = 0 \quad [y(0) = y_0, y'(0) = y_1] \rightarrow Y(s) = \frac{(as+b)y_0 + ay_1}{as^2 + bs + c}$$

Differential Equation with Initial Conditions

$$y''(t) + 6\,y'(t) + 2\,y(t) = 0 \qquad (\; y_0 = 1 \quad y_1 = 0 \;)$$

Laplace transform of y(t) using the expression on top

$$Y_L(s) = \frac{s+6}{s^2+6\,s+2}$$

Laplace transform of y(t) as partial fractions

$$Y_L(s) = \frac{s+6}{s^2+6\,s+2}$$

Solution using inverse Laplace

$$y_L(t) = \mathrm{e}^{-3t-\sqrt{7}t}\left(\frac{\mathrm{e}^{2\sqrt{7}t}}{2} + \frac{3\sqrt{7}\,\mathrm{e}^{2\sqrt{7}t}}{14} - \frac{3\sqrt{7}}{14} + \frac{1}{2}\right)$$

asymptotically stable system

Example # 3.9

```
abc=[1,-3,2];
```

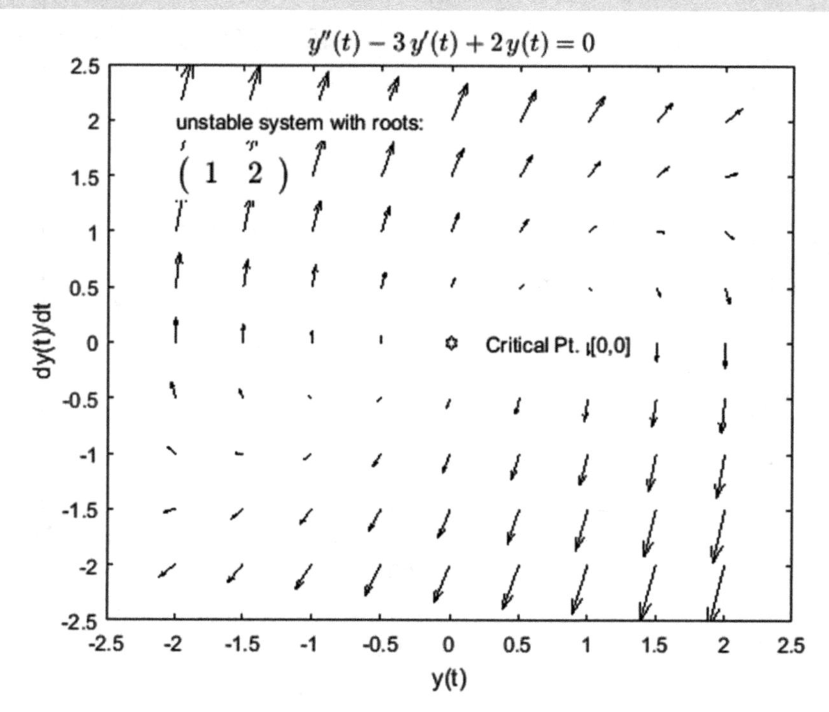

$$y''(t) - 3\,y'(t) + 2\,y(t) = 0$$

Characteristic Equation $\quad r^2 - 3r + 2 = 0$

Roots (r_1 & r_2) $\quad \begin{pmatrix} 1 & 2 \end{pmatrix}$

One/both roots: + ve; unstable system

------ General Solution using the roots $y(t) = c_1 e^{r_1 t} + c_2 e^{r_2 t}$ ----

$$y(t) = c_1\,e^{t} + c_2\,e^{2\,t}$$

------ General Solution using dsolve(.) obtained directly ---

$$y_s(t) = b_2\,e^{t} + b_1\,e^{2\,t}$$

No Initial Conditions Given

Example # 3.10

```
abc=[1,-2,1,1,0];
```

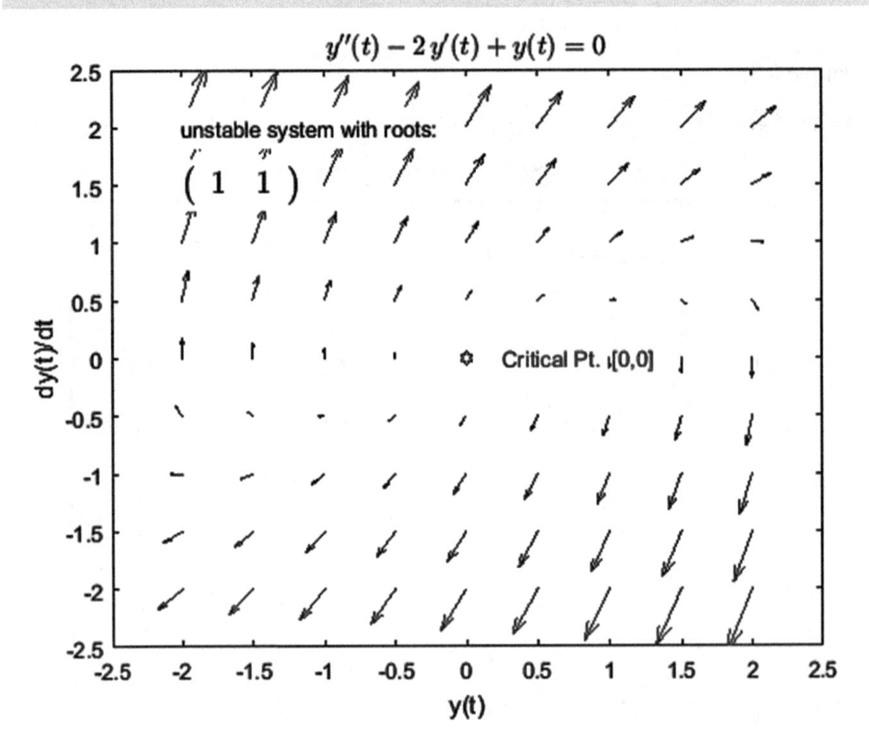

$$y''(t) - 2\,y'(t) + y(t) = 0 \qquad [\,y(0) = 1,\, y'(0) = 0\,]$$

Characteristic Equation $\quad r^2 - 2r + 1 = 0$

Roots (r_1 & r_2) $\quad \begin{pmatrix} 1 & 1 \end{pmatrix}$

Equal Real roots (r_1, r_2 = r): + ve; unstable system

------ General Solution using the roots $y(t) = e^{rt}\,[c_1 + c_2 t]$ -----

$$y(t) = \mathrm{e}^t\,(c_2 + c_1\,t)$$

Differentiate y(t) & apply initial conditions ==> [y(0) = 1, y'(0) = 0]

$$\begin{pmatrix} c_2 = 1 \\ c_1 + c_2 = 0 \end{pmatrix} \quad ===> \quad \begin{pmatrix} c_1 = -1 \\ c_2 = 1 \end{pmatrix}$$

----- Solution after substitution (c_1 & c_2) in the general solution -----

$$y(t) = -\mathrm{e}^t\,(t-1)$$

$$y''(t) - 2\,y'(t) + y(t) = 0 \qquad [\,y(0) = 1,\, y'(0) = 0\,]$$

unstable system

solution (Laplace Transform) details on next page $\quad y_L(t) = -\mathrm{e}^t\,(t-1)$

solution (dsolve) $\quad y_d(t) = -\mathrm{e}^t\,(t-1)$

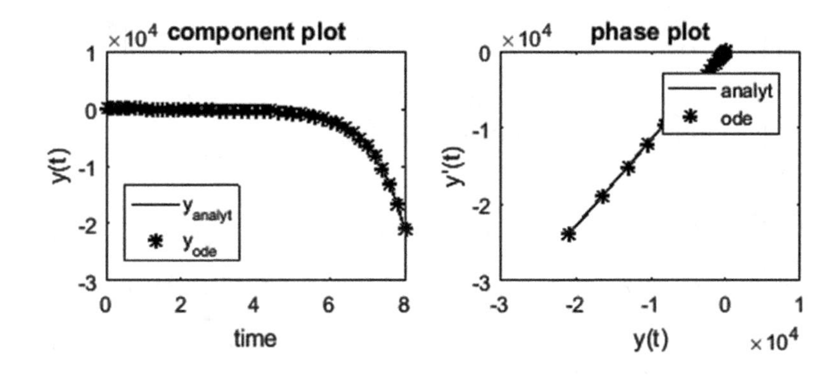

Inverse Laplace based solution of a 2nd order Differential Equation

$$ay''(t) + by'(t) + cy(t) = 0 \quad [y(0) = y_0, y'(0) = y_1] \rightarrow Y(s) = \frac{(as+b)y_0 + ay_1}{as^2 + bs + c}$$

Differential Equation with Initial Conditions

$$y''(t) - 2y'(t) + y(t) = 0 \qquad (y_0 = 1 \quad y_1 = 0)$$

Laplace transform of y(t) using the expression on top

$$Y_L(s) = \frac{s-2}{(s-1)^2}$$

Laplace transform of y(t) as partial fractions

$$Y_L(s) = \frac{1}{s-1} - \frac{1}{(s-1)^2}$$

Solution using inverse Laplace

$$y_L(t) = -e^t\,(t-1)$$

unstable system

Example # 3.11

```
abc=[1,-2,0,1,1];
```

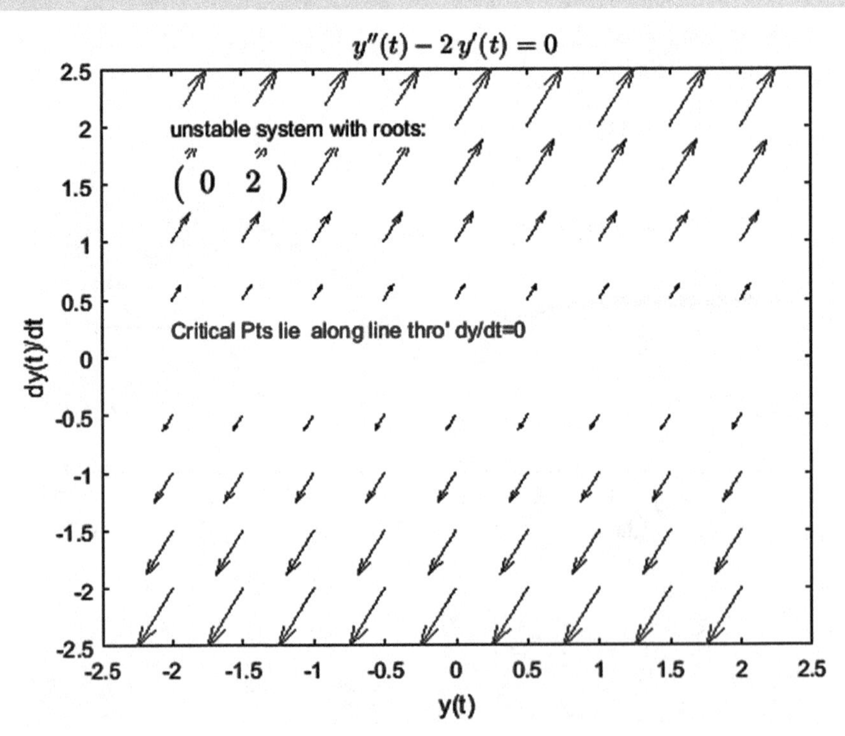

$$y''(t) - 2\,y'(t) = 0 \qquad [\,y(0) = 1,\ y'(0) = 1\,]$$

Characteristic Equation $\quad r^2 - 2\,r = 0$

Roots (r_1 & r_2) $\quad \begin{pmatrix} 0 & 2 \end{pmatrix}$

Roots are +ve or zero; r_1=0 r_2=r_0; unstable system

------ General Solution using the roots $y(t) = [c_1 + c_2 e^{r_0 t}]$ -----

$$y(t) = c_1 + c_2\,e^{2t}$$

Differentiate y(t) & apply initial conditions ==> [y(0) = 1, y'(0) = 1]

$$\begin{pmatrix} c_1 + c_2 = 1 \\ 2\,c_2 = 1 \end{pmatrix} \quad ===> \quad \begin{pmatrix} c_1 = \frac{1}{2} \\ c_2 = \frac{1}{2} \end{pmatrix}$$

----- Solution after substitution (c_1 & c_2) in the general solution -----

$$y(t) = \frac{e^{2t}}{2} + \frac{1}{2}$$

$$y''(t) - 2\,y'(t) = 0 \qquad [\,y(0) = 1,\ y'(0) = 1\,]$$

unstable system

solution (Laplace Transform) $\quad y_L(t) = \frac{e^{2t}}{2} + \frac{1}{2}$

details on next page

solution (dsolve) $\quad y_d(t) = \frac{e^{2t}}{2} + \frac{1}{2}$

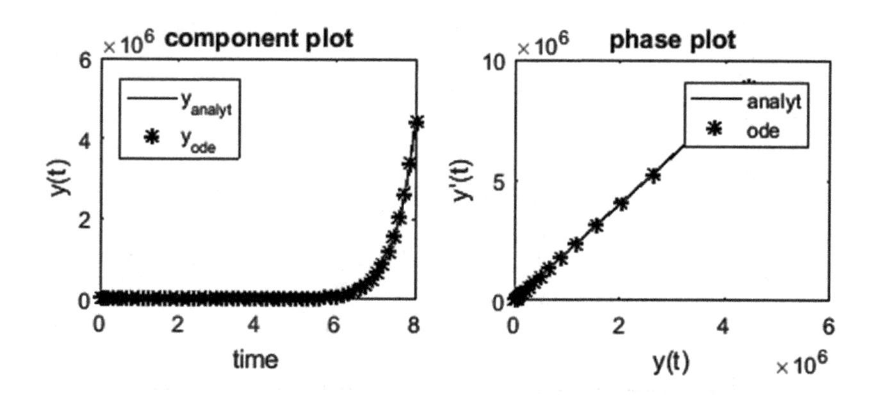

Inverse Laplace based solution of a 2nd order Differential Equation

$$ay''(t) + by'(t) + cy(t) = 0 \quad [y(0) = y_0, y'(0) = y_1] \rightarrow Y(s) = \frac{(as+b)y_0 + ay_1}{as^2 + bs + c}$$

Differential Equation with Initial Conditions

$$y''(t) - 2\,y'(t) = 0 \qquad (\ y_0 = 1 \quad y_1 = 1\ )$$

Laplace transform of y(t) using the expression on top

$$Y_L(s) = \frac{1}{2\,(s-2)} + \frac{1}{2\,s}$$

Laplace transform of y(t) as partial fractions

$$Y_L(s) = \frac{1}{2\,(s-2)} + \frac{1}{2\,s}$$

Solution using inverse Laplace

$$y_L(t) = \frac{e^{2t}}{2} + \frac{1}{2}$$

unstable system

Example # 3.12

```
abc=[1,4,0,1,1];
```

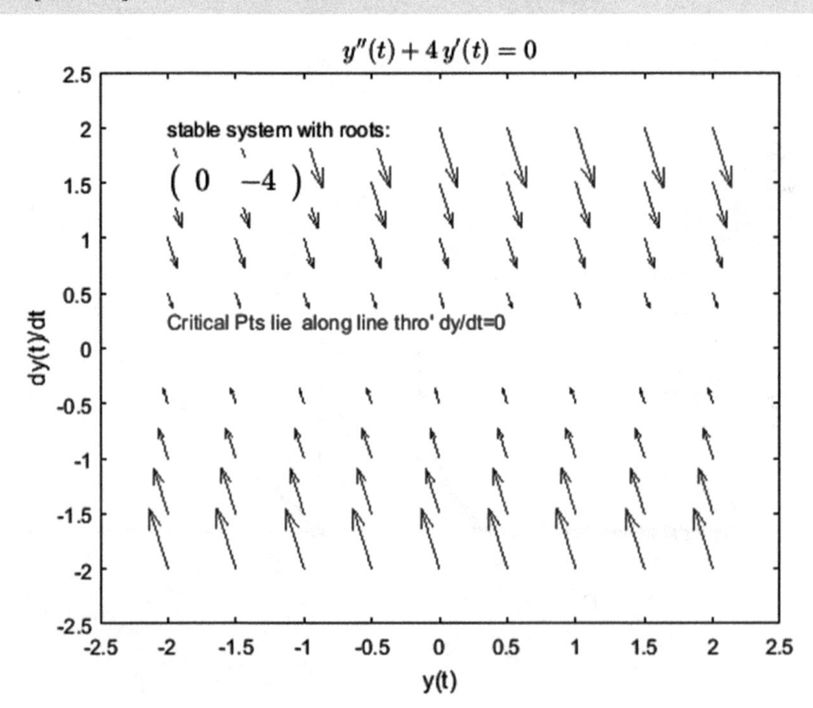

$$y''(t) + 4\,y'(t) = 0$$ [y(0) = 1, y'(0) = 1]

Characteristic Equation $r^2 + 4\,r = 0$

Roots (r_1 & r_2) $\begin{pmatrix} 0 & -4 \end{pmatrix}$

Roots are -ve or zero; r_1=0 r_2=r_0; stable system

------ General Solution using the roots y(t) = [c_1+$c_2$$e^{r_0 t}$] -----

$$y(t) = c_1 + c_2\,\mathrm{e}^{-4\,t}$$

Differentiate y(t) & apply initial conditions ==> [y(0) = 1, y'(0) = 1]

$$\begin{pmatrix} c_1 + c_2 = 1 \\ -4\,c_2 = 1 \end{pmatrix} \quad ===> \quad \begin{pmatrix} c_1 = \frac{5}{4} \\ c_2 = -\frac{1}{4} \end{pmatrix}$$

----- Solution after substitution (c_1 & c_2) in the general solution -----

$$y(t) = \frac{5}{4} - \frac{\mathrm{e}^{-4\,t}}{4}$$

$$y''(t) + 4\,y'(t) = 0$$ [y(0) = 1, y'(0) = 1]

stable system

solution
(Laplace Transform) $y_L(t) = \frac{5}{4} - \frac{\mathrm{e}^{-4t}}{4}$
details on next page

solution
(dsolve) $y_d(t) = \frac{5}{4} - \frac{\mathrm{e}^{-4t}}{4}$

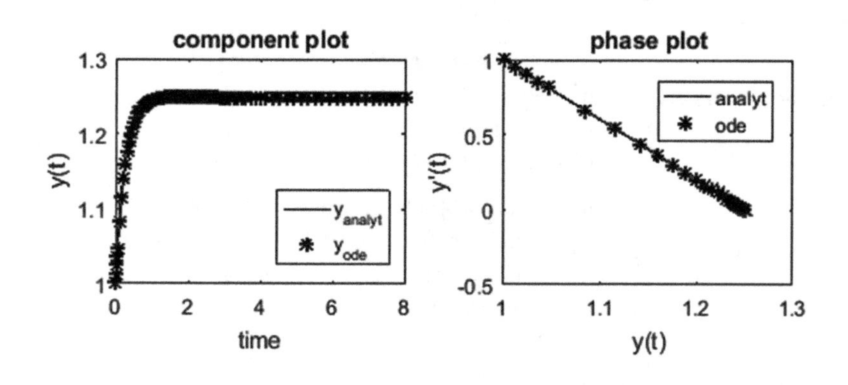

Inverse Laplace based solution of a 2nd order Differential Equation

$$ay''(t)+by'(t)+cy(t)=0 \quad [y(0)=y_0, y'(0)=y_1] \rightarrow Y(s)=\frac{(as+b)y_0+ay_1}{as^2+bs+c}$$

Differential Equation with Initial Conditions

$$y''(t)+4y'(t)=0 \qquad (y_0=1 \quad y_1=1)$$

Laplace transform of y(t) using the expression on top

$$Y_L(s)=\frac{5}{4s}-\frac{1}{4(s+4)}$$

Laplace transform of y(t) as partial fractions

$$Y_L(s)=\frac{5}{4s}-\frac{1}{4(s+4)}$$

Solution using inverse Laplace

$$y_L(t)=\frac{5}{4}-\frac{e^{-4t}}{4}$$

stable system

Example # 3.13

```
abc=[1,0,0];
```

Linear 2nd Order Differential Equation with Constant Coefficients

Homogeneous Differential Equation: $a\frac{d^2y}{dt^2}+b\frac{dy}{dt}+cy=0, \quad a\neq 0$

Characteristic Equation: $ar^2+br+c=0$

Roots of Characteristic Equation or Eigenvalues: $r_{1,2}=r_\pm=\frac{-b\pm\sqrt{b^2-4ac}}{2a}$

Properties of the Roots ⇒
- $b^2-4ac>0, \quad r_1\neq r_2\ (real\ roots)$
- $b^2-4ac=0, \quad r_1=r_2=r_0\ (real\ root)$
- $c=0, \quad r_1=0, r_2\neq 0\ or\ r_1\neq 0, r_2=0$
- $b^2-4ac<0, \quad r_{1,2}=r_\pm=\alpha\pm j\beta$

Stability of the System ⇒
- [r_1 > 0 or r_2 > 0] or [$r_0 \geq 0$] or [α > 0], unstable
- [r_1 < 0 and r_2 < 0] or [r_0<0] or [α < 0], asymptotically stable
- [r_1 = 0, r_2 < 0] or [r_1 < 0, r_2 = 0], stable
- [α = 0], oscillatory

Solution y(t) ⇒
- $c_1e^{r_1t}+c_2e^{r_2t}, \quad r_1\neq r_2$
- $[c_1+c_2t]\,e^{r_0t}, \quad r_1=r_2=r_0$
- $[c_1cos(\beta t)+c_2sin(\beta t)]\,e^{\alpha t}, \quad r_\pm=\alpha\pm j\beta$

Special Case ⇒ $\frac{d^2y}{dt^2}=0, \quad r_1=r_2=0, \quad y(t)=c_1+c_2t$

3.3 Non-homogeneous Differential Equations: Particular Solutions and Complete Solutions

Methods to solve the nonhomogeneous differential equation in eqn. (3.3) can now be described. The solution to the nonhomogeneous differential equation can be expressed as

$$y(t) = y_h(t) + y_p(t). \tag{3.53}$$

In eqn. (3.53), $y_h(t)$ is the solution to the homogeneous part given in eqn. (3.9) and $y_p(t)$ is identified as the particular solution of the nonhomogeneous differential equation in eqn. (3.3). Before examining the two approaches to finding the particular solution $y_p(t)$, it should be noted that the homogeneous solution can be expressed as a linear combination of two solutions $y_1(t)$ and $y_2(t)$ as

$$y_h(t) = c_1 y_1(t) + c_2 y_2(t) = \begin{bmatrix} c_1 & c_2 \end{bmatrix} \begin{bmatrix} y_1(t) \\ y_2(t) \end{bmatrix}. \tag{3.54}$$

The two solutions $y_1(t)$ and $y_2(t)$ can be obtained from eqn. (3.9) as

$$\vec{y}(t) = \begin{bmatrix} y_1(t) \\ y_2(t) \end{bmatrix} = \begin{cases} \begin{bmatrix} e^{r_1 t} \\ e^{r_2 t} \end{bmatrix} & \Rightarrow \; r_1 \neq r_2 \,(real) \\ \begin{bmatrix} e^{rt} \\ te^{rt} \end{bmatrix} & \Rightarrow \; r_1 = r_2 = r\,(real) \\ e^{\alpha t} \begin{bmatrix} \cos(\beta t) \\ \sin(\beta t) \end{bmatrix} & \Rightarrow r_1 = \alpha + j\beta, r_2 = \alpha - j\beta \end{cases}. \tag{3.55}$$

The representation of the primary (or basic or basis) solutions (also known as the fundamental solutions), $y_1(t)$ and $y_2(t)$, formed from the roots of the characteristic equation play a crucial role in obtaining the particular solution.

Two different approaches exist for obtaining the particular solution. The first one is the method of undetermined coefficients and the second one is the method of variation of parameters and both methods rely on $y_1(t)$ and $y_2(t)$. There are advantages and disadvantages and limitations to each of these methods. These methods are described in the next section along with a comparison and approaches for obtaining the complete solution of a non-homogeneous differential equation. We will start with the particular solution before exploring methods to obtain the complete solution to the differential equation expressed as the sum of the solution to the homogeneous differential equation and the particular solution as given in eqn. (3.53).

3.3.1 Particular solution

3.3.1.1 Method of undetermined coefficients

This method relies on an initial guess of the particular solution by examining the relationship of the roots of the characteristic function and the forcing function g(t), thus indirectly exploring its relationship to the primary solutions $y_1(t)$ and $y_2(t)$. The initial guess is made on the basis of hypothesizing the possible forms of the solutions that can satisfy the differential equation and then creating a linear combination of these. Once the initial guess is chosen, unknown constants that are used to combine the various possible solutions are determined by substitution of the particular solution in the differential equation and solving for them. The limitation of this method arises from the approach itself, which is based on the association between the solution of the homogeneous differential equation (more precisely the roots of the characteristic equation) and the forcing function. Because the homogeneous solutions always appear as functions of sin(.), cos(.), t and exp(rt) only, the method is applicable only when the forcing function g(t) takes one of the two forms

$$g(t) = Ct^m e^{qt} \tag{3.56}$$

$$g(t) = \begin{cases} Ct^m e^{\mu t} \cos(\omega t) \\ Ct^m e^{\mu t} \sin(\omega t) \end{cases} \tag{3.57}$$

In eqns. (3.56) and (3.57), C is a constant and m is an integer. This suggests that the method of undetermined coefficients is not applicable for forcing functions of the type h(t)/f(t), tan(t) or non-integer powers in t. Note that it is possible to consider cases where g(t) contains a power series in t as

$$t^m \to a_n t^n + a_{n-1} t^{n-1} + \ldots a_0, n = 0,1,2,\ldots \tag{3.58}$$

The method of undetermined coefficients provides the particular solution for the case in eqn. (3.56) as

$$y_p(t) = t^s \left[A_m t^m + A_{m-1} t^{m-1} + \ldots A_0 \right] e^{qt}. \tag{3.59}$$

The value of s in eqn. (3.59) is

$$s = \begin{cases} 0, & q \neq r_1 \neq r_2 \\ 1, & q = r_1 \; or \; q = r_2 \\ 2, & q = r, r_1 = r_2 \; (double \; root) \end{cases}. \tag{3.60}$$

The particular solution for the case in eqn. (3.57) is

$$\begin{aligned} y_p(t) = & \left[A_m t^m + A_{m-1} t^{m-1} + \ldots A_0 \right] t^s e^{\mu t} \cos(\omega t) + \\ & \left[B_m t^m + B_{m-1} t^{m-1} + \ldots B_0 \right] t^s e^{\mu t} \sin(\omega t) \end{aligned}. \tag{3.61}$$

Equation (3.61) can be written using two unknown constants P and Q in a slightly more compact form as

$$y_p(t) = \left[A_m t^m + A_{m-1} t^{m-1} + \ldots A_0 \right] e^{\mu t} t^s \left[P\cos(\omega t) + Q\sin(\omega t) \right]. \tag{3.62}$$

The value of s in eqn. (3.61) is

$$s = \begin{cases} 0, & \mu + j\omega \neq \alpha + j\beta \\ 1, & \mu + j\omega = \alpha + j\beta \end{cases} \tag{3.63}$$

All the unknown constants can be determined by inserting $y_p(t)$ for y(t) in eqn. (3.3) and solving for them.

If the initial conditions are given, a complete solution can be obtained by solving for c_1 and c_2 in eqn. (3.53).

3.3.1.2 Method of variation of parameters

As mentioned in the previous section, the method of variation of parameters is not limited to specific forms of the forcing function. It relies on the Wronskian W associated with the differential equation, given by

$$W = \begin{bmatrix} y_1(t) & y_2(t) \\ y_1'(t) & y_2'(t) \end{bmatrix}. \tag{3.64}$$

The particular solution is given by

$$y_p(t) = \begin{bmatrix} y_1(t) & y_2(t) \end{bmatrix} \int \vec{v}(t)\,dt. \tag{3.65}$$

The vectorial function $\vec{v}(t)$ in eqn. (3.65) is related to the inverse of the Wronskian as

$$\vec{v}(t) = W^{-1} \begin{bmatrix} 0 \\ \dfrac{g(t)}{a} \end{bmatrix}. \tag{3.66}$$

While the particular solution from both approaches match, it should be noted that the method of variation of parameters even works when the forcing function is of the type h(t)/f(t). The limitation of the method arises from the fact that explicit integration is required (which might not be possible in all cases), while the method of undetermined coefficients only requires algebraic manipulation and substitution.

3.3.2 Complete solution

The complete solution to a non-homogeneous equation can be obtained from the homogeneous and particular solutions. If initial conditions are given, solution can also be obtained using Laplace transforms and Runge-Kutta methods.

3.3.2.1 Complete solution from homogeneous and particular solutions

The complete solution of the non-homogeneous differential equation can be written using eqns. (3.53) and (3.54), as

$$y(t) = c_1 y_1(t) + c_2 y_2(t) + y_p(t). \tag{3.67}$$

Given initial conditions y(0) and y'(0) the unknown constants, c_1 and c_2 can be evaluated and the complete (exact) solution is obtained.

3.3.2.2 Complete solution using laplace transforms

If the initial conditions are given, the method based on Laplace transforms can also be used to obtain the complete solution. Taking the Laplace transform of the differential equation in eqn. (3.3) leads to

$$a\left[s^2Y(s)-sy_0-y_1\right]+b\left[sY(s)-y_0\right]+cY(s)=G(s). \tag{3.68}$$

In eqn. (3.68), G(s) is the Laplace transform of g(t). Solving for Y(s), the Laplace transform of y(t) becomes

$$Y(s)=\frac{(as+b)y_0+ay_1}{as^2+bs+c}+\frac{G(s)}{as^2+bs+c}. \tag{3.69}$$

The complete solution is obtained by evaluating the inverse Laplace transform of y(t). Details are provided in Appendix B.

3.3.2.3 Complete solution using Runge-Kutta methods

If the initial conditions are given, the Runge-Kutta method can be applied for further verification of the results. For the case of the non-homogeneous differential equation in eqn. (3.3), the coupled first order equations become

$$\begin{bmatrix}\frac{dy}{dt}\\ \frac{dz}{dt}\end{bmatrix}=\begin{bmatrix}0 & 1\\ -\frac{c}{a} & -\frac{b}{a}\end{bmatrix}\begin{bmatrix}y\\ z\end{bmatrix}+\begin{bmatrix}0\\ \frac{g(t)}{a}\end{bmatrix}\Rightarrow \vec{x}'=A\vec{x}+\vec{p}. \tag{3.70}$$

In eqn. (3.70), $\vec{p}$ is a R^2 vector.

$$\vec{p}=\begin{bmatrix}0\\ \frac{g(t)}{a}\end{bmatrix}. \tag{3.71}$$

As an example (Example # 3.14), consider the following second order differential equation,

$$y''(t)-2y'(t)=t^2e^{2t}. \tag{3.72}$$

The characteristic equation is

$$r^2-2r=0. \tag{3.73}$$

The roots are 0 and 2 and therefore, the homogeneous solution is

$$y_h(t)=c_1+c_2e^{2t}. \tag{3.74}$$

We will look at the method of undetermined coefficients first. The forcing function is the product of two terms, a quadratic term in t and exponential term containing (2t). The scaling factor in the exponent matches one of the roots of the characteristic equation, namely 2. This means that the initial guess of the particular solution must involve scaling by t. In other words, one of the terms in the initial guess of the particular solution will be

$$y_{p1}(t) = te^{2t}. \tag{3.75}$$

The other term in the initial guess of the particular solution will be a quadratic equation containing t^2, t and a constant,

$$y_{p2}(t) = At^2 + Bt + C. \tag{3.76}$$

In eqn. (3.76), A, B, and C are constants to be determined. The initial guess of the particular solution will be the product of eqns. (3.75) and (3.76)

$$y_p(t) = y_{p1}(t)y_{p2}(t) = te^{2t}\left(At^2 + Bt + C\right). \tag{3.77}$$

If $y_p(t)$ is a valid solution, it must satisfy the differential equation in eqn. (3.72). Taking the first and second derivatives of the particular solution in eqn. (3.77) and substituting in eqn. (3.72), an equation relating A, B and C is obtained as

$$2B + 2C + 6At + 4Bt + 6At^2 = t^2. \tag{3.78}$$

Collecting matching terms,

$$6A = 1 \Rightarrow A = \frac{1}{6} \tag{3.79}$$

$$6A + 4B = 0 \Rightarrow B = -\frac{1}{4} \tag{3.80}$$

$$2B + 2C = 0 \Rightarrow C = \frac{1}{4} \tag{3.81}$$

Substituting for A, B, and C, the particular solution in eqn. (3.77) becomes

$$y_p(t) = te^{2t}\left(\frac{t^2}{6} - \frac{t}{4} + \frac{1}{4}\right) \tag{3.82}$$

Now, we can explore the solution using the method of variation of parameters. Because the roots of the characteristic function are 0 and 2, the two primary solutions of the homogeneous part are

$$\begin{aligned} y_1(t) &= 1 \\ y_2(t) &= e^{2t} \end{aligned}. \tag{3.83}$$

It is easy to observe that the solution to the homogeneous part is a linear combination of $y_1(t)$ and $y_2(t)$ in eqn. (3.83). This means that the Wronskian W is

$$W = \begin{bmatrix} y_1(t) & y_2(t) \\ y_1^{'}(t) & y_1^{'}(t) \end{bmatrix} = \begin{bmatrix} 1 & e^{2t} \\ 0 & 2e^{2t} \end{bmatrix}. \tag{3.84}$$

The inverse (Appendix D) of the Wronskian is

$$W^{-1} = \begin{bmatrix} 1 & -\frac{1}{2} \\ 0 & \frac{1}{2}e^{2t} \end{bmatrix}. \tag{3.85}$$

The intermediate vector $\vec{v}(t)$ in eqn. (3.66) is

$$\vec{v}(t) = \begin{bmatrix} -\frac{1}{2}t^2 \exp(2t) \\ \frac{1}{2}t^2 \end{bmatrix}. \tag{3.86}$$

The particular solution is obtained using eqn. (3.65) as

$$y_p(t) = \left[\frac{1}{6}t^2 - \frac{1}{4}t + \frac{1}{4}\right]te^{2t} - \frac{1}{8}e^{2t}. \tag{3.87}$$

It might appear that the particular solution obtained using the method of variation of parameters contains an extra term $-\frac{1}{8}e^{2t}$. But this is not an error because it is a scaled version of one of the solutions, namely

$$y_2(t) = e^{2t}. \tag{3.88}$$

This term can always be absorbed into the homogeneous solution. To see this, let us write the complete solution using the method of undetermined coefficients first,

$$y(t) = y_h(t) + y_p(t) = c_1 + c_2e^{2t} + te^{2t}\left(\frac{t^2}{6} - \frac{t}{4} + \frac{1}{4}\right). \tag{3.89}$$

The complete solution using the method of variation of parameters is

$$y(t) = c_1 + c_2e^{2t} + te^{2t}\left(\frac{t^2}{6} - \frac{t}{4} + \frac{1}{4}\right) - \frac{1}{8}e^{2t} = c_1 + c_3e^{2t} + te^{2t}\left(\frac{t^2}{6} - \frac{t}{4} + \frac{1}{4}\right) \tag{3.90}$$

It is now easy to see that equations (3.89) and (3.90) are a match except for the unknown constants and without any loss of significance, c_3 can be replaced by c_2. If the initial conditions are given, the constants c_1 and c_2 can be evaluated.

Consider another example (Example # 3.15). The second order differential equation is

$$y''(t) + 16y(t) = t\cos(4t), y(0) = -1, y'(0) = 1 \tag{3.91}$$

The characteristic equation is

$$r^2 + 16 = 0. \tag{3.92}$$

The roots are $\pm 4j$ resulting in the homogeneous solution of

$$y_h(t) = c_1\cos(4t) + c_2\sin(4t) \tag{3.93}$$

The forcing function is

$$g(t) = t\cos(4t) \cdot \tag{3.94}$$

Because the forcing function contains the term cos(4t), the initial guess of the solution based on the method of undetermined coefficients is

$$y_p(t) = \underbrace{[A_1 t + A_0]}_{\text{corresponds to } t} \underbrace{t[P\cos(4t) + Q\sin(4t)]}_{\text{corresponds to } \cos(4t)} . \tag{3.95}$$

Note that the scaling by t arises from the match of one of the solutions with the cos(.) term in the forcing function. The quantity in the first bracket is the term from the presence of the term t in the forcing function and the term arises from cos(4t). Rewriting eqn. (3.95) in terms of unknowns B_1, B_2, B_3, B_4, we have

$$y_p(t) = B_1 t^2 \cos(4t) + B_2 t^2 \sin(4t) + B_3 t\cos(4t) + B_4 t\sin(4t). \tag{3.96}$$

Substituting eqn. (3.96) in the left hand side of eqn. (3.91), we get

$$2B_1\cos(4t) + 8B_4\cos(4t) + 2B_2\sin(4t) - 8B_3\sin(4t) + 16B_2 t\cos(4t) - 16B_1 t\sin(4t) = t\cos(4t) \tag{3.97}$$

Collecting the appropriate terms and equating their coefficients, we have

$$16B_2 = 1 . \tag{3.98}$$

$$16B_1 = 0 \tag{3.99}$$

$$2B_1 + 8B_4 = 0 \tag{3.100}$$

$$2B_2 - 8B_3 = 0 \tag{3.101}$$

From equations (3.98)–(3.101), we have

$$B_2 = \frac{1}{16} \tag{3.102}$$

$$B_3 = \frac{1}{64} \tag{3.103}$$

$$B_1 = B_4 = 0 \tag{3.104}$$

Substituting the values of B_1, B_2, B_3, B_4 in eqn. (3.96), the particular solution in eqn. (3.96) becomes

$$y_p(t) = \frac{1}{16}t^2\sin(4t) + \frac{1}{64}t\cos(4t). \tag{3.105}$$

Using the method of variation of parameters, the Wronskian becomes

$$W = \begin{bmatrix} \cos(4t) & \sin(4t) \\ -4\sin(4t) & 4\cos(4t) \end{bmatrix}. \tag{3.106}$$

The particular solution becomes

$$y_p(t) = \frac{1}{16}t^2\sin(4t) + \frac{1}{64}t\cos(4t) - \frac{1}{256}\sin(4t). \tag{3.107}$$

It is seen that eqn. (3.107) contains the term sin(4t) which is part of the homogeneous solution and can be discarded when the complete solution is written. The complete solution becomes

$$y(t) = c_1\cos(4t) + c_2\sin(4t) + \frac{1}{16}t^2\sin(4t) + \frac{1}{64}t\cos(4t). \tag{3.108}$$

Applying the initial conditions,

$$\begin{aligned} c_1 &= -1 \\ c_2 &= \frac{63}{256} \end{aligned}. \tag{3.109}$$

Following the substitution of the unknown constants c_1 and c_2, the solution becomes

$$y(t) = -\cos(4t) + \frac{63}{256}\sin(4t) + \frac{1}{16}t^2\sin(4t) + \frac{1}{64}t\cos(4t). \tag{3.110}$$

Taking the Laplace transform of the differential equation and collecting the terms in Y(s), we have

$$Y(s) = \frac{1}{(s^2+16)^2} - \frac{32}{(s^2+16)^3} - \frac{s-1}{(s^2+16)}. \tag{3.111}$$

Taking the inverse Laplace transform, eqn. (3.111) leads to the solution in eqn. (3.108).

3.3.3 Additional examples

A MATLAB script was written to obtain the complete solution to a second order linear non-homogeneous differential equation with constant coefficients. The goal was to create a document consisting of theory, steps involved, solutions using multiple methods, comparison of the solution, etc. All these were accomplished through a simple input consisting of a vector of values of a, b, c, initial conditions, and another input consisting of the forcing function. The results obtained consist of the following displays:

1. Description of the Method of Variation of parameters (MVP).
2. Description of the method of undetermined coefficients (MUC).

3. Step-by-step derivation of the particular solution using MVP.
4. Step-by-step derivation of the particular solution using MUC. If scaling by t or t^2 is required, it will be indicated, giving the reasons.
5. The particular solution obtained using the method of variation of parameters may contain copies of the components of the homogeneous solution. To check this, comparison of the particular solution, explanation of any discrepancy between the two (MVP vs. MUC) and mitigation of the discrepancy are provided.
6. General solution shown if no initial conditions (determined by the number of input values).
7. If initial conditions are given, the unknown constants are evaluated (with all the steps shown).
8. Theory of Laplace transform based solution, Laplace transform of y(t) and solution obtained using the Laplace transform is also shown.
9. Comparison of all results (overview) including those obtained using numerical techniques shown.

Example # 3.17

```
abc=[1,0,4];gt='cos(2*t)';
```

Method of Variation of Parameters

$$ay''(t) + by'(t) + cy(t) = g(t)$$

Roots of the characteristic equation, $ar^2+br+c=0 \Rightarrow r_1$ and r_2

Solutions $y_1(t)$ and $y_2(t)$ of the homogeneous differential equation:

$y_1(t) = e^{r_1 t}; y_2(t) = e^{r_2 t}$, $\quad r_1 \neq r_2$ & REAL

$y_1(t) = e^{rt}; y_2(t) = t\,e^{rt}$, $\quad r_1 = r_2 = r$

$y_1(t) = e^{\alpha t}\cos(\beta t);\ y_2(t) = e^{\alpha t}\sin(\beta t)$, $\quad r_{1,2} = \alpha \pm j\beta$

WRONSKIAN $\quad W = \begin{bmatrix} y_1 & y_2 \\ y_1' & y_2' \end{bmatrix}$

$$\vec{v} = [W]^{-1}\begin{bmatrix} 0 \\ \frac{g(t)}{a} \end{bmatrix}$$

Particular Solution $\quad y_p(t) = [y_1\ \ y_2]\int \vec{v}\,dt$

Method of Undetermined Coefficients

$$ay''(t) + by'(t) + cy(t) = Ct^m e^{qt}$$

Roots of the characteristic equation, $ar^2+br+c=0$ are r_1 and r_2 (Real)

s=0, if $q \neq r_1$, $q \neq r_2$, $r_1 \neq r_2$

s=1 if $q=r_1$ or $q=r_2$, $r_1 \neq r_2$

s=2 if $q=r_1=r_2=r$ (double root)

$$y_p(t) = t^s(A_m t^m + A_{m-1}t^{m-1} + \ldots + A_2 t^2 + A_1 t + A_0)e^{qt}$$

$$ay''(t) + by'(t) + cy(t) = \begin{cases} Ct^m \; e^{\mu t}cos(\omega t) \\ Ct^m \; e^{\mu t}sin(\omega t) \end{cases}$$

Roots of the characteristic equation, $ar^2+br+c=0$, $r_{\pm} = \alpha \pm j\beta$ (Complex)

s = 0, if $\mu+j\omega \neq \alpha+j\beta$

s = 1, if $\mu+j\omega = \alpha+j\beta$

$$y_p(t) = t^s(A_m t^m + A_{m-1}t^{m-1} + \ldots + A_2 t^2 + A_1 t + A_0)e^{\mu t}cos(\omega t) + t^s(B_m t^m + B_{m-1}t^{m-1} + \ldots + B_2 t^2 + B_1 t + B_0)e^{\mu t}sin(\omega t)$$

$$y_p(t) = t^s(A_m t^m + \ldots + A_1 t + A_0)e^{\mu t}\,[P\,cos(\omega t) + Q\,sin(\omega t)]$$

Particular Solution: Variation of Parameters

$$y''(t) + 4\,y(t) = \cos(2\,t)$$

Roots of the Characteristic or Auxiliary Equation: $r^2 + 4 = 0$

$$(\; 2\mathrm{i} \quad -2\mathrm{i} \;)$$

General solutions of the Homogen. Diff. EQN: $y''(t) + 4\,y(t) = 0$

$$(\; y_1(t) = \cos(2\,t) \quad y_2(t) = \sin(2\,t) \;)$$

WRONSKIAN [W] given below

$$\begin{pmatrix} \cos(2\,t) & \sin(2\,t) \\ -2\,\sin(2\,t) & 2\,\cos(2\,t) \end{pmatrix}$$

Particular solution obtained may contain $y_1(t)$ or $y_2(t)$ which can be included in the homogeneous solution with appropriate scaling factors

$$y_p(t) = \frac{\cos(2\,t)}{16} + \frac{t\,\sin(2\,t)}{4}$$

Complete Solution (with unknown constants b_1 & b_2)

$$y(t) = y_p(t) + b_2\,\sin(2\,t) + b_1\,\cos(2\,t)$$

Particular solution: Method of Undetermined Coefficients

$$y''(t) + 4\,y(t) = \cos(2\,t)$$

Roots of the Characteristic or Auxiliary Equation

$$(\ 2\,i \quad -2\,i\ )$$

Initial guess for the particular solution (Root present; scaling by t)

$y_p(t)$ = t[P*cos(2*t) + Q*sin(2*t)]

Particular Solution Obtained

$$y_p(t) = \frac{t\,\sin(2\,t)}{4}$$

Solution of the Homogeneous Differential Equation (c_1 & c_2 unknown)

$$y_h(t) = c_1\,\cos(2\,t) - c_2\,\sin(2\,t)$$

Complete Solution

$$y(t) = y_p(t) + y_h(t)$$

Comparison of Particular Solutions

$$y''(t) + 4\,y(t) = \cos(2\,t)$$

Roots of the Characteristic or Auxiliary Equation: $(\ 2\,i \quad -2\,i\ )$

General solutions of the Homogen. Diff. EQN: $y''(t) + 4\,y(t) = 0$

$$(\ y_1(t) = \cos(2\,t) \quad y_2(t) = \sin(2\,t)\ )$$

Particular solution (Variation of parameters)

$$y_p(t) = \frac{\cos(2\,t)}{16} + \frac{t\,\sin(2\,t)}{4}$$

Contains term below (scaled version of general solution), $y_1(t) = \cos(2\,t)$

$$\frac{\cos(t)^2}{8} - \frac{1}{16}$$

Particular solution (Variation of Parameters) after removing the term above

$$y_p(t) = \frac{t\,\sin(2\,t)}{4}$$

Particular solution (Undetermined Coefficients)

$$y_p(t) = \frac{t\,\sin(2\,t)}{4}$$

Particular solutions are now a match

Particular solution: Method of Undetermined Coefficients

$$y''(t) + 4\,y(t) = \cos(2\,t)$$

Roots of the Characteristic or Auxiliary Equation

$$\begin{pmatrix} 2\mathrm{i} & -2\mathrm{i} \end{pmatrix}$$

Initial guess for the particular solution (Root present; scaling by t)

$y_p(t) = t[P*\cos(2*t) + Q*\sin(2*t)]$

Particular Solution Obtained

$$y_p(t) = \frac{t\,\sin(2\,t)}{4}$$

Solution of the Homogeneous Differential Equation (c_1 & c_2 unknown)

$$y_h(t) = c_1\,\cos(2\,t) - c_2\,\sin(2\,t)$$

Complete Solution

$$y(t) = y_p(t) + y_h(t)$$

Comparison of Particular Solutions

$$y''(t) + 4\,y(t) = \cos(2\,t)$$

Roots of the Characteristic or Auxiliary Equation: $\begin{pmatrix} 2\mathrm{i} & -2\mathrm{i} \end{pmatrix}$

General solutions of the Homogen. Diff. EQN: $y''(t) + 4\,y(t) = 0$

$$\begin{pmatrix} y_1(t) = \cos(2\,t) & y_2(t) = \sin(2\,t) \end{pmatrix}$$

Particular solution (Variation of parameters)

$$y_p(t) = \frac{\cos(2\,t)}{16} + \frac{t\,\sin(2\,t)}{4}$$

Contains term below (scaled version of general solution), $y_1(t) = \cos(2\,t)$

$$\frac{\cos(t)^2}{8} - \frac{1}{16}$$

Particular solution (Variation of Parameters) after removing the term above

$$y_p(t) = \frac{t\,\sin(2\,t)}{4}$$

Particular solution (Undetermined Coefficients)

$$y_p(t) = \frac{t\,\sin(2\,t)}{4}$$

Particular solutions are now a match

Example # 3.18

```
abc=[1,-2,0];gt='t^2*exp(2*t)';
```

Method of Variation of Parameters

$$ay''(t) + by'(t) + cy(t) = g(t)$$

Roots of the characteristic equation, $ar^2+br+c=0 \Rightarrow r_1$ and r_2

Solutions $y_1(t)$ and $y_2(t)$ of the homogeneous differential equation:

$y_1(t) = e^{r_1 t}$; $y_2(t) = e^{r_2 t}$, $\quad r_1 \neq r_2$ & REAL

$y_1(t) = e^{rt}$; $y_2(t) = t\,e^{rt}$, $\quad r_1 = r_2 = r$

$y_1(t) = e^{\alpha t}\cos(\beta t)$; $y_2(t) = e^{\alpha t}\sin(\beta t)$, $\quad r_{1,2} = \alpha \pm j\beta$

WRONSKIAN $$W = \begin{bmatrix} y_1 & y_2 \\ y_1' & y_2' \end{bmatrix}$$

$$\vec{v} = [W]^{-1} \begin{bmatrix} 0 \\ \frac{g(t)}{a} \end{bmatrix}$$

Particular Solution $$y_p(t) = [y_1 \;\; y_2] \int \vec{v}\, dt$$

Method of Undetermined Coefficients

$$ay''(t) + by'(t) + cy(t) = Ct^m e^{qt}$$

Roots of the characteristic equation, $ar^2+br+c=0$ are r_1 and r_2 (Real)

$s=0$, if $q \neq r_1$, $q \neq r_2$, $r_1 \neq r_2$

$s=1$ if $q=r_1$ or $q=r_2$, $r_1 \neq r_2$

$s=2$ if $q=r_1=r_2=r$ (double root)

$$y_p(t) = t^s(A_m t^m + A_{m-1}t^{m-1} + \ldots + A_2 t^2 + A_1 t + A_0)e^{qt}$$

$$ay''(t) + by'(t) + cy(t) = \begin{cases} Ct^m\, e^{\mu t}cos(\omega t) \\ Ct^m\, e^{\mu t}sin(\omega t) \end{cases}$$

Roots of the characteristic equation, $ar^2+br+c=0$, $r_\pm = \alpha \pm j\beta$ (Complex)

$s = 0$, if $\mu + j\omega \neq \alpha + j\beta$

$s = 1$, if $\mu + j\omega = \alpha + j\beta$

$$\begin{aligned} y_p(t) &= t^s(A_m t^m + A_{m-1}t^{m-1} + \ldots + A_2 t^2 + A_1 t + A_0)e^{\mu t}cos(\omega t) \\ &+ t^s(B_m t^m + B_{m-1}t^{m-1} + \ldots + B_2 t^2 + B_1 t + B_0)e^{\mu t}sin(\omega t) \end{aligned}$$

$$y_p(t) = t^s(A_m t^m + \ldots + A_1 t + A_0)e^{\mu t}\,[P\; cos(\omega t) + Q\; sin(\omega t)]$$

Particular Solution: Variation of Parameters

$$y''(t) - 2\,y'(t) = t^2\,e^{2t}$$

Roots of the Characteristic or Auxiliary Equation: $r^2 - 2r = 0$

$$\begin{pmatrix} 0 & 2 \end{pmatrix}$$

General solutions of the Homogen. Diff. EQN: $y''(t) - 2\,y'(t) = 0$

$$\begin{pmatrix} y_1(t) = 1 & y_2(t) = e^{2t} \end{pmatrix}$$

WRONSKIAN [W] given below

$$\begin{pmatrix} 1 & e^{2t} \\ 0 & 2e^{2t} \end{pmatrix}$$

Particular solution obtained may contain $y_1(t)$ or $y_2(t)$ which can be included in the homogeneous solution with appropriate scaling factors

$$y_p(t) = \frac{e^{2t}\left(4t^3 - 6t^2 + 6t - 3\right)}{24}$$

Complete Solution (with unknown constants b_1 & b_2)

$$y(t) = b_1 + y_p(t) + b_2\,e^{2t}$$

Particular solution: Method of Undetermined Coefficients

$$y''(t) - 2\,y'(t) = t^2\,e^{2t}$$

Roots of the Characteristic or Auxiliary Equation

$$\begin{pmatrix} 0 & 2 \end{pmatrix}$$

Initial guess for the particular solution (One root matches; scaling by t)

$y_p(t) = t\ [A_2t^2 + A_1t + A_0]\ \exp(2*t)$

Particular Solution Obtained

$$y_p(t) = t\,e^{2t}\left(\frac{t^2}{6} - \frac{t}{4} + \frac{1}{4}\right)$$

Solution of the Homogeneous Differential Equation (c_1 & c_2 unknown)

$$y_h(t) = c_1 + c_2\,e^{2t}$$

Complete Solution

$$y(t) = y_p(t) + y_h(t)$$

Comparison of Particular Solutions

$$y''(t) - 2y'(t) = t^2 e^{2t}$$

Roots of the Characteristic or Auxiliary Equation: $\begin{pmatrix} 0 & 2 \end{pmatrix}$

General solutions of the Homogen. Diff. EQN: $y''(t) - 2y'(t) = 0$

$$\begin{pmatrix} y_1(t) = 1 & y_2(t) = e^{2t} \end{pmatrix}$$

Particular solution (Variation of parameters)

$$y_p(t) = \frac{e^{2t}\left(4t^3 - 6t^2 + 6t - 3\right)}{24}$$

Contains term below (scaled version of general solution), $y_2(t) = e^{2t}$

$$-\frac{e^{2t}}{8}$$

Particular solution (Variation of Parameters) after removing the term above

$$y_p(t) = \frac{t e^{2t}\left(2t^2 - 3t + 3\right)}{12}$$

Particular solution (Undetermined Coefficients)

$$y_p(t) = t e^{2t}\left(\frac{t^2}{6} - \frac{t}{4} + \frac{1}{4}\right)$$

Particular solutions are now a match

Example # 3.19

```
abc=[1,2,1];gt='4*exp(t)';
```

Method of Variation of Parameters

$$ay''(t) + by'(t) + cy(t) = g(t)$$

Roots of the characteristic equation, $ar^2+br+c=0 \Rightarrow r_1$ and r_2

Solutions $y_1(t)$ and $y_2(t)$ of the homogeneous differential equation:

$y_1(t) = e^{r_1 t}$; $y_2(t) = e^{r_2 t}$,	$r_1 \neq r_2$ & REAL
$y_1(t) = e^{rt}$; $y_2(t) = t\,e^{rt}$,	$r_1 = r_2 = r$
$y_1(t) = e^{\alpha t}\cos(\beta t)$; $y_2(t) = e^{\alpha t}\sin(\beta t)$,	$r_{1,2} = \alpha \pm j\beta$

WRONSKIAN $\quad W = \begin{bmatrix} y_1 & y_2 \\ y_1' & y_2' \end{bmatrix}$

$$\vec{v} = [W]^{-1}\begin{bmatrix} 0 \\ \frac{g(t)}{a} \end{bmatrix}$$

Particular Solution $\quad y_p(t) = [y_1 \;\; y_2]\int \vec{v}\,dt$

Method of Undetermined Coefficients

$$ay''(t) + by'(t) + cy(t) = Ct^m e^{qt}$$

Roots of the characteristic equation, $ar^2+br+c=0$ are r_1 and r_2 (Real)

s=0, if $q \neq r_1, q \neq r_2, r_1 \neq r_2$

s=1 if $q=r_1$ or $q=r_2, r_1 \neq r_2$

s=2 if $q=r_1=r_2=r$ (double root)

$$y_p(t) = t^s(A_m t^m + A_{m-1}t^{m-1} + \ldots + A_2t^2 + A_1t + A_0)e^{qt}$$

$$ay''(t) + by'(t) + cy(t) = \begin{cases} Ct^m\ e^{\mu t}cos(\omega t) \\ Ct^m\ e^{\mu t}sin(\omega t) \end{cases}$$

Roots of the characteristic equation, $ar^2+br+c=0$, $r_\pm = \alpha \pm j\beta$ (Complex)

s = 0, if $\mu+j\omega \neq \alpha+j\beta$

s = 1, if $\mu+j\omega = \alpha+j\beta$

$$y_p(t) = t^s(A_m t^m + A_{m-1}t^{m-1} + \ldots + A_2t^2 + A_1t + A_0)e^{\mu t}cos(\omega t)$$
$$+ t^s(B_m t^m + B_{m-1}t^{m-1} + \ldots + B_2t^2 + B_1t + B_0)e^{\mu t}sin(\omega t)$$

$$y_p(t) = t^s(A_m t^m + \ldots + A_1t + A_0)e^{\mu t}\ [P\ cos(\omega t) + Q\ sin(\omega t)]$$

Particular Solution: Variation of Parameters

$$y''(t) + 2\,y'(t) + y(t) = 4\,e^t$$

Roots of the Characteristic or Auxiliary Equation: $r^2 + 2r + 1 = 0$

$$(\ -1 \quad -1\ )$$

General solutions of the Homogen. Diff. EQN: $y''(t) + 2\,y'(t) + y(t) = 0$

$$(\ y_1(t) = e^{-t} \quad y_2(t) = t\,e^{-t}\ )$$

WRONSKIAN [W] given below

$$\begin{pmatrix} e^{-t} & t\,e^{-t} \\ -e^{-t} & e^{-t} - t\,e^{-t} \end{pmatrix}$$

Particular solution obtained may contain $y_1(t)$ or $y_2(t)$ which can be included in the homogeneous solution with appropriate scaling factors

$$y_p(t) = e^t$$

Complete Solution (with unknown constants b_1 & b_2)

$$y(t) = y_p(t) + b_1\,e^{-t} + b_2\,t\,e^{-t}$$

Particular solution: Method of Undetermined Coefficients

$$y''(t) + 2\,y'(t) + y(t) = 4\,e^{t}$$

Roots of the Characteristic or Auxiliary Equation

$$(\ -1 \quad -1\ )$$

Initial guess for the particular solution (Root absent: no scaling by t)

$y_p(t) = A_0 \exp(t)$

Particular Solution Obtained

$$y_p(t) = e^{t}$$

Solution of the Homogeneous Differential Equation (c_1 & c_2 unknown)

$$y_h(t) = e^{-t}\,(c_2 + c_1\,t)$$

Complete Solution

$$y(t) = y_p(t) + y_h(t)$$

Comparison of Particular Solutions

$$y''(t) + 2\,y'(t) + y(t) = 4\,e^{t}$$

Roots of the Characteristic or Auxiliary Equation: $(\ -1 \quad -1\ )$

General solutions of the Homogen. Diff. EQN: $y''(t) + 2\,y'(t) + y(t) = 0$

$$(\ y_1(t) = e^{-t} \quad y_2(t) = t\,e^{-t}\ )$$

Particular solution (Variation of parameters)

$$y_p(t) = e^{t}$$

Particular solution (Undetermined Coefficients)

$$y_p(t) = e^{t}$$

Particular solutions match

Example # 3.20

```
abc=[1,-3,-4,1,-1];gt='2*exp(-t)';
```

Method of Variation of Parameters

$$ay''(t) + by'(t) + cy(t) = g(t)$$

Roots of the characteristic equation, $ar^2+br+c=0 \Rightarrow r_1$ and r_2

Solutions $y_1(t)$ and $y_2(t)$ of the homogeneous differential equation:

$y_1(t) = e^{r_1 t};\ y_2(t) = e^{r_2 t},$ $\quad r_1 \neq r_2$ & REAL

$y_1(t) = e^{rt};\ y_2(t) = t\,e^{rt},$ $\quad r_1 = r_2 = r$

$y_1(t) = e^{\alpha t}\cos(\beta t);\ y_2(t) = e^{\alpha t}\sin(\beta t),$ $\quad r_{1,2} = \alpha \pm j\beta$

WRONSKIAN $\quad W = \begin{bmatrix} y_1 & y_2 \\ y_1' & y_2' \end{bmatrix}$

$$\vec{v} = [W]^{-1}\begin{bmatrix} 0 \\ \frac{g(t)}{a} \end{bmatrix}$$

Particular Solution $\quad y_p(t) = [y_1 \ \ y_2]\int \vec{v}\,dt$

Method of Undetermined Coefficients

$$ay''(t) + by'(t) + cy(t) = Ct^m e^{qt}$$

Roots of the characteristic equation, $ar^2+br+c=0$ are r_1 and r_2 (Real)

$s=0$, if $q \neq r_1,\ q \neq r_2,\ r_1 \neq r_2$

$s=1$ if $q=r_1$ or $q=r_2,\ r_1 \neq r_2$

$s=2$ if $q=r_1=r_2=r$ (double root)

$$y_p(t) = t^s(A_m t^m + A_{m-1}t^{m-1} + \ldots + A_2 t^2 + A_1 t + A_0)e^{qt}$$

$$ay''(t) + by'(t) + cy(t) = \begin{cases} Ct^m\ e^{\mu t}cos(\omega t) \\ Ct^m\ e^{\mu t}sin(\omega t) \end{cases}$$

Roots of the characteristic equation, $ar^2+br+c=0$, $r_{\pm} = \alpha \pm j\beta$ (Complex)

$s = 0$, if $\mu + j\omega \neq \alpha + j\beta$

$s = 1$, if $\mu + j\omega = \alpha + j\beta$

$$\begin{aligned} y_p(t) = {} & t^s(A_m t^m + A_{m-1}t^{m-1} + \ldots + A_2 t^2 + A_1 t + A_0)e^{\mu t}cos(\omega t) \\ & + t^s(B_m t^m + B_{m-1}t^{m-1} + \ldots + B_2 t^2 + B_1 t + B_0)e^{\mu t}sin(\omega t) \end{aligned}$$

$$y_p(t) = t^s(A_m t^m + \ldots + A_1 t + A_0)e^{\mu t}\,[P\ cos(\omega t) + Q\ sin(\omega t)]$$

Particular Solution: Variation of Parameters

$$y''(t) - 3\,y'(t) - 4\,y(t) = 2\,e^{-t}$$

Roots of the Characteristic or Auxiliary Equation: $r^2 - 3r - 4 = 0$

$$(\; -1 \quad 4 \;)$$

General solutions of the Homogen. Diff. EQN: $y''(t) - 3\,y'(t) - 4\,y(t) = 0$

$$(\; y_1(t) = e^{-t} \quad y_2(t) = e^{4t} \;)$$

WRONSKIAN [W] given below

$$\begin{pmatrix} e^{-t} & e^{4t} \\ -e^{-t} & 4\,e^{4t} \end{pmatrix}$$

Particular solution obtained may contain $y_1(t)$ or $y_2(t)$ which can be included in the homogeneous solution with appropriate scaling factors

$$y_p(t) = -\frac{2e^{-t}\,(5\,t+1)}{25}$$

Complete Solution (with unknown constants b_1 & b_2)

$$y(t) = y_p(t) + b_1\,e^{-t} + b_2\,e^{4t}$$

Particular solution: Method of Undetermined Coefficients

$$y''(t) - 3\,y'(t) - 4\,y(t) = 2\,e^{-t}$$

Roots of the Characteristic or Auxiliary Equation

$$(\; -1 \quad 4 \;)$$

Initial guess for the particular solution (One root matches; scaling by t)

$y_p(t) = t\ A_0\ \exp(-t)$

Particular Solution Obtained

$$y_p(t) = -\frac{2\,t\,e^{-t}}{5}$$

Solution of the Homogeneous Differential Equation (c_1 & c_2 unknown)

$$y_h(t) = c_1\,e^{-t} + c_2\,e^{4t}$$

Complete Solution

$$y(t) = y_p(t) + y_h(t)$$

Comparison of Particular Solutions

$$y''(t) - 3\,y'(t) - 4\,y(t) = 2\,e^{-t}$$

Roots of the Characteristic or Auxiliary Equation: $\begin{pmatrix} -1 & 4 \end{pmatrix}$

General solutions of the Homogen. Diff. EQN: $y''(t) - 3\,y'(t) - 4\,y(t) = 0$

$$\begin{pmatrix} y_1(t) = e^{-t} & y_2(t) = e^{4t} \end{pmatrix}$$

Particular solution (Variation of parameters)

$$y_p(t) = -\frac{2e^{-t}\,(5\,t+1)}{25}$$

Contains term below (scaled version of general solution), $y_1(t) = e^{-t}$

$$-\frac{2e^{-t}}{25}$$

Particular solution (Variation of Parameters) after removing the term above

$$y_p(t) = -\frac{2\,t\,e^{-t}}{5}$$

Particular solution (Undetermined Coefficients)

$$y_p(t) = -\frac{2\,t\,e^{-t}}{5}$$

Particular solutions are now a match

Complete solution using the Particular Solution

Differential Equation

$$y''(t) - 3\,y'(t) - 4\,y(t) = 2\,e^{-t}$$

Initial conditions (IC)

[y(0)=1, y'(0)= -1]

General solution (Particular Solution plus Homogeneous solution)

$$y(t) = c_1\,e^{-t} - \frac{2\,t\,e^{-t}}{5} + c_2\,e^{4t}$$

Equations for c_1 and c_2 after the application of Initial Conditions

$$\begin{pmatrix} c_1 + c_2 = 1 \\ 4\,c_2 - c_1 - \frac{2}{5} = -1 \end{pmatrix} \quad ===> \quad \begin{pmatrix} c_1 = \frac{23}{25} \\ c_2 = \frac{2}{25} \end{pmatrix}$$

Analytical Solution from Particular Solution & Homogeneous solution

$$y(t) = \frac{e^{-t}\left(2\,e^{5t} - 10\,t + 23\right)}{25}$$

Complete Solution using Inverse Laplace Transform

$$a\,y''(t)+b\,y'(t)+cy=g(t) \quad [y(0)=y_0,\,y'(0)=y_1] \;\rightarrow Y_L(s)=\frac{(as+b)y_0+ay_1+G(s)}{(as^2+bs+c)}$$

$$y''(t)-3\,y'(t)-4\,y(t)=2\,\mathrm{e}^{-t} \qquad (\;y_0=1 \quad y_1=-1\;)$$

Laplace transform Y(s) of y(t)

$$Y_L(s)=-\frac{-s^2+3\,s+2}{(s+1)^2\,(s-4)}$$

Solution using inverse Laplace

$$y_L(t)=\frac{\mathrm{e}^{-t}\left(2\,\mathrm{e}^{5t}-10\,t+23\right)}{25}$$

Solution directly using dsolve(.)

$$y(t)=\frac{\mathrm{e}^{-t}\left(2\,\mathrm{e}^{5t}-10\,t+23\right)}{25}$$

Overview of Results

The Differential Equation

$$y''(t)-3\,y'(t)-4\,y(t)=2\,\mathrm{e}^{-t} \qquad (\;y_0=1 \quad y_1=-1\;)$$

Particular solution

$$y_p(t)=-\frac{2\,t\,\mathrm{e}^{-t}}{5}$$

Complete Solution

$$y(t)=\frac{\mathrm{e}^{-t}\left(2\,\mathrm{e}^{5t}-10t+23\right)}{25}$$

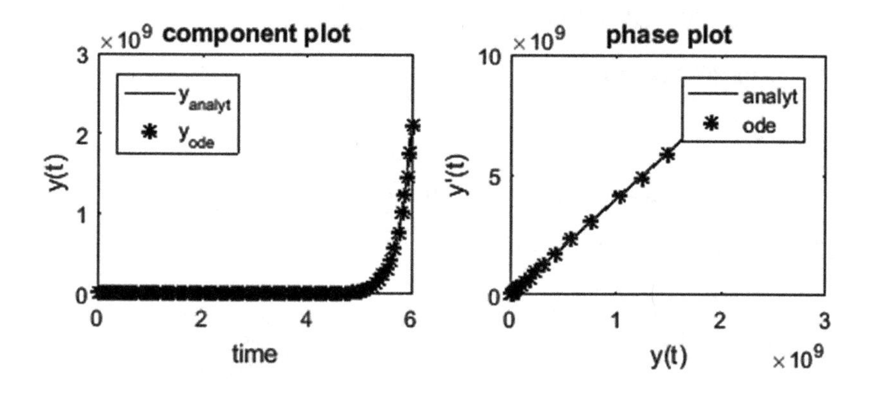

For the remaining examples the displays on their theory are not shown.

Example # 3.21

```
abc=[1,0,4,-1,1];gt='cos(2*t)';
```

Particular Solution: Variation of Parameters

$$y''(t) + 4\,y(t) = \cos(2\,t)$$

Roots of the Characteristic or Auxiliary Equation: $r^2 + 4 = 0$

$$\begin{pmatrix} 2i & -2i \end{pmatrix}$$

General solutions of the Homogen. Diff. EQN: $y''(t) + 4\,y(t) = 0$

$$\begin{pmatrix} y_1(t) = \cos(2t) & y_2(t) = \sin(2t) \end{pmatrix}$$

WRONSKIAN [W] given below

$$\begin{pmatrix} \cos(2\,t) & \sin(2t) \\ -2\,\sin(2\,t) & 2\,\cos(2\,t) \end{pmatrix}$$

Particular solution obtained may contain $y_1(t)$ or $y_2(t)$ which can be included in the homogeneous solution with appropriate scaling factors

$$y_p(t) = \frac{\cos(2\,t)}{16} + \frac{t\,\sin(2\,t)}{4}$$

Complete Solution (with unknown constants b_1 & b_2)

$$y(t) = y_p(t) + b_2\,\sin(2t) + b_1\,\cos(2\,t)$$

Particular solution: Method of Undetermined Coefficients

$$y''(t) + 4\,y(t) = \cos(2\,t)$$

Roots of the Characteristic or Auxiliary Equation

$$\begin{pmatrix} 2i & -2i \end{pmatrix}$$

Initial guess for the particular solution (Root present; scaling by t)

$y_p(t)$ = t[P*cos(2*t) + Q*sin(2*t)]

Particular Solution Obtained

$$y_p(t) = \frac{t\,\sin(2\,t)}{4}$$

Solution of the Homogeneous Differential Equation (c_1 & c_2 unknown)

$$y_h(t) = c_1\,\cos(2\,t) - c_2\,\sin(2t)$$

Complete Solution

$$y(t) = y_p(t) + y_h(t)$$

Comparison of Particular Solutions

$$y''(t) + 4\,y(t) = \cos(2\,t)$$

Roots of the Characteristic or Auxiliary Equation: $\begin{pmatrix} 2i & -2i \end{pmatrix}$

General solutions of the Homogen. Diff. EQN: $y''(t) + 4\,y(t) = 0$

$$\begin{pmatrix} y_1(t) = \cos(2t) & y_2(t) = \sin(2t) \end{pmatrix}$$

Particular solution (Variation of parameters)

$$y_p(t) = \frac{\cos(2\,t)}{16} + \frac{t\,\sin(2\,t)}{4}$$

Contains term below (scaled version of general solution), $y_1(t) = \cos(2\,t)$

$$\frac{\cos(t)^2}{8} - \frac{1}{16}$$

Particular solution (Variation of Parameters) after removing the term above

$$y_p(t) = \frac{t\,\sin(2\,t)}{4}$$

Particular solution (Undetermined Coefficients)

$$y_p(t) = \frac{t\,\sin(2\,t)}{4}$$

Particular solutions are now a match

Complete solution using the Particular Solution

Differential Equation | Initial conditions (IC)

$$y''(t) + 4\,y(t) = \cos(2\,t)$$

[y(0)=-1, y'(0)= 1]

General solution (Particular Solution plus Homogeneous solution)

$$y(t) = \frac{t\,\sin(2\,t)}{4} - c_2\,\sin(2\,t) + c_1\,\cos(2\,t)$$

Equations for c_1 and c_2 after the application of Initial Conditions

$$\begin{pmatrix} c_1 = -1 \\ -2\,c_2 = 1 \end{pmatrix} \quad ===> \quad \begin{pmatrix} c_1 = -1 \\ c_2 = -\frac{1}{2} \end{pmatrix}$$

Analytical Solution from Particular Solution & Homogeneous solution

$$y(t) = \frac{\sin(2\,t)}{2} - \cos(2\,t) + \frac{t\,\sin(2\,t)}{4}$$

Complete Solution using Inverse Laplace Transform

$$a\,y''(t) + b\,y'(t) + c\,y = g(t) \quad [y(0) = y_0, y'(0) = y_1] \rightarrow Y_L(s) = \frac{(as+b)y_0 + ay_1 + G(s)}{(as^2+bs+c)}$$

$$y''(t) + 4\,y(t) = \cos(2\,t) \qquad (\; y_0 = -1 \quad y_1 = 1 \;)$$

Laplace transform Y(s) of y(t)

$$Y_L(s) = -\frac{s^3 - s^2 + 3\,s - 4}{(s^2+4)^2}$$

Solution using inverse Laplace

$$y_L(t) = \frac{\sin(2\,t)}{2} - \cos(2\,t) + \frac{t\,\sin(2\,t)}{4}$$

Solution directly using dsolve(.)

$$y(t) = \frac{\sin(2\,t)}{2} - \cos(2\,t) + \frac{t\,\sin(2\,t)}{4}$$

Overview of Results

The Differential Equation

$$y''(t) + 4\,y(t) = \cos(2\,t) \qquad (\; y_0 = -1 \quad y_1 = 1 \;)$$

Particular solution

$$y_p(t) = \frac{t\,\sin(2\,t)}{4}$$

Complete Solution

$$y(t) = \frac{\sin(2\,t)}{2} - \cos(2\,t) + \frac{t\,\sin(2\,t)}{4}$$

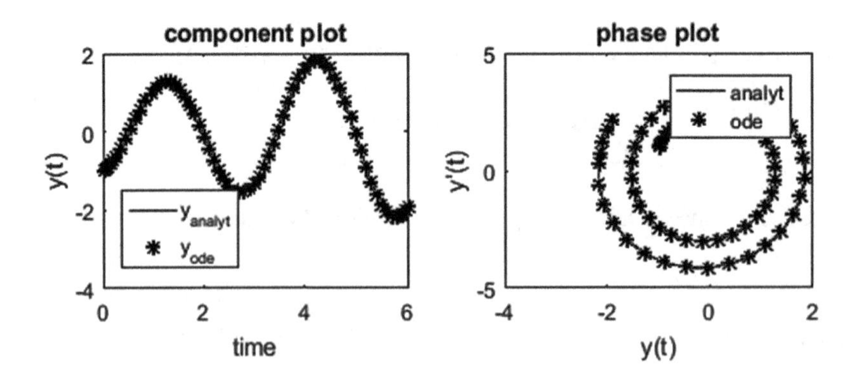

Example # 3.22

```
abc=[1,2,1,0,1];gt='4*t*exp(t)';
```

Particular Solution: Variation of Parameters

$$y''(t) + 2\,y'(t) + y(t) = 4\,t\,e^{t}$$

Roots of the Characteristic or Auxiliary Equation: $r^2 + 2r + 1 = 0$

$$\begin{pmatrix} -1 & -1 \end{pmatrix}$$

General solutions of the Homogen. Diff. EQN: $y''(t) + 2\,y'(t) + y(t) = 0$

$$\begin{pmatrix} y_1(t) = e^{-t} & y_2(t) = t\,e^{-t} \end{pmatrix}$$

WRONSKIAN [W] given below

$$\begin{pmatrix} e^{-t} & t\,e^{-t} \\ -e^{-t} & e^{-t} - t\,e^{-t} \end{pmatrix}$$

Particular solution obtained may contain $y_1(t)$ or $y_2(t)$ which can

be included in the homogeneous solution with appropriate scaling factors

$$y_p(t) = e^{t}\ (t-1)$$

Complete Solution (with unknown constants b_1 & b_2)

$$y(t) = y_p(t) + b_1\,e^{-t} + b_2\,t\,e^{-t}$$

Particular solution: Method of Undetermined Coefficients

$$y''(t) + 2\,y'(t) + y(t) = 4\,t\,e^{t}$$

Roots of the Characteristic or Auxiliary Equation

$$\begin{pmatrix} -1 & -1 \end{pmatrix}$$

Initial guess for the particular solution (Root absent: no scaling by t)

$y_p(t) =$ $[A_1 t + A_0]$ exp(t)

Particular Solution Obtained

$$y_p(t) = e^{t}\ (t-1)$$

Solution of the Homogeneous Differential Equation (c_1 & c_2 unknown)

$$y_h(t) = e^{-t}\ (c_2 + c_1\,t)$$

Complete Solution

$$y(t) = y_p(t) + y_h(t)$$

Comparison of Particular Solutions

$$y''(t) + 2\,y'(t) + y(t) = 4\,t\,e^t$$

Roots of the Characteristic or Auxiliary Equation: $\begin{pmatrix} -1 & -1 \end{pmatrix}$

General solutions of the Homogen. Diff. EQN: $y''(t) + 2\,y'(t) + y(t) = 0$

$$\begin{pmatrix} y_1(t) = e^{-t} & y_2(t) = t\,e^{-t} \end{pmatrix}$$

Particular solution (Variation of parameters)

$$y_p(t) = e^t\ (t-1)$$

Particular solution (Undetermined Coefficients)

$$y_p(t) = e^t\ (t-1)$$

Particular solutions match

Complete solution using the Particular Solution

Differential Equation

$$y''(t) + 2\,y'(t) + y(t) = 4\,t\,e^t$$

Initial conditions (IC)

[y(0)=0, y'(0)= 1]

General solution (Particular Solution plus Homogeneous solution)

$$y(t) = e^t\ (t-1) + e^{-t}\ (c_2 + c_1\,t)$$

Equations for c_1 and c_2 after the application of Initial Conditions

$$\begin{pmatrix} c_2 - 1 = 0 \\ c_1 - c_2 = 1 \end{pmatrix} \quad ===> \quad \begin{pmatrix} c_1 = 2 \\ c_2 = 1 \end{pmatrix}$$

Analytical Solution from Particular Solution & Homogeneous solution

$$y(t) = e^t\ (t-1) + e^{-t}\ (2\,t + 1)$$

Complete Solution using Inverse Laplace Transform

$$\begin{array}{l} a\,y''(t) + b\,y'(t) + c\,y = g(t) \\ [y(0) = y_0, y'(0) = y_1] \end{array} \rightarrow Y_L(s) = \frac{(as+b)y_0 + ay_1 + G(s)}{(as^2 + bs + c)}$$

$$y''(t) + 2\,y'(t) + y(t) = 4\,t\,e^t \qquad (\; y_0 = 0 \quad y_1 = 1 \;)$$

Laplace transform Y(s) of y(t)

$$Y_L(s) = \frac{s^2 - 2\,s + 5}{(s^2 - 1)^2}$$

Solution using inverse Laplace

$$y_L(t) = e^{-t}\,\left(2\,t - e^{2\,t} + t\,e^{2\,t} + 1\right)$$

Solution directly using dsolve(.)

$$y(t) = e^{-t}\,\left(2\,t - e^{2\,t} + t\,e^{2\,t} + 1\right)$$

Overview of Results

The Differential Equation

$$y''(t) + 2\,y'(t) + y(t) = 4\,t\,e^t \qquad (\; y_0 = 0 \quad y_1 = 1 \;)$$

Particular solution

$$y_p(t) = e^t\,(t - 1)$$

Complete Solution

$$y(t) = e^{-t}\,\left(2\,t - e^{2\,t} + t\,e^{2\,t} + 1\right)$$

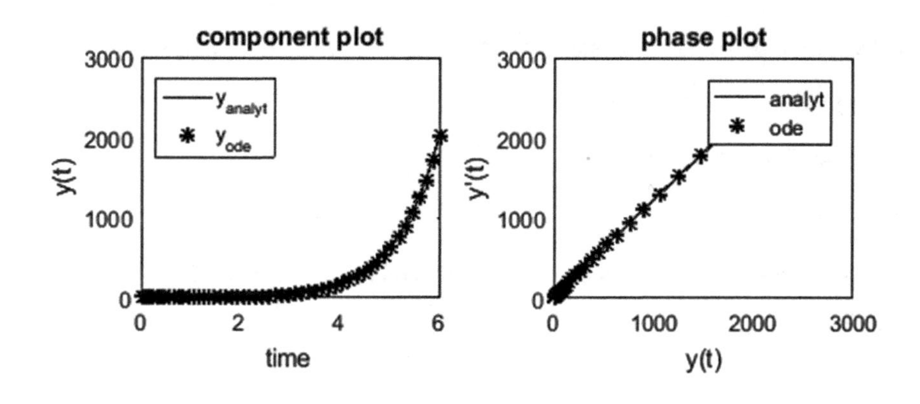

Example # 3.23

```
abc=[1,-2,1];gt='4*t*exp(t)';
```

Particular Solution: Variation of Parameters

$$y''(t) - 2\,y'(t) + y(t) = 4\,t\,e^t$$

Roots of the Characteristic or Auxiliary Equation: $r^2 - 2r + 1 = 0$

$$\begin{pmatrix} 1 & 1 \end{pmatrix}$$

General solutions of the Homogen. Diff. EQN: $y''(t) - 2\,y'(t) + y(t) = 0$

$$\begin{pmatrix} y_1(t) = e^t & y_2(t) = t\,e^t \end{pmatrix}$$

WRONSKIAN [W] given below

$$\begin{pmatrix} e^t & t\,e^t \\ e^t & e^t + t\,e^t \end{pmatrix}$$

Particular solution obtained may contain $y_1(t)$ or $y_2(t)$ which can be included in the homogeneous solution with appropriate scaling factors

$$y_p(t) = \frac{2\,t^3\,e^t}{3}$$

Complete Solution (with unknown constants b_1 & b_2)

$$y(t) = y_p(t) + b_1\,e^t + b_2\,t\,e^t$$

Particular solution: Method of Undetermined Coefficients

$$y''(t) - 2\,y'(t) + y(t) = 4\,t\,e^t$$

Roots of the Characteristic or Auxiliary Equation

$$\begin{pmatrix} 1 & 1 \end{pmatrix}$$

Initial guess for the particular solution (Double roots match; scaling by t^2)

$$y_p(t) = t^2\,[A_1 t + A_0]\ \exp(t)$$

Particular Solution Obtained

$$y_p(t) = \frac{2\,t^3\,e^t}{3}$$

Solution of the Homogeneous Differential Equation (c_1 & c_2 unknown)

$$y_h(t) = e^t\,(c_2 + c_1\,t)$$

Complete Solution

$$y(t) = y_p(t) + y_h(t)$$

Comparison of Particular Solutions

$$y''(t) - 2y'(t) + y(t) = 4t\,e^t$$

Roots of the Characteristic or Auxiliary Equation: $\begin{pmatrix} 1 & 1 \end{pmatrix}$

General solutions of the Homogen. Diff. EQN: $y''(t) - 2y'(t) + y(t) = 0$

$$\begin{pmatrix} y_1(t) = e^t & y_2(t) = t\,e^t \end{pmatrix}$$

Particular solution (Variation of parameters)

$$y_p(t) = \frac{2t^3 e^t}{3}$$

Particular solution (Undetermined Coefficients)

$$y_p(t) = \frac{2t^3 e^t}{3}$$

Particular solutions match

Example # 3.24

```
abc=[1,0,0,0,-1];gt='t^2+t+1';
```

Particular Solution: Variation of Parameters

$$y''(t) = t^2 + t + 1$$

Roots of the Characteristic or Auxiliary Equation: $r^2 = 0$

$$\begin{pmatrix} 0 & 0 \end{pmatrix}$$

General solutions of the Homogen. Diff. EQN: $y''(t) = 0$

$$\begin{pmatrix} y_1(t) = 1 & y_2(t) = t \end{pmatrix}$$

WRONSKIAN [W] given below

$$\begin{pmatrix} 1 & t \\ 0 & 1 \end{pmatrix}$$

Particular solution obtained may contain $y_1(t)$ or $y_2(t)$ which can be included in the homogeneous solution with appropriate scaling factors

$$y_p(t) = \frac{t^2\left(t^2+2t+6\right)}{12}$$

Complete Solution (with unknown constants b_1 & b_2)

$$y(t) = b_1 + y_p(t) + b_2\,t$$

Particular solution: Method of Undetermined Coefficients

$$y''(t) = t^2 + t + 1$$

Roots of the Characteristic or Auxiliary Equation

$$(\ 0 \quad 0\ )$$

Initial guess for the particular solution (Double roots match; scaling by t^2)

$$y_p(t) = t^2[A_2 t^2 + A_1 t + A_0]$$

Particular Solution Obtained

$$y_p(t) = t^2 \left(\frac{t^2}{12} + \frac{t}{6} + \frac{1}{2}\right)$$

Solution of the Homogeneous Differential Equation (c_1 & c_2 unknown)

$$y_h(t) = c_2 + c_1\, t$$

Complete Solution

$$y(t) = y_p(t) + y_h(t)$$

Comparison of Particular Solutions

$$y''(t) = t^2 + t + 1$$

Roots of the Characteristic or Auxiliary Equation: $(\ 0 \quad 0\ )$

General solutions of the Homogen. Diff. EQN: $y''(t) = 0$

$$(\ y_1(t) = 1 \quad y_2(t) = t\ )$$

Particular solution (Variation of parameters)

$$y_p(t) = \frac{t^2\left(t^2+2\,t+6\right)}{12}$$

Particular solution (Undetermined Coefficients)

$$y_p(t) = t^2 \left(\frac{t^2}{12} + \frac{t}{6} + \frac{1}{2}\right)$$

Particular solutions match

Complete solution using the Particular Solution

Differential Equation

$$y''(t) = t^2 + t + 1$$

Initial conditions (IC)

[y(0)=0, y'(0)= -1]

General solution (Particular Solution plus Homogeneous solution)

$$y(t) = c_2 + c_1\,t + t^2\left(\frac{t^2}{12} + \frac{t}{6} + \frac{1}{2}\right)$$

Equations for c_1 and c_2 after the application of Initial Conditions

$$\begin{pmatrix} c_2 = 0 \\ c_1 = -1 \end{pmatrix} \quad ===> \quad \begin{pmatrix} c_1 = -1 \\ c_2 = 0 \end{pmatrix}$$

Analytical Solution from Particular Solution & Homogeneous solution

$$y(t) = \frac{t^2\left(t^2+2\,t+6\right)}{12} - t$$

Complete Solution using Inverse Laplace Transform

$$\begin{array}{l} a\,y''(t) + b\,y'(t) + c\,y = g(t) \\ [y(0) = y_0, y'(0) = y_1] \end{array} \rightarrow Y_L(s) = \frac{(as+b)y_0 + ay_1 + G(s)}{(as^2+bs+c)}$$

$$y''(t) = t^2 + t + 1 \qquad (\; y_0 = 0 \quad y_1 = -1 \;)$$

Laplace transform Y(s) of y(t)

$$Y_L(s) = -\frac{(s-2)\left(s^2+s+1\right)}{s^5}$$

Solution using inverse Laplace

$$y_L(t) = \frac{t\left(t^3+2\,t^2+6\,t-12\right)}{12}$$

Solution directly using dsolve(.)

$$y(t) = \frac{t\left(t^3+2\,t^2+6\,t-12\right)}{12}$$

Overview of Results

The Differential Equation

$$y''(t) = t^2 + t + 1 \qquad (\; y_0 = 0 \quad y_1 = -1 \;)$$

Particular solution

$$y_p(t) = t^2 \left(\frac{t^2}{12} + \frac{t}{6} + \frac{1}{2}\right)$$

Complete Solution

$$y(t) = \frac{t\left(t^3 + 2\,t^2 + 6\,t - 12\right)}{12}$$

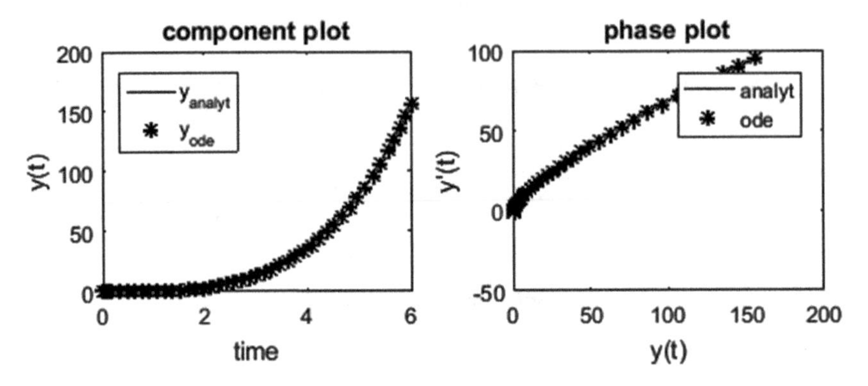

Example # 3.25

```
abc=[1,1,2,0,-1];gt='exp(t)*sin(t)';
```

Particular Solution: Variation of Parameters

$$y''(t) + y'(t) + 2\,y(t) = e^t \sin(t)$$

Roots of the Characteristic or Auxiliary Equation: $\quad r^2 + r + 2 = 0$

$$\left(-\frac{1}{2} + \frac{\sqrt{7}\,i}{2} \quad -\frac{1}{2} - \frac{\sqrt{7}\,i}{2} \right)$$

General solutions of the Homogen. Diff. EQN: $y''(t) + y'(t) + 2\,y(t) = 0$

$$\left(y_1(t) = e^{-\frac{t}{2}} \cos\!\left(\frac{\sqrt{7}\,t}{2}\right) \quad y_2(t) = e^{-\frac{t}{2}} \sin\!\left(\frac{\sqrt{7}\,t}{2}\right) \right)$$

WRONSKIAN [W] given below

$$\begin{pmatrix} e^{-\frac{t}{2}} \cos\!\left(\frac{\sqrt{7}\,t}{2}\right) & e^{-\frac{t}{2}} \sin\!\left(\frac{\sqrt{7}\,t}{2}\right) \\ -\frac{e^{-\frac{t}{2}} \cos\left(\frac{\sqrt{7}\,t}{2}\right)}{2} - \frac{\sqrt{7}\,e^{-\frac{t}{2}} \sin\left(\frac{\sqrt{7}\,t}{2}\right)}{2} & \frac{\sqrt{7}\,e^{-\frac{t}{2}} \cos\left(\frac{\sqrt{7}\,t}{2}\right)}{2} - \frac{e^{-\frac{t}{2}} \sin\left(\frac{\sqrt{7}\,t}{2}\right)}{2} \end{pmatrix}$$

Particular solution obtained may contain $y_1(t)$ or $y_2(t)$ which can be included in the homogeneous solution with appropriate scaling factors

$$y_p(t) = -\frac{\sqrt{2}\,e^t \cos\left(t + \frac{\pi}{4}\right)}{6}$$

Complete Solution (with unknown constants b_1 & b_2)

$$y(t) = y_p(t) + b_1\,e^{-\frac{t}{2}} \cos\!\left(\frac{\sqrt{7}\,t}{2}\right) + b_2\,e^{-\frac{t}{2}} \sin\!\left(\frac{\sqrt{7}\,t}{2}\right)$$

Particular solution: Method of Undetermined Coefficients

$$y''(t) + y'(t) + 2\,y(t) = e^t \sin(t)$$

Roots of the Characteristic or Auxiliary Equation

$$\left(-\tfrac{1}{2} + \tfrac{\sqrt{7}\,i}{2} \quad -\tfrac{1}{2} - \tfrac{\sqrt{7}\,i}{2} \right)$$

Initial guess for the particular solution (Root absent: no scaling by t)

$$y_p(t) = [P^*\cos(t) + Q^*\sin(t)]\, e^t$$

Particular Solution Obtained

$$y_p(t) = -e^t \left(\frac{\cos(t)}{6} - \frac{\sin(t)}{6} \right)$$

Solution of the Homogeneous Differential Equation (c_1 & c_2 unknown)

$$y_h(t) = e^{-\frac{t}{2}} \left(c_1 \cos\!\left(\frac{\sqrt{7}t}{2}\right) - c_2 \sin\!\left(\frac{\sqrt{7}t}{2}\right) \right)$$

Complete Solution

$$y(t) = y_p(t) + y_h(t)$$

Comparison of Particular Solutions

$$y''(t) + y'(t) + 2\,y(t) = e^t \sin(t)$$

Roots of the Characteristic or Auxiliary Equation: $\left(-\tfrac{1}{2} + \tfrac{\sqrt{7}\,i}{2} \quad -\tfrac{1}{2} - \tfrac{\sqrt{7}\,i}{2} \right)$

General solutions of the Homogen. Diff. EQN: $y''(t) + y'(t) + 2\,y(t) = 0$

$$\left(y_1(t) = e^{-\frac{t}{2}} \cos\!\left(\frac{\sqrt{7}t}{2}\right) \quad y_2(t) = e^{-\frac{t}{2}} \sin\!\left(\frac{\sqrt{7}t}{2}\right) \right)$$

Particular solution (Variation of parameters)

$$y_p(t) = -\frac{\sqrt{2}\, e^t \cos\!\left(t + \frac{\pi}{4}\right)}{6}$$

Particular solution (Undetermined Coefficeints)

$$y_p(t) = -e^t \left(\frac{\cos(t)}{6} - \frac{\sin(t)}{6} \right)$$

Particular solutions match

Complete solution using the Particular Solution

Differential Equation

$y''(t) + y'(t) + 2\,y(t) = e^t \sin(t)$

Initial conditions (IC)

[y(0)=0, y'(0)= -1]

General solution (Particular Solution plus Homogeneous solution)

$$y(t) = e^{-\frac{t}{2}}\left(c_1 \cos\left(\frac{\sqrt{7}\,t}{2}\right) - c_2 \sin\left(\frac{\sqrt{7}\,t}{2}\right)\right) - e^t\left(\frac{\cos(t)}{6} - \frac{\sin(t)}{6}\right)$$

Equations for c_1 and c_2 after the application of Initial Conditions

$$\begin{pmatrix} c_1 - \frac{1}{6} = 0 \\ -\frac{c_1}{2} - \frac{\sqrt{7}\,c_2}{2} = -1 \end{pmatrix} \quad ===> \quad \begin{pmatrix} c_1 = \frac{1}{6} \\ c_2 = \frac{11\sqrt{7}}{42} \end{pmatrix}$$

Analytical Solution from Particular Solution & Homogeneous solution

$$y(t) = e^{-\frac{t}{2}}\left(\frac{\cos\left(\frac{\sqrt{7}\,t}{2}\right)}{6} - \frac{11\sqrt{7}\sin\left(\frac{\sqrt{7}\,t}{2}\right)}{42}\right) - e^t\left(\frac{\cos(t)}{6} - \frac{\sin(t)}{6}\right)$$

Complete Solution using Inverse Laplace Transform

$$\begin{matrix} a\,y''(t) + b\,y'(t) + cy = g(t) \\ [y(0) = y_0, y'(0) = y_1] \end{matrix} \rightarrow Y_L(s) = \frac{(as+b)y_0 + ay_1 + G(s)}{(as^2 + bs + c)}$$

$y''(t) + y'(t) + 2\,y(t) = e^t \sin(t)$ $\quad (y_0 = 0 \quad y_1 = -1)$

Laplace transform Y(s) of y(t)

$$Y_L(s) = \frac{\frac{1}{(s-1)^2+1} - 1}{s^2+s+2}$$

Solution using inverse Laplace

$$y_L(t) = \frac{e^{-\frac{t}{2}}\left(\cos\left(\frac{\sqrt{7}\,t}{2}\right) - \frac{11\sqrt{7}\sin\left(\frac{\sqrt{7}\,t}{2}\right)}{7}\right)}{6} - \frac{e^t(\cos(t) - \sin(t))}{6}$$

Solution directly using dsolve(.)

$$y(t) = \frac{e^{-\frac{t}{2}}\left(7\cos\left(\frac{\sqrt{7}\,t}{2}\right) - 11\sqrt{7}\sin\left(\frac{\sqrt{7}\,t}{2}\right) - 7e^{\frac{3t}{2}}\cos(t) + 7e^{\frac{3t}{2}}\sin(t)\right)}{42}$$

Overview of Results

The Differential Equation

$$y''(t) + y'(t) + 2\,y(t) = e^t \sin(t) \qquad (\ y_0 = 0 \quad y_1 = -1\ )$$

Particular solution

$$y_p(t) = -e^t \left(\frac{\cos(t)}{6} - \frac{\sin(t)}{6}\right)$$

Complete Solution

$$y(t) = \frac{e^{-\frac{t}{2}}\left(7\cos\left(\frac{\sqrt{7}\,t}{2}\right) - 11\sqrt{7}\sin\left(\frac{\sqrt{7}\,t}{2}\right) - 7\,e^{\frac{3t}{2}}\cos(t) + 7\,e^{\frac{3t}{2}}\sin(t)\right)}{42}$$

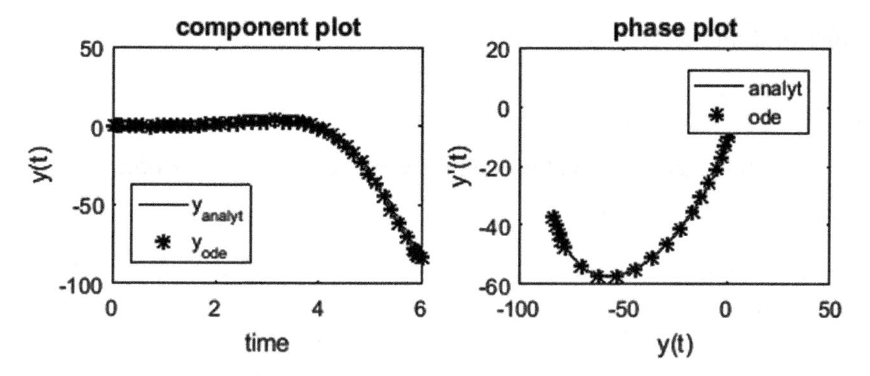

Example # 3.26

```
abc=[1,-3,-4,1,-1];gt='2*t*exp(-t)';
```

Particular Solution: Variation of Parameters

$$y''(t) - 3\,y'(t) - 4\,y(t) = 2\,t\,e^{-t}$$

Roots of the Characteristic or Auxiliary Equation: $r^2 - 3r - 4 = 0$

$$(\ -1 \quad 4\ )$$

General solutions of the Homogen. Diff. EQN: $y''(t) - 3\,y'(t) - 4\,y(t) = 0$

$$(\ y_1(t) = e^{-t} \quad y_2(t) = e^{4t}\ )$$

WRONSKIAN [W] given below

$$\begin{pmatrix} e^{-t} & e^{4t} \\ -e^{-t} & 4\,e^{4t} \end{pmatrix}$$

Particular solution obtained may contain y_1(t) or y_2(t) which can be included in the homogeneous solution with appropriate scaling factors

$$y_p(t) = -\frac{e^{-t}\left(25\,t^2 + 10\,t + 2\right)}{125}$$

Complete Solution (with unknown constants b_1 & b_2)

$$y(t) = y_p(t) + b_1\,e^{-t} + b_2\,e^{4t}$$

Particular solution: Method of Undetermined Coefficients

$$y''(t) - 3\,y'(t) - 4\,y(t) = 2\,t\,\mathrm{e}^{-t}$$

Roots of the Characteristic or Auxiliary Equation

$$\begin{pmatrix} -1 & 4 \end{pmatrix}$$

Initial guess for the particular solution (One root matches; scaling by t)

$y_p(t) = t\ [A_1 t + A_0]\ \exp(-t)$

Particular Solution Obtained

$$y_p(t) = -t\,\mathrm{e}^{-t}\,\left(\frac{t}{5} + \frac{2}{25}\right)$$

Solution of the Homogeneous Differential Equation (c_1 & c_2 unknown)

$$y_h(t) = c_1\,\mathrm{e}^{-t} + c_2\,\mathrm{e}^{4t}$$

Complete Solution

$$y(t) = y_p(t) + y_h(t)$$

Comparison of Particular Solutions

$$y''(t) - 3\,y'(t) - 4\,y(t) = 2\,t\,\mathrm{e}^{-t}$$

Roots of the Characteristic or Auxiliary Equation: $\begin{pmatrix} -1 & 4 \end{pmatrix}$

General solutions of the Homogen. Diff. EQN: $y''(t) - 3\,y'(t) - 4\,y(t) = 0$

$$\begin{pmatrix} y_1(t) = \mathrm{e}^{-t} & y_2(t) = \mathrm{e}^{4t} \end{pmatrix}$$

Particular solution (Variation of parameters)

$$y_p(t) = -\frac{\mathrm{e}^{-t}\left(25\,t^2 + 10\,t + 2\right)}{125}$$

Contains term below (scaled version of general solution), $y_1(t) = \mathrm{e}^{-t}$

$$-\frac{2\mathrm{e}^{-t}}{125}$$

Particular solution (Variation of Parameters) after removing the term above

$$y_p(t) = -\frac{t\mathrm{e}^{-t}\,(5\,t + 2)}{25}$$

Particular solution (Undetermined Coefficients)

$$y_p(t) = -t\,\mathrm{e}^{-t}\,\left(\frac{t}{5} + \frac{2}{25}\right)$$

Particular solutions are now a match

Complete solution using the Particular Solution

Differential Equation

$$y''(t) - 3\,y'(t) - 4\,y(t) = 2\,t\,e^{-t}$$

Initial conditions (IC)

[y(0)=1, y'(0)= -1]

General solution (Particular Solution plus Homogeneous solution)

$$y(t) = c_1\,e^{-t} + c_2\,e^{4t} - t\,e^{-t}\,\left(\frac{t}{5} + \frac{2}{25}\right)$$

Equations for c_1 and c_2 after the application of Initial Conditions

$$\begin{pmatrix} c_1 + c_2 = 1 \\ 4\,c_2 - c_1 - \frac{2}{25} = -1 \end{pmatrix} \quad ===> \quad \begin{pmatrix} c_1 = \frac{123}{125} \\ c_2 = \frac{2}{125} \end{pmatrix}$$

Analytical Solution from Particular Solution & Homogeneous solution

$$y(t) = -\frac{e^{-t}\left(10\,t - 2\,e^{5t} + 25\,t^2 - 123\right)}{125}$$

Complete Solution using Inverse Laplace Transform

$$\begin{matrix} a\,y''(t) + b\,y'(t) + c\,y = g(t) \\ [y(0) = y_0, y'(0) = y_1] \end{matrix} \rightarrow Y_L(s) = \frac{(as+b)y_0 + ay_1 + G(s)}{(as^2 + bs + c)}$$

$$y''(t) - 3\,y'(t) - 4\,y(t) = 2\,t\,e^{-t} \qquad (\; y_0 = 1 \quad y_1 = -1 \;)$$

Laplace transform Y(s) of y(t)

$$Y_L(s) = -\frac{-s^3 + 2\,s^2 + 7\,s + 2}{(s+1)^3\,(s-4)}$$

Solution using inverse Laplace

$$y_L(t) = -\frac{e^{-t}\left(10\,t - 2\,e^{5t} + 25\,t^2 - 123\right)}{125}$$

Solution directly using dsolve(.)

$$y(t) = -\frac{e^{-t}\left(10\,t - 2\,e^{5t} + 25\,t^2 - 123\right)}{125}$$

Overview of Results

The Differential Equation

$$y''(t) - 3\,y'(t) - 4\,y(t) = 2\,t\,\mathrm{e}^{-t} \qquad (\; y_0 = 1 \quad y_1 = -1 \;)$$

Particular solution

$$y_p(t) = -t\,\mathrm{e}^{-t}\,\left(\frac{t}{5} + \frac{2}{25}\right)$$

Complete Solution

$$y(t) = -\frac{\mathrm{e}^{-t}\left(10\,t - 2\,\mathrm{e}^{5t} + 25\,t^2 - 123\right)}{125}$$

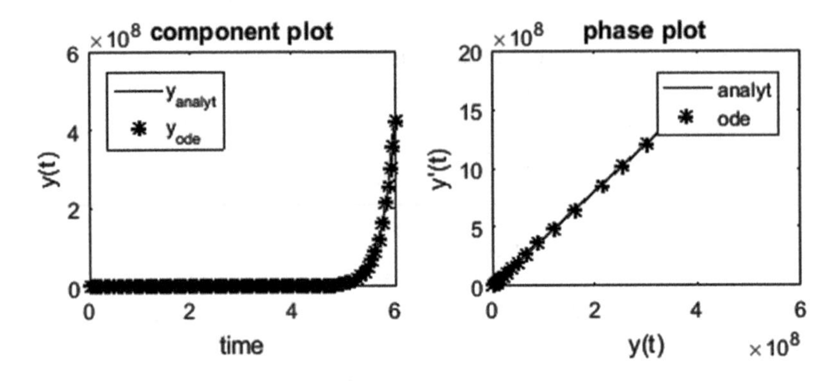

Example # 3.27

```
abc=[1,8,16];gt='2*t*exp(-4*t)';
```

Particular Solution: Variation of Parameters

$$y''(t) + 8\,y'(t) + 16\,y(t) = 2\,t\,\mathrm{e}^{-4\,t}$$

Roots of the Characteristic or Auxiliary Equation: $r^2 + 8r + 16 = 0$

$$(\; -4 \quad -4 \;)$$

General solutions of the Homogen. Diff. EQN: $y''(t) + 8\,y'(t) + 16\,y(t) = 0$

$$(\; y_1(t) = \mathrm{e}^{-4\,t} \quad y_2(t) = t\,\mathrm{e}^{-4\,t} \;)$$

WRONSKIAN [W] given below

$$\begin{pmatrix} \mathrm{e}^{-4\,t} & t\,\mathrm{e}^{-4\,t} \\ -4\,\mathrm{e}^{-4t} & \mathrm{e}^{-4t} - 4\,t\,\mathrm{e}^{-4t} \end{pmatrix}$$

Particular solution obtained may contain $y_1(t)$ or $y_2(t)$ which can be included in the homogeneous solution with appropriate scaling factors

$$y_p(t) = \frac{t^3\,\mathrm{e}^{-4\,t}}{3}$$

Complete Solution (with unknown constants b_1 & b_2)

$$y(t) = y_p(t) + b_1\,\mathrm{e}^{-4t} + b_2\,t\,\mathrm{e}^{-4\,t}$$

Particular solution: Method of Undetermined Coefficients

$$y''(t) + 8\,y'(t) + 16\,y(t) = 2\,t\,e^{-4\,t}$$

Roots of the Characteristic or Auxiliary Equation

$$(\ -4 \quad -4\ )$$

Initial guess for the particular solution (Double roots match; scaling by t^2)

$y_p(t) = t^2\,[A_1 t + A_0]\ \exp(-4*t)$

Particular Solution Obtained

$$y_p(t) = \frac{t^3\,e^{-4t}}{3}$$

Solution of the Homogeneous Differential Equation (c_1 & c_2 unknown)

$$y_h(t) = e^{-4\,t}\,(c_2 + c_1\,t)$$

Complete Solution

$$y(t) = y_p(t) + y_h(t)$$

Comparison of Particular Solutions

$$y''(t) + 8\,y'(t) + 16\,y(t) = 2\,t\,e^{-4\,t}$$

Roots of the Characteristic or Auxiliary Equation: $(\ -4 \quad -4\ )$

General solutions of the Homogen. Diff. EQN: $y''(t) + 8\,y'(t) + 16\,y(t) = 0$

$$\left(\ y_1(t) = e^{-4\,t} \quad y_2(t) = t\,e^{-4\,t}\ \right)$$

Particular solution (Variation of parameters)

$$y_p(t) = \frac{t^3\,e^{-4t}}{3}$$

Particular solution (Undetermined Coefficients)

$$y_p(t) = \frac{t^3\,e^{-4t}}{3}$$

Particular solutions match

Example # 3.28

```
abc=[1,8,16];gt='2*t^2*exp(-4*t)';
```

Particular Solution: Variation of Parameters

$$y''(t) + 8\,y'(t) + 16\,y(t) = 2t^2\,e^{-4t}$$

Roots of the Characteristic or Auxiliary Equation: $r^2 + 8r + 16 = 0$

$$\begin{pmatrix} -4 & -4 \end{pmatrix}$$

General solutions of the Homogen. Diff. EQN: $y''(t) + 8\,y'(t) + 16\,y(t) = 0$

$$\begin{pmatrix} y_1(t) = e^{-4t} & y_2(t) = t\,e^{-4t} \end{pmatrix}$$

WRONSKIAN [W] given below

$$\begin{pmatrix} e^{-4t} & t\,e^{-4t} \\ -4\,e^{-4t} & e^{-4t} - 4t\,e^{-4t} \end{pmatrix}$$

Particular solution obtained may contain $y_1(t)$ or $y_2(t)$ which can

be included in the homogeneous solution with appropriate scaling factors

$$y_p(t) = \frac{t^4\,e^{-4t}}{6}$$

Complete Solution (with unknown constants b_1 & b_2)

$$y(t) = y_p(t) + b_1\,e^{-4t} + b_2\,t\,e^{-4t}$$

Particular solution: Method of Undetermined Coefficients

$$y''(t) + 8\,y'(t) + 16\,y(t) = 2t^2\,e^{-4t}$$

Roots of the Characteristic or Auxiliary Equation

$$\begin{pmatrix} -4 & -4 \end{pmatrix}$$

Initial guess for the particular solution (Double roots match; scaling by t^2)

$y_p(t) = t^2\,[A_2 t^2 + A_1 t + A_0]\ \exp(-4*t)$

Particular Solution Obtained

$$y_p(t) = \frac{t^4\,e^{-4t}}{6}$$

Solution of the Homogeneous Differential Equation (c_1 & c_2 unknown)

$$y_h(t) = e^{-4t}\,(c_2 + c_1\,t)$$

Complete Solution

$$y(t) = y_p(t) + y_h(t)$$

Comparison of Particular Solutions

$$y''(t) + 8\,y'(t) + 16\,y(t) = 2t^2\,e^{-4t}$$

Roots of the Characteristic or Auxiliary Equation: $\begin{pmatrix} -4 & -4 \end{pmatrix}$

General solutions of the Homogen. Diff. EQN: $y''(t) + 8\,y'(t) + 16\,y(t) = 0$

$$\begin{pmatrix} y_1(t) = e^{-4t} & y_2(t) = t\,e^{-4t} \end{pmatrix}$$

Particular solution (Variation of parameters)

$$y_p(t) = \frac{t^4\,e^{-4t}}{6}$$

Particular solution (Undetermined Coefficients)

$$y_p(t) = \frac{t^4\,e^{-4t}}{6}$$

Particular solutions match

Example # 3.29

```
gt='6';abc=[1,1,0];
```

Particular Solution: Variation of Parameters

$$y''(t) + y'(t) = 6$$

Roots of the Characteristic or Auxiliary Equation: $r^2 + r = 0$

$$\begin{pmatrix} 0 & -1 \end{pmatrix}$$

General solutions of the Homogen. Diff. EQN: $y''(t) + y'(t) = 0$

$$\begin{pmatrix} y_1(t) = 1 & y_2(t) = e^{-t} \end{pmatrix}$$

WRONSKIAN [W] given below

$$\begin{pmatrix} 1 & e^{-t} \\ 0 & -e^{-t} \end{pmatrix}$$

Particular solution obtained may contain $y_1(t)$ or $y_2(t)$ which can be included in the homogeneous solution with appropriate scaling factors

$$y_p(t) = 6\,t - 6$$

Complete Solution (with unknown constants b_1 & b_2)

$$y(t) = b_1 + y_p(t) + b_2\,e^{-t}$$

Particular solution: Method of Undetermined Coefficients

$$y''(t) + y'(t) = 6$$

Roots of the Characteristic or Auxiliary Equation

$$\begin{pmatrix} 0 & -1 \end{pmatrix}$$

Initial guess for the particular solution (One root matches; scaling by t)

$y_p(t) = t\ [A_0]$

Particular Solution Obtained

$$6\,t$$

Solution of the Homogeneous Differential Equation (c_1 & c_2 unknown)

$$y_h(t) = c_1 + c_2\,e^{-t}$$

Complete Solution

$$y(t) = y_p(t) + y_h(t)$$

Comparison of Particular Solutions

$$y''(t) + y'(t) = 6$$

Roots of the Characteristic or Auxiliary Equation: $\begin{pmatrix} 0 & -1 \end{pmatrix}$

General solutions of the Homogen. Diff. EQN: $y''(t) + y'(t) = 0$

$$\begin{pmatrix} y_1(t) = 1 & y_2(t) = e^{-t} \end{pmatrix}$$

Particular solution (Variation of parameters)

$$y_p(t) = 6t - 6$$

Contains term below (scaled version of general solution), $y_1(t) = 1$

$$-6$$

Particular solution (Variation of Parameters) after removing the term above

$$y_p(t) = 6t$$

Particular solution (Undetermined Coefficients)

$$y_p(t) = 6t$$

Particular solutions are now a match

Example # 3.30

```
gt='t*exp(2*t)*cos(2*t)';abc=[1,0,4];
```

Particular Solution: Variation of Parameters

$$y''(t) + 4\,y(t) = t\,\cos(2t)\,e^{2t}$$

Roots of the Characteristic or Auxiliary Equation: $r^2 + 4 = 0$

$$\begin{pmatrix} 2i & -2i \end{pmatrix}$$

General solutions of the Homogen. Diff. EQN: $y''(t) + 4\,y(t) = 0$

$$\begin{pmatrix} y_1(t) = \cos(2t) & y_2(t) = \sin(2t) \end{pmatrix}$$

WRONSKIAN [W] given below

$$\begin{pmatrix} \cos(2t) & \sin(2t) \\ -2\sin(2t) & 2\cos(2t) \end{pmatrix}$$

Particular solution obtained may contain y_1(t) or y_2(t) which can be included in the homogeneous solution with appropriate scaling factors

$$y_p(t) = -\frac{e^{2t}\,(\cos(2t) + 7\sin(2t) - 5t\cos(2t) - 10t\sin(2t))}{100}$$

Complete Solution (with unknown constants b_1 & b_2)

$$y(t) = y_p(t) + b_2\,\sin(2t) + b_1\,\cos(2t)$$

Particular solution: Method of Undetermined Coefficients

$$y''(t) + 4\,y(t) = t\,\cos(2t)\,e^{2t}$$

Roots of the Characteristic or Auxiliary Equation

$$\begin{pmatrix} 2i & -2i \end{pmatrix}$$

Initial guess for the particular solution (Root absent: no scaling by t)

$y_p(t) = [A_1 t + A_0]\,[P*\cos(2*t) + Q*\sin(2*t)]\,e^{2*t}$

Particular Solution Obtained

$$y_p(t) = -e^{2t}\left(\frac{\cos(2t)}{100} + \frac{7\sin(2t)}{100} - \frac{t\cos(2t)}{20} - \frac{t\sin(2t)}{10}\right)$$

Solution of the Homogeneous Differential Equation (c_1 & c_2 unknown)

$$y_h(t) = c_1\,\cos(2t) - c_2\,\sin(2t)$$

Complete Solution

$$y(t) = y_p(t) + y_h(t)$$

Comparison of Particular Solutions

$$y''(t) + 4\,y(t) = t\,\cos(2t)\,e^{2t}$$

Roots of the Characteristic or Auxiliary Equation: $(\,2i \quad -2i\,)$

General solutions of the Homogen. Diff. EQN: $y''(t) + 4\,y(t) = 0$

$$(\; y_1(t) = \cos(2t) \quad y_2(t) = \sin(2t) \;)$$

Particular solution (Variation of parameters)

$$y_p(t) = -\frac{e^{2t}\,(\cos(2t) + 7\,\sin(2t) - 5t\,\cos(2t) - 10\,t\,\sin(2t))}{100}$$

Particular solution (Undetermined Coefficeints)

$$y_p(t) = -e^{2t}\left(\frac{\cos(2t)}{100} + \frac{7\,\sin(2t)}{100} - \frac{t\,\cos(2t)}{20} - \frac{t\,\sin(2t)}{10}\right)$$

Particular solutions match

Example # 3.31

```
gt='t^2*cos(2*t)';abc=[1,0,4];
```

Particular Solution: Variation of Parameters

$$y''(t) + 4\,y(t) = t^2\,\cos(2\,t)$$

Roots of the Characteristic or Auxiliary Equation: $r^2 + 4 = 0$

$$(\,2i \quad -2i\,)$$

General solutions of the Homogen. Diff. EQN: $y''(t) + 4\,y(t) = 0$

$$(\; y_1(t) = \cos(2t) \quad y_2(t) = \sin(2t) \;)$$

WRONSKIAN [W] given below

$$\begin{pmatrix} \cos(2\,t) & \sin(2\,t) \\ -2\,\sin(2\,t) & 2\,\cos(2\,t) \end{pmatrix}$$

Particular solution obtained may contain y_1(t) or y_2(t) which can be included in the homogeneous solution with appropriate scaling factors

$$y_p(t) = \frac{t^2\,\cos(2\,t)}{16} - \frac{t\,\sin(2\,t)}{32} - \frac{\cos(2\,t)}{128} + \frac{t^3\,\sin(2\,t)}{12}$$

Complete Solution (with unknown constants b_1 & b_2)

$$y(t) = y_p(t) + b_2\,\sin(2t) + b_1\,\cos(2t)$$

Particular solution: Method of Undetermined Coefficient

$$y''(t) + 4\,y(t) = t^2 \cos(2\,t)$$

Roots of the Characteristic or Auxiliary Equation

$$(\ 2i \quad -2i\ )$$

Initial guess for the particular solution (Root present; scaling by t)

$y_p(t) = t\,[A_2t^2 + A_1t + A_0]\,[P*\cos(2*t) + Q*\sin(2*t)]$

Particular Solution Obtained

$$y_p(t) = t\left(\frac{t\cos(2t)}{16} - \frac{\sin(2\,t)}{32} + \frac{t^2\sin(2\,t)}{12}\right)$$

Solution of the Homogeneous Differential Equation (c_1 & c_2 unknown)

$$y_h(t) = c_1\cos(2\,t) - c_2\sin(2\,t)$$

Complete Solution

$$y(t) = y_p(t) + y_h(t)$$

Comparison of Particular Solutions

$$y''(t) + 4\,y(t) = t^2 \cos(2\,t)$$

Roots of the Characteristic or Auxiliary Equation: $(\ 2i \quad -2i\ )$

General solutions of the Homogen. Diff. EQN: $y''(t) + 4\,y(t) = 0$

$$\left(\ y_1(t) = \cos(2\,t) \quad y_2(t) = \sin(2\,t)\ \right)$$

Particular solution (Variation of parameters)

$$y_p(t) = \frac{t^2\cos(2\,t)}{16} - \frac{t\sin(2\,t)}{32} - \frac{\cos(2\,t)}{128} + \frac{t^3\sin(2\,t)}{12}$$

Contains term below (scaled version of general solution), $y_1(t) = \cos(2\,t)$

$$\frac{\sin(t)^2}{64} - \frac{1}{128}$$

Particular solution (Variation of Parameters) after removing the term above

$$y_p(t) = \frac{t^2\cos(2\,t)}{16} - \frac{t\sin(2\,t)}{32} + \frac{t^3\sin(2\,t)}{12}$$

Particular solution (Undetermined Coefficients)

$$y_p(t) = t\left(\frac{t\cos(2t)}{16} - \frac{\sin(2\,t)}{32} + \frac{t^2\sin(2\,t)}{12}\right)$$

Particular solutions are now a match

Example # 3.32

```
gt='t';abc=[1,1,0];
```

Particular Solution: Variation of Parameters

$$y''(t) + y'(t) = t$$

Roots of the Characteristic or Auxiliary Equation: $r^2 + r = 0$

$$\begin{pmatrix} 0 & -1 \end{pmatrix}$$

General solutions of the Homogen. Diff. EQN: $y''(t) + y'(t) = 0$

$$\begin{pmatrix} y_1(t) = 1 & y_2(t) = e^{-t} \end{pmatrix}$$

WRONSKIAN [W] given below

$$\begin{pmatrix} 1 & e^{-t} \\ 0 & -e^{-t} \end{pmatrix}$$

Particular solution obtained may contain $y_1(t)$ or $y_2(t)$ which can be included in the homogeneous solution with appropriate scaling factors

$$y_p(t) = \frac{t^2}{2} - t + 1$$

Complete Solution (with unknown constants b_1 & b_2)

$$y(t) = b_1 + y_p(t) + b_2\, e^{-t}$$

Particular solution: Method of Undetermined Coefficients

$$y''(t) + y'(t) = t$$

Roots of the Characteristic or Auxiliary Equation

$$\begin{pmatrix} 0 & -1 \end{pmatrix}$$

Initial guess for the particular solution (One root matches; scaling by t)

$$y_p(t) = t\ [A_1 t + A_0]$$

Particular Solution Obtained

$$y_p(t) = t\ \left(\frac{t}{2} - 1\right)$$

Solution of the Homogeneous Differential Equation (c_1 & c_2 unknown)

$$y_h(t) = c_1 + c_2\, e^{-t}$$

Complete Solution

$$y(t) = y_p(t) + y_h(t)$$

Comparison of Particular Solutions

$$y''(t) + y'(t) = t$$

Roots of the Characteristic or Auxiliary Equation: $\begin{pmatrix} 0 & -1 \end{pmatrix}$

General solutions of the Homogen. Diff. EQN: $y''(t) + y'(t) = 0$

$$\begin{pmatrix} y_1(t) = 1 & y_2(t) = \mathrm{e}^{-t} \end{pmatrix}$$

Particular solution (Variation of parameters)

$$y_p(t) = \frac{t^2}{2} - t + 1$$

Contains term below (scaled version of general solution), $y_1(t) = 1$

$$1$$

Particular solution (Variation of Parameters) after removing the term above

$$y_p(t) = \frac{t\,(t-2)}{2}$$

Particular solution (Undetermined Coefficients)

$$y_p(t) = t\,\left(\frac{t}{2} - 1\right)$$

Particular solutions are now a match

Example # 3.33

```
gt='t^2*exp(-2*t)';abc=[1,2,0];
```

Particular Solution: Variation of Parameters

$$y''(t) + 2\,y'(t) = t^2\,\mathrm{e}^{-2\,t}$$

Roots of the Characteristic or Auxiliary Equation: $r^2 + 2\,r = 0$

$$\begin{pmatrix} 0 & -2 \end{pmatrix}$$

General solutions of the Homogen. Diff. EQN: $y''(t) + 2\,y'(t) = 0$

$$\begin{pmatrix} y_1(t) = 1 & y_2(t) = \mathrm{e}^{-2\,t} \end{pmatrix}$$

WRONSKIAN [W] given below

$$\begin{pmatrix} 1 & \mathrm{e}^{-2\,t} \\ 0 & -2\,\mathrm{e}^{-2\,t} \end{pmatrix}$$

Particular solution obtained may contain $y_1(t)$ or $y_2(t)$ which can be included in the homogeneous solution with appropriate scaling factors

$$y_p(t) = -\frac{\mathrm{e}^{-2\,t}\left(4\,t^3 + 6\,t^2 + 6\,t + 3\right)}{24}$$

Complete Solution (with unknown constants b_1 & b_2)

$$y(t) = b_1 + y_p(t) + b_2\,\mathrm{e}^{-2\,t}$$

Particular solution: Method of Undetermined Coefficients

$$y''(t) + 2y'(t) = t^2 e^{-2t}$$

Roots of the Characteristic or Auxiliary Equation

$$(0 \quad -2)$$

Initial guess for the particular solution (One root matches; scaling by t)

$y_p(t) = t\ [A_2 t^2 + A_1 t + A_0]\ \exp(-2^*t)$

Particular Solution Obtained

$$y_p(t) = -t\, e^{-2t} \left(\frac{t^2}{6} + \frac{t}{4} + \frac{1}{4} \right)$$

Solution of the Homogeneous Differential Equation (c_1 & c_2 unknown)

$$y_h(t) = c_1 + c_2\, e^{-2t}$$

Complete Solution

$$y(t) = y_p(t) + y_h(t)$$

Comparison of Particular Solutions

$$y''(t) + 2y'(t) = t^2 e^{-2t}$$

Roots of the Characteristic or Auxiliary Equation: $(0 \quad -2)$

General solutions of the Homogen. Diff. EQN: $y''(t) + 2y'(t) = 0$

$$(y_1(t) = 1 \quad y_2(t) = e^{-2t})$$

Particular solution (Variation of parameters)

$$y_p(t) = -\frac{e^{-2t}\left(4t^3 + 6t^2 + 6t + 3\right)}{24}$$

Contains term below (scaled version of general solution), $y_2(t) = e^{-2t}$

$$-\frac{e^{-2t}}{8}$$

Particular solution (Variation of Parameters) after removing the term above

$$y_p(t) = -\frac{t e^{-2t}\left(2t^2 + 3t + 3\right)}{12}$$

Particular solution (Undetermined Coefficients)

$$y_p(t) = -t\, e^{-2t} \left(\frac{t^2}{6} + \frac{t}{4} + \frac{1}{4} \right)$$

Particular solutions are now a match

Example # 3.34

```
gt='t';abc=[1,-1,-1];
```

Particular Solution: Variation of Parameters

$$y''(t) - y'(t) - y(t) = t$$

Roots of the Characteristic or Auxiliary Equation: $r^2 - r - 1 = 0$

$$\left(\frac{1}{2} - \frac{\sqrt{5}}{2} \quad \frac{\sqrt{5}}{2} + \frac{1}{2} \right)$$

General solutions of the Homogen. Diff. EQN: $y''(t) - y'(t) - y(t) = 0$

$$\left(y_1(t) = e^{-t\left(\frac{\sqrt{5}}{2} - \frac{1}{2}\right)} \quad y_2(t) = e^{t\left(\frac{\sqrt{5}}{2} + \frac{1}{2}\right)} \right)$$

WRONSKIAN [W] given below

$$\begin{pmatrix} e^{-t\left(\frac{\sqrt{5}}{2} - \frac{1}{2}\right)} & e^{t\left(\frac{\sqrt{5}}{2} + \frac{1}{2}\right)} \\ -e^{-t\left(\frac{\sqrt{5}}{2} - \frac{1}{2}\right)} \left(\frac{\sqrt{5}}{2} - \frac{1}{2}\right) & e^{t\left(\frac{\sqrt{5}}{2} + \frac{1}{2}\right)} \left(\frac{\sqrt{5}}{2} + \frac{1}{2}\right) \end{pmatrix}$$

Particular solution obtained may contain $y_1(t)$ or $y_2(t)$ which can be included in the homogeneous solution with appropriate scaling factors

$$y_p(t) = 1 - t$$

Complete Solution (with unknown constants b_1 & b_2)

$$y(t) = y_p(t) + b_1 e^{-t\left(\frac{\sqrt{5}}{2} - \frac{1}{2}\right)} + b_2 e^{t\left(\frac{\sqrt{5}}{2} + \frac{1}{2}\right)}$$

Particular solution: Method of Undetermined Coefficients

$$y''(t) - y'(t) - y(t) = t$$

Roots of the Characteristic or Auxiliary Equation

$$\left(\frac{1}{2} - \frac{\sqrt{5}}{2} \quad \frac{\sqrt{5}}{2} + \frac{1}{2} \right)$$

Initial guess for the particular solution (Root absent: no scaling by t)

$$y_p(t) = [A_1 t + A_0]$$

Particular Solution Obtained

$$y_p(t) = 1 - t$$

Solution of the Homogeneous Differential Equation (c_1 & c_2 unknown)

$$y_h(t) = c_1 e^{-t\left(\frac{\sqrt{5}}{2} - \frac{1}{2}\right)} + c_2 e^{t\left(\frac{\sqrt{5}}{2} + \frac{1}{2}\right)}$$

Complete Solution

$$y(t) = y_p(t) + y_h(t)$$

Comparison of Particular Solutions

$$y''(t) - y'(t) - y(t) = t$$

Roots of the Characteristic or Auxiliary Equation: $\left(\frac{1}{2} - \frac{\sqrt{5}}{2} \quad \frac{\sqrt{5}}{2} + \frac{1}{2} \right)$

General solutions of the Homogen. Diff. EQN: $y''(t) - y'(t) - y(t) = 0$

$$\left(y_1(t) = e^{-t\left(\frac{\sqrt{5}}{2} - \frac{1}{2}\right)} \quad y_2(t) = e^{t\left(\frac{\sqrt{5}}{2} + \frac{1}{2}\right)} \right)$$

Particular solution (Variation of parameters)

$$y_p(t) = 1 - t$$

Particular solution (Undetermined Coefficients)

$$y_p(t) = 1 - t$$

Particular solutions match

Example # 3.35

```
gt='t*exp(t)';abc=[1,-3,2];
```

Particular Solution: Variation of Parameters

$$y''(t) - 3\,y'(t) + 2\,y(t) = t\,e^t$$

Roots of the Characteristic or Auxiliary Equation: $r^2 - 3r + 2 = 0$

$$\left(1 \quad 2 \right)$$

General solutions of the Homogen. Diff. EQN: $y''(t) - 3\,y'(t) + 2\,y(t) = 0$

$$\left(y_1(t) = e^t \quad y_2(t) = e^{2t} \right)$$

WRONSKIAN [W] given below

$$\begin{pmatrix} e^t & e^{2t} \\ e^t & 2e^{2t} \end{pmatrix}$$

Particular solution obtained may contain $y_1(t)$ or $y_2(t)$ which can be included in the homogeneous solution with appropriate scaling factors

$$y_p(t) = -\frac{e^t\left(t^2+2t+2\right)}{2}$$

Complete Solution (with unknown constants b_1 & b_2)

$$y(t) = y_p(t) + b_1\,e^t + b_2\,e^{2t}$$

Particular solution: Method of Undetermined Coefficients

$$y''(t) - 3y'(t) + 2y(t) = t\,e^t$$

Roots of the Characteristic or Auxiliary Equation

$$(\ 1 \quad 2\ )$$

Initial guess for the particular solution (One root matches; scaling by t)

$$y_p(t) = t\ [A_1 t + A_0]\ \exp(t)$$

Particular Solution Obtained

$$y_p(t) = -t\,e^t\ \left(\tfrac{t}{2} + 1\right)$$

Solution of the Homogeneous Differential Equation (c_1 & c_2 unknown)

$$y_h(t) = c_1\,e^t + c_2\,e^{2t}$$

Complete Solution

$$y(t) = y_p(t) + y_h(t)$$

Comparison of Particular Solutions

$$y''(t) - 3y'(t) + 2y(t) = t\,e^t$$

Roots of the Characteristic or Auxiliary Equation: $(\ 1 \quad 2\ )$

General solutions of the Homogen. Diff. EQN: $y''(t) - 3y'(t) + 2y(t) = 0$

$$\left(\ y_1(t) = e^t \quad y_2(t) = e^{2t}\ \right)$$

Particular solution (Variation of parameters)

$$y_p(t) = -\frac{e^t\left(t^2+2t+2\right)}{2}$$

Contains term below (scaled version of general solution), $y_1(t) = e^t$

$$-e^t$$

Particular solution (Variation of Parameters) after removing the term above

$$y_p(t) = -\frac{t e^t\,(t+2)}{2}$$

Particular solution (Undetermined Coefficients)

$$y_p(t) = -t\,e^t\ \left(\tfrac{t}{2} + 1\right)$$

Particular solutions are now a match

Example # 3.36

```
gt='t^2+t+2';abc=[1,0,0];
```

Particular Solution: Variation of Parameters

$$y''(t) = t^2 + t + 2$$

Roots of the Characteristic or Auxiliary Equation: $r^2 = 0$

$$\begin{pmatrix} 0 & 0 \end{pmatrix}$$

General solutions of the Homogen. Diff. EQN: $y''(t) = 0$

$$\begin{pmatrix} y_1(t) = 1 & y_2(t) = t \end{pmatrix}$$

WRONSKIAN [W] given below

$$\begin{pmatrix} 1 & t \\ 0 & 1 \end{pmatrix}$$

Particular solution obtained may contain $y_1(t)$ or $y_2(t)$ which can be included in the homogeneous solution with appropriate scaling factors

$$y_p(t) = \frac{t^2\left(t^2+2t+12\right)}{12}$$

Complete Solution (with unknown constants b_1 & b_2)

$$y(t) = b_1 + y_p(t) + b_2\, t$$

Particular solution: Method of Undetermined Coefficients

$$y''(t) = t^2 + t + 2$$

Roots of the Characteristic or Auxiliary Equation

$$\begin{pmatrix} 0 & 0 \end{pmatrix}$$

Initial guess for the particular solution (Double roots match; scaling by t^2)

$$y_p(t) = t^2[A_2t^2 + A_1t + A_0]$$

Particular Solution Obtained

$$y_p(t) = t^2\left(\frac{t2}{12} + \frac{t}{6} + 1\right)$$

Solution of the Homogeneous Differential Equation (c_1 & c_2 unknown)

$$y_h(t) = c_2 + c_1\, t$$

Complete Solution

$$y(t) = y_p(t) + y_h(t)$$

Comparison of Particular Solutions

$$y''(t) = t^2 + t + 2$$

Roots of the Characteristic or Auxiliary Equation: $\begin{pmatrix} 0 & 0 \end{pmatrix}$

General solutions of the Homogen. Diff. EQN: $y''(t) = 0$

$$\begin{pmatrix} y_1(t) = 1 & y_2(t) = t \end{pmatrix}$$

Particular solution (Variation of parameters)

$$y_p(t) = \frac{t^2\left(t^2+2t+12\right)}{12}$$

Particular solution (Undetermined Coefficients)

$$y_p(t) = t^2\left(\frac{t^2}{12} + \frac{t}{6} + 1\right)$$

Particular solutions match

Example # 3.37

```
abc=[1,3,2];; gt='t^2*exp(-2*t)';
```

Particular Solution: Variation of Parameters

$$y''(t) + 3\,y'(t) + 2\,y(t) = t^2\,e^{-2t}$$

Roots of the Characteristic or Auxiliary Equation: $r^2 + 3r + 2 = 0$

$$\begin{pmatrix} -2 & -1 \end{pmatrix}$$

General solutions of the Homogen. Diff. EQN: $y''(t) + 3\,y'(t) + 2\,y(t) = 0$

$$\begin{pmatrix} y_1(t) = e^{-2t} & y_2(t) = e^{-t} \end{pmatrix}$$

WRONSKIAN [W] given below

$$\begin{pmatrix} e^{-2t} & e^{-t} \\ -2\,e^{-2t} & -e^{-t} \end{pmatrix}$$

Particular solution obtained may contain y_1(t) or y_2(t) which can

be included in the homogeneous solution with appropriate scaling factors

$$y_p(t) = -\frac{e^{-2t}\left(t^3+3t^2+6t+6\right)}{3}$$

Complete Solution (with unknown constants b_1 & b_2)

$$y(t) = y_p(t) + b_1\,e^{-2t} + b_2\,e^{-t}$$

Particular solution: Method of Undetermined Coefficients

$$y''(t) + 3\,y'(t) + 2\,y(t) = t^2\,e^{-2\,t}$$

Roots of the Characteristic or Auxiliary Equation

$$\left(-2 \quad -1 \right)$$

Initial guess for the particular solution (One root matches; scaling by t)

$y_p(t) = t\ [A_2 t^2 + A_1 t + A_0]\ \exp(-2*t)$

Particular Solution Obtained

$$y_p(t) = -t\,e^{-2t}\left(\frac{t^2}{3} + t + 2\right)$$

Solution of the Homogeneous Differential Equation (c_1 & c_2 unknown)

$$y_h(t) = c_1\,e^{-2\,t} + c_2\,e^{-t}$$

Complete Solution

$$y(t) = y_p(t) + y_h(t)$$

Comparison of Particular Solutions

$$y''(t) + 3\,y'(t) + 2\,y(t) = t^2\,e^{-2\,t}$$

Roots of the Characteristic or Auxiliary Equation: $\left(-2 \quad -1 \right)$

General solutions of the Homogen. Diff. EQN: $y''(t) + 3\,y'(t) + 2\,y(t) = 0$

$$\left(y_1(t) = e^{-2\,t} \quad y_2(t) = e^{-t} \right)$$

Particular solution (Variation of parameters)

$$y_p(t) = -\frac{e^{-2\,t}\left(t^3+3\,t^2+6\,t+6\right)}{3}$$

Contains term below (scaled version of general solution), $y_1(t) = e^{-2t}$

$$-2\,e^{-2\,t}$$

Particular solution (Variation of Parameters) after removing the term above

$$y_p(t) = -\frac{t\,e^{-2\,t}\left(t^2+3\,t+6\right)}{3}$$

Particular solution (Undetermined Coefficients)

$$y_p(t) = -t\,e^{-2t}\left(\frac{t^2}{3} + t + 2\right)$$

Particular solutions are now a match

Example # 3.38

```
gt='t*exp(-2*t)';abc=[1,4,4];
```

Particular Solution: Variation of Parameters

$$y''(t) + 4\,y'(t) + 4\,y(t) = t\,e^{-2t}$$

Roots of the Characteristic or Auxiliary Equation: $r^2 + 4r + 4 = 0$

$$\begin{pmatrix} -2 & -2 \end{pmatrix}$$

General solutions of the Homogen. Diff. EQN: $y''(t) + 4\,y'(t) + 4\,y(t) = 0$

$$\begin{pmatrix} y_1(t) = e^{-2t} & y_2(t) = t\,e^{-2t} \end{pmatrix}$$

WRONSKIAN [W] given below

$$\begin{pmatrix} e^{-2t} & t\,e^{-2t} \\ -2\,e^{-2t} & e^{-2t} - 2t\,e^{-2t} \end{pmatrix}$$

Particular solution obtained may contain $y_1(t)$ or $y_2(t)$ which can

be included in the homogeneous solution with appropriate scaling factors

$$y_p(t) = \frac{t^3\,e^{-2t}}{6}$$

Complete Solution (with unknown constants b_1 & b_2)

$$y(t) = y_p(t) + b_1\,e^{-2t} + b_2\,t\,e^{-2t}$$

Particular solution: Method of Undetermined Coefficients

$$y''(t) + 4\,y'(t) + 4\,y(t) = t\,e^{-2t}$$

Roots of the Characteristic or Auxiliary Equation

$$\begin{pmatrix} -2 & -2 \end{pmatrix}$$

Initial guess for the particular solution (Double roots match; scaling by t^2)

$y_p(t) = t^2\,[A_1 t + A_0]$ exp(-2*t)

Particular Solution Obtained

$$y_p(t) = \frac{t^3\,e^{-2t}}{6}$$

Solution of the Homogeneous Differential Equation (c_1 & c_2 unknown)

$$y_h(t) = e^{-2t}\,(c_2 + c_1\,t)$$

Complete Solution

$$y(t) = y_p(t) + y_h(t)$$

Comparison of Particular Solutions

$$y''(t) + 4\,y'(t) + 4\,y(t) = t\,e^{-2t}$$

Roots of the Characteristic or Auxiliary Equation: $\begin{pmatrix} -2 & -2 \end{pmatrix}$

General solutions of the Homogen. Diff. EQN: $y''(t) + 4\,y'(t) + 4\,y(t) = 0$

$$\begin{pmatrix} y_1(t) = e^{-2t} & y_2(t) = t\,e^{-2t} \end{pmatrix}$$

Particular solution (Variation of parameters)

$$y_p(t) = \frac{t^3\,e^{-2t}}{6}$$

Particular solution (Undetermined Coefficients)

$$y_p(t) = \frac{t^3\,e^{-2t}}{6}$$

Particular solutions match

Example # 3.39

```
gt='6*sin(t)';abc=[1,0,1];
```

Particular Solution: Variation of Parameters

$$y''(t) + y(t) = 6\,\sin(t)$$

Roots of the Characteristic or Auxiliary Equation: $r^2 + 1 = 0$

$$\begin{pmatrix} i & -i \end{pmatrix}$$

General solutions of the Homogen. Diff. EQN: $y''(t) + y(t) = 0$

$$\begin{pmatrix} y_1(t) = \cos(t) & y_2(t) = \sin(t) \end{pmatrix}$$

WRONSKIAN [W] given below

$$\begin{pmatrix} \cos(t) & \sin(t) \\ -\sin(t) & \cos(t) \end{pmatrix}$$

Particular solution obtained may contain $y_1(t)$ or $y_2(t)$ which can be included in the homogeneous solution with appropriate scaling factors

$$y_p(t) = -3\,t\,\cos(t)$$

Complete Solution (with unknown constants b_1 & b_2)

$$y(t) = y_p(t) + b_1\,\cos(t) + b_2\,\sin(t)$$

Particular solution: Method of Undetermined Coefficients

$$y''(t) + y(t) = 6\ \sin(t)$$

Roots of the Characteristic or Auxiliary Equation

$$(\mathrm{i} \quad -\mathrm{i})$$

Initial guess for the particular solution (Root present; scaling by t)

$y_p(t) = t\,[P*\cos(t) + Q*\sin(t)]$

Particular Solution Obtained

$$y_p(t) = -3\,t\ \cos(t)$$

Solution of the Homogeneous Differential Equation (c_1 & c_2 unknown)

$$y_h(t) = c_1\ \cos(t) - c_2\ \sin(t)$$

Complete Solution

$$y(t) = y_p(t) + y_h(t)$$

Comparison of Particular Solutions

$$y''(t) + y(t) = 6\ \sin(t)$$

Roots of the Characteristic or Auxiliary Equation: $(\mathrm{i} \quad -\mathrm{i})$

General solutions of the Homogen. Diff. EQN: $y''(t) + y(t) = 0$

$$\left(\ y_1(t) = \cos(t) \quad y_2(t) = \sin(t) \ \right)$$

Particular solution (Variation of parameters)

$$y_p(t) = -3\,t\ \cos(t)$$

Particular solution (Undetermined Coefficients)

$$y_p(t) = -3\,t\ \cos(t)$$

Particular solutions match

Example # 3.40

```
gt='t^2*exp(-2*t)*sin(3*t)';abc=[1,4,13];
```

Particular Solution: Variation of Parameters

$$y''(t)+4\,y'(t)+13\,y(t)=t^2\,\sin(3\,t)\,e^{-2t}$$

Roots of the Characteristic or Auxiliary Equation: $r^2+4r+13=0$

$$\left(\,-2+3i\quad -2-3i\,\right)$$

General solutions of the Homogen. Diff. EQN: $y''(t)+4\,y'(t)+13\,y(t)=0$

$$\left(\,y_1(t)=\cos(3t)\,e^{-2t}\quad y_2(t)=\sin(3t)\,e^{-2t}\,\right)$$

WRONSKIAN [W] given below

$$\begin{pmatrix} \cos(3\,t)\,e^{-2t} & \sin(3\,t)\,e^{-2t} \\ -2\cos(3\,t)\,e^{-2t}-3\sin(3\,t)\,e^{-2t} & 3\cos(3\,t)\,e^{-2t}-2\sin(3\,t)\,e^{-2t} \end{pmatrix}$$

Particular solution obtained may contain $y_1(t)$ or $y_2(t)$ which can be included in the homogeneous solution with appropriate scaling factors

$$y_p(t)=-\frac{e^{-2t}\left(\sin(3\,t)-6\,t\,\cos(3\,t)+36\,t^3\,\cos(3\,t)-18\,t^2\,\sin(3\,t)\right)}{648}$$

Complete Solution (with unknown constants b_1 & b_2)

$$y(t)=y_p(t)+b_1\,\cos(3\,t)\,e^{-2t}+b_2\,\sin(3\,t)\,e^{-2t}$$

Particular solution: Method of Undetermined Coefficients

$$y''(t)+4\,y'(t)+13\,y(t)=t^2\,\sin(3\,t)\,e^{-2t}$$

Roots of the Characteristic or Auxiliary Equation

$$\left(\,-2+3i\quad -2-3i\,\right)$$

Initial guess for the particular solution (Root present; scaling by t)

$y_p(t)$ = t[A_2t^2 + A_1t + A_0] [P*cos(3*t) + Q*sin(3*t)] e^{-2*t}

Particular Solution Obtained

$$y_p(t)=t\,e^{-2t}\left(\frac{\cos(3\,t)}{108}+\frac{t\,\sin(3\,t)}{36}-\frac{t^2\,\cos(3\,t)}{18}\right)$$

Solution of the Homogeneous Differential Equation (c_1 & c_2 unknown)

$$y_h(t)=-e^{-2t}\,(c_2\,\sin(3\,t)-c_1\,\cos(3\,t))$$

Complete Solution

$$y(t)=y_p(t)+y_h(t)$$

Comparison of Particular Solutions

$$y''(t) + 4\,y'(t) + 13\,y(t) = t^2 \sin(3\,t)\ e^{-2\,t}$$

Roots of the Characteristic or Auxiliary Equation: $\left(-2+3\,i \quad -2-3\,i \right)$

General solutions of the Homogen. Diff. EQN: $y''(t) + 4\,y'(t) + 13\,y(t) = 0$

$$\left(y_1(t) = \cos(3\,t)\ e^{-2\,t} \quad y_2(t) = \sin(3\,t)\ e^{-2\,t} \right)$$

Particular solution (Variation of parameters)

$$y_p(t) = -\frac{e^{-2\,t}\left(\sin(3\,t) - 6\,t\,\cos(3\,t) + 36\,t^3\,\cos(3\,t) - 18\,t^2\,\sin(3\,t)\right)}{648}$$

Contains term below (scaled version of general solution), $y_2(t) = \sin(3\,t)\ e^{-2\,t}$

$$-\frac{\sin(3\,t)\,e^{-2\,t}}{648}$$

Particular solution (Variation of Parameters) after removing the term above

$$y_p(t) = \frac{t\,e^{-2\,t}\left(\cos(3\,t) + 3\,t\,\sin(3\,t) - 6\,t^2\,\cos(3\,t)\right)}{108}$$

Particular solution (Undetermined Coefficients)

$$y_p(t) = t\,e^{-2\,t}\left(\frac{\cos(3\,t)}{108} + \frac{t\,\sin(3\,t)}{36} - \frac{t^2\,\cos(3\,t)}{18}\right)$$

Particular solutions are now a match

3.4 Summary

This chapter is devoted to the detailed study of second order differential equations with constant coefficients. The methodology based on the roots of the characteristic equation has been the primary focus for the determination of the solution. Unlike the conventional pedagogic formats where other approaches for obtaining the solutions are presented separately either in the Appendix or in other chapters, all the methods to obtain the solutions are presented together here. These methods include breaking the second order differential equation into a pair of coupled first order differential equations, Laplace transforms, as well as numerical methods based on Runge-Kutta using MATLAB. This allows the reader to compare the results from different approaches. The results also include detailed analysis of the phase portraits which help explain the stability of the systems modeled through these differential equations. While homogeneous differential equations are solved through the multiple ways described above, non-homogeneous differential equations are solved through the use of method of variation of parameters as well as the method of undetermined coefficients in order to obtain the particular solution. The results from these two approaches for obtaining the particular solutions are compared and explanations are provided appropriately in cases where the particular solutions appear different. Even in the case of the non-homogeneous differential equations, Laplace transforms as well as ODE based methods are employed as additional verification steps. The examples cover a substantial number of different cases and the solution in each case is prepared to be self-contained with appropriate theory.

3.5 Exercises

1. Examine the phase portraits for the set of second order homogeneous differential equations with constant coefficients with [a,b,c] as a=1, b=[-2,-1,-.5,0,0.5,1,2,3], c=[-2,-1,-.5,0,0.5,1,2,3].

2. For all the cases above, obtain the general solutions starting with the roots of the characteristic equation and formulating the solution from the roots. Verify your results directly using MATLAB.

   [a,b,c]
   [1,-4,-4]
   [1,-4,-3]
   [1,-4,2]
   [1,-4,1]
   [1,-3,-4]
   [1,-3,-3]
   [1,-3,0]
   [1,-3,2]
   [1,-3,1]
   [1,-2,-4]
   [1,-2,-3]
   [1,-2,2]
   [1,-2,1]
   [1,4,-4]
   [1,4,-3]
   [1,4,2]
   [1,4,1]
   [1,5,-4]
   [1,5,-3]
   [1,5,0]
   [1,5,2]
   [1,5,1]
   [1,6,-4]
   [1,6,-3]
   [1,6,0]
   [1,6,2]
   [1,6,1]

3. For the set of differential equations with [a, b, c, y(0), y'(0)], obtain the complete solution using (a) roots (b) Laplace transforms and (c) using Runge-Kutta methods and verify that the results match.

   [1,-4,-4,-1,1]

   [1,-4,-3,-1,1]

   [1,-4,0,1,-1]

   [1,-4,2,1,-1]

   [1,-4,1,1,-1]

   [1,-3,-4,1,-1]

   [1,-3,-3,1,-1]

   [1,-3,0,1,-1]

   [1,-3,2,1,-1]

   [1,-3,1,1,-1]

   [1,-2,-4,1,-1]

   [1,-2,-3,1,-1]

   [1,-2,2,1,-1]

   [1,-2,1,-1,1]

   [1,4,-4,-1,1]

   [1,4,-3,-1,1]

   [1,4,2,-1,1]

   [1,4,1,-1,1]

   [1,5,-4,-1,1]

   [1,5,-3,-1,1]

   [1,5,0,-1,1]

   [1,5,2,-1,1]

   [1,5,1,-1,1]

   [1,6,-4,-1,0]

   [1,6,-3,1,0]

   [1,6,0,-1,0]

   [1,6,2,0,-1]

   [1,6,1,0,1]

4. For the following sets of differential equations identified by the coefficients a, b, c as a row vector and the corresponding forcing functions, obtain expressions for the particular solution obtained using the method of variation of parameters and the method of undetermined coefficients.

   A=[1,2,1]; g(t)=t exp(-t)

   A=[1,2,1]; g(t)=tcos(t);

A=[1,-2,-3]; g(t)=t2exp(3t)
A=[1,4,0]; g(t)=t2+t+2
A=[1,4,4];g(t)=texp(-2t)
A=[1,0,0];g(t)=t+5;
A=[1,0,-9]; g(t)=t exp(2t)
A=[1,0,25]; g(t)=t sin(5t)
A=[1,-2,2]; g(t)=exp(t) cos(t)
A=[1,2,2]; g(t)=exp(-t)*sin(t);

5. Obtain the complete solutions for the problem in 5.5 with the initial conditions given. Verify your results using Laplace transforms and plot the results obtained using ode45 and the roots.

   Initial conditions constitute the vector [y(0);y'(0)].

   A=[1,2,1];g(t)=t exp(-t);IC=[1;-1]
   A=[1,2,1];g(t)=tcos(t));IC=[1;-1]
   A=[1,-2,-3]; g(t)=t2exp(3t);IC=[1;-1]
   A=[1,4,0];g(t)=t2+t+2;IC=[1;-1];
   A=[1,4,4];g(t)=texp(-2t);IC=[-1;1];
   A=[1,0,0];g(t)=t+5;IC=[2;1];
   A=[1,0,-9];g(t)=texp(2t);IC=[1,-2];
   A=[1,0,25];g(t)=tsin(5t);IC=[-1,1];
   A=[1,-2,2];g(t)=exp(t)cos(t);IC=[1,0]
   A=[1, 2,2];g(t)=exp(-t)sin(t);IC=[1,0]

CHAPTER 4

Linear Higher Order Differential Equations with Constant Coefficients

4.1 Introduction 181
4.2 Homogeneous Differential Equations (n<5) 182
4.3 Non-homogeneous Differential Equations and Particular Solutions (n<5) 185
4.4 Additional Methods of Obtaining the Solution and Verification 186
4.4.1 Laplace transform 186
4.4.2 Numerical techniques 187
4.5 Higher Order Differential Equations (n>4) 188
4.6 Examples 188
4.7 Summary 233
4.8 Exercises 234

4.1 Introduction

The concept of a second order linear differential equation with constant coefficients can be extended to higher order differential equations. A general higher order linear differential equation with constant coefficients can be written as

$$A_n y^{(n)}(t) + A_{n-1} y^{(n-1)}(t) + \dots A_0 y(t) = g(t). \tag{4.1}$$

In eqn. (4.1), A_0, A_1,.., A_n are constants and

$$y^{(n)}(t) = \frac{d^n y(t)}{dt^n}. \tag{4.2}$$

When the excitation function or the external forcing function g(t)=0, eqn. (4.1) becomes a homogeneous type and g(t) ≠ 0, eqn. (4.1) becomes a non-homogeneous type. The solution to the differential equation (4.1) may be obtained using procedures similar to the ones presented in connection with the second order ones as described in Chapter 3. These include the use of the characteristic equation or the characteristic polynomial, variation of parameters for obtaining the particular solution, Laplace transforms and numerical ones such as the Runge-Kutta method.

One of the difficulties with the use of the roots of the characteristic equation arises from the breadth and scope of the relationships among the various roots. For example with n=2, roots are either real and distinct, real and equal or complex. With n=3, the roots may be distinct, two of them may be equal or all three of them may be equal. It is also possible that one of the roots is real while the other two form a complex pair. One can therefore envision the various possibilities that might exist in terms of the relationships among the roots when n exceeds 4. Because of this, the initial discussion is limited to n<5. The cases of higher order differential equations with n>4 are treated separately.

4.2 Homogeneous Differential Equations (n<5)

The characteristic equation associated with the nth order homogeneous differential equation is given by

$$A_n r^n + A_{n-1} r^{n-1} + \dots A_1 r + A_0 = 0. \tag{4.3}$$

Following the approach and arguments put forth in obtaining the solution for a second order homogeneous differential equation with constant coefficients, the basis solution vector becomes

$$\vec{Y}(t) = \begin{bmatrix} y_1(t) \\ y_2(t) \\ \cdot \\ y_{n-1}(t) \\ y_n(t) \end{bmatrix}. \tag{4.4}$$

Note that the solution vector in eqn. (4.4) may also be written as row vector of size [1 x n]

$$\vec{Y}(t) = \begin{bmatrix} y_1(t) & y_2(t) & \cdot & y_{n-1}(t) & y_n(t) \end{bmatrix}. \tag{4.5}$$

The general solution for the homogeneous differential equation becomes

$$y_h(t) = \sum_{k=1}^{n} c_k y_k(t). \tag{4.6}$$

In eqn. (4.6), the coefficients c_k may be obtained if initial conditions are provided and the fundamental solution set consisting of the basic solutions, $y_1(t)$, $y_2(t)$,..,$y_n(t)$, are formed from the roots of the characteristic equation in eqn. (4.3).

The nature of the solution set is more complicated than the three forms associated with the case of a second order differential equation because the interrelationships (if any) among the roots of eqn. (4.3) determine the complexity of the solution set. Therefore, one needs to consider different values of n (2, 3, 4,..) separately to formulate the solution set. While the solution set for n=2 is described in Chapter 3, the case of n=3, 4 will be presented in detail now and higher order cases will be discussed in general terms.

The characteristic equation associated with a 3rd order differential equation is

$$A_3r^3 + A_2r^2 + A_1r + A_0 = 0. \tag{4.7}$$

If the three roots are identified by r_k, k=1,2,3, depending on whether roots are distinct, equal or complex conjugates, the solution set may take different forms. If the roots are distinct and real, the solution vector is

$$\vec{Y}(t) = \begin{bmatrix} y_1(t) \\ y_2(t) \\ y_3(t) \end{bmatrix} = \begin{bmatrix} e^{r_1t} \\ e^{r_2t} \\ e^{r_3t} \end{bmatrix}, \quad r_1 \neq r_2 \neq r_3 \tag{4.8}$$

If all the roots are identical, the solution vector is

$$\vec{Y}(t) = \begin{bmatrix} y_1(t) \\ y_2(t) \\ y_3(t) \end{bmatrix} = \begin{bmatrix} e^{rt} \\ te^{rt} \\ t^2e^{rt} \end{bmatrix}, \quad r = r_1 = r_2 = r_3. \tag{4.9}$$

If the two roots are equal and real, the solution vector is

$$\vec{Y}(t) = \begin{bmatrix} y_1(t) \\ y_2(t) \\ y_3(t) \end{bmatrix} = \begin{bmatrix} e^{rt} \\ e^{r_1t} \\ te^{r_1t} \end{bmatrix}, \quad r \neq r_1 = r_2 \tag{4.10}$$

The remaining possibility is that one root is real while the other two form a complex conjugate pair with real valued constants α and β. The solution vector is

$$\vec{Y}(t) = \begin{bmatrix} y_1(t) \\ y_2(t) \\ y_3(t) \end{bmatrix} = \begin{bmatrix} e^{rt} \\ e^{\alpha t}\cos(\beta t) \\ e^{\alpha t}\sin(\beta t) \end{bmatrix}, \quad r \neq r_{1,2} = \alpha \pm j\beta. \tag{4.11}$$

The nature of the roots can now be expanded to provide the solution set for differential equations of order 4. We identify the four roots as r_k, k=1, 2, 3, 4. If all roots are real and distinct, the solution vector is

$$\vec{Y}(t) = \begin{bmatrix} y_1(t) \\ y_2(t) \\ y_3(t) \\ y_4(t) \end{bmatrix} = \begin{bmatrix} e^{r_1t} \\ e^{r_2t} \\ e^{r_3t} \\ e^{r_4t} \end{bmatrix}. \tag{4.12}$$

When all four roots are real and equal, the solution vector becomes

$$\vec{Y}(t) = \begin{bmatrix} y_1(t) \\ y_2(t) \\ y_3(t) \\ y_4(t) \end{bmatrix} = \begin{bmatrix} e^{rt} \\ te^{rt} \\ t^2e^{rt} \\ t^3e^{rt} \end{bmatrix}, r = r_{1,2,3,4} \tag{4.13}$$

It is possible to proceed similarly when two of the roots are equal in pairs or one is distinct and three are equal. The other cases will be when two are distinct and the other two form a complex pair, and when two real roots are equal and the other pair is complex. The final case will correspond to the case of two pairs of equal complex conjugate sets.

$$\vec{Y}(t)=\begin{bmatrix} y_1(t) \\ y_2(t) \\ y_3(t) \\ y_4(t) \end{bmatrix}=\begin{bmatrix} e^{r_1 t} \\ e^{rt} \\ te^{rt} \\ t^2 e^{rt} \end{bmatrix},\ r_1;\ r=r_{2,3,4} \tag{4.14}$$

$$\vec{Y}(t)=\begin{bmatrix} y_1(t) \\ y_2(t) \\ y_3(t) \\ y_4(t) \end{bmatrix}=\begin{bmatrix} e^{r_1 t} \\ e^{r_2 t} \\ e^{rt} \\ te^{rt} \end{bmatrix},r_1,r_2;\quad r=r_{3,4} \tag{4.15}$$

$$\vec{Y}(t)=\begin{bmatrix} y_1(t) \\ y_2(t) \\ y_3(t) \\ y_4(t) \end{bmatrix}=\begin{bmatrix} e^{r_1 t} \\ te^{r_1 t} \\ e^{r_3 t} \\ te^{r_3 t} \end{bmatrix},r_1=r_{1,2},r_3=r_3,r_4 \tag{4.16}$$

$$\vec{Y}(t)=\begin{bmatrix} y_1(t) \\ y_2(t) \\ y_3(t) \\ y_4(t) \end{bmatrix}=\begin{bmatrix} e^{r_1 t} \\ e^{r_2 t} \\ e^{\alpha t}\cos(\beta t) \\ e^{\alpha t}\sin(\beta t) \end{bmatrix},\ real\ r_1,r_2\ and\quad r_{3,4}=\alpha\pm j\beta \tag{4.17}$$

$$\vec{Y}(t)=\begin{bmatrix} y_1(t) \\ y_2(t) \\ y_3(t) \\ y_4(t) \end{bmatrix}=\begin{bmatrix} e^{rt} \\ te^{rt} \\ e^{\alpha t}\cos(\beta t) \\ e^{\alpha t}\sin(\beta t) \end{bmatrix},\ real\ r=r_{1,2}\ and\quad r_{3,4}=\alpha\pm j\beta \tag{4.18}$$

$$\vec{Y}(t)=\begin{bmatrix} y_1(t) \\ y_2(t) \\ y_3(t) \\ y_4(t) \end{bmatrix}=\begin{bmatrix} e^{\alpha t}\cos(\beta t) \\ e^{\alpha t}\sin(\beta t) \\ te^{\alpha t}\cos(\beta t) \\ te^{\alpha t}\sin(\beta t) \end{bmatrix},r_{1,2}=r_{3,4}=\alpha\pm j\beta \tag{4.19a}$$

$$\vec{Y}(t)=\begin{bmatrix} y_1(t) \\ y_2(t) \\ y_3(t) \\ y_4(t) \end{bmatrix}=\begin{bmatrix} e^{\alpha_1 t}\cos(\beta_1 t) \\ e^{\alpha_1 t}\sin(\beta_1 t) \\ e^{\alpha_2 t}\cos(\beta_2 t) \\ e^{\alpha_2 t}\sin(\beta_2 t) \end{bmatrix},r_{1,2}=\alpha_1\pm j\beta_1,r_{3,4}=\alpha_2\pm j\beta_2 \tag{4.19b}$$

It can easily be seen that as the order of the differential equation increases, the possible grouping of the basic solutions becomes more diverse making it difficult to form a solution. A general approach can be stated as follows:

If the multiplicity of the roots is m ($m \le n$, m roots are equal to r) and if the roots are real, the solutions can be expressed as

$$e^{rt}, te^{rt}, t^2e^{rt}, \ldots, t^{m-1}e^{rt}. \tag{4.20}$$

The remaining roots, (n-m) will be distinct, each with a unique solution. When n>3, possibilities exist that conjugate complex pair of roots might also display multiplicity. The multiplicity of m results in 2m solutions as

$$\begin{aligned} &e^{\alpha t}\cos(\beta t), te^{\alpha t}\cos(\beta t), t^2e^{\alpha t}\cos(\beta t), \ldots, t^{m-1}e^{\alpha t}\cos(\beta t) \\ &e^{\alpha t}\sin(\beta t), te^{\alpha t}\sin(\beta t), t^2e^{\alpha t}\sin(\beta t), \ldots, t^{m-1}e^{\alpha t}\sin(\beta t) \end{aligned} \tag{4.21}$$

The remaining (n-2m) might be distinct in which case, there will be distinct solutions. If multiplicities exist among (n-2m) real roots, the solutions will follow the format in eqn. (4.20).

For reasons mentioned above, solutions for higher order differential equations with n>4 may be obtained easily using a method based on Laplace transforms discussed below and verified using the Runge-Kutta method.

4.3 Non-homogeneous Differential Equations and Particular Solutions (n<5)

Just as it was done for the case of second order non-homogeneous equations of constant coefficients, the solution can now be expressed as

$$y(t) = y_h(t) + y_p(t) = \sum_{k=1}^{n} c_k y_k(t) + y_p(t). \tag{4.22}$$

The particular solution resulting from the excitation force represented by g(t) in eqn. (4.1) is represented by $y_p(t)$.

While the particular solution may be obtained using the method of undetermined coefficients as it was done for the case of second order systems, the process of obtaining the solution is simplified and more formal through the use of the method of variation of parameters. This approach requires the use of the Wronskian associated with the nth order differential equation. Proceeding similarly as in the case of a second order differential equation, the Wronskian W becomes

$$W = \begin{bmatrix} y_1 & y_2 & \cdot & \cdot & y_n \\ y_1^{(1)} & y_2^{(1)} & \cdot & \cdot & y_n^{(1)} \\ \cdot & \cdot & \cdot & \cdot & \cdot \\ \cdot & \cdot & \cdot & \cdot & \cdot \\ y_1^{(n-1)} & y_2^{(n-1)} & \cdot & \cdot & y_n^{(n-1)} \end{bmatrix}. \tag{4.23}$$

The particular solution $y_p(t)$ is given by

$$y_p(t) = \left[y_1(t)y_1(t)\ldots y_n(t)\right]\int \vec{v}(t)dt. \tag{4.24}$$

In eqn. (4.24),

$$\vec{v}(t) = W^{-1}\begin{bmatrix} 0 \\ 0 \\ \cdot \\ \cdot \\ \frac{g(t)}{A_n} \end{bmatrix}. \tag{4.25}$$

In eqn. (4.25), W^{-1} is the inverse of the Wronskian and the matrix on the right hand side is of size [n x 1] with the first (n-1) elements being zeros. If initial conditions IC are given, the unknown coefficients $c_1, c_2, .., c_n$ in eqn. (4.22) can be determined,

$$IC = \left[y(0)\, y^{(1)}(0) \ldots y^{(n-1)}(0) \right] = \left[y_0\ y_1 \cdots y_{n-1} \right]. \tag{4.26}$$

Once the unknown coefficients are evaluated, the complete solution of the differential equation in eqn. (4.1) can be obtained.

4.4 Additional Methods of Obtaining the Solution and Verification

Two methods can be implemented to obtain the solution to the higher order differential equation. The first one is the method based on the Laplace transform and the other is the use of numerical techniques. The numerical techniques use ODE solvers in MATLAB® and this requires that the higher order differential equation be broken down into a set of n-first order differential equations. Details of both techniques are provided in the Appendix.

4.4.1 Laplace transform

As shown in Appendix B, the Laplace transform based approach is a simple and direct way of obtaining the solution to a set of differential equations or a higher order differential equation. The solution is obtained only when initial conditions are available.

Taking the Laplace transform of the differential equation in eqn. (4.1), it is possible to write

$$A(s) = \frac{y_0\left[A_n s^{n-1} + A_{n-1}s^{n-2} + \ldots + A_1 \right] + y_1\left[A_n s^{n-2} + A_{n-1}s^{n-3} + \ldots + A_2 \right] + \ldots + y_{n-1}A_n}{A_n s^n + A_{n-1}s^{n-1} + \ldots + A_1 s + A_0} \tag{4.27}$$

$$B(s) = \frac{G(s)}{A_n s^n + A_{n-1}s^{n-1} + \ldots + A_1 s + A_0}. \tag{4.28}$$

In eqn. (4.28) G(s) is the Laplace transform of g(t). If Y(s) is the Laplace transform of the solution y(t),

$$Y(s) = A(s) + B(s). \tag{4.29}$$

The solution y(t) is obtained by taking the inverse Laplace transform of eqn. (4.29).

4.4.2 Numerical techniques

If the initial conditions are available, it is also possible to obtain the solution numerically using the Runge-Kutta method described in Appendix A. The first step in the solution is the creation of n-first order differential equations. We define

$$\begin{aligned} Z_1 &= y(t) \\ Z_2 &= y^{(1)}(t) \\ Z_3 &= y^{(2)}(t) \\ &\dots \\ Z_n &= y^{(n-1)}(t) \end{aligned} \tag{4.30}$$

Using eqn. (4.30), it is possible to write n-first order equations as

$$\begin{aligned} Z_1^{(1)} &= Z_2 \\ Z_2^{(1)} &= Z_3 \\ Z_3^{(1)} &= Z_4 \\ &\dots \\ Z_{n-1}^{(1)} &= Z_n \\ Z_n^{(1)} &= -\frac{A_{n-1}}{A_n}Z_n - \frac{A_{n-2}}{A_n}Z_{n-1} - \dots - \frac{A_1}{A_n}Z_2 - \frac{A_0}{A_n}Z_1 + \frac{g(t)}{A_n} \end{aligned} \tag{4.31}$$

Equation (4.31) can be expressed in matrix form as

$$\begin{bmatrix} Z_1^{(1)} \\ Z_2^{(1)} \\ Z_3^{(1)} \\ \vdots \\ Z_{n-1}^{(1)} \\ Z_n^{(1)} \end{bmatrix} = \begin{bmatrix} 0 & 1 & 0 & \dots & 0 & 0 \\ 0 & 0 & 1 & \dots & 0 & 0 \\ 0 & 0 & 0 & \dots & 0 & 0 \\ . & . & . & \dots & . & . \\ 0 & 0 & 0 & \dots & 0 & 1 \\ -\frac{A_0}{A_n} & -\frac{A_1}{A_n} & -\frac{A_2}{A_n} & \dots & -\frac{A_{n-2}}{A_n} & -\frac{A_{n-1}}{A_n} \end{bmatrix} \begin{bmatrix} Z_1 \\ Z_2 \\ Z_3 \\ \vdots \\ Z_{n-1} \\ Z_n \end{bmatrix} + \begin{bmatrix} 0 \\ 0 \\ 0 \\ \vdots \\ 0 \\ \frac{g(t)}{A_n} \end{bmatrix} \tag{4.32}$$

Equation (4.32) can be simplified by defining the following:

$$a_{n-1} = -\frac{A_{n-1}}{A_n}, a_{n-2} = -\frac{A_{2-1}}{A_n}, \dots, a_0 = -\frac{A_0}{A_n} \tag{4.33}$$

$$h(t) = \frac{g(t)}{A_n} \tag{4.34}$$

$$\vec{Z}' = \begin{bmatrix} Z_1^{(1)} \\ Z_2^{(1)} \\ Z_3^{(1)} \\ \vdots \\ Z_{n-1}^{(1)} \\ Z_n^{(1)} \end{bmatrix} \tag{4.35}$$

$$A = \begin{bmatrix} 0 & 1 & 0 & \dots & 0 & 0 \\ 0 & 0 & 1 & \dots & 0 & 0 \\ 0 & 0 & 0 & \dots & 0 & 0 \\ . & . & . & \dots & . & . \\ 0 & 0 & 0 & \dots & 0 & 1 \\ -\frac{A_0}{A_n} & -\frac{A_1}{A_n} & -\frac{A_2}{A_n} & \dots & -\frac{A_{n-2}}{A_n} & -\frac{A_{n-1}}{A_n} \end{bmatrix} = \begin{bmatrix} 0 & 1 & 0 & \dots & 0 & 0 \\ 0 & 0 & 1 & \dots & 0 & 0 \\ 0 & 0 & 0 & \dots & 0 & 0 \\ \dots & & & & & \\ 0 & 0 & 0 & \dots & 0 & 1 \\ a_0 & a_1 & a_2 & \dots & a_{n-2} & a_{n-1} \end{bmatrix} \tag{4.36}$$

$$\vec{b}(t) = \begin{bmatrix} 0 \\ 0 \\ 0 \\ \vdots \\ 0 \\ h(t) \end{bmatrix} \tag{4.37}$$

The higher order differential equations can now be written in the form of a matrix equation as

$$\vec{Z}'(t) = A\vec{Z}(t) + \vec{b}(t). \tag{4.38}$$

It should be noted that numerical evaluation of the solution requires the availability of the initial conditions given in eqn. (4.26).

4.5 Higher Order Differential Equations (n>4)

Because of the possible existence of varied relationships among the roots when n>4, solutions are obtained easily using Laplace transforms as shown in Section 4.4.1. Results can be verified using the numerical techniques described in Section 4.4.2.

4.6 Examples

Example # 4.1 Consider a 3rd order differential equation,

$$y'''(t) + y'(t) = 4 + \sin(t). \tag{4.39}$$

The initial conditions are

$$\begin{aligned} y(0) &= 0 \\ y'(0) &= 1 \\ y''(0) &= -1 \end{aligned} \quad . \tag{4.40}$$

The characteristic equation associated with the differential eqn. (4.39) is

$$r^3 + r = 0. \tag{4.41}$$

The roots of the characteristic equation are (one real and a pair of complex ones)

$$\begin{aligned} &0 \\ &-j \\ &j \end{aligned} \,. \tag{4.42}$$

The solution set associated with the homogeneous differential equation is

$$\begin{aligned} y_1(t) &= t \\ y_2(t) &= \cos(t) \\ y_3(t) &= \sin(t) \end{aligned} \,. \tag{4.43}$$

The solution to the homogeneous differential equation is

$$y_h(t) = b_1 y_1(t) + b_2 y_2(t) + b_3 y_3(t) = b_1 t + b_2 \cos(t) + b_3 \sin(t). \tag{4.44}$$

Using the solution set in eqn. (4.43), the Wronskian W in eqn. (4.23) becomes

$$W = \begin{bmatrix} 1 & \cos(t) & \sin(t) \\ 0 & -\sin(t) & \cos(t) \\ 0 & -\cos(t) & -\sin(t) \end{bmatrix}. \tag{4.45}$$

The vector $\vec{v}(t)$ in eqn. (4.25) is

$$\vec{v}(t) = \begin{bmatrix} \sin(t) + 4 \\ -\cos(t)\left[\sin(t) + 4\right] \\ -\sin(t)\left[\sin(t) + 4\right] \end{bmatrix}. \tag{4.46}$$

The particular solution can be obtained using eqn. (4.24) and is

$$y_p(t) = 4t - \cos(t) - \frac{t}{2}\sin(t). \tag{4.47}$$

The complete solution is

$$y(t) = y_h(t) + y_p(t) = b_1 t + b_2 \cos(t) + b_3 \sin(t) + 4t - \cos(t) - \frac{t}{2}\sin(t). \tag{4.48}$$

Applying the initial conditions in eqn. (4.40) to the solution in eqn. (4.48), the constants b_1, b_2, b_3 are obtained as

$$b_1 = 0 \quad b_2 = 1 \quad b_3 = -3. \tag{4.49}$$

Substituting the values of the constants in eqn. (4.48), the solution to the 3rd order differential equation in eqn. (4.39) becomes

$$y(t) = 4t - 3\sin(t) - \frac{t}{2}\sin(t). \tag{4.50}$$

The Laplace transform of the solution can be obtained using eqn. (4.29) as

$$Y(s) = \frac{s^4 - s^3 + 5s^2 + 4}{\left(s^3 + s\right)^2}. \tag{4.51}$$

Taking inverse Laplace transforms, eqn. (4.51) leads to a solution matching the one in eqn. (4.50).

Example # 4.2 Consider another 3rd order differential equation,

$$y'''(t) - y''(t) - y'(t) + y(t) = t^2 + t. \tag{4.52}$$

The initial conditions are

$$\begin{aligned} y(0) &= -1 \\ y'(0) &= 0 \\ y''(0) &= 0 \end{aligned}. \tag{4.53}$$

The characteristic equation associated with the differential eqn. (4.39) is

$$r^3 - r^2 - r + 1 = 0. \tag{4.54}$$

The roots of the characteristic equation are

$$\begin{matrix} -1 \\ 1 \\ 1 \end{matrix}. \tag{4.55}$$

The roots are not distinct, with one set being equal to unity while the other is equal to –1 leading to the set associated with the homogeneous differential equation as

$$\begin{aligned} y_1(t) &= e^{-t} \\ y_2(t) &= e^{t} \\ y_3(t) &= te^{t} \end{aligned}. \tag{4.56}$$

The solution to the homogeneous differential equation is

$$y_h(t) = b_1 y_1(t) + b_2 y_2(t) + b_3 y_3(t) = b_1 e^{-t} + b_2 e^{t} + b_3 te^{t}. \tag{4.57}$$

Using the solution set in eqn. (4.56), the Wronskian W in eqn. (4.23) becomes

$$W = \begin{bmatrix} e^{-t} & e^{t} & te^{t} \\ -e^{-t} & e^{t} & e^{t}(t+1) \\ e^{-t} & e^{t} & e^{t}(t+2) \end{bmatrix}. \tag{4.58}$$

The vector $\vec{v}(t)$ in eqn. (4.25) is

$$\vec{v}(t) = \begin{bmatrix} \frac{t}{4}e^{t}(t+1) \\ -\frac{t}{4}e^{-t}(t+1)(2t+1) \\ \frac{t}{2}e^{-t}(t+1) \end{bmatrix}. \quad (4.59)$$

The particular solution can be obtained using eqn. (4.24) and is

$$y_p(t) = t^2 + 3t + 5. \quad (4.60)$$

The complete solution is

$$y(t) = y_h(t) + y_p(t) = b_1e^{-t} + b_2e^{t} + b_3te^{t} + t^2 + 3t + 5. \quad (4.61)$$

Applying the initial conditions in eqn. (4.53) to the solution in eqn. (4.61), the constants b_1, b_2, b_3 are obtained as

$$b_1 = -\frac{1}{2} \quad b_2 = -\frac{11}{2} \quad b_3 = 2. \quad (4.62)$$

Substituting the values of the constants in eqn. (4.62), the solution to the 3rd order differential equation in eqn. (4.61) becomes

$$y(t) = 3t - \frac{1}{2}e^{-t} - \frac{11}{2}e^{t} + 2te^{t} + t^2 + 5. \quad (4.63)$$

The Laplace transform of the solution can be obtained using eqn. (4.29) as

$$Y(s) = \frac{-s^5 + s^4 + s^3 + s + 2}{s^3(s-1)^2(s+1)}. \quad (4.64)$$

Taking inverse Laplace transform, eqn. (4.64) leads to a solution matching the one in eqn. (4.63).

Example # 4.3 Consider a 4th order differential equation,

$$y''''(t) - y(t) = e^{t} + e^{-t}. \quad (4.65)$$

The initial conditions are

$$\begin{aligned} y(0) &= 1 \\ y'(0) &= 1 \\ y''(0) &= -1 \\ y'''(0) &= -1 \end{aligned} \quad (4.66)$$

The characteristic equation associated with the differential eqn. (4.65) is

$$r^4 - 1 = 0. \quad (4.67)$$

The roots of the characteristic equation are

$$\begin{matrix} -1 \\ 1 \\ -j \\ j \end{matrix}. \tag{4.68}$$

The solution set associated with the homogeneous differential equation is

$$\begin{aligned} y_1(t) &= e^{-t} \\ y_2(t) &= e^{t} \\ y_3(t) &= \cos(t) \\ y_4(t) &= -\sin(t) \end{aligned}. \tag{4.69}$$

Note that use of –sin(t) or sin(t) in eqn. (4.69) will not change the result. The solution to the homogeneous differential equation is

$$y_h(t) = b_1 y_1(t) + b_2 y_2(t) + b_3 y_3(t) + b_4 y_4(t) = b_1 e^{-t} + b_2 e^{t} + b_3 \cos(t) - b_4 \sin(t). \tag{4.70}$$

Using the solution set in eqn. (4.69), the Wronskian W in eqn. (4.23) becomes

$$W = \begin{bmatrix} e^{-t} & e^{t} & \cos(t) & -\sin(t) \\ -e^{-t} & e^{t} & -\sin(t) & -\cos(t) \\ e^{-t} & e^{t} & -\cos(t) & \sin(t) \\ -e^{-t} & e^{t} & \sin(t) & \cos(t) \end{bmatrix}. \tag{4.71}$$

The vector $\vec{v}(t)$ in eqn. (4.25) is

$$\vec{v}(t) = \begin{bmatrix} -\frac{1}{4}\left(1+e^{2t}\right) \\ \frac{1}{4}\left(1+e^{2t}\right) \\ \sin(t)\cosh(t) \\ \cos(t)\cosh(t) \end{bmatrix}. \tag{4.72}$$

The particular solution can be obtained using eqn. (4.24) and it is

$$y_p(t) = -\frac{1}{8}\left(2t + 3e^{2t} - 2te^{2t} + 3\right). \tag{4.73}$$

The complete solution is

$$y(t) = b_1 e^{-t} + b_2 e^{t} + b_3 \cos(t) - b_4 \sin(t) - \frac{1}{8}\left(2t + 3e^{2t} - 2te^{2t} + 3\right). \tag{4.74}$$

Applying the initial conditions in eqn. (4.66) to the solution in eqn. (4.74) the constants b_1, b_2, b_3, b_4 are obtained as

$$b_1 = \frac{1}{8} \quad b_2 = \frac{1}{8} \quad b_3 = \frac{3}{2} \quad b_4 = -1. \tag{4.75}$$

Substituting the values of the constants in eqn. (4.74), the solution to the 4th order differential equation in eqn. (4.65) becomes

$$y(t) = \frac{3}{2}\cos(t) + \sin(t) - \frac{1}{2}\cosh(t) + \frac{1}{2}t\sinh(t). \tag{4.76}$$

The Laplace transform of the solution can be obtained using eqn. (4.29) as

$$Y(s) = \frac{s^5 + s^4 - 2s^3 - 2s^2 + 3s + 1}{(s^2-1)(s^4-1)}. \tag{4.77}$$

Taking the inverse Laplace transform, eqn. (4.77) leads to a solution matching the one in eqn. (4.76).

Example # 4.4 Consider another 4th order differential equation,

$$y''''(t) + y'(t) = te^{-t} + 4. \tag{4.78}$$

The initial conditions are

$$\begin{aligned} y(0) &= -1 \\ y'(0) &= 0 \\ y''(0) &= 0 \\ y'''(0) &= 1 \end{aligned}. \tag{4.79}$$

The characteristic equation associated with the differential eqn. (4.78) is

$$r^4 + r = 0. \tag{4.80}$$

The roots of the characteristic equation are

$$\begin{matrix} 0 \\ -1 \\ \dfrac{1-j\sqrt{3}}{2} \\ \dfrac{1+j\sqrt{3}}{2} \end{matrix}. \tag{4.81}$$

The solution set associated with the homogeneous differential equation is

$$\begin{aligned} y_1(t) &= 1 \\ y_2(t) &= e^{-t} \\ y_3(t) &= e^{\frac{t}{2}}\cos\left(\frac{\sqrt{3}}{2}t\right) \\ y_4(t) &= -e^{\frac{t}{2}}\sin\left(\frac{\sqrt{3}}{2}t\right) \end{aligned}. \tag{4.82}$$

The solution to the homogeneous differential equation is

$$y_h(t) = b_1 y_1(t) + b_2 y_2(t) + b_3 y_3(t) + b_4 y_4(t) = b_1 + b_2 e^{-t} + b_3 e^{\frac{t}{2}} \cos\left(\frac{\sqrt{3}}{2}t\right) - b_4 e^{\frac{t}{2}} \sin\left(\frac{\sqrt{3}}{2}t\right). \tag{4.83}$$

Using the solution set in eqn. (4.82), the Wronskian W in eqn. (4.23) becomes

$$W = \begin{bmatrix} 1 & e^{-t} & e^{\frac{t}{2}} \cos\left(\frac{\sqrt{3}}{2}t\right) & -e^{\frac{t}{2}} \sin\left(\frac{\sqrt{3}}{2}t\right) \\ 0 & -e^{-t} & e^{\frac{t}{2}} \cos\left(\frac{\pi}{3} + \frac{\sqrt{3}}{2}t\right) & -e^{\frac{t}{2}} \sin\left(\frac{\pi}{3} + \frac{\sqrt{3}}{2}t\right) \\ 0 & e^{-t} & -e^{\frac{t}{2}} \sin\left(\frac{\pi}{6} + \frac{\sqrt{3}}{2}t\right) & -e^{\frac{t}{2}} \cos\left(\frac{\pi}{6} + \frac{\sqrt{3}}{2}t\right) \\ 0 & -e^{-t} & -e^{\frac{t}{2}} \cos\left(\frac{\sqrt{3}}{2}t\right) & e^{\frac{t}{2}} \sin\left(\frac{\sqrt{3}}{2}t\right) \end{bmatrix}. \tag{4.84}$$

The vector $\vec{v}(t)$ in eqn. (4.25) is

$$\vec{v}(t) = \begin{bmatrix} 4 + te^{-t} \\ -\frac{1}{3}\left(t + 4e^{t}\right) \\ -\frac{2}{3} e^{-\frac{1}{2}t} \cos\left(\frac{\sqrt{3}}{2}t\right)\left(t + 4e^{t}\right) \\ \frac{2}{3} e^{-\frac{1}{2}t} \sin\left(\frac{\sqrt{3}}{2}t\right)\left(t + 4e^{t}\right) \end{bmatrix}. \tag{4.85}$$

The particular solution can be obtained using eqn. (4.24) and is

$$y_p(t) = -\frac{e^{-t}}{18}\left(12t - 72te^{2t} + 3t^2 + 16\right). \tag{4.86}$$

The complete solution is

$$y(t) = b_1 + b_2 e^{-t} + b_3 e^{\frac{t}{2}} \cos\left(\frac{\sqrt{3}}{2}t\right) - b_4 e^{\frac{t}{2}} \sin\left(\frac{\sqrt{3}}{2}t\right) - \frac{e^{-t}}{18}\left(12t - 72te^{2t} + 3t^2 + 16\right). \tag{4.87}$$

Applying the initial conditions in eqn. (4.79) to the solution in eqn. (4.87), the constants b_1, b_2, b_3, b_4 are obtained as

$$b_1 = 1 \quad b_2 = 1 \quad b_3 = -\frac{19}{9} \quad b_4 = \frac{13\sqrt{3}}{9}. \tag{4.88}$$

Substituting the values of the constants in eqn. (4.88), the solution to the 4th order differential equation in eqn. (4.87) becomes

$$y(t) = 1 + 4t + \frac{1}{9}e^{-t} - \frac{2}{3}te^{-t} - \frac{1}{6}t^2e^{-t} - \frac{19}{9}e^{\frac{t}{2}}\cos\left(\frac{\sqrt{3}}{2}t\right) - \frac{13\sqrt{3}}{9}e^{\frac{t}{2}}\sin\left(\frac{\sqrt{3}}{2}t\right). \quad (4.89)$$

The Laplace transform of the solution can be obtained using eqn. (4.29) as

$$Y(s) = \frac{\frac{4}{s} - s^3 + \left(\frac{1}{1+s}\right)^2}{s\left(s^3+1\right)}. \quad (4.90)$$

Taking the inverse Laplace transform, eqn. (4.90) leads to a solution matching the one in eqn. (4.89).

When analytical approaches based on the roots of the characteristic equation become cumbersome (as the order of the differential equation increases), the convenient option is to rely on the Laplace transform based approach. In all cases, additional verification is possible using numerical techniques based on Runge-Kutta methods as described in Appendix A and in Section 4.4.2. The MATLAB script for the case of a 3rd order differential equation and results are given first.

Example # 4.5 Consider the case of the following 3rd order differential equation,

$$y'''(t) + y''(t) - 2y'(t) = t + e^t, \quad y(0) = 1, y'(0) = 1, y''(0) = -1. \quad (4.91)$$

The MATLAB script created for solving the differential equation and the results are shown.

```
function higherorder_numerical_example
% june 2017 solution of a higher order differential equation using Laplace transform
% and verification using Runge-Kutta method P M Shankar
% y'''(t)+y''(t)-2y'(t)=t+exp(t), y(0)=1, y'(0)=1,y''(0)=-1
syms t s
A3=1;A2=1;A1=-2;A0=0;
y0=1;y1=1;y2=-1; % initial conditions
gt=t+exp(t); % forcing functions
Dr=(A3*s^3+A2*s^2+A1*s+A0); % Denominator of the Laplace transform
As=(y0*(A3*s*s+A2*s+A1)+A3*y2+y1*(A2+A3*s))/Dr;
Bs=laplace(gt,t,s)/Dr; % Laplace transform of the forcing function
yL=ilaplace(As+Bs,s,t); % inverse Laplace transform to get the solution
simplify(yL,'steps',100); % analytical solution using Laplace
fy=MATLABFunction(yL);
syms z(t) % use z as a placement or dummay variable
D3=diff(z,3);
D2=diff(z,2);
D1=diff(z,1);
f=[A3*D3+A2*D2+A1*D1+A0*z==t+exp(t)];
V = odeToVectorField(f);% break higher order DE into 3 first order ones
F = MATLABFunction(V,'vars', {'t','Y'});% create inline Function for ODE
tspan=[0 5];
[T,yode]=ode45(F,tspan,[1;1;-1]);
% numerical solution; only use the first column of the solution yode
```

```
figure,plot(T,fy(T),'r-',T,yode(:,1),'k*')
xlabel('time t'),ylabel('solution')
legend('analyt','numerical')
syms y(t)
yy=[y(t)==yL]; % create the expression for the title
title(['$' latex(yy) '$'],'interpreter','latex','fontsize',14,'color','b')

end
```

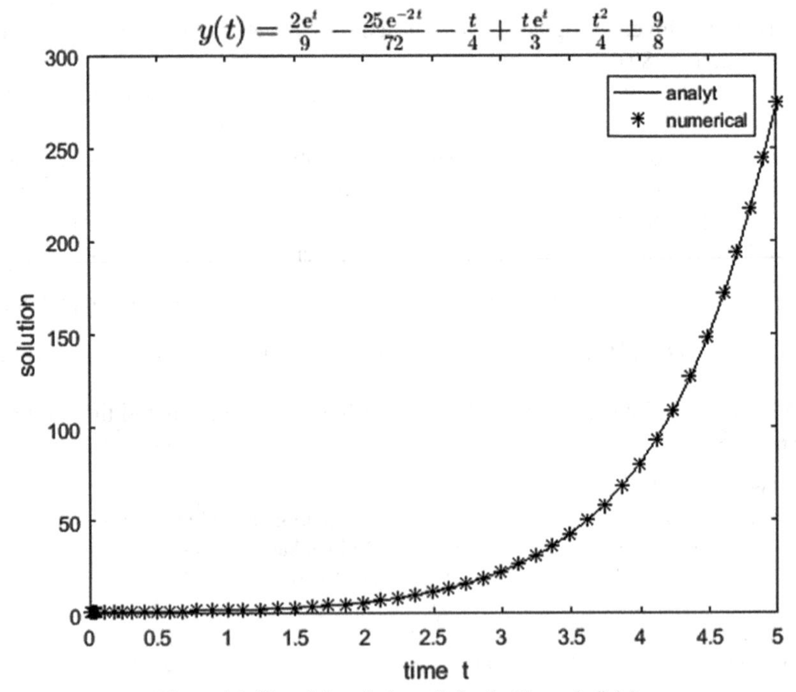

Figure 4.1 Plot of the solution y(t) for the Example # 4.5.

The approach based on combining Laplace transform and Runge-Kutta methods can also be used to solve differential equations in orders higher than the 4th. As the order increases, the increasing number of roots leads to difficulties in forming the solutions set for the corresponding homogeneous differential equation. The Laplace transform offers a simple means to obtain an analytical expression for the solution and the verification of the solution can be accomplished through the use of the Runge-Kutta method implemented in MATLAB.

Example # 4.6 Consider a 5th order differential equation

$$y'''''(t)+y'''(t)+y''(t)+y(t)=e^t, \quad y(0)=1, y'(0)=y''(0)=y'''(0)=0, y''''(0)=1 \tag{4.92}$$

The characteristic equation associated with differential equation is

$$r^5+r^3+r^2+r=0. \tag{4.93}$$

The roots of the characteristic equation are

$$\begin{matrix} -1 \\ -j \\ j \\ \frac{1}{2}\left(1-j\sqrt{3}\right) \\ \frac{1}{2}\left(1+j\sqrt{3}\right) \end{matrix}. \tag{4.94}$$

As suggested, the simpler way to obtain the analytical solution is to use the Laplace transform based approach. Taking the Laplace transform of the differential equation in eqn. (4.92), the expression for the Laplace transform of the solution becomes

$$Y(s) = \frac{s^3}{s^4 - 1}. \tag{4.95}$$

Taking the inverse Laplace transform of eqn. (4.95), the solution to the differential equation becomes

$$y(t) = \frac{\cos(t) + \cosh(t)}{2}. \tag{4.96}$$

The results of verifying the solution are shown in Figure 4.2.

Example # 4.7 Consider the case of a 6th order differential equation,

$$y''''''(t) + y'''''(t) = te^{-t},\ y(0) = 1, y'(0) = y''(0) = y'''(0) = 0, y''''(0) = 1, y'''''(0) = 0. \tag{4.97}$$

The characteristic equation and its roots are

$$r^6 + r^5 = 0. \tag{4.98}$$

$$\begin{matrix} 0 \\ 0 \\ 0 \\ 0 \\ 0 \\ -1 \end{matrix}. \tag{4.99}$$

Taking the Laplace transform of the differential equation in eqn. (4.97), the Laplace transform of the solution is

$$Y(s) = \frac{s + \frac{1}{(s+1)^2} + s^4 + s^5 + 1}{s^5(s+1)}. \tag{4.100}$$

Taking the inverse Laplace transform of eqn. (4.100) leads to the solution

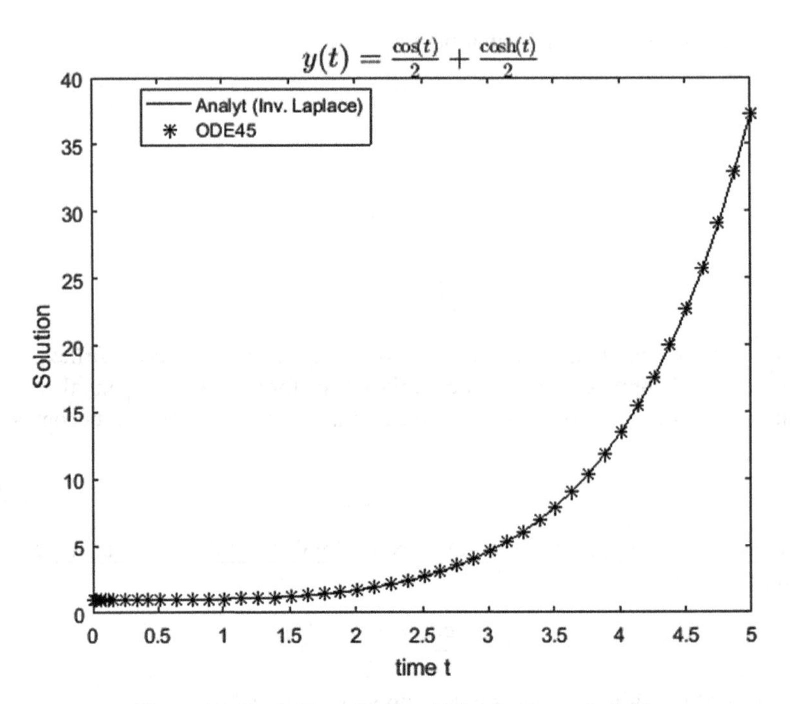

Figure 4.2 Plot of the solution y(t) for the Example # 4.6.

$$y(t) = 16 - 10t + 3t^2 - \frac{t^3}{2} + \frac{t^4}{12} - 15e^{-t} - 5te^{-t} - \frac{1}{2}t^2e^{-t}. \tag{4.101}$$

The verification is provided in Figure 4.3.

Example 4.8 Consider the case of a 7th order differential equation

$$\begin{aligned} &y'''''''(t) - y'''''(t) + y''''(t) - y''(t) = t, \\ &\quad y(0) = 1, y'(0) = y''(0) = y'''(0) = y''''(0) = y'''''(0) = 0, y''''''(0) = 1 \end{aligned}. \tag{4.102}$$

The characteristic equation and the roots of the characteristic equation are

$$r^7 - r^5 + r^4 - r^2 = 0. \tag{4.103}$$

$$\begin{matrix} 0 \\ 0 \\ -1 \\ -1 \\ 1 \\ \frac{1}{2}\left(1 - j\sqrt{3}\right) \\ \frac{1}{2}\left(1 + j\sqrt{3}\right) \end{matrix}. \tag{4.104}$$

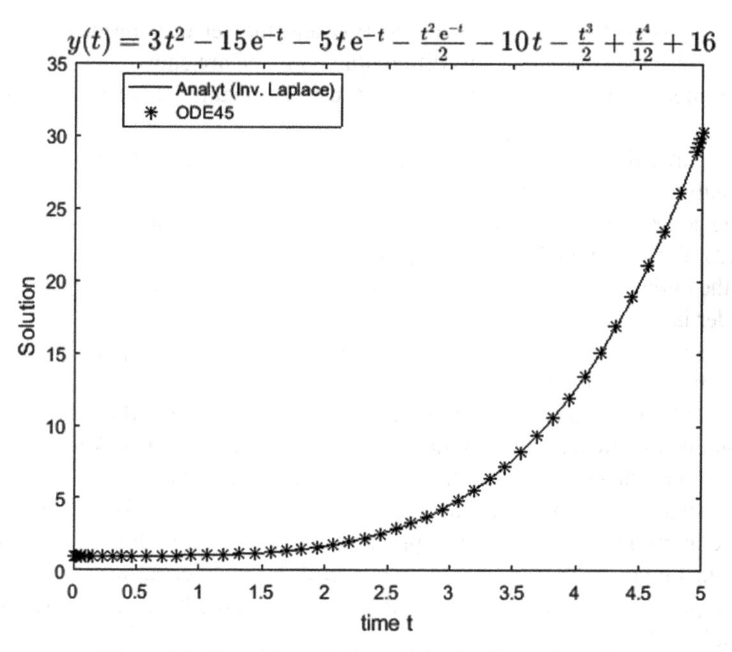

Figure 4.3 Plot of the solution y(t) for the Example # 4.7.

Taking the Laplace transform of the differential equation in eqn. (4.102), the Laplace transform of the solution is

$$Y(s) = \frac{s^8 - s^6 + s^5 - s^3 + s^2 + 1}{s^4(s-1)(s+1)^2(s^2 - s + 1)}. \tag{4.105}$$

Taking the inverse transform of eqn. (4.105), the solution to the differential equation becomes

$$y(t) = 2 - 2t + \frac{1}{2}e^t - \frac{3}{2}e^{-t} - \frac{1}{3}te^{-t} - \frac{1}{6}t^3 + \frac{2\sqrt{3}}{6}e^{\frac{t}{2}} \sin\left(\frac{\sqrt{3}}{2}t\right). \tag{4.106}$$

The verification is given in Figure 4.4.

In keeping up with the theme of this manuscript, theoretical aspects are provided with every example given. Of specific mention is the fact that explicit values of the fundamental solution set based on the roots are provided for every example, for all types of relationships among the roots as mentioned in Section 4.4.2.

The MATLAB script generates the following displays.

1. Based on the vector A representing the coefficients, a determination is made regarding the order of the differential equation and the general form of the equation, the characteristic equation, roots and the form of the general solution is provided. While the form of the general solution is simple for order 2, with order 3 and order 4, more possibilities exist for the general solution based on the relationships among the roots of the characteristic function. All these possibilities are automatically shown.

2. The roots of the characteristic equation and the set of solutions forming the solution to the homogeneous differential equation are shown next.
3. The method of variation of parameters used to obtain the particular solution is shown.
4. The particular solution is then obtained, tested by substitution in the differential equation and the validation is shown.
5. The general solution incorporating the particular one is shown and the analysis ends if initial conditions are not provided.
6. If the initial conditions are provided, the applicable theory for the specific highest order is shown.
7. The Laplace transform of the solution y(t) is shown along with the solution to the differential equation.
8. The solution is then verified by applying the initial conditions to the general solution obtained using the roots displaying the values of the unknown constants as well as the complete solution obtained from the roots.
9. The solution obtained using dsolve(.) is also shown for the sake of completeness.
10. The numerical computational approach is shown next with the decomposition of the higher order into several first order differential equations.
11. The plots of the solutions obtained using various methods are shown as in the final display.

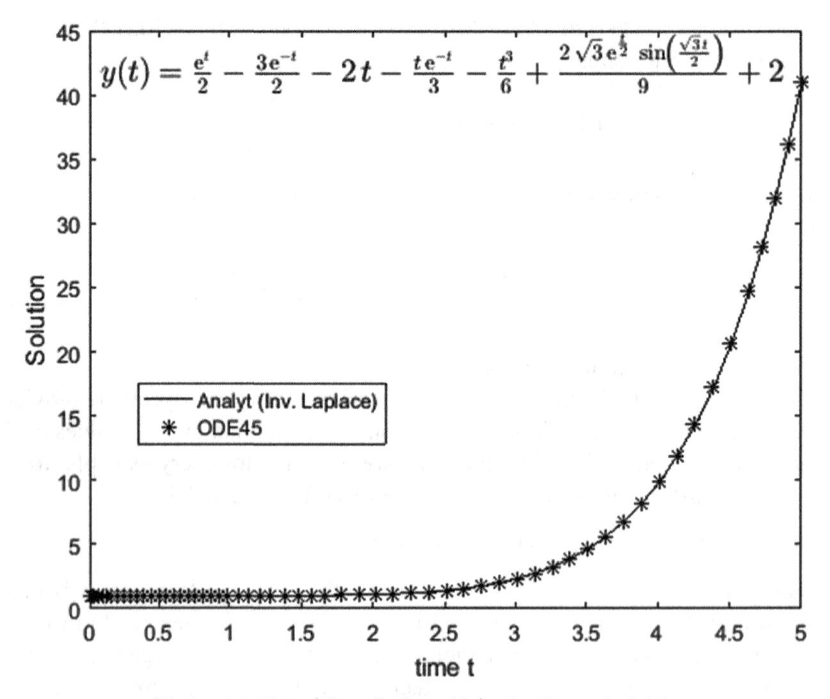

Figure 4.4 Plot of the solution y(t) for the Example # 4.8.

Example # 4.9

```
A=[1 0 1 0];IC=[0,1,-1];gt='cos(t)+sin(t)';
```

3rd Order Homogeneous Differential Equations with Constant Coefficients

Differential Equation $A_3y'''(t) + A_2y''(t) + A_1y'(t) + A_0y(t) = 0$

Characteristic Equation/Roots $A_3r^3 + A_2r^2 + A_1r + A_0 = 0 \Rightarrow r_1,\ r_2,\ r_3,$

Solution set and solution y(t) $Y = [y_1(t), y_2(t), y_3(t)] \Rightarrow y(t) = \sum_{k=1}^{k=3} b_k y_k(t)$

$$Y(t) \Rightarrow \begin{cases} e^{r_1 t} & e^{r_2 t} & e^{r_3 t} & r_{1,2,3}\ real \\ e^{rt} & te^{rt} & e^{r_3 t} & r_{1,2} = r \\ e^{rt} & te^{rt} & t^2e^{rt} & r_{1,2,3} = r \\ e^{r_1 t} & e^{\alpha t}cos(\beta t) & e^{\alpha t}sin(\beta t) & r_{2,3} = \alpha \pm j\beta \end{cases}$$

Set of Solutions: Homogeneous case

Particular and complete solutions
Higher order differential equations

Differential Equation and Initial Conditions

$$y'''(t) + y'(t) = \cos(t) + \sin(t)$$
$$(\ y(0) = 0 \quad y'(0) = 1 \quad y''(0) = -1\ )$$

Characteristic equation: $r^3 + r = 0$

Roots of the Characteristic equation

$$\begin{pmatrix} 0 \\ -\mathrm{i} \\ \mathrm{i} \end{pmatrix}$$

Homogeneous solutions set

$$y_1(t) = 1$$
$$y_2(t) = \cos(t)$$
$$y_3(t) = \sin(t)$$

Roots $\Rightarrow r_{\pm} = \alpha \pm j\beta, r_3$

$y(t) = e^{\alpha t}cos(\beta t),\ y(t) = e^{\alpha t}sin(\beta t)$ Solutions Set ⇑

$y(t) = e^{r_3 t}$

Particular Solution using the Method of Variation of Parameters

Forcing Function $g(t) = \cos(t) + \sin(t)$

WRONSKIAN $$W = \begin{bmatrix} y_1(t) & y_2(t) & y_3(t) \\ y_1'(t) & y_2'(t) & y_3'(t) \\ y_1''(t) & y_2''(t) & y_3''(t) \end{bmatrix}$$

$$\vec{v}(t) = [W]^{-1} \begin{bmatrix} 0 \\ 0 \\ \frac{g(t)}{A_3} \end{bmatrix}$$, A_3 is the coefficient of $y'''(t)$

Particular Solution $$y_p(t) = [y_1(t)\, y_2(t)\, y_3(t)] \int \vec{v}(t)\, dt$$

Particular Solution using Variation of Parameters

$$y_p(t) = \frac{3\sin(t)}{4} - \frac{3\cos(t)}{4} - \frac{t\cos(t)}{2} - \frac{t\sin(t)}{2}$$

Test the validity by substitution in Differential EQN

VALID/PASS

Details of Wronskian

Wronskian W ⇒

$$\begin{pmatrix} 1 & \cos(t) & \sin(t) \end{pmatrix}$$
$$\begin{pmatrix} 0 & -\sin(t) & \cos(t) \end{pmatrix}$$
$$\begin{pmatrix} 0 & -\cos(t) & -\sin(t) \end{pmatrix}$$

$$\sqrt{2}\sin\left(t + \frac{\pi}{4}\right)$$

$\vec{v}(t) \Rightarrow$

$$-\frac{\sqrt{2}\sin\left(2t+\frac{\pi}{4}\right)}{2} - \frac{1}{2}$$

$$\frac{\sqrt{2}\cos\left(2t+\frac{\pi}{4}\right)}{2} - \frac{1}{2}$$

Inverse Laplace based solution: Theory and Results

Differential Equation and Initial Conditions (General Case)

$$A_3\, y'''(t) + A_2\, y''(t) + A_1\, y'(t) + A_0\, y = g(t)$$

$$[y(0) = y_0, y'(0) = y_1, y''(0) = y_2]$$

Laplace Transform Y(s)=A(s)+B(s)

$$A(s) = \frac{y_0\left(A_3\, s^2 + A_2\, s + A_1\right) + A_3\, y_2 + y_1\,(A_2 + A_3\, s)}{A_3\, s^3 + A_2\, s^2 + A_1\, s + A_0}$$

$$B(s) = \frac{G(s)}{A_3\, s^3 + A_2\, s^2 + A_1\, s + A_0}$$

Differential Equation and Initial Conditions (given)

$$y'''(t) + y'(t) = \cos(t) + \sin(t)$$

$$(\ y(0) = 0 \quad y'(0) = 1 \quad y''(0) = -1\ )$$

Laplace Transform Y(s) of y(t)

$$Y_L(s) = \frac{s^2 - s + 2}{(s^2+1)^2}$$

Solution y(t) using inverse Laplace

$$y(t) = \frac{3\sin(t)}{2} - \frac{t\cos(t)}{2} - \frac{t\sin(t)}{2}$$

Complete Solution and Validation

Solution to the Homogeneous part using roots (unknown constants, $b_1, b_2, \ldots$)

$$y_h(t) = b_1\, y_1(t) + b_2\, y_2(t) + b_3\, y_3(t)$$

Complete solution $y(t)=y_h(t)+y_p(t)$

$$y(t) = \frac{3\sin(t)}{4} - \frac{3\cos(t)}{4} - \frac{t\cos(t)}{2} + b_1\, y_1(t) + b_2\, y_2(t) + b_3\, y_3(t) - \frac{t\sin(t)}{2}$$

Obtain unknown constants (b_1 b_2 b_3) by applying the initial conditions

$$(b_1\ b_2\ b_3) \Rightarrow \qquad \left(0 \quad \tfrac{3}{4} \quad \tfrac{3}{4}\right)$$

Complete solution using roots (replace unknown constants)

$$y(t) = \frac{3\sin(t)}{2} - \frac{t\cos(t)}{2} - \frac{t\sin(t)}{2}$$

Complete solution using inverse Laplace transforms

$$y(t) = \frac{3\sin(t)}{2} - \frac{t\cos(t)}{2} - \frac{t\sin(t)}{2}$$

Complete solution using dsolve(.) in Matlab

$$y(t) = \frac{3\sin(t)}{2} - \frac{t\cos(t)}{2} - \frac{t\sin(t)}{2}$$

ODE45 and comparison

Differential Equation and Initial Conditions

$$y'''(t) + y'(t) = \cos(t) + \sin(t)$$
$$(\ y(0) = 0 \quad y'(0) = 1 \quad y''(0) = -1\ )$$

Corresponding first order differential equations for ODE [$Y_1 \equiv y(t)$] vector

$$\begin{pmatrix} Y_1'(t) = Y_2 \\ Y_2'(t) = Y_3 \\ Y_3'(t) = \cos(t) + \sin(t) - Y_2 \end{pmatrix}$$

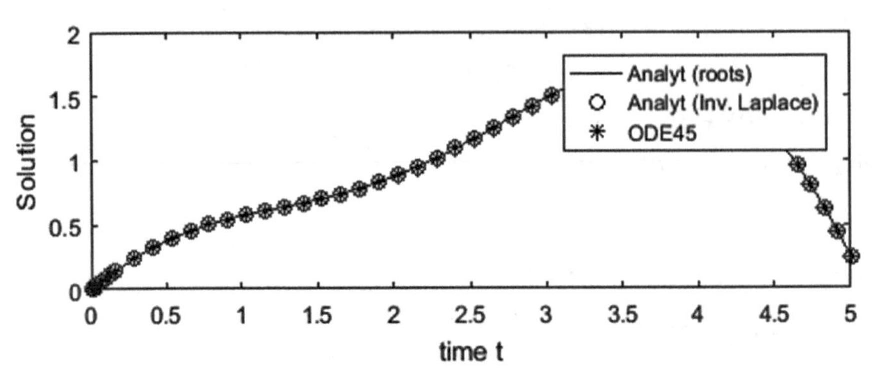

Example # 4.10

```
A=[1 -1 -1 1];IC=[-1,0,1 ];gt='exp(t)-t*exp(-t)';
```

3rd Order Homogeneous Differential Equations with Constant Coefficients

Differential Equation $A_3y'''(t) + A_2y''(t) + A_1y'(t) + A_0y(t) = 0$

Characteristic Equation/Roots $A_3r^3 + A_2r^2 + A_1r + A_0 = 0 \Rightarrow r_1,\ r_2,\ r_3,$

Solution set and solution y(t) $Y = [y_1(t), y_2(t), y_3(t)] \Rightarrow y(t) = \sum_{k=1}^{k=3} b_k y_k(t)$

$$Y(t) \Rightarrow \begin{cases} e^{r_1t} & e^{r_2t} & e^{r_3t} & r_{1,2,3}\ real \\ e^{rt} & te^{rt} & e^{r_3t} & r_{1,2} = r \\ e^{rt} & te^{rt} & t^2e^{rt} & r_{1,2,3} = r \\ e^{r_1t} & e^{\alpha t}cos(\beta t) & e^{\alpha t}sin(\beta t) & r_{2,3} = \alpha \pm j\beta \end{cases}$$

Set of Solutions: Homogeneous case

Particular and complete solutions
Higher order differential equations

Differential Equation and Initial Conditions

$$y'''(t) - y''(t) - y'(t) + y(t) = e^t - t\,e^{-t}$$

$$(\ y(0) = -1 \quad y'(0) = 0 \quad y''(0) = 1\ )$$

Characteristic equation: $r^3 - r^2 - r + 1 = 0$

Roots of the Characteristic equation

$$\begin{pmatrix} -1 \\ 1 \\ 1 \end{pmatrix}$$

Homogeneous solutions set

$$y_1(t) = e^{-t}$$

$$y_2(t) = e^t$$

$$y_3(t) = t\,e^t$$

Roots $\Rightarrow r, r(equal), and\ p$

$y(t) = e^{rt}, y(t) = te^{rt}$

$y(t) = e^{pt}$

Solutions Set ⇑

Particular Solution using the Method of Variation of Parameters

Forcing Function $g(t) = e^t - t\,e^{-t}$

WRONSKIAN $W = \begin{bmatrix} y_1(t) & y_2(t) & y_3(t) \\ y_1'(t) & y_2'(t) & y_3'(t) \\ y_1''(t) & y_2''(t) & y_3''(t) \end{bmatrix}$

$$\vec{v}(t) = [W]^{-1} \begin{bmatrix} 0 \\ 0 \\ \frac{g(t)}{A_3} \end{bmatrix}, \text{ A}_3 \text{ is the coefficient of } y'''(t)$$

Particular Solution $y_p(t) = [y_1(t)\, y_2(t)\, y_3(t)] \int \vec{v}(t)\, dt$

Particular Solution using Variation of Parameters

$$y_p(t) = -\frac{e^{-t}\left(4t - 2e^{2t} + 4te^{2t} - 4t^2e^{2t} + 2t^2 + 3\right)}{16}$$

Test the validity by substitution in Differential EQN

VALID/PASS

Details of Wronskian

Wronskian W ⇒
$$\begin{pmatrix} e^{-t} & e^{t} & t\,e^{t} \\ -e^{-t} & e^{t} & e^{t}\,(t+1) \\ e^{-t} & e^{t} & e^{t}\,(t+2) \end{pmatrix}$$

$\vec{v}(t) \Rightarrow$
$$\frac{e^{2t}}{4} - \frac{t}{4}$$
$$\frac{(2t+1)\left(t\,e^{-2t}-1\right)}{4}$$
$$\frac{1}{2} - \frac{t\,e^{-2t}}{2}$$

Inverse Laplace based solution: Theory and Results

Differential Equation and Initial Conditions (General Case)

$$A_3\,y'''(t) + A_2\,y''(t) + A_1\,y'(t) + A_0\,y = g(t)$$
$$[y(0) = y_0, y'(0) = y_1, y''(0) = y_2]$$

Laplace Transform Y(s)=A(s)+B(s)

$$A(s) = \frac{y_0\left(A_3\,s^2+A_2\,s+A_1\right)+A_3\,y_2+y_1\,(A_2+A_3\,s)}{A_3\,s^3+A_2\,s^2+A_1\,s+A_0}$$
$$B(s) = \frac{G(s)}{A_3\,s^3+A_2\,s^2+A_1\,s+A_0}$$

Differential Equation and Initial Conditions (given)

$$y'''(t) - y''(t) - y'(t) + y(t) = e^{t} - t\,e^{-t}$$
$$(\ y(0) = -1 \quad y'(0) = 0 \quad y''(0) = 1\ )$$

Laplace Transform Y(s) of y(t)

$$Y_L(s) = \frac{s\left(-s^4+4\,s^2+3\,s-2\right)}{(s^2-1)^3}$$

Solution y(t) using inverse Laplace

$$y(t) = -\frac{e^{-t}\left(4t+11\,e^{2t}-10\,t\,e^{2t}-4t^2\,e^{2t}+2\,t^2+5\right)}{16}$$

Complete Solution and Validation

Solution to the Homogeneous part using roots (unknown constants, b_1,b_2,...)

$$y_h(t) = b_1\, y_1(t) + b_2\, y_2(t) + b_3\, y_3(t)$$

Complete solution $y(t)=y_h(t)+y_p(t)$

$$y(t) = b_1\, y_1(t) - \frac{e^{-t}\left(4t - 2e^{2t} + 4te^{2t} - 4t^2e^{2t} + 2t^2 + 3\right)}{16} + b_2\, y_2(t) + b_3\, y_3(t)$$

Obtain unknown constants (b_1 b_2 b_3) by applying the initial conditions

$$(b_1\ b_2\ b_3) \Rightarrow \quad \left(\, -\tfrac{1}{8} \quad -\tfrac{13}{16} \quad \tfrac{7}{8}\, \right)$$

Complete solution using roots (replace unknown constants)

$$y(t) = -\frac{e^{-t}\left(4t + 11e^{2t} - 10te^{2t} - 4t^2e^{2t} + 2t^2 + 5\right)}{16}$$

Complete solution using inverse Laplace transforms

$$y(t) = -\frac{e^{-t}\left(4t + 11e^{2t} - 10te^{2t} - 4t^2e^{2t} + 2t^2 + 5\right)}{16}$$

Complete solution using dsolve(.) in Matlab

$$y(t) = -\frac{e^{-t}\left(4t + 11e^{2t} - 10te^{2t} - 4t^2e^{2t} + 2t^2 + 5\right)}{16}$$

ODE45 and comparison

Differential Equation and Initial Conditions

$$y'''(t) - y''(t) - y'(t) + y(t) = e^t - t\,e^{-t}$$
$$\left(\ y(0) = -1 \quad y'(0) = 0 \quad y''(0) = 1\ \right)$$

Corresponding first order differential equations for ODE [$Y_1 \equiv y(t)$] vector

$$\begin{pmatrix} Y_1'(t) = Y_2 \\ Y_2'(t) = Y_3 \\ Y_3'(t) = e^t - t\,e^{-t} - Y_1 + Y_2 + Y_3 \end{pmatrix}$$

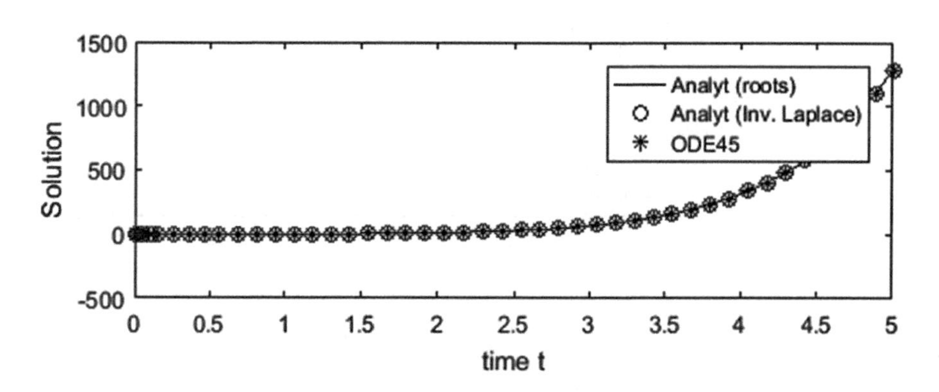

Example # 4.11

```
A=[1 0 0 0 -1];IC=[1,-1,0,-1 ];gt='t+cos(t)';
```

4th Order Homogeneous Differential Equations with Constant Coefficients

Differential Equation $A_4 y''''(t) + A_3 y'''(t) + A_2 y''(t) + A_1 y'(t) + A_0 y(t) = 0$

Characteristic Equation/Roots $A_4 r^4 + A_3 r^3 + A_2 r^2 + A_1 r + A_0 = 0 \Rightarrow r_1,\ r_2,\ r_3,\ r_4$

Solution set and solution y(t) $Y = [y_1(t), y_2(t), y_3(t), y_4(t)] \Rightarrow y(t) = \sum_{k=1}^{k=4} C_k y_k(t)$

$$Y(t) \Rightarrow \begin{cases} e^{r_1 t} & e^{r_2 t} & e^{r_3 t} & e^{r_4 t} & r_{1,2,3,4}\ real \\ e^{rt} & te^{rt} & e^{r_3 t} & e^{r_4 t} & r_{1,2} = r\ real \\ e^{rt} & te^{rt} & t^2 e^{rt} & e^{r_4 t} & r_{1,2,3} = r\ real \\ e^{rt} & te^{rt} & t^2 e^{rt} & t^3 e^{rt} & r_{1,2,3,4} = r\ real \\ e^{r_1 t} & te^{r_1 t} & e^{r_2 t} & te^{r_3 t} & r_{1,2} = r_1, r_{3,4} = r_2\ real \\ e^{r_1 t} & e^{r_2 t} & e^{\alpha t} cos(\beta t) & e^{\alpha t} sin(\beta t) & r_{3,4} = \alpha \pm j\beta \\ e^{rt} & te^{rt} & e^{\alpha t} cos(\beta t) & e^{\alpha t} sin(\beta t) & r_{1,2} = r,\ r_{3,4} = \alpha \pm j\beta \\ e^{\alpha_1 t} cos(\beta_1 t) & e^{\alpha_1} sin(\beta_1 t) & e^{\alpha_2 t} cos(\beta_2 t) & e^{\alpha_2 t} sin(\beta_2 t) & \begin{bmatrix} r_{1,2} = \alpha_1 \pm j\beta_1 \\ r_{3,4} = \alpha_2 \pm j\beta_2 \end{bmatrix} \\ e^{\alpha t} cos(\beta t) & e^{\alpha} sin(\beta t) & te^{\alpha t} cos(\beta t) & te^{\alpha t} sin(\beta t) & r_{1,2} = r_{3,4} = \alpha \pm j\beta \end{cases}$$

Set of Solutions: Homogeneous case

Particular and complete solutions
Higher order differential equations

Differential Equation and Initial Conditions

$$y''''(t) - y(t) = t + \cos(t)$$

$$(\ y(0) = 1 \quad y'(0) = -1 \quad y''(0) = 0 \quad y'''(0) = -1\ )$$

Characteristic equation: $r^4 - 1 = 0$

Roots of the Characteristic equation

$$\begin{pmatrix} -1 \\ 1 \\ -\mathrm{i} \\ \mathrm{i} \end{pmatrix}$$

Homogeneous solutions set

$y_1(t) = \mathrm{e}^{-t}$

$y_2(t) = \mathrm{e}^{t}$

$y_3(t) = \cos(t)$

$y_4(t) = -\sin(t)$

Roots $\Rightarrow r_1 \neq r_2,\ r_{3,4} = \alpha \pm j\beta$

$y_k(t) = e^{rt}, k = 1, 2$

$y_3(t) = e^{\alpha t} cos(\beta t),\ y_4(t) = e^{\alpha t} sin(\beta t)$

Solutions Set ⇑

Particular Solution using the Method of Variation of Parameters

Forcing Function $g(t) = t + \cos(t)$

WRONSKIAN $$W = \begin{bmatrix} y_1(t) & y_2(t) & y_3(t) & y_4(t) \\ y_1'(t) & y_2'(t) & y_3'(t) & y_4'(t) \\ y_1''(t) & y_2''(t) & y_3''(t) & y_4'' \\ y_1'''(t) & y_2'''(t) & y_3'''(t) & y_4''' \end{bmatrix}$$

$$\vec{v}(t) = [W]^{-1} \begin{bmatrix} 0 \\ 0 \\ 0 \\ \frac{g(t)}{A_4} \end{bmatrix}, \text{ A}_4 \text{ is the coefficient of } y''''(t)$$

Particular Solution $$y_p(t) = [y_1(t)\, y_2(t)\, y_3(t)\, y_4(t)] \int \vec{v}(t)\, dt$$

Particular Solution using Variation of Parameters

$$y_p(t) = -t - \frac{\cos(t)}{2} - \frac{t\,\sin(t)}{4}$$

Test the validity by substitution in Differential EQN

VALID/PASS

Details of Wronskian

Wronskian W ⇒
$$\begin{pmatrix} e^{-t} & e^{t} & \cos(t) & -\sin(t) \\ -e^{-t} & e^{t} & -\sin(t) & -\cos(t) \\ e^{-t} & e^{t} & -\cos(t) & \sin(t) \\ -e^{-t} & e^{t} & \sin(t) & \cos(t) \end{pmatrix}$$

$\vec{v}(t) \Rightarrow$
$$-\frac{e^{t}\,(t+\cos(t))}{4}$$

$$\frac{e^{-t}\,(t+\cos(t))}{4}$$

$$\frac{\sin(t)\,(t+\cos(t))}{2}$$

$$\frac{\cos(t)\,(t+\cos(t))}{2}$$

Inverse Laplace based solution: Theory and Results

Differential Equation and Initial Conditions (General Case)

$$A_4\, y''''(t) + A_3\, y'''(t) + A_2\, y''(t) + A_1\, y'(t) + A_0\, y = g(t)$$

$$[y(0) = y_0, y'(0) = y_1, y''(0) = y_2, y'''(0) = y_3]$$

Laplace Transform Y(s)=A(s)+B(s)

$$A(s) = \frac{y_1\left(A_4\, s^2 + A_3\, s + A_2\right) + A_4\, y_3 + y_0\left(A_4\, s^3 + A_3\, s^2 + A_2\, s + A_1\right) + y_2\,(A_3 + A_4\, s)}{A_4\, s^4 + A_3\, s^3 + A_2\, s^2 + A_1\, s + A_0}$$

$$B(s) = \frac{G(s)}{A_4\, s^4 + A_3\, s^3 + A_2\, s^2 + A_1\, s + A_0}$$

Differential Equation and Initial Conditions (given)

$$y''''(t) - y(t) = t + \cos(t)$$

$$(\ y(0) = 1 \quad y'(0) = -1 \quad y''(0) = 0 \quad y'''(0) = -1\ )$$

Laplace Transform Y(s) of y(t)

$$Y_L(s) = \frac{\frac{s}{s^2+1} + \frac{1}{s^2} - s^2 + s^3 - 1}{s^4 - 1}$$

Solution y(t) using inverse Laplace

$$y(t) = \frac{5\,e^{-t}}{8} - t + \frac{\cos(t)}{4} + \frac{e^t}{8} + \frac{\sin(t)}{2} - \frac{t\,\sin(t)}{4}$$

Complete Solution and Validation

Solution to the Homogeneous part using roots (unknown constants, b_1,b_2,...)

$$y_h(t) = b_1\, y_1(t) + b_2\, y_2(t) + b_3\, y_3(t) + b_4\, y_4(t)$$

Complete solution $y(t)=y_h(t)+y_p(t)$

$$y(t) = b_1\, y_1(t) - \frac{\cos(t)}{2} - t + b_2\, y_2(t) + b_3\, y_3(t) + b_4\, y_4(t) - \frac{t\,\sin(t)}{4}$$

Obtain unknown constants (b_1 b_2 b_3 b_4) by applying the initial conditions

(b_1 b_2 b_3 b_4) $\Rightarrow$ $\left(\ \frac{5}{8} \quad \frac{1}{8} \quad \frac{3}{4} \quad -\frac{1}{2}\ \right)$

Complete solution using roots (replace unknown constants)

$$y(t) = \frac{5e^{-t}}{8} - t + \frac{\cos(t)}{4} + \frac{e^t}{8} + \frac{\sin(t)}{2} - \frac{t\,\sin(t)}{4}$$

Complete solution using inverse Laplace transforms

$$y(t) = \frac{5e^{-t}}{8} - t + \frac{\cos(t)}{4} + \frac{e^t}{8} + \frac{\sin(t)}{2} - \frac{t\,\sin(t)}{4}$$

Complete solution using dsolve(.) in Matlab

$$y(t) = \frac{5e^{-t}}{8} - t + \frac{\cos(t)}{4} + \frac{e^t}{8} + \frac{\sin(t)}{2} - \frac{t\,\sin(t)}{4}$$

ODE45 and comparison

Differential Equation and Initial Conditions

$$y''''(t) - y(t) = t + \cos(t)$$

$$(\ y(0) = 1 \quad y'(0) = -1 \quad y''(0) = 0 \quad y'''(0) = -1\ )$$

Corresponding first order differential equations for ODE [$Y_1 \equiv y(t)$] vector

$$\begin{pmatrix} Y_1'(t) = Y_2 \\ Y_2'(t) = Y_3 \\ Y_3'(t) = Y_4 \\ Y_4'(t) = t + \cos(t) + Y_1 \end{pmatrix}$$

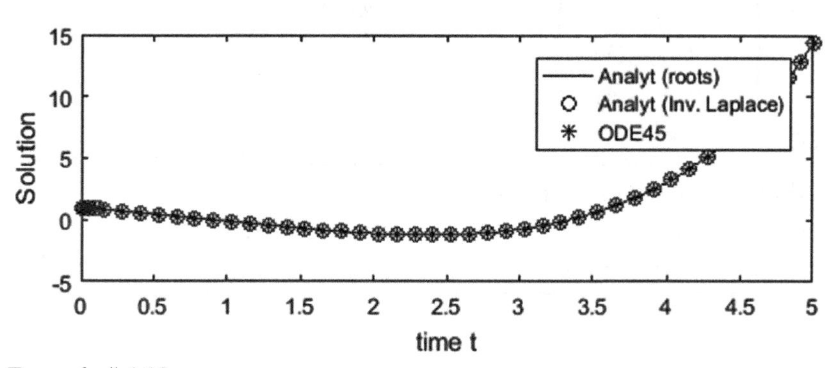

Example # 4.12

```
A=[1 0 0 -1 0];IC=[-1,1,0,1];gt='exp(-t)';
```

4th Order Homogeneous Differential Equations with Constant Coefficients

Differential Equation $A_4y''''(t) + A_3y'''(t) + A_2y''(t) + A_1y'(t) + A_0y(t) = 0$

Characteristic Equation/Roots $A_4r^4 + A_3r^3 + A_2r^2 + A_1r + A_0 = 0 \Rightarrow r_1,\ \ r_2,\ \ r_3,\ \ r_4$

Solution set and solution y(t) $Y = [y_1(t), y_2(t), y_3(t), y_4(t)] \Rightarrow y(t) = \sum_{k=1}^{k=4} C_k y_k(t)$

$$Y(t) \Rightarrow \begin{cases} e^{r_1t} & e^{r_2t} & e^{r_3t} & e^{r_4t} & r_{1,2,3,4}\ real \\ e^{rt} & te^{rt} & e^{r_3t} & e^{r_4t} & r_{1,2} = r\ real \\ e^{rt} & te^{rt} & t^2e^{rt} & e^{r_4t} & r_{1,2,3} = r\ real \\ e^{rt} & te^{rt} & t^2e^{rt} & t^3e^{rt} & r_{1,2,3,4} = r\ real \\ e^{r_1t} & te^{r_1t} & e^{r_2t} & te^{r_3t} & r_{1,2} = r_1, r_{3,4} = r_2\ real \\ e^{r_1t} & e^{r_2t} & e^{\alpha t}cos(\beta t) & e^{\alpha t}sin(\beta t) & r_{3,4} = \alpha \pm j\beta \\ e^{rt} & te^{rt} & e^{\alpha t}cos(\beta t) & e^{\alpha t}sin(\beta t) & r_{1,2} = r,\ r_{3,4} = \alpha \pm j\beta \\ e^{\alpha_1 t}cos(\beta_1 t) & e^{\alpha_1}sin(\beta_1 t) & e^{\alpha_2 t}cos(\beta_2 t) & e^{\alpha_2 t}sin(\beta_2 t) & \begin{bmatrix} r_{1,2} = \alpha_1 \pm j\beta_1 \\ r_{3,4} = \alpha_2 \pm j\beta_2 \end{bmatrix} \\ e^{\alpha t}cos(\beta t) & e^{\alpha}sin(\beta t) & te^{\alpha t}cos(\beta t) & te^{\alpha t}sin(\beta t) & r_{1,2} = r_{3,4} = \alpha \pm j\beta \end{cases}$$

Set of Solutions: Homogeneous case

Particular and complete solutions
Higher order differential equations

Differential Equation and Initial Conditions

$$y''''(t) - y'(t) = e^{-t}$$

$$(\ y(0) = -1 \quad y'(0) = 1 \quad y''(0) = 0 \quad y'''(0) = 1\ )$$

Characteristic equation: $r^4 - r = 0$

Roots of the Characteristic equation

$$\begin{pmatrix} 0 \\ 1 \\ -\frac{1}{2} - \frac{\sqrt{3}\,i}{2} \\ -\frac{1}{2} + \frac{\sqrt{3}\,i}{2} \end{pmatrix}$$

Homogeneous solutions set

$$y_1(t) = 1$$

$$y_2(t) = e^{t}$$

$$y_3(t) = e^{-\frac{t}{2}} \cos\left(\frac{\sqrt{3}\,t}{2}\right)$$

$$y_4(t) = -e^{-\frac{t}{2}} \sin\left(\frac{\sqrt{3}\,t}{2}\right)$$

Roots $\Rightarrow r_1 \neq r_2,\ r_{3,4} = \alpha \pm j\beta$

$y_k(t) = e^{rt}, k = 1, 2$

$y_3(t) = e^{\alpha t} cos(\beta t),\ y_4(t) = e^{\alpha t} sin(\beta t)$

Solutions Set ⇑

Particular Solution using the Method of Variation of Parameters

Forcing Function $g(t) = e^{-t}$

WRONSKIAN
$$W = \begin{bmatrix} y_1(t) & y_2(t) & y_3(t) & y_4(t) \\ y_1'(t) & y_2'(t) & y_3'(t) & y_4'(t) \\ y_1''(t) & y_2''(t) & y_3''(t) & y_4'' \\ y_1'''(t) & y_2'''(t) & y_3'''(t) & y_4''' \end{bmatrix}$$

$$\vec{v}(t) = [W]^{-1} \begin{bmatrix} 0 \\ 0 \\ 0 \\ \frac{g(t)}{A_4} \end{bmatrix},\ A_4 \text{ is the coefficient of } y''''(t)$$

Particular Solution
$$y_p(t) = [y_1(t)\, y_2(t)\, y_3(t)\, y_4(t)] \int \vec{v}(t)\, dt$$

Particular Solution using Variation of Parameters

$$y_p(t) = \frac{e^{-t}}{2}$$

Test the validity by substitution in Differential EQN

VALID/PASS

Details of Wronskian

Wronskian W ⇒

$$\begin{pmatrix} 1 & e^{t} & e^{-\frac{t}{2}}\cos\left(\frac{\sqrt{3}t}{2}\right) & -e^{-\frac{t}{2}}\sin\left(\frac{\sqrt{3}t}{2}\right) \end{pmatrix}$$
$$\begin{pmatrix} 0 & e^{t} & -e^{-\frac{t}{2}}\cos\left(\frac{\pi}{3}-\frac{\sqrt{3}t}{2}\right) & -e^{-\frac{t}{2}}\cos\left(\frac{\pi}{6}+\frac{\sqrt{3}t}{2}\right) \end{pmatrix}$$
$$\begin{pmatrix} 0 & e^{t} & -e^{-\frac{t}{2}}\cos\left(\frac{\pi}{3}+\frac{\sqrt{3}t}{2}\right) & e^{-\frac{t}{2}}\sin\left(\frac{\pi}{3}+\frac{\sqrt{3}t}{2}\right) \end{pmatrix}$$
$$\begin{pmatrix} 0 & e^{t} & e^{-\frac{t}{2}}\cos\left(\frac{\sqrt{3}t}{2}\right) & -e^{-\frac{t}{2}}\sin\left(\frac{\sqrt{3}t}{2}\right) \end{pmatrix}$$

$$-e^{-t}$$

$\vec{v}(t) \Rightarrow$

$$\frac{e^{-2t}}{3}$$

$$\frac{2e^{-\frac{t}{2}}\cos\left(\frac{\sqrt{3}t}{2}\right)}{3}$$

$$-\frac{2e^{-\frac{t}{2}}\sin\left(\frac{\sqrt{3}t}{2}\right)}{3}$$

Inverse Laplace based solution: Theory and Results

Differential Equation and Initial Conditions (General Case)

$$A_4\, y''''(t) + A_3\, y'''(t) + A_2\, y''(t) + A_1\, y'(t) + A_0\, y = g(t)$$
$$[y(0) = y_0, y'(0) = y_1, y''(0) = y_2, y'''(0) = y_3]$$

Laplace Transform Y(s)=A(s)+B(s)

$$A(s) = \frac{y_1\left(A_4 s^2+A_3 s+A_2\right)+A_4 y_3+y_0\left(A_4 s^3+A_3 s^2+A_2 s+A_1\right)+y_2\left(A_3+A_4 s\right)}{A_4 s^4+A_3 s^3+A_2 s^2+A_1 s+A_0}$$
$$B(s) = \frac{G(s)}{A_4 s^4+A_3 s^3+A_2 s^2+A_1 s+A_0}$$

Differential Equation and Initial Conditions (given)

$$y''''(t) - y'(t) = e^{-t}$$
$$(\ y(0) = -1 \quad y'(0) = 1 \quad y''(0) = 0 \quad y'''(0) = 1\ )$$

Laplace Transform Y(s) of y(t)

$$Y_L(s) = -\frac{\frac{1}{s+1}+s^2-s^3+2}{s-s^4}$$

Solution y(t) using inverse Laplace

$$y(t) = \frac{e^{-t}}{2} + \frac{5e^{t}}{6} + \frac{2e^{-\frac{t}{2}}\left(\cos\left(\frac{\sqrt{3}t}{2}\right)+\sqrt{3}\sin\left(\frac{\sqrt{3}t}{2}\right)\right)}{3} - 3$$

Complete Solution and Validation

Solution to the Homogeneous part using roots (unknown constants, $b_1, b_2, \ldots$)

$$y_h(t) = b_1\, y_1(t) + b_2\, y_2(t) + b_3\, y_3(t) + b_4\, y_4(t)$$

Complete solution $y(t)=y_h(t)+y_p(t)$

$$y(t) = \frac{e^{-t}}{2} + b_1\, y_1(t) + b_2\, y_2(t) + b_3\, y_3(t) + b_4\, y_4(t)$$

Obtain unknown constants (b_1 b_2 b_3 b_4) by applying the initial conditions

$$(b_1\ b_2\ b_3\ b_4) \Rightarrow \qquad \begin{pmatrix} -3 & \frac{5}{6} & \frac{2}{3} & -\frac{2\sqrt{3}}{3} \end{pmatrix}$$

Complete solution using roots (replace unknown constants)

$$y(t) = \frac{e^{-t}}{2} + \frac{5\,e^{t}}{6} + \frac{2e^{-\frac{t}{2}}\cos\left(\frac{\sqrt{3}\,t}{2}\right)}{3} + \frac{2\sqrt{3}\,e^{-\frac{t}{2}}\sin\left(\frac{\sqrt{3}\,t}{2}\right)}{3} - 3$$

Complete solution using inverse Laplace transforms

$$y(t) = \frac{e^{-t}}{2} + \frac{5\,e^{t}}{6} + \frac{2e^{-\frac{t}{2}}\left(\cos\left(\frac{\sqrt{3}\,t}{2}\right)+\sqrt{3}\sin\left(\frac{\sqrt{3}\,t}{2}\right)\right)}{3} - 3$$

Complete solution using dsolve(.) in Matlab

$$y(t) = \frac{e^{-t}}{2} + \frac{5\,e^{t}}{6} + \frac{2e^{-\frac{t}{2}}\cos\left(\frac{\sqrt{3}\,t}{2}\right)}{3} + \frac{2\sqrt{3}\,e^{-\frac{t}{2}}\sin\left(\frac{\sqrt{3}\,t}{2}\right)}{3} - 3$$

ODE45 and comparison

Differential Equation and Initial Conditions

$$y''''(t) - y'(t) = e^{-t}$$

$$\left(\ y(0) = -1 \quad y'(0) = 1 \quad y''(0) = 0 \quad y'''(0) = 1\ \right)$$

Corresponding first order differential equations for ODE [$Y_1 \equiv y(t)$] vector

$$\begin{pmatrix} Y_1'(t) = Y_2 \\ Y_2'(t) = Y_3 \\ Y_3'(t) = Y_4 \\ Y_4'(t) = e^{-t} + Y_2 \end{pmatrix}$$

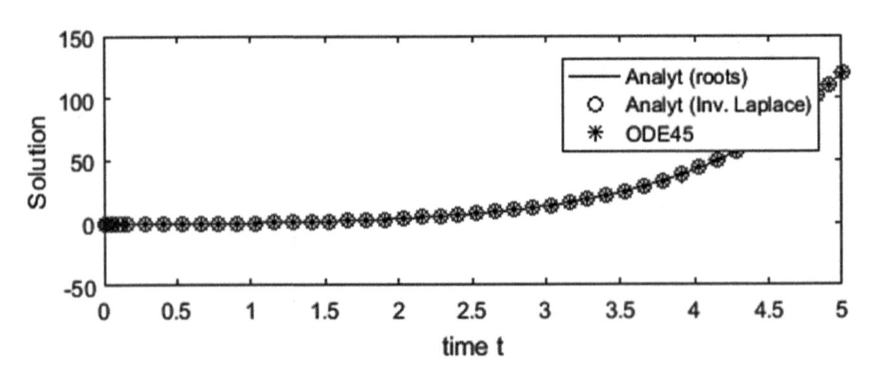

Example # 4.13

```
A=[1,1,-2,0];IC=[1,1,-1];gt='t+exp(t)';
```

3rd Order Homogeneous Differential Equations with Constant Coefficients

Differential Equation $A_3y'''(t) + A_2y''(t) + A_1y'(t) + A_0y(t) = 0$

Characteristic Equation/Roots $A_3r^3 + A_2r^2 + A_1r + A_0 = 0 \Rightarrow r_1,\ r_2,\ r_3,$

Solution set and solution y(t) $Y = [y_1(t), y_2(t), y_3(t)] \Rightarrow y(t) = \sum_{k=1}^{k=3} b_k y_k(t)$

$$Y(t) \Rightarrow \begin{cases} e^{r_1t} & e^{r_2t} & e^{r_3t} & r_{1,2,3}\ real \\ e^{rt} & te^{rt} & e^{r_3t} & r_{1,2} = r \\ e^{rt} & te^{rt} & t^2e^{rt} & r_{1,2,3} = r \\ e^{r_1t} & e^{\alpha t}cos(\beta t) & e^{\alpha t}sin(\beta t) & r_{2,3} = \alpha \pm j\beta \end{cases}$$

Set of Solutions: Homogeneous case

Particular and complete solutions
Higher order differential equations

Differential Equation and Initial Conditions

$$y'''(t) + y''(t) - 2y'(t) = t + e^t$$

$$(\ y(0) = 1 \quad y'(0) = 1 \quad y''(0) = -1\ )$$

Characteristic equation: $r^3 + r^2 - 2r = 0$

Roots of the Characteristic equation

$$\begin{pmatrix} 0 \\ -2 \\ 1 \end{pmatrix}$$

Homogeneous solutions set

$$y_1(t) = 1$$

$$y_2(t) = e^{-2t}$$

$$y_3(t) = e^t$$

Roots $\Rightarrow r_1 \neq r_2 \neq r_3$

$y_k(t) = e^{r_kt}, k = 1, 2, 3$ Solutions Set ⇑

Particular Solution using the Method of Variation of Parameters

Forcing Function $g(t) = t + e^t$

WRONSKIAN $W = \begin{bmatrix} y_1(t) & y_2(t) & y_3(t) \\ y_1'(t) & y_2'(t) & y_3'(t) \\ y_1''(t) & y_2''(t) & y_3''(t) \end{bmatrix}$

$$\vec{v}(t) = [W]^{-1} \begin{bmatrix} 0 \\ 0 \\ \frac{g(t)}{A_3} \end{bmatrix}, \text{ A}_3 \text{ is the coefficient of } y'''(t)$$

Particular Solution $y_p(t) = [y_1(t)\, y_2(t)\, y_3(t)] \int \vec{v}(t)\, dt$

Particular Solution using Variation of Parameters

$$y_p(t) = \frac{t\,e^t}{3} - \frac{4\,e^t}{9} - \frac{t}{4} - \frac{t^2}{4} - \frac{3}{8}$$

Test the validity by substitution in Differential EQN

VALID/PASS

Details of Wronskian

Wronskian W ⇒

$$\begin{pmatrix} 1 & e^{-2t} & e^t \end{pmatrix}$$
$$\begin{pmatrix} 0 & -2\,e^{-2t} & e^t \end{pmatrix}$$
$$\begin{pmatrix} 0 & 4e^{-2t} & e^t \end{pmatrix}$$

$\vec{v}(t)$ ⇒

$$-\frac{t}{2} - \frac{e^t}{2}$$

$$\frac{e^{2t}\,(t+e^t)}{6}$$

$$\frac{t\,e^{-t}}{3} + \frac{1}{3}$$

Inverse Laplace based solution: Theory and Results

Differential Equation and Initial Conditions (General Case)

$$A_3\, y'''(t) + A_2\, y''(t) + A_1\, y'(t) + A_0\, y = g(t)$$

$$[y(0) = y_0, y'(0) = y_1, y''(0) = y_2]$$

Laplace Transform Y(s)=A(s)+B(s)

$$A(s) = \frac{y_0\left(A_3\, s^2 + A_2\, s + A_1\right) + A_3\, y_2 + y_1\,(A_2 + A_3\, s)}{A_3\, s^3 + A_2\, s^2 + A_1\, s + A_0}$$

$$B(s) = \frac{G(s)}{A_3\, s^3 + A_2\, s^2 + A_1\, s + A_0}$$

Differential Equation and Initial Conditions (given)

$$y'''(t) + y''(t) - 2\, y'(t) = t + \mathrm{e}^t$$

$$(\; y(0) = 1 \quad y'(0) = 1 \quad y''(0) = -1 \;)$$

Laplace Transform Y(s) of y(t)

$$Y_L(s) = \frac{s^5 + s^4 - 4\, s^3 + 3\, s^2 + s - 1}{s^3\,(s-1)^2\,(s+2)}$$

Solution y(t) using inverse Laplace

$$y(t) = \frac{2\,\mathrm{e}^t}{9} - \frac{25\,\mathrm{e}^{-2t}}{72} - \frac{t}{4} + \frac{t\,\mathrm{e}^t}{3} - \frac{t^2}{4} + \frac{9}{8}$$

Complete Solution and Validation

Solution to the Homogeneous part using roots (unknown constants, b_1,b_2,...)

$$y_h(t) = b_1\, y_1(t) + b_2\, y_2(t) + b_3\, y_3(t)$$

Complete solution $y(t)=y_h(t)+y_p(t)$

$$y(t) = \frac{t\,e^t}{3} - \frac{4e^t}{9} - \frac{t}{4} + b_1\, y_1(t) + b_2\, y_2(t) + b_3\, y_3(t) - \frac{t^2}{4} - \frac{3}{8}$$

Obtain unknown constants (b_1 b_2 b_3) by applying the initial conditions

(b_1 b_2 b_3) $\Rightarrow$ $\left(\frac{3}{2} \quad -\frac{25}{72} \quad \frac{2}{3}\right)$

Complete solution using roots (replace unknown constants)

$$y(t) = \frac{2e^t}{9} - \frac{25e^{-2t}}{72} - \frac{t}{4} + \frac{t\,e^t}{3} - \frac{t^2}{4} + \frac{9}{8}$$

Complete solution using inverse Laplace transforms

$$y(t) = \frac{2e^t}{9} - \frac{25e^{-2t}}{72} - \frac{t}{4} + \frac{t\,e^t}{3} - \frac{t^2}{4} + \frac{9}{8}$$

Complete solution using dsolve(.) in Matlab

$$y(t) = \frac{2e^t}{9} - \frac{25e^{-2t}}{72} - \frac{t}{4} + \frac{t\,e^t}{3} - \frac{t^2}{4} + \frac{9}{8}$$

ODE45 and comparison

Differential Equation and Initial Conditions

$$y'''(t) + y''(t) - 2\,y'(t) = t + e^t$$

$$(\ y(0) = 1 \quad y'(0) = 1 \quad y''(0) = -1\ )$$

Corresponding first order differential equations for ODE [$Y_1 \equiv y(t)$] vector

$$\begin{pmatrix} Y_1'(t) = Y_2 \\ Y_2'(t) = Y_3 \\ Y_3'(t) = t + e^t + 2\,Y_2 - Y_3 \end{pmatrix}$$

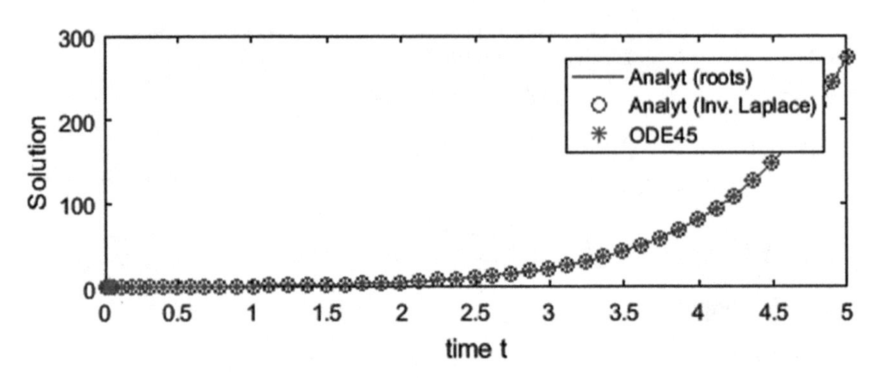

Example # 4.14

```
A=[1,1,0,0,0,0];IC=[1;0;1;0;1];gt='t';
```

Theory of Laplace Transforms: Higher order differential Equations

The Differential Equation

$$A_n y^{(n)}(t) + A_{n-1} y^{(n-1)}(t) + \cdots + A_1 y^{(1)}(t) + A_0 y(t) = g(t)$$

The Initial Conditions (IC)

$$y(0) = y_0, y^{(1)}(0) = y_1, y^{(2)}(0) = y_2, \cdots, y^{(n-1)}(0) = y_{n-1}$$

The Laplace Transform Y(s)

$$\frac{(A_n s^{n-1} + A_{n-1} s^{n-2} + \cdots + A_2 s + A_1) y_0 + (A_n s^{n-2} + A_{n-1} s^{n-3} + \cdots + A_2) y_1 + \cdots + y_{n-1} A_n}{(A_n s^n + A_{n-1} s^{n-1} + \cdots + A_1 s + A_0)}$$

$$+$$

$$\frac{G(s)}{(A_n s^n + A_{n-1} s^{n-1} + \cdots + A_1 s + A_0)}$$

Inverse Laplace based solution

Differential Equation and Initial Conditions

$$y''''(t) + y'''(t) = t$$

$$(\ y(0) = 1 \quad y'(0) = 0 \quad y''(0) = 1 \quad y'''(0) = 0 \quad y''''(0) = 1\ )$$

Laplace Transform $Y_L(s)$ of y(t)

$$Y_L(s) = \frac{(s^2+1)(s^4+s^3+1)}{s^6(s+1)}$$

Solution y(t) using inverse Laplace transform of $Y_L(s)$

$$y(t) = 2t + 2e^{-t} - \frac{t^2}{2} + \frac{t^3}{3} - \frac{t^4}{24} + \frac{t^5}{120} - 1$$

Solution y(t) using dsolve(.) in Matlab

$$y(t) = 2t + 2e^{-t} - \frac{t^2}{2} + \frac{t^3}{3} - \frac{t^4}{24} + \frac{t^5}{120} - 1$$

ODE45 and comparison

Differential Equation and Initial Conditions

$$y''''(t) + y'''(t) = t$$

$$(\ y(0) = 1 \quad y'(0) = 0 \quad y''(0) = 1 \quad y'''(0) = 0 \quad y''''(0) = 1\ )$$

Vector: 1st first order differential equations for ODE [$Y_1 \equiv y(t)$] ⇒

$$\begin{pmatrix} Y_1'(t) = Y_2 \\ Y_2'(t) = Y_3 \\ Y_3'(t) = Y_4 \\ Y_4'(t) = Y_5 \\ Y_5'(t) = t - Y_5 \end{pmatrix}$$

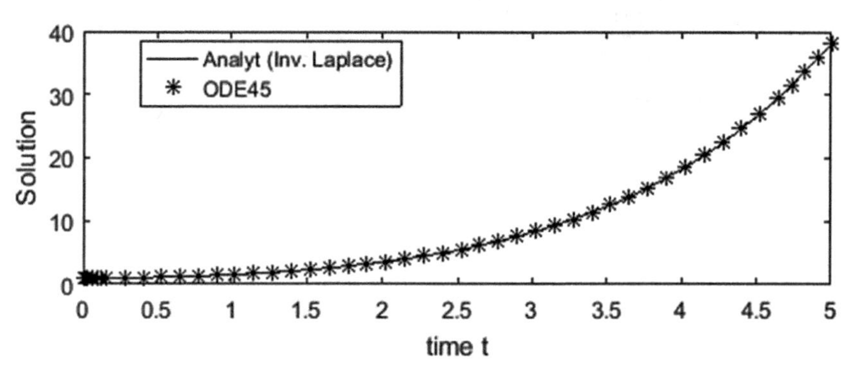

Example # 4.15

```
A=[1,0,1,1,0,1];IC=[1;0;0;0;1];gt='exp(t)';
```

Theory of Laplace Transforms: Higher order differential Equations

The Differential Equation

$$A_n y^{(n)}(t) + A_{n-1} y^{(n-1)}(t) + \cdots + A_1 y^{(1)}(t) + A_0 y(t) = g(t)$$

The Initial Conditions (IC)

$$y(0) = y_0, y^{(1)}(0) = y_1, y^{(2)}(0) = y_2, \cdots, y^{(n-1)}(0) = y_{n-1}$$

The Laplace Transform Y(s)

$$\frac{(A_n s^{n-1} + A_{n-1} s^{n-2} + \cdots + A_2 s + A_1) y_0 + (A_n s^{n-2} + A_{n-1} s^{n-3} + \cdots + A_2) y_1 + \cdots + y_{n-1} A_n}{(A_n s^n + A_{n-1} s^{n-1} + \cdots + A_1 s + A_0)}$$

$$+$$

$$\frac{G(s)}{(A_n s^n + A_{n-1} s^{n-1} + \cdots + A_1 s + A_0)}$$

Inverse Laplace based solution

Differential Equation and Initial Conditions

$$y''''(t) + y'''(t) + y''(t) + y(t) = e^t$$

$$(\; y(0) = 1 \quad y'(0) = 0 \quad y''(0) = 0 \quad y'''(0) = 0 \quad y''''(0) = 1 \;)$$

Laplace Transform $Y_L(s)$ of y(t)

$$Y_L(s) = \frac{s^3}{s^4 - 1}$$

Solution y(t) using inverse Laplace transform of $Y_L(s)$

$$y(t) = \frac{\cos(t)}{2} + \frac{\cosh(t)}{2}$$

Solution y(t) using dsolve(.) in Matlab

$$y(t) = \frac{\cos(t)}{2} + \frac{\cosh(t)}{2}$$

ODE45 and comparison

Differential Equation and Initial Conditions

$$y''''(t) + y'''(t) + y''(t) + y(t) = e^t$$

$$(\; y(0) = 1 \quad y'(0) = 0 \quad y''(0) = 0 \quad y'''(0) = 0 \quad y''''(0) = 1 \;)$$

Vector: 1st first order differential equations for ODE [$Y_1 \equiv y(t)$] ⇒

$$\begin{pmatrix} Y_1'(t) = Y_2 \\ Y_2'(t) = Y_3 \\ Y_3'(t) = Y_4 \\ Y_4'(t) = Y_5 \\ Y_5'(t) = e^t - Y_1 - Y_3 - Y_4 \end{pmatrix}$$

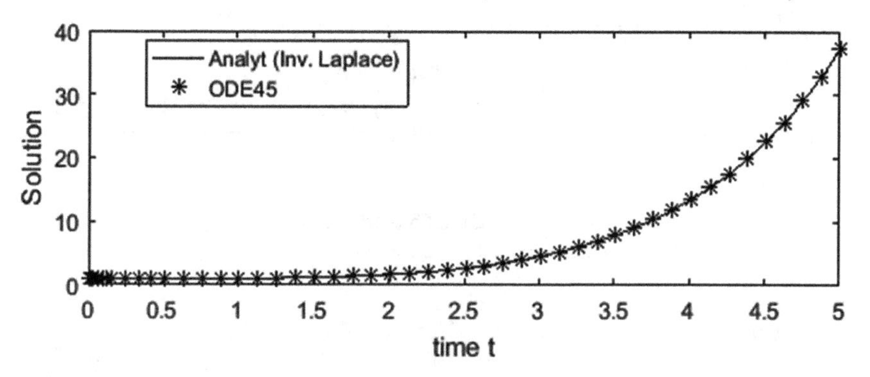

Example # 4.16

```
A=[1,0,0,-1,0,0];IC=[1;0;1;0;0];gt='exp(-t)';
```

Theory of Laplace Transforms: Higher order differential Equations

The Differential Equation

$$A_n y^{(n)}(t) + A_{n-1} y^{(n-1)}(t) + \cdots + A_1 y^{(1)}(t) + A_0 y(t) = g(t)$$

The Initial Conditions (IC)

$$y(0) = y_0, y^{(1)}(0) = y_1, y^{(2)}(0) = y_2, \cdots, y^{(n-1)}(0) = y_{n-1}$$

The Laplace Transform Y(s)

$$\frac{(A_n s^{n-1} + A_{n-1} s^{n-2} + \cdots + A_2 s + A_1) y_0 + (A_n s^{n-2} + A_{n-1} s^{n-3} + \cdots + A_2) y_1 + \cdots + y_{n-1} A_n}{(A_n s^n + A_{n-1} s^{n-1} + \cdots + A_1 s + A_0)}$$

$$+$$

$$\frac{G(s)}{(A_n s^n + A_{n-1} s^{n-1} + \cdots + A_1 s + A_0)}$$

Inverse Laplace based solution

Differential Equation and Initial Conditions

$$y''''(t) - y''(t) = e^{-t}$$

$$(\ y(0) = 1 \quad y'(0) = 0 \quad y''(0) = 1 \quad y'''(0) = 0 \quad y''''(0) = 0\ )$$

Laplace Transform $Y_L(s)$ of y(t)

$$Y_L(s) = -\frac{\frac{1}{s+1} - s + s^2 + s^4}{s^2 - s^5}$$

Solution y(t) using inverse Laplace transform of $Y_L(s)$

$$y(t) = \sinh(t) - t - \frac{4\sqrt{3}\,e^{-\frac{t}{2}}\sin\left(\frac{\pi}{3} + \frac{\sqrt{3}t}{2}\right)}{3} + 2$$

Solution y(t) using dsolve(.) in Matlab

$$y(t) = \sinh(t) - t - 2\,e^{-\frac{t}{2}}\cos\left(\frac{\sqrt{3}t}{2}\right) - \frac{2\sqrt{3}\,e^{-\frac{t}{2}}\sin\left(\frac{\sqrt{3}t}{2}\right)}{3} + 2$$

ODE45 and comparison

Differential Equation and Initial Conditions

$$y''''(t) - y''(t) = e^{-t}$$

$$(\ y(0) = 1 \quad y'(0) = 0 \quad y''(0) = 1 \quad y'''(0) = 0 \quad y''''(0) = 0\ )$$

Vector: 1st first order differential equations for ODE [$Y_1 \equiv y(t)$] ⇒

$$\begin{pmatrix} Y_1'(t) = Y_2 \\ Y_2'(t) = Y_3 \\ Y_3'(t) = Y_4 \\ Y_4'(t) = Y_5 \\ Y_5'(t) = e^{-t} + Y_3 \end{pmatrix}$$

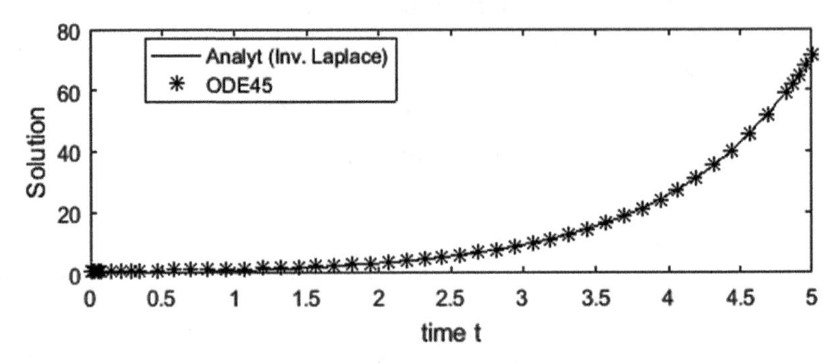

Example # 4.17

```
A=[1,1,0,0,0,0,0];IC=[1;0;0;0;1;0];gt='exp(-t)*t';
```

Theory of Laplace Transforms: Higher order differential Equations

The Differential Equation

$$A_n y^{(n)}(t) + A_{n-1} y^{(n-1)}(t) + \cdots + A_1 y^{(1)}(t) + A_0 y(t) = g(t)$$

The Initial Conditions (IC)

$$y(0) = y_0, y^{(1)}(0) = y_1, y^{(2)}(0) = y_2, \cdots, y^{(n-1)}(0) = y_{n-1}$$

The Laplace Transform Y(s)

$$\frac{(A_n s^{n-1} + A_{n-1} s^{n-2} + \cdots + A_2 s + A_1) y_0 + (A_n s^{n-2} + A_{n-1} s^{n-3} + \cdots + A_2) y_1 + \cdots + y_{n-1} A_n}{(A_n s^n + A_{n-1} s^{n-1} + \cdots + A_1 s + A_0)}$$

$$+$$

$$\frac{G(s)}{(A_n s^n + A_{n-1} s^{n-1} + \cdots + A_1 s + A_0)}$$

Inverse Laplace based solution

Differential Equation and Initial Conditions

$$y''''''(t) + y'''''(t) = t\,e^{-t}$$

$$(\ y(0) = 1 \quad y'(0) = 0 \quad y''(0) = 0 \quad y'''(0) = 0 \quad y''''(0) = 1 \quad y'''''(0) = 0\ )$$

Laplace Transform $Y_L(s)$ of y(t)

$$Y_L(s) = \frac{s + \frac{1}{(s+1)^2} + s^4 + s^5 + 1}{s^5\,(s+1)}$$

Solution y(t) using inverse Laplace transform of $Y_L(s)$

$$y(t) = 3\,t^2 - 15\,e^{-t} - 5\,t\,e^{-t} - \frac{t^2\,e^{-t}}{2} - 10\,t - \frac{t^3}{2} + \frac{t^4}{12} + 16$$

Solution y(t) using dsolve(.) in Matlab

$$y(t) = 3\,t^2 - 15\,e^{-t} - 5\,t\,e^{-t} - \frac{t^2\,e^{-t}}{2} - 10\,t - \frac{t^3}{2} + \frac{t^4}{12} + 16$$

ODE45 and comparison

Differential Equation and Initial Conditions

$$y''''''(t) + y'''''(t) = t\,\mathrm{e}^{-t}$$

$$(\ y(0) = 1 \quad y'(0) = 0 \quad y''(0) = 0 \quad y'''(0) = 0 \quad y''''(0) = 1 \quad y'''''(0) = 0\ )$$

Vector: 1st first order differential equations for ODE [$Y_1 \equiv y(t)$] ⇒

$$\begin{pmatrix} Y_1'(t) = Y_2 \\ Y_2'(t) = Y_3 \\ Y_3'(t) = Y_4 \\ Y_4'(t) = Y_5 \\ Y_5'(t) = Y_6 \\ Y_6'(t) = t\mathrm{e}^{-t} - Y_6 \end{pmatrix}$$

Analyt (Inv. Laplace)
ODE45
Solution
time t

Example # 4.18

```
A=[1,0,0,1,0,0,0];IC=[0;0;0;0;0;1];gt='t+5';
```

Theory of Laplace Transforms: Higher order differential Equations

The Differential Equation

$$A_n y^{(n)}(t) + A_{n-1} y^{(n-1)}(t) + \cdots + A_1 y^{(1)}(t) + A_0 y(t) = g(t)$$

The Initial Conditions (IC)

$$y(0) = y_0, y^{(1)}(0) = y_1, y^{(2)}(0) = y_2, \cdots, y^{(n-1)}(0) = y_{n-1}$$

The Laplace Transform Y(s)

$$\frac{(A_n s^{n-1} + A_{n-1} s^{n-2} + \cdots + A_2 s + A_1) y_0 + (A_n s^{n-2} + A_{n-1} s^{n-3} + \cdots + A_2) y_1 + \cdots + y_{n-1} A_n}{(A_n s^n + A_{n-1} s^{n-1} + \cdots + A_1 s + A_0)}$$

$$+$$

$$\frac{G(s)}{(A_n s^n + A_{n-1} s^{n-1} + \cdots + A_1 s + A_0)}$$

Inverse Laplace based solution

Differential Equation and Initial Conditions

$$y'''''(t) + y'''(t) = t + 5$$

$$(\; y(0) = 0 \quad y'(0) = 0 \quad y''(0) = 0 \quad y'''(0) = 0 \quad y''''(0) = 0 \quad y'''''(0) = 1 \;)$$

Laplace Transform $Y_L(s)$ of y(t)

$$Y_L(s) = \frac{s^2+5\,s+1}{s^5\,(s^3+1)}$$

Solution y(t) using inverse Laplace transform of $Y_L(s)$

$$y(t) = \mathrm{e}^{-t} - t + 4\,\mathrm{e}^{\frac{t}{2}}\,\cos\left(\frac{\sqrt{3}\,t}{2}\right) + \frac{t^2}{2} + \frac{5\,t^3}{6} + \frac{t^4}{24} - 5$$

Solution y(t) using dsolve(.) in Matlab

$$y(t) = \mathrm{e}^{-t} - t + 4\,\mathrm{e}^{\frac{t}{2}}\,\cos\left(\frac{\sqrt{3}\,t}{2}\right) + \frac{t^2}{2} + \frac{5\,t^3}{6} + \frac{t^4}{24} - 5$$

ODE45 and comparison

Differential Equation and Initial Conditions

$$y'''''(t) + y'''(t) = t + 5$$

$$(\; y(0) = 0 \quad y'(0) = 0 \quad y''(0) = 0 \quad y'''(0) = 0 \quad y''''(0) = 0 \quad y'''''(0) = 1 \;)$$

Vector: 1st first order differential equations for ODE [$Y_1 \equiv y(t)$] $\Rightarrow$

$$\begin{pmatrix} Y_1'(t) = Y_2 \\ Y_2'(t) = Y_3 \\ Y_3'(t) = Y_4 \\ Y_4'(t) = Y_5 \\ Y_5'(t) = Y_6 \\ Y_6'(t) = t - Y_4 + 5 \end{pmatrix}$$

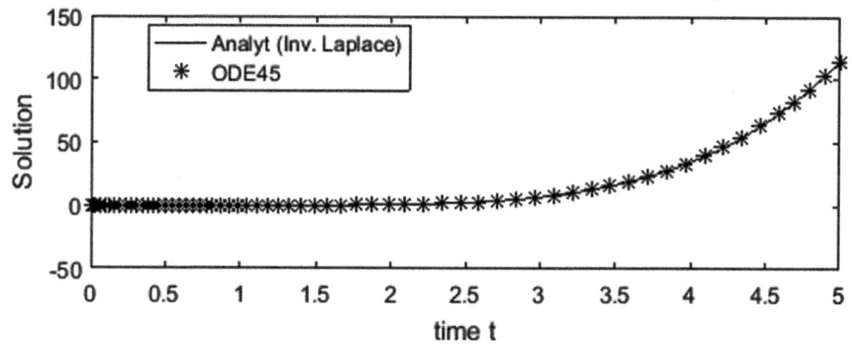

Example # 4.19

```
A=[1,0,0,1,0,0,0,0];IC=[1;0;0;0;0;0;1];gt='t';
```

Theory of Laplace Transfoms: Higher order differential Equations

The Differential Equation

$$A_n y^{(n)}(t) + A_{n-1} y^{(n-1)}(t) + \cdots + A_1 y^{(1)}(t) + A_0 y(t) = g(t)$$

The Initial Conditions (IC)

$$y(0) = y_0, y^{(1)}(0) = y_1, y^{(2)}(0) = y_2, \cdots, y^{(n-1)}(0) = y_{n-1}$$

The Laplace Transform Y(s)

$$\frac{(A_n s^{n-1} + A_{n-1} s^{n-2} + \cdots + A_2 s + A_1) y_0 + (A_n s^{n-2} + A_{n-1} s^{n-3} + \cdots + A_2) y_1 + \cdots + y_{n-1} A_n}{(A_n s^n + A_{n-1} s^{n-1} + \cdots + A_1 s + A_0)}$$

$$+$$

$$\frac{G(s)}{(A_n s^n + A_{n-1} s^{n-1} + \cdots + A_1 s + A_0)}$$

Inverse Laplace based solution

Differential Equation and Initial Conditions

$$y''''''(t) + y''''(t) = t$$

$$(\ y(0) = 1 \quad y'(0) = 0 \quad y''(0) = 0 \quad y'''(0) = 0 \quad y''''(0) = 0 \quad y'''''(0) = 0 \quad y''''''(0) = 1\ )$$

Laplace Transform $Y_L(s)$ of y(t)

$$Y_L(s) = \frac{s^8 + s^5 + s^2 + 1}{s^6 (s^3 + 1)}$$

Solution y(t) using inverse Laplace transform of $Y_L(s)$

$$y(t) = \frac{2e^{-t}}{3} + \frac{2e^{\frac{t}{2}} \sin\left(\frac{\pi}{6} + \frac{\sqrt{3}t}{2}\right)}{3} - \frac{t^2}{2} + \frac{t^3}{6} + \frac{t^5}{120}$$

Solution y(t) using dsolve(.) in Matlab

$$y(t) = \frac{2e^{-t}}{3} + \frac{e^{\frac{t}{2}} \cos\left(\frac{\sqrt{3}t}{2}\right)}{3} - \frac{t^2}{2} + \frac{t^3}{6} + \frac{t^5}{120} + \frac{\sqrt{3}e^{\frac{t}{2}} \sin\left(\frac{\sqrt{3}t}{2}\right)}{3}$$

ODE45 and comparison

Differential Equation and Initial Conditions

$$y''''''(t) + y''''(t) = t$$

$$(\ y(0) = 1 \quad y'(0) = 0 \quad y''(0) = 0 \quad y'''(0) = 0 \quad y''''(0) = 0 \quad y'''''(0) = 0 \quad y''''''(0) = 1\ )$$

Vector: 1st first order differential equations for ODE [$Y_1 \equiv y(t)$] ⇒

$$\begin{pmatrix} Y_1'(t) = Y_2 \\ Y_2'(t) = Y_3 \\ Y_3'(t) = Y_4 \\ Y_4'(t) = Y_5 \\ Y_5'(t) = Y_6 \\ Y_6'(t) = Y_7 \\ Y_7'(t) = t - Y_5 \end{pmatrix}$$

Solution
30
20
10
0
0 0.5 1 1.5 2 2.5 3 3.5 4 4.5 5
time t
Analyt (Inv. Laplace)
ODE45

Example # 4.20

```
A=[1,0,0,1,0,0,0,0];IC=[1;0;0;-1;0;0;1];gt='t*exp(-t)';
```

Theory of Laplace Transforms: Higher order differential Equations

The Differential Equation

$$A_n y^{(n)}(t) + A_{n-1} y^{(n-1)}(t) + \cdots + A_1 y^{(1)}(t) + A_0 y(t) = g(t)$$

The Initial Conditions (IC)

$$y(0) = y_0, y^{(1)}(0) = y_1, y^{(2)}(0) = y_2, \cdots, y^{(n-1)}(0) = y_{n-1}$$

The Laplace Transform Y(s)

$$\frac{(A_n s^{n-1} + A_{n-1} s^{n-2} + \cdots + A_2 s + A_1) y_0 + (A_n s^{n-2} + A_{n-1} s^{n-3} + \cdots + A_2) y_1 + \cdots + y_{n-1} A_n}{(A_n s^n + A_{n-1} s^{n-1} + \cdots + A_1 s + A_0)}$$

$$+$$

$$\frac{G(s)}{(A_n s^n + A_{n-1} s^{n-1} + \cdots + A_1 s + A_0)}$$

Inverse Laplace based solution

Differential Equation and Initial Conditions

$$y''''''(t) + y''''(t) = t\,e^{-t}$$

$$(\; y(0)=1 \quad y'(0)=0 \quad y''(0)=0 \quad y'''(0)=-1 \quad y''''(0)=0 \quad y'''''(0)=0 \quad y''''''(0)=1 \;)$$

Laplace Transform $Y_L(s)$ of y(t)

$$Y_L(s) = \frac{\frac{1}{(s+1)^2}+s^6}{s^4\,(s^3+1)}$$

Solution y(t) using inverse Laplace transform of $Y_L(s)$

$$y(t) = 3\,t + \frac{47\,e^{-t}}{9} + \frac{5\,t\,e^{-t}}{3} + \frac{t^2\,e^{-t}}{6} - t^2 + \frac{t^3}{6} + \frac{7\,e^{\frac{t}{2}}\left(\cos\left(\frac{\sqrt{3}\,t}{2}\right)+\frac{\sqrt{3}\,\sin\left(\frac{\sqrt{3}\,t}{2}\right)}{7}\right)}{9} - 5$$

Solution y(t) using dsolve(.) in Matlab

$$y(t) = 3\,t + \frac{47\,e^{-t}}{9} + \frac{5\,t\,e^{-t}}{3} + \frac{7\,e^{\frac{t}{2}}\,\cos\left(\frac{\sqrt{3}\,t}{2}\right)}{9} + \frac{t^2\,e^{-t}}{6} - t^2 + \frac{t^3}{6} + \frac{\sqrt{3}\,e^{\frac{t}{2}}\,\sin\left(\frac{\sqrt{3}\,t}{2}\right)}{9} - 5$$

ODE45 and comparison

Differential Equation and Initial Conditions

$$y''''''(t) + y''''(t) = t\,e^{-t}$$

$$(\; y(0)=1 \quad y'(0)=0 \quad y''(0)=0 \quad y'''(0)=-1 \quad y''''(0)=0 \quad y'''''(0)=0 \quad y''''''(0)=1 \;)$$

Vector: 1st first order differential equations for ODE [$Y_1 \equiv y(t)$] ⇒

$$\begin{pmatrix} Y_1'(t) = Y_2 \\ Y_2'(t) = Y_3 \\ Y_3'(t) = Y_4 \\ Y_4'(t) = Y_5 \\ Y_5'(t) = Y_6 \\ Y_6'(t) = Y_7 \\ Y_7'(t) = t\,e^{-t} - Y_5 \end{pmatrix}$$

Example # 4.21

```
A=[1,0,-1,1,0,-1,0,0];IC=[1;0;0;0;0;0;1];gt='t';
```

Theory of Laplace Transforms: Higher order differential Equations

The Differential Equation

$$A_n y^{(n)}(t) + A_{n-1} y^{(n-1)}(t) + \cdots + A_1 y^{(1)}(t) + A_0 y(t) = g(t)$$

The Initial Conditions (IC)

$$y(0) = y_0, y^{(1)}(0) = y_1, y^{(2)}(0) = y_2, \cdots, y^{(n-1)}(0) = y_{n-1}$$

The Laplace Transform Y(s)

$$\frac{(A_n s^{n-1} + A_{n-1} s^{n-2} + \cdots + A_2 s + A_1) y_0 + (A_n s^{n-2} + A_{n-1} s^{n-3} + \cdots + A_2) y_1 + \cdots + y_{n-1} A_n}{(A_n s^n + A_{n-1} s^{n-1} + \cdots + A_1 s + A_0)}$$

$$+$$

$$\frac{G(s)}{(A_n s^n + A_{n-1} s^{n-1} + \cdots + A_1 s + A_0)}$$

Inverse Laplace based solution

Differential Equation and Initial Conditions

$$y''''''(t) - y''''(t) + y'''(t) - y''(t) = t$$

$$\left(y(0) = 1 \quad y'(0) = 0 \quad y''(0) = 0 \quad y'''(0) = 0 \quad y''''(0) = 0 \quad y'''''(0) = 0 \quad y''''''(0) = 1 \right)$$

Laplace Transform $Y_L(s)$ of y(t)

$$Y_L(s) = \frac{s^8 - s^6 + s^5 - s^3 + s^2 + 1}{s^4\,(s-1)\,(s+1)^2\,(s^2 - s + 1)}$$

Solution y(t) using inverse Laplace transform of $Y_L(s)$

$$y(t) = \frac{e^t}{2} - \frac{3\,e^{-t}}{2} - 2\,t - \frac{t\,e^{-t}}{3} - \frac{t^3}{6} + \frac{2\sqrt{3}\,e^{\frac{t}{2}}\,\sin\left(\frac{\sqrt{3}\,t}{2}\right)}{9} + 2$$

Solution y(t) using dsolve(.) in Matlab

$$y(t) = \frac{e^t}{2} - \frac{3\,e^{-t}}{2} - 2\,t - \frac{t\,e^{-t}}{3} - \frac{t^3}{6} + \frac{2\sqrt{3}\,e^{\frac{t}{2}}\,\sin\left(\frac{\sqrt{3}\,t}{2}\right)}{9} + 2$$

ODE45 and comparison

Differential Equation and Initial Conditions

$$y''''''(t) - y'''''(t) + y'''(t) - y''(t) = t$$

$$(\; y(0) = 1 \quad y'(0) = 0 \quad y''(0) = 0 \quad y'''(0) = 0 \quad y''''(0) = 0 \quad y'''''(0) = 0 \quad y''''''(0) = 1 \;)$$

Vector: 1st first order differential equations for ODE [$Y_1 \equiv y(t)$] ⇒

$$\begin{pmatrix} Y_1'(t) = Y_2 \\ Y_2'(t) = Y_3 \\ Y_3'(t) = Y_4 \\ Y_4'(t) = Y_5 \\ Y_5'(t) = Y_6 \\ Y_6'(t) = Y_7 \\ Y_7'(t) = t + Y_3 - Y_5 + Y_6 \end{pmatrix}$$

Example # 4.22

```
A=[1,0,-1,1,0,-1,0,0];IC=[1;0;0;0;0;0;1];gt='t+exp(-t)';
```

Theory of Laplace Transforms: Higher order differential Equations

The Differential Equation

$$A_n y^{(n)}(t) + A_{n-1} y^{(n-1)}(t) + \cdots + A_1 y^{(1)}(t) + A_0 y(t) = g(t)$$

The Initial Conditions (IC)

$$y(0) = y_0, y^{(1)}(0) = y_1, y^{(2)}(0) = y_2, \cdots, y^{(n-1)}(0) = y_{n-1}$$

The Laplace Transform Y(s)

$$\frac{(A_n s^{n-1} + A_{n-1} s^{n-2} + \cdots + A_2 s + A_1) y_0 + (A_n s^{n-2} + A_{n-1} s^{n-3} + \cdots + A_2) y_1 + \cdots + y_{n-1} A_n}{(A_n s^n + A_{n-1} s^{n-1} + \cdots + A_1 s + A_0)}$$

$$+$$

$$\frac{G(s)}{(A_n s^n + A_{n-1} s^{n-1} + \cdots + A_1 s + A_0)}$$

Inverse Laplace based solution

Differential Equation and Initial Conditions

$$y''''''(t) - y'''''(t) + y''''(t) - y''(t) = t + \mathrm{e}^{-t}$$

$$\left(\ y(0) = 1 \quad y'(0) = 0 \quad y''(0) = 0 \quad y'''(0) = 0 \quad y''''(0) = 0 \quad y'''''(0) = 0 \quad y''''''(0) = 1\ \right)$$

Laplace Transform $Y_L(s)$ of y(t)

$$Y_L(s) = \frac{s^9+s^8-s^7+s^5-s^4+2\,s^2+s+1}{s^4\,(s-1)\,(s+1)^3\,(s^2-s+1)}$$

Solution y(t) using inverse Laplace transform of $Y_L(s)$

$$y(t) = \frac{5\,\mathrm{e}^t}{8} - \frac{197\,\mathrm{e}^{-t}}{72} - 3\,t - \frac{11\,t\,\mathrm{e}^{-t}}{12} - \frac{t^2\,\mathrm{e}^{-t}}{12} - \frac{t^3}{6} + \frac{\mathrm{e}^{\frac{t}{2}}\left(\cos\left(\frac{\sqrt{3}\,t}{2}\right) + 3\,\sqrt{3}\,\sin\left(\frac{\sqrt{3}\,t}{2}\right)\right)}{9} + 3$$

Solution y(t) using dsolve(.) in Matlab

$$y(t) = \frac{5\,\mathrm{e}^t}{8} - \frac{197\,\mathrm{e}^{-t}}{72} - 3\,t - \frac{11\,t\,\mathrm{e}^{-t}}{12} + \frac{\mathrm{e}^{\frac{t}{2}}\,\cos\left(\frac{\sqrt{3}\,t}{2}\right)}{9} - \frac{t^2\,\mathrm{e}^{-t}}{12} - \frac{t^3}{6} + \frac{\sqrt{3}\,\mathrm{e}^{\frac{t}{2}}\,\sin\left(\frac{\sqrt{3}\,t}{2}\right)}{3} + 3$$

ODE45 and comparison

Differential Equation and Initial Conditions

$$y''''''(t) - y'''''(t) + y''''(t) - y''(t) = t + \mathrm{e}^{-t}$$

$$\left(\ y(0) = 1 \quad y'(0) = 0 \quad y''(0) = 0 \quad y'''(0) = 0 \quad y''''(0) = 0 \quad y'''''(0) = 0 \quad y''''''(0) = 1\ \right)$$

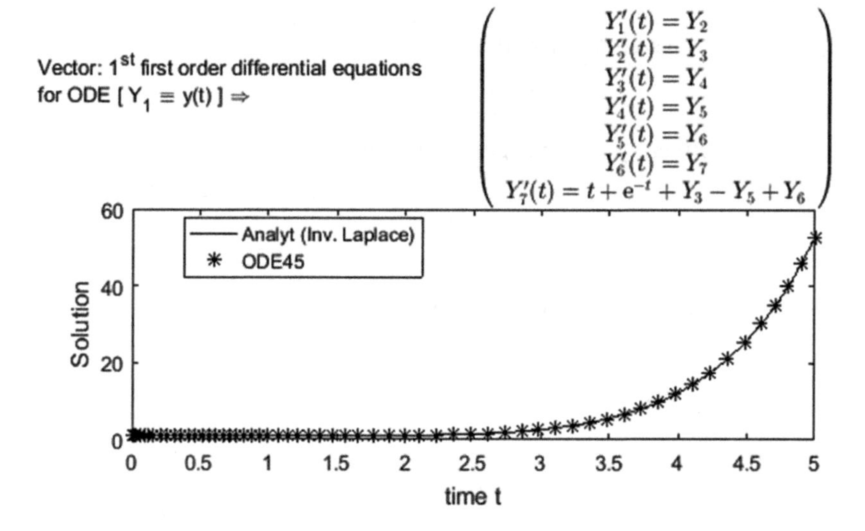

Example # 4.23

```
A=[1,0,-1,1,0,-1,0,0];IC=[1;0;0;0;0;0;1];gt='exp(-t)';
```

Theory of Laplace Transforms: Higher order differential Equations

The Differential Equation

$$A_n y^{(n)}(t) + A_{n-1} y^{(n-1)}(t) + \cdots + A_1 y^{(1)}(t) + A_0 y(t) = g(t)$$

The Initial Conditions (IC)

$$y(0) = y_0, y^{(1)}(0) = y_1, y^{(2)}(0) = y_2, \cdots, y^{(n-1)}(0) = y_{n-1}$$

The Laplace Transform Y(s)

$$\frac{(A_n s^{n-1} + A_{n-1} s^{n-2} + \cdots + A_2 s + A_1) y_0 + (A_n s^{n-2} + A_{n-1} s^{n-3} + \cdots + A_2) y_1 + \cdots + y_{n-1} A_n}{(A_n s^n + A_{n-1} s^{n-1} + \cdots + A_1 s + A_0)}$$

$$+$$

$$\frac{G(s)}{(A_n s^n + A_{n-1} s^{n-1} + \cdots + A_1 s + A_0)}$$

Inverse Laplace based solution

Differential Equation and Initial Conditions

$$y''''''(t) - y''''(t) + y'''(t) - y''(t) = e^{-t}$$

$$(\ y(0) = 1 \quad y'(0) = 0 \quad y''(0) = 0 \quad y'''(0) = 0 \quad y''''(0) = 0 \quad y'''''(0) = 0 \quad y''''''(0) = 1\ )$$

Laplace Transform $Y_L(s)$ of y(t)

$$Y_L(s) = \frac{s^7 + s^6 - s^5 + s^3 - s^2 + 2}{s^2\,(s-1)\,(s+1)^3\,(s^2 - s + 1)}$$

Solution y(t) using inverse Laplace transform of $Y_L(s)$

$$y(t) = \frac{3\,e^t}{8} - \frac{131\,e^{-t}}{72} - 2\,t - \frac{3\,t\,e^{-t}}{4} - \frac{t^2\,e^{-t}}{12} + \frac{4\,e^{\frac{t}{2}}\left(\cos\left(\frac{\sqrt{3}\,t}{2}\right) + \frac{\sqrt{3}\sin\left(\frac{\sqrt{3}\,t}{2}\right)}{2}\right)}{9} + 2$$

Solution y(t) using dsolve(.) in Matlab

$$y(t) = \frac{3\,e^t}{8} - \frac{131\,e^{-t}}{72} - 2\,t - \frac{3\,t\,e^{-t}}{4} + \frac{4\,e^{\frac{t}{2}}\cos\left(\frac{\sqrt{3}\,t}{2}\right)}{9} - \frac{t^2\,e^{-t}}{12} + \frac{2\sqrt{3}\,e^{\frac{t}{2}}\sin\left(\frac{\sqrt{3}\,t}{2}\right)}{9} + 2$$

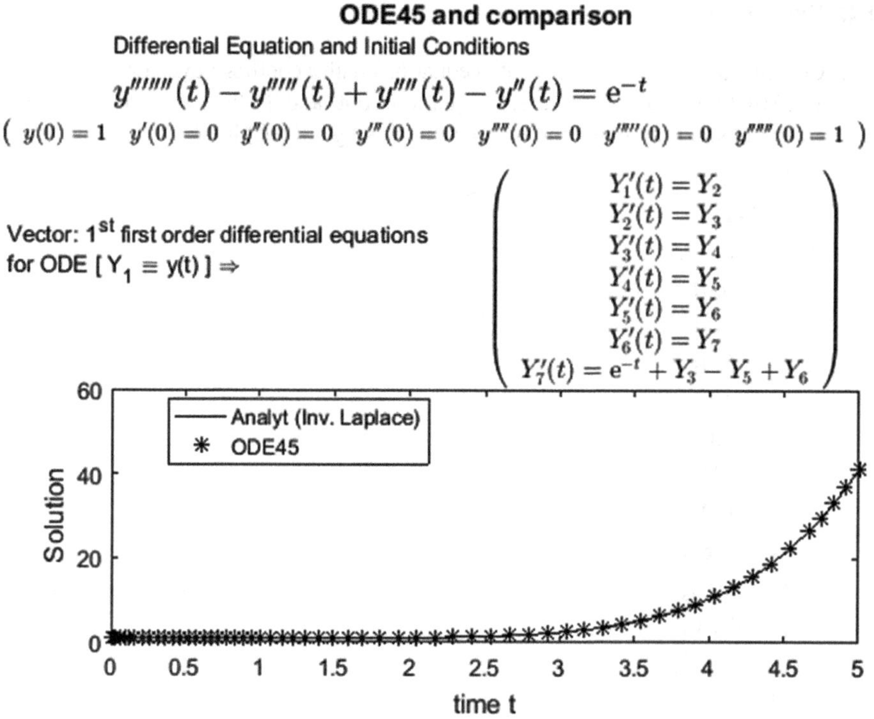

ODE45 and comparison

Differential Equation and Initial Conditions

$$y'''''''(t) - y'''''(t) + y''''(t) - y''(t) = e^{-t}$$

$$(\; y(0) = 1 \quad y'(0) = 0 \quad y''(0) = 0 \quad y'''(0) = 0 \quad y''''(0) = 0 \quad y'''''(0) = 0 \quad y''''''(0) = 1 \;)$$

Vector: 1st first order differential equations for ODE [$Y_1 \equiv y(t)$] ⇒

$$\begin{pmatrix} Y_1'(t) = Y_2 \\ Y_2'(t) = Y_3 \\ Y_3'(t) = Y_4 \\ Y_4'(t) = Y_5 \\ Y_5'(t) = Y_6 \\ Y_6'(t) = Y_7 \\ Y_7'(t) = e^{-t} + Y_3 - Y_5 + Y_6 \end{pmatrix}$$

Analyt (Inv. Laplace)
ODE45
Solution
time t

4.7 Summary

In this chapter, we explored ways of solving higher order differential equations with constant coefficients. The characteristic equation is used to formulate the solution to the homogeneous differential equation while the concept of the Wronskian introduced in Chapter 3 is invoked to get the particular solution. The difficulties encountered in using the roots of the characteristic equation of the higher order differential equations is clearly articulated in terms of the relationships that might exist among the various roots making it difficult to proceed as the order increases beyond 4. As has been done throughout, the solution obtained through the use of roots and the Wronskian is compared to the one obtained using the theory of Laplace transforms. Additional confirmation is provided through the use of numerical techniques based on the Runge-Kutta method in MATLAB. Differential equations with orders larger than the 4th are solved using Laplace transforms and results are compared using the Runge-Kutta method.

4.8 Exercises

1. Consider a higher order differential equations with a coefficient vector A=[An,An-1,..,A0]. Obtain the characteristic equation, roots and the homogeneous solution. Does the solution match the one obtained using solve(.) in MATLAB.

   A=[1,-3,-1,0]
   A=[1,0,-2,-1]
   A=[1,-1,0,-4];
   A=[1,-1,0,0];
   A=[1,0,0,-3];
   A=[1,0,-3,0];
   A=[1,-1,1,0];
   A=[1,0,0,0];
   A=[1,0,0,1];
   A=[1,-4,0,0];
   A=[1,0,0,-8,0];
   A=[1,4,0,0,0];
   A=[1,0,0,1,0];
   A=[1,0,0,-1,0];
   A=[1,-1,0,0,0];

2. A number of third order differential equations are identified by the coefficient vector A=[A3,A2,A1,A0]. For each of the cases, a forcing function g(t) is provided. Obtain the particular solution in each case. Provide the homogeneous solution set and Wronskian in each case.

   A=[1,1,0,0]; g(t)=texp(-t)+t
   A=[1,-1,0,0];g(t)=t;
   A=[1,0,4,0];g(t)=cos(2t)
   A=[2,-5,2,0]=g(t)=exp(2t)
   A=[1,0,-9,0];g(t)=cosh(3t);
   A=[1,-4,-5,0]; g(t)=t+t^2exp(5*t)
   A=[2,0,-2,0];g(t)=sinh(t)
   A=[1,2,0,-1];g(t)=cos(t);
   A=[2,1,0,0];g(t)=t^2;
   A=[1,0,0,-8,0];g(t)=t+exp(2t)
   A=[1,4,0,0,0];g(t)=t*exp(-4t)
   A=[1,0,0,1,0];g(t)=exp(-t)
   A=[1,0,0,-1,0];g(t)=5t+exp(t)
   A=[1,-1,0,0,0];g(t)=t+t^2
   A=[1,-1,-1,0,0];g(t)=t;

3. For the given initial conditions, obtain the complete solutions. Verify the results using Laplace transforms and ode45.

```
A=[1,1,0,0];g(t)=t*exp(-t)+t;IC=[1,-1,0];
A=[1,-1,0,0];g(t)=t;IC=[1,1,0];
A=[1,0,4,0];g(t)=cos(2*t);IC=[0,-1,0];
A=[2,-5,2,0];g(t)=exp(2*t);IC=[-1,-1,0];
A=[1,0,-9,0];g(t)=cosh(3*t);IC=[0,-1,1];
A=[1,-4,-5,0];g(t)=t+t^2*exp(5*t);IC=[-1,-1,1];
A=[2,0,-2,0];g(t)=sinh(t);IC=[0,-1,-1];
A=[1,2,0,-1];g(t)=cos(t);IC=[-1,1,0];
A=[2,1,0,0];g(t)=t^2;IC=[0,-1,1];
A=[1,0,0,-8,0];g(t)=t+exp(2*t);IC=[1,0,-1,1];
A=[1,4,0,0,0];g(t)=t*exp(-4*t);IC=[1,0,1,1];
A=[1,0,0,1,0];g(t)='exp(-t)';IC=[1,1,0,1];
A=[1,0,0,-1,0];g(t)='5*t+exp(t)';IC=[1,0,-1,0];
A=[1,-1,0,0,0];g(t)=t+t^2;IC=[1,1,0,0];
A=[1,-1,-1,0,0];g(t)=t;IC=[1,0,0,0];
```

CHAPTER 5

First Order Coupled Differential Equations with Constant Coefficients

5.1 Introduction 236
5.2 A Pair of Coupled Differential Equations 237
5.2.1 Homogeneous systems 237
5.2.2 First order non-homogeneous systems 268
5.3 Multiple First Order Coupled Differential Equations of Constant Coefficients 279
5.3.1 Solution using eigenvalues and eigenvectors 280
5.3.2 Solution using Laplace transforms 281
5.3.3 Examples 282
5.4 Numerical Solutions 295
5.5 Examples 298
5.6 Summary 381
5.7 Exercises 381

5.1 Introduction

Systems often comprise more than one component or rely on two or more components linked together for efficient operations. Examples include electrical, mechanical and chemical flow systems, and traffic networks. Such systems require two or more dependent variables for modeling and the behavior or operation of the systems can be described in terms of a set of coupled first order differential equations. The analysis presented here examines only linear systems with constant coefficients. We start with the case of a pair of coupled equations before we explore coupled systems with three or more variables.

5.2 A Pair of Coupled Differential Equations

An example of a system with two variables $x_1(t)$ and $x_2(t)$ can be written as

$$\begin{aligned}\frac{dx_1}{dt} &= A_{11}x_1 + A_{12}x_2 \\ \frac{dx_2}{dt} &= A_{21}x_1 + A_{22}x_2\end{aligned} \tag{5.1}$$

Equation (5.1) can be written as a matrix equation (Appendix D)

$$\frac{d\vec{x}}{dt} = \begin{bmatrix} \frac{dx_1}{dt} \\ \frac{dx_2}{dt} \end{bmatrix} = \begin{bmatrix} A_{11} & A_{12} \\ A_{21} & A_{22} \end{bmatrix}\begin{bmatrix} x_1 \\ x_2 \end{bmatrix} = A\vec{x} \tag{5.2}$$

In eqn. (5.2), A is the coefficient matrix and $\vec{x}$ is the vector

$$\vec{x} = \begin{bmatrix} x_1 \\ x_2 \end{bmatrix}. \tag{5.3}$$

While eqn. (5.1) is homogeneous, the non-homogeneous set of differential equations can be expressed as

$$\frac{d\vec{x}}{dt} = \begin{bmatrix} \frac{dx_1}{dt} \\ \frac{dx_2}{dt} \end{bmatrix} = A\vec{x} + \vec{g}(t). \tag{5.4}$$

In eqn. (5.4),

$$\vec{g}(t) = \begin{bmatrix} g_1(t) \\ g_2(t) \end{bmatrix}. \tag{5.5}$$

In eqn. (5.5), $g_1(t)$ and $g_2(t)$ are the forcing functions. If the initial conditions are available, they can be expressed in vectorial form as

$$\vec{x}(0) = \begin{bmatrix} x_1(0) \\ x_2(0) \end{bmatrix}. \tag{5.6}$$

We will now consider the homogeneous and non-homogeneous differential equations separately.

5.2.1 Homogeneous systems

5.2.1.1 Solution using eigenvalues and eigenvectors

Consider the case of a pair of first order homogeneous equations expressed as

$$\frac{d\vec{x}}{dt} = \begin{bmatrix} \frac{dx_1}{dt} \\ \frac{dx_2}{dt} \end{bmatrix} = A\vec{x}, \quad \vec{x}(0) = \begin{bmatrix} x_1(0) \\ x_2(0) \end{bmatrix}. \tag{5.7}$$

In eqn. (5.7),

$$A = \begin{bmatrix} A_{11} & A_{12} \\ A_{21} & A_{22} \end{bmatrix} \tag{5.8}$$

The solution to the first order homogeneous (coupled) system can be obtained by examining the coefficient matrix A in eqn. (5.8) and its properties in terms of the eigenvalues and eigenvectors. The concepts of eigenvalues and eigenvectors are discussed in Appendix D. From the analysis presented in Appendix D, three distinct possibilities exist for the two sets of eigenvectors: real and distinct, real and equal or form a complex conjugate pair.

Distinct eigenvectors

When the coefficient matrix is not defective (it has two distinct eigenvectors), two possibilities exist, either the eigenvalues are distinct or equal. Consider the case where the two eigenvalues are real and distinct. If the two eigenvalues associated with A are λ_1 and λ_2 with corresponding eigenvectors $\vec{v}_1$ and $\vec{v}_2$, the two linearly independent solution vectors are

$$\begin{aligned} \vec{X}_1(t) &= \vec{v}_1 e^{\lambda_1 t} \\ \vec{X}_2(t) &= \vec{v}_2 e^{\lambda_2 t} \end{aligned}. \tag{5.9}$$

The fundamental matrix X(t) associated with the system is the [2 x 2] matrix

$$X(t) = \begin{bmatrix} \vec{X}_1 & \vec{X}_2 \end{bmatrix} \tag{5.10}$$

The general solution of the homogeneous system is the superposition of the two solutions in eqn. (5.9) as

$$\vec{x}(t) = X\vec{c} = c_1\vec{X}_1(t) + c_2\vec{X}_2(t) = \sum_{k=1}^{2} c_k \vec{X}_k(t). \tag{5.11}$$

In eqn. (5.11), $\vec{c}$ is the vector with two elements c_1 and c_2 representing the two unknown constants or scaling factors.

The two unknown coefficients can be evaluated from the initial conditions $\vec{x}(0)$ in eqn. (5.7). The analysis and the solution given in eqn. (5.9) and eqn. (5.11) do not change even when $\lambda_1 = \lambda_2 = \lambda$, if the two eigenvectors are distinct because the set of equations in eqn. (5.9) represents two linearly independent solutions.

Complex eigenvectors

When eigenvectors are complex, eigenvalues are also complex and vice versa. If the two eigenvalues are

$$\lambda_{1,2} = \lambda_{\pm} = \alpha \pm j\beta. \tag{5.12}$$

In eqn. (5.12), α and β are real. The two eigenvectors are

$$\vec{v}_{1,2} = \vec{v}_{\pm} = \vec{u} \pm j\vec{w}. \tag{5.13}$$

The two solution vectors are

$$\begin{aligned}\vec{X}_1(t) &= e^{\alpha t}\left[\vec{u}\cos(\beta t) - \vec{w}\sin(\beta t)\right] \\ \vec{X}_2(t) &= e^{\alpha t}\left[\vec{u}\sin(\beta t) + \vec{w}\cos(\beta t)\right]\end{aligned}. \tag{5.14}$$

The general solution is once again given by eqn. (5.11) with the solution vectors given in eqn. (5.14).

Equal eigenvectors

When the eigenvectors are equal (this certainly implies that eigenvalues are equal), the coefficient matrix is defective and two independent solution vectors do not exist. This case requires the generation of a pair of generalized eigenvectors as described in Appendix D. While it is possible to keep the single eigenvector obtained and get an extra single generalized eigenvector and use these two, the approach described here does not use the original eigenvector. Instead, a pair of generalized eigenvectors is obtained from the solution of

$$(A - \lambda I_2)^2 \vec{v} = 0 . \tag{5.15}$$

In eqn. (5.15), I_2 is the 2 x 2 identity matrix and if the pair of generalized eigenvectors are $\vec{v}_{g_1}$ and $\vec{v}_{g_2}$, the solution vectors become (Appendix D)

$$\begin{aligned}\vec{X}_1(t) &= e^{\lambda t}\left[\vec{v}_{g_1} + t(A - \lambda I_2)\vec{v}_{g_1}\right] \\ \vec{X}_2(t) &= e^{\lambda t}\left[\vec{v}_{g_2} + t(A - \lambda I_2)\vec{v}_{g_2}\right]\end{aligned}. \tag{5.16}$$

The general solution is once again given by eqn. (5.11) with the solution vectors given in eqn. (5.16). It should be noted that for the case of a 2 x 2 defective matrix, the generalized eigenvectors will be (see appendix D)

$$\begin{aligned}\vec{v}_{g_1} &= \begin{bmatrix}1\\0\end{bmatrix} \\ \vec{v}_{g_2} &= \begin{bmatrix}0\\1\end{bmatrix}\end{aligned} \tag{5.17}$$

5.2.1.2 Solution without using eigenvectors

While the solutions to a set of first order coupled homogeneous differential equations with constant coefficients is simple and straightforward, it creates some difficulties when the coefficient matrix is defective. It is possible to obtain the solutions using the characteristic equation approach outlined in Chapter 3 in connection with the solutions of second order homogeneous differential equations with constant coefficients. The same approach can be implemented here by converting a pair of homogeneous first order equations into a second order differential equation of constant coefficients. Consider the first differential equation in x_1 in eqn. (5.1). Its differentiation leads to

$$x_1'' = A_{11}x_1' + A_{12}x_2' . \tag{5.18}$$

Equation (5.18) can be rewritten using the differential equation for x_2 from eqn. (5.1) and creating a second order differential equation in x_1 as

$$x_1'' = A_{11}x_1' + A_{12}\left[A_{21}x_1 + A_{22}x_2\right] = A_{11}x_1' + A_{12}A_{21}x_1 + A_{22}\left(x_1' - A_{11}x_1\right) \quad (5.19)$$

Simplifying eqn. (5.19) leads to

$$x_1'' - \left(A_{11} + A_{22}\right)x_1' + \left(A_{11}A_{22} - A_{12}A_{21}\right)x_1 = 0 \quad (5.20)$$

Using the method based on the roots of the characteristic equation developed in Chapter 3, it is possible to get a general solution for $x_1(t)$. Substituting $x_1(t)$ in the first differential equation eqn. (5.1), a general solution for $x_2(t)$ can be obtained. Using the initial conditions, the unknown coefficients can be obtained completing the solution.

In some of the instances, $A_{12}=0$ and $A_{21}\neq 0$. In this case, the first differential equation eqn. (5.1) is used to obtain a second order differential equation in $x_2(t)$ and the procedure is repeated to obtain the solution for $x_1(t)$. In the special case of $A_{12}= A_{21}=0$, the two differential equations are uncoupled and they can easily be solved.

5.2.1.3 Solution using Laplace transforms

If the initial conditions are given, the solutions set can be obtained using the concept of Laplace and inverse Laplace transforms. The solutions vector can be written as

$$\begin{bmatrix} x_1(t) \\ x_2(t) \end{bmatrix} = L^{-1}\left\{\left[sI_2 - A\right]^{-1}\begin{bmatrix} x_1(0) \\ x_2(0) \end{bmatrix}\right\}. \quad (5.21)$$

In eqn. (5.21), I_2 is a [2 x 2] identity matrix and $[sI_2 - A]^{-1}$ is the inverse of the matrix $[sI_2 - A]$. Inversion of a matrix is described in Appendix D. Solutions may also be obtained using numerical techniques outlined in Appendix A.

5.2.1.4 Analysis of coupled first order homogeneous systems with constant coefficients

An analysis on a set of homogeneous first order coupled differential equations of constant coefficients will be incomplete without a discussion of the critical point. The critical point or the equilibrium point is the solution to

$$\frac{d\vec{x}}{dt} = 0 = A\vec{x}\ . \quad (5.22)$$

Because we have a homogeneous system, the critical point will be [0. 0] or the origin. The critical point and the behavior of the solutions of the differential equations can be used to characterize the stability of the system described by the set of differential equations. The phase plane plots or phase portraits (Appendix C) show the behavior of the solutions and hence the behavior of the system. A number of specific coefficient matrices have been analyzed to demonstrate the usefulness of the phase portrait and its use in understanding the behavior of the system described by the coupled first order system. While the detailed procedure is described in Appendix C, the analytical solutions are not plotted alongside the quiver(.) plot.

```
clear;close all;
% generate phase plots for several coefficient matrices
% P M Shankar, July 2017
for k=1:12
   if k==1
A=[-1,2;2,1];
   elseif k==2
   A=[-1,2;-4,1];
   elseif k==3
A=[-1,-1;1,-1];
   elseif k==4
      A=[2,1;-1,2];
   elseif k==5
      A=[2,-2;1,-1];
   elseif k==6
      A=[-2,2;1,-1];
   elseif k==7
     A=[2,0;1,1];
   elseif k==8
     A=[2,0;0,2];
  elseif k==9
    A=[-1,0;-1,-1];
  elseif k==10
   A=[-2,0;0,-2];
   elseif k==11
 A=[-2,1;0,-2];
  elseif k==12
   A=[2,2;-2,-2];
 end;
[x1,x2]=meshgrid(-3:.5:3,-3:.5:3);
DX1=A(1,1)*x1+A(1,2)*x2;
DX2=A(2,1)*x1+A(2,2)*x2;
figure
quiver(x1,x2,DX1,DX2,1.5,'color','r'),xlabel('x_1(t)'),ylabel('x_2(t)')
xlim([-3,3]),ylim([-3,3])
lambda=eig(sym(A));
title(['EigenValues\Rightarrow',' [',num2str(double(lambda')),']'])
end;
```

The results are presented below. Consider the case of the coefficient matrix

$$A = \begin{bmatrix} -1 & 2 \\ 2 & 1 \end{bmatrix}. \tag{5.23}$$

The eigenvalues are real with one of them being positive and the other being negative resulting in the phase portrait in Figure 5.1. The solution moves away from the critical point indicating that the system is unstable.

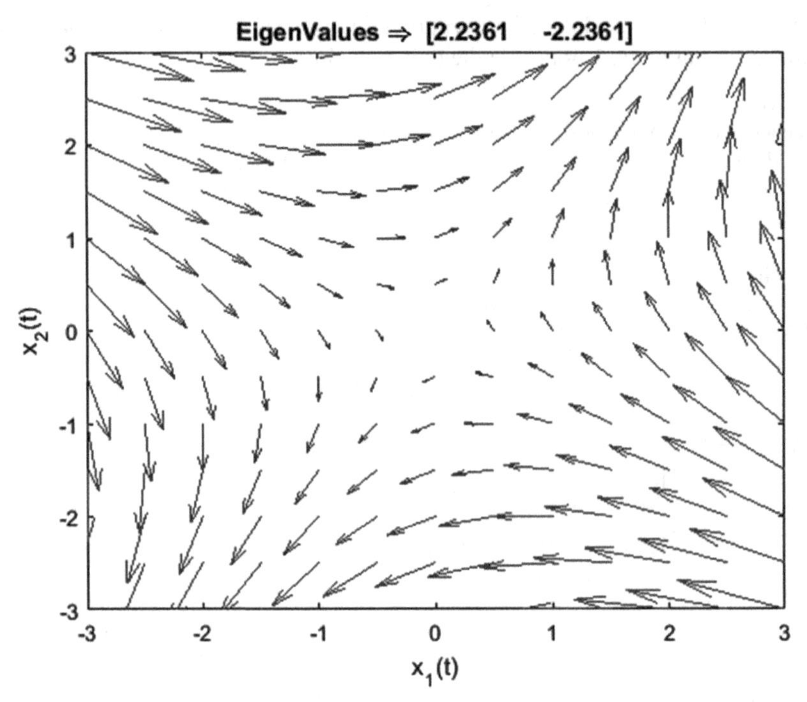

Figure 5.1 The eigenvalues are real and have opposite signs. The solutions move away from the critical point [0,0] indicating that the system is unstable.

Consider the case of the coefficient matrix

$$A = \begin{bmatrix} -1 & 2 \\ -4 & 1 \end{bmatrix}. \tag{5.24}$$

The eigenvalues are purely imaginary (complex conjugate pair) resulting in the phase portrait in Figure 5.2. The solutions continously encircle the critical point [0,0] indicating that the system is oscillatory, because the solutions are sines and cosines.

Consider the case of the coefficient matrix

$$A = \begin{bmatrix} -1 & -1 \\ 1 & -1 \end{bmatrix}. \tag{5.25}$$

The eigenvalues form a complex conjugate pair with the real part being negative resulting in the phase portrait in Figure 5.3. The solutions continously move towards the critical point [0,0] indicating that the system is asymptotically stable.

Consider the case of the coefficient matrix

$$A = \begin{bmatrix} 2 & 1 \\ -1 & 2 \end{bmatrix}. \tag{5.26}$$

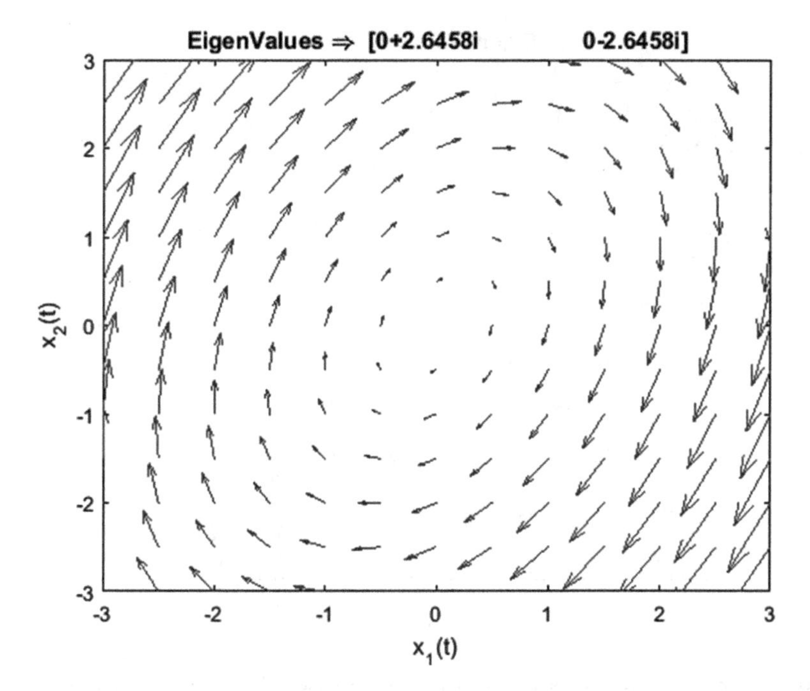

Figure 5.2 The eigenvalues are purely imaginary. The solutions continously encircle the critical point [0,0], indicating that the system is oscillatory.

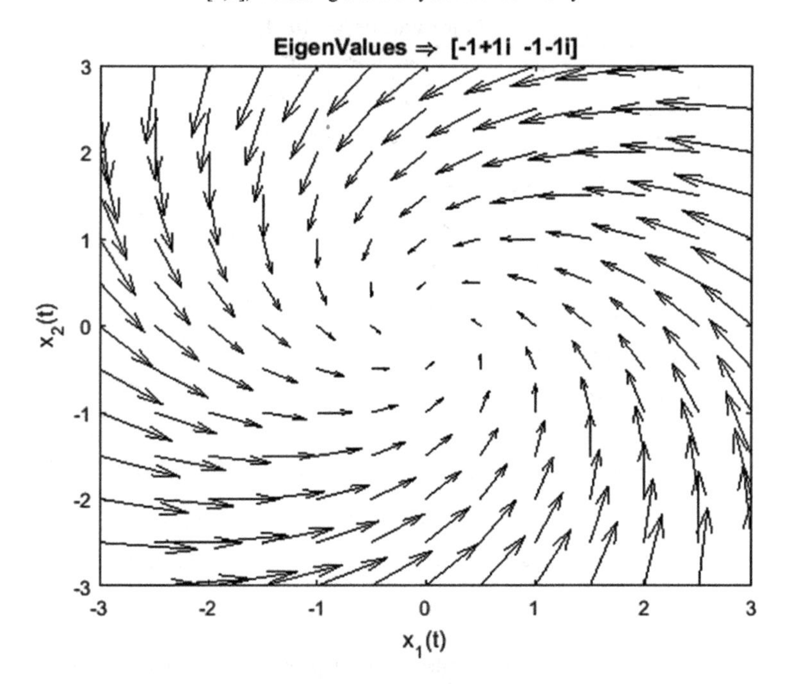

Figure 5.3 The eigenvalues are complex with a negative real part. The solutions move towards the critical point [0,0] indicating that the system is asymptotically stable.

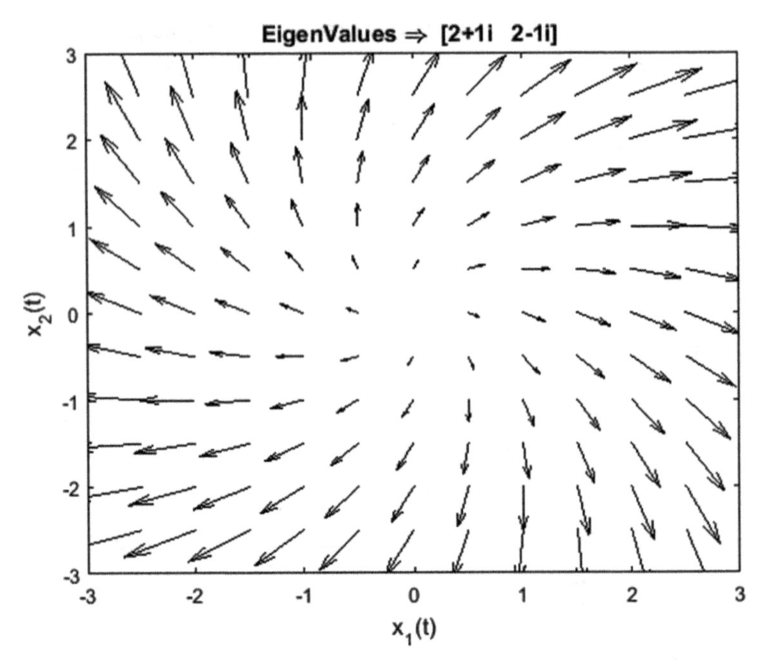

Figure 5.4 The eigenvalues are complex with a positive real part. The solutions move away from the critical point [0,0] indicating that the system is unstable.

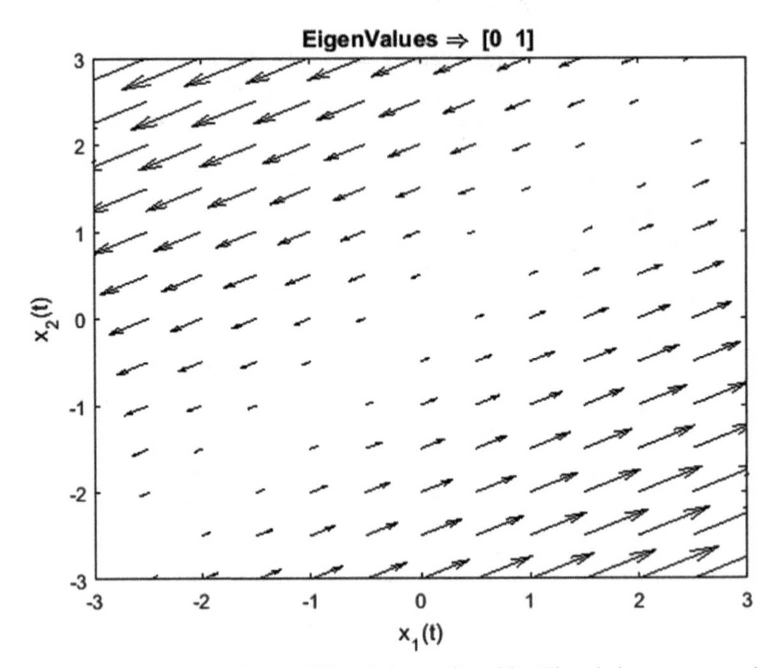

Figure 5.5 One of the eigenvalues is zero while the other one is positive. The solutions move away (opposite direction) from a diagonal line going through the critical point [0,0] indicating that the system is unstable.

The eigenvalues form a complex conjugate pair with the real part being positive resulting in the phase portrait in Figure 5.4. The solutions continously move away from the critical point [0,0] indicating that the system is unstable.
Consider the case of the coefficient matrix

$$A = \begin{bmatrix} 2 & -2 \\ 1 & -1 \end{bmatrix}. \tag{5.27}$$

The eigenvalues are real with one of them being a zero and the other one being positive resulting in the phase portrait in Figure 5.5. The solutions move away (opposite direction) from a diagonal line through critical point [0,0] indicating that the system is unstable. The unstable nature in this case being defined w.r.t. a line is due to the existence of an eigenvalue of 0.

Consider the case of the coefficient matrix

$$A = \begin{bmatrix} -2 & 2 \\ 1 & -1 \end{bmatrix}. \tag{5.28}$$

The eigenvalues are real with one of them being a zero and the other one being negative resulting in the phase portrait in Figure 5.6. The solutions move towards the diagonal line through the critical point [0,0], indicating that the system is stable. The stability in this case being defined w.r.t. a line is due to the existence of an eigenvalue of 0.

Consider the case of the coefficient matrix

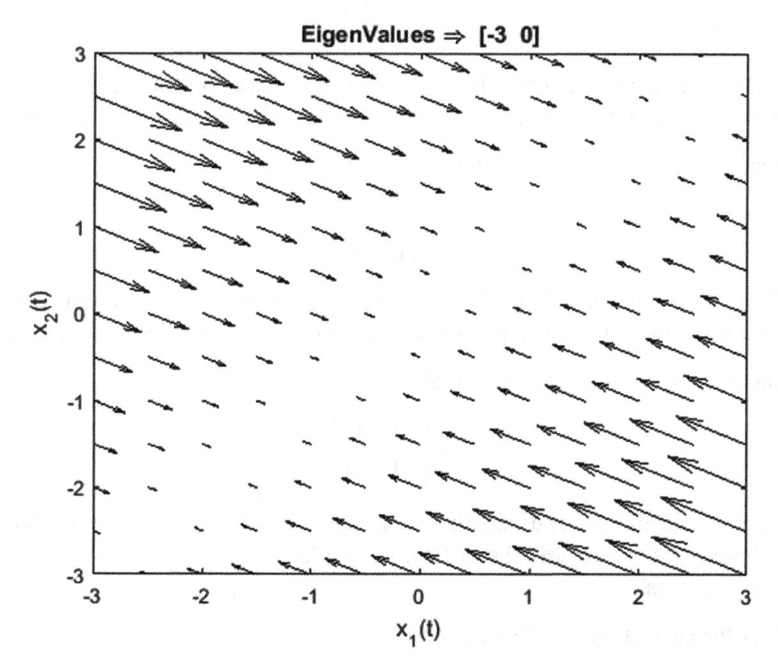

Figure 5.6 One of the eigenvalues is zero while the other one is negative. The solutions point to the a diagonal line through the critical point [0,0] indicating that the system is stable.

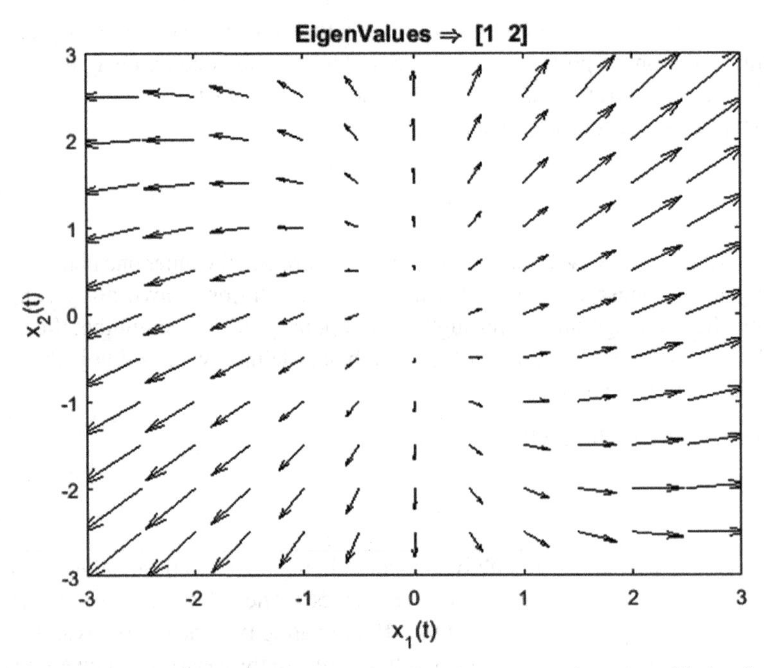

Figure 5.7 Both eigenvalues are positive. The solutions move away from the critical point [0,0], indicating that the system is unstable.

$$A = \begin{bmatrix} 2 & 0 \\ 1 & 1 \end{bmatrix}. \tag{5.29}$$

The eigenvalues are real and positive resulting in the phase portrait in Figure 5.7. The solutions move away from the critical point [0,0] indicating that the system is unstable.

Consider the case of the coefficient matrix

$$A = \begin{bmatrix} 2 & 0 \\ 0 & 2 \end{bmatrix}. \tag{5.30}$$

The eigenvalues are real and positive resulting in the phase portrait in Figure 5.8. The solutions move away from the critical point [0,0], indicating that the system is unstable.

Consider the case of the coefficient matrix

$$A = \begin{bmatrix} -1 & 0 \\ -1 & -1 \end{bmatrix}. \tag{5.31}$$

The eigenvalues are real and negative resulting in the phase portrait in Figure 5.9. The solutions move towards the critical point [0,0], indicating that the system is asymptotically stable.

Consider the case of the coefficient matrix

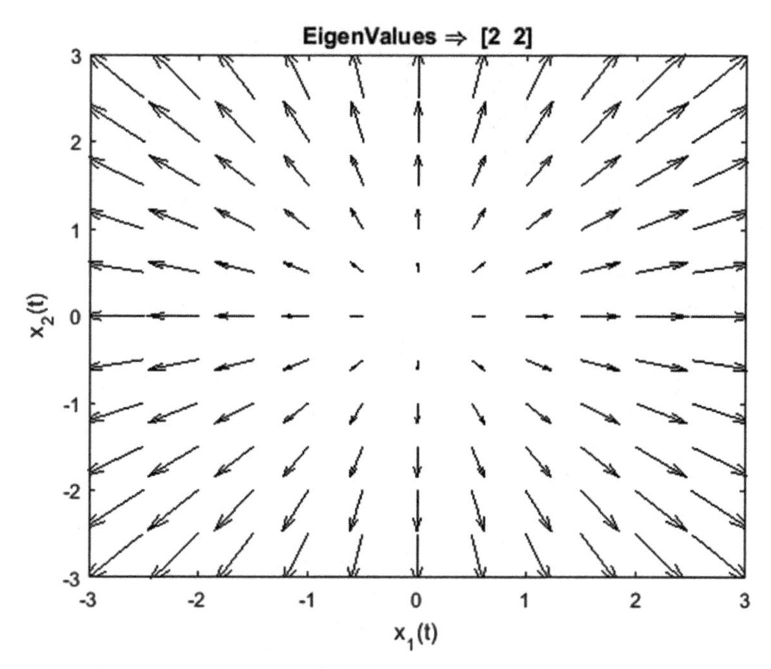

Figure 5.8 Both eigenvalues are equal and positive. The solutions move away from the critical point [0,0], indicating that the system is unstable.

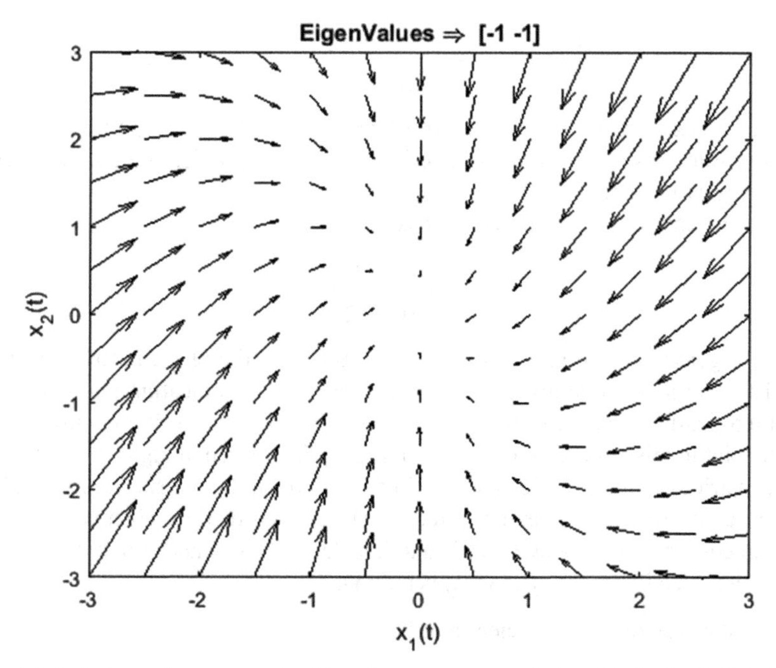

Figure 5.9 Both eigenvalues are negative. The solutions move towards the critical point [0,0], indicating that the system is asymptotically stable.

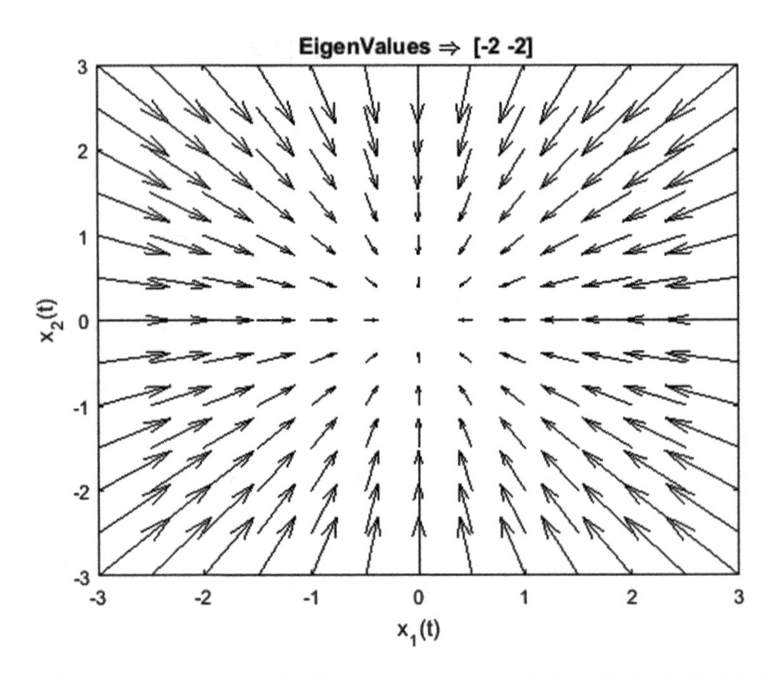

Figure 5.10 Both eigenvalues are negative. The solutions move towards the critical point [0,0] indicating that the system is asymptotically stable.

$$A = \begin{bmatrix} -2 & 0 \\ 0 & -2 \end{bmatrix}. \tag{5.32}$$

The eigenvalues are real and negative resulting in the phase portrait in Figure 5.10. The solutions move towards the critical point [0,0] indicating that the system is asymptotically stable.
Consider the case of the coefficient matrix

$$A = \begin{bmatrix} -2 & 1 \\ 0 & -2 \end{bmatrix}. \tag{5.33}$$

The eigenvalues are real and negative resulting in the phase portrait in Figure 5.11. The solutions move towards the critical point [0,0] indicating that the system is asymptotically stable. Note the difference in how the solutions move towards the critical point in this case vs. what is seen in Figure 5.10. Even though the eigenvalues for the matrices in eqns. (5.32) and (5.33) are identical, the matrix in eqn. (5.32) is symmetric and therefore it is not defective (it has two distinct eigenvectors). The matrix in eqn. (5.33) is not symmetric and it is defective because it only has a single eigenvector. Details can be seen in Appendix D.

Consider the case of the coefficient matrix

$$A = \begin{bmatrix} 2 & 2 \\ -2 & -2 \end{bmatrix}. \tag{5.34}$$

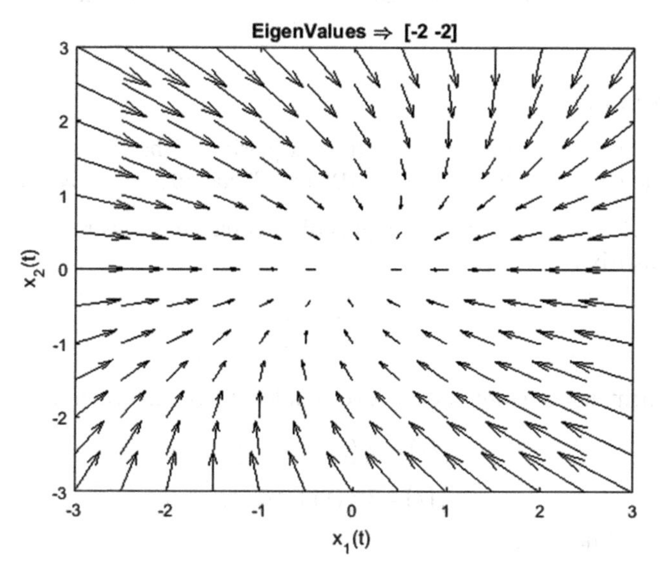

Figure 5.11 Both eigenvalues are negative. The solutions move towards the critical point [0,0], indicating that the system is asymptotically stable. Note the difference in how the solutions move towards the critical point in this case vs. what is seen in Figure 5.10. This is due to the fact that the matrix is defective in this case while the matrix is not defective for the case depicted in Figure 5.10.

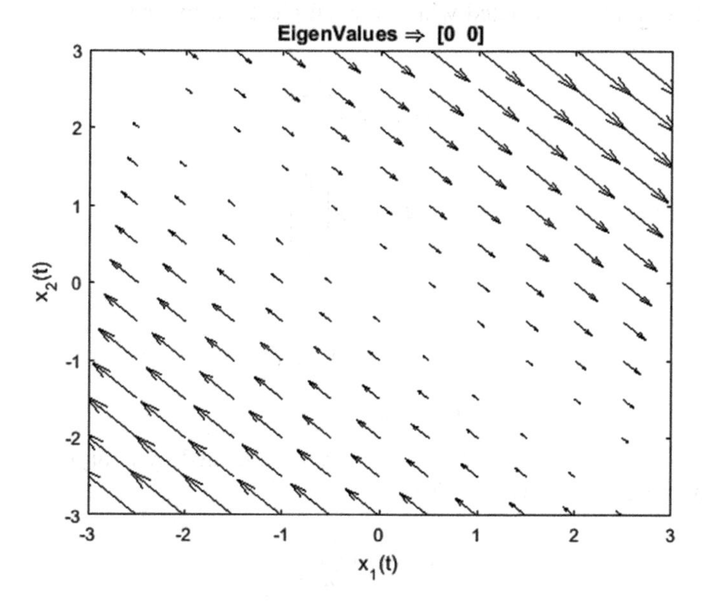

Figure 5.12 Both eigenvalues are zeros. The solutions move in opposite directions with respect to the line through the critical point [0,0] indicating that the system is unstable.

The eigenvalues are real and both of these are 0's resulting in the phase portrait in Figure 5.12. The solutions move in opposite directions with respect to the line through the critical point [0,0], indicating that the system is unstable.

The phase plane plots and the stability criteria associated with a pair of first order homogeneous differential equations with constant coefficients is summarized in Table 5.1

5.2.1.5 Examples

A set of examples demonstrating the different pairing of the eigenvalues described above will now be given before MATLAB® based examples are provided.

Example # 5.1 Consider a pair of coupled differential equations represented by the coefficient matrix

$$A=\begin{bmatrix}1 & 2\\ 3 & 2\end{bmatrix}. \tag{5.35}$$

This coefficient matrix represents the two coupled differential equations,

$$\begin{aligned}x_1'(t)&=x_1(t)+2x_2(t)\\ x_2'(t)&=3x_1(t)+2x_2(t)\end{aligned}. \tag{5.36}$$

The initial conditions are

$$\begin{bmatrix}x_1(0)\\ x_2(0)\end{bmatrix}=\begin{bmatrix}1\\ 0\end{bmatrix}. \tag{5.37}$$

The two eigenvalues associated with the coefficient matrix and the corresponding eigenvectors are

$$\begin{aligned}\lambda_1&=-1 \quad \vec{v}_1=\begin{bmatrix}-1\\ 1\end{bmatrix}\\ \lambda_2&=4 \quad \vec{v}_2=\begin{bmatrix}\frac{2}{3}\\ 1\end{bmatrix}\end{aligned}. \tag{5.38}$$

The solution set can now be written as

$$\begin{aligned}x_1(t)&=-c_1e^{-t}+\frac{2}{3}c_2e^{4t}\\ x_2(t)&=c_1e^{-t}+c_2e^{4t}\end{aligned}. \tag{5.39}$$

The two unknowns constants c_1 and c_2 may be obtained by applying the initial conditions in eqn. (5.37). This leads to

$$\begin{aligned}1&=-c_1+\frac{2}{3}c_2\\ 0&=c_1+c_2\end{aligned}. \tag{5.40}$$

Table 5.1. Eigenvalues, eigenvectors and stability.

Eigenvalues	Eigenvectors	Type of Critical Point	Stability	Gallery of solutions
$\lambda_1,\ \lambda_2 > 0$	Real and distinct	Node	Unstable	Not straight lines
$\lambda_1,\ \lambda_2 < 0$	Real and distinct	Node	Asymptotically stable	Not straight lines
$\lambda_1 < 0 < \lambda_2$	Real and distinct	Saddlepoint	Unstable	Not straight lines
$\lambda_1 = \lambda_2 > 0$	Real and distinct	Node	Unstable	Straight lines
$\lambda_1 = \lambda_2 < 0$	Real and distinct	Node	Asymptotically stable	Straight lines
$\lambda_1 = \lambda_2 > 0$	Real, but not distinct	Node	Unstable	Not straight lines
$\lambda_1 = \lambda_2 < 0$	Real, but not distinct	Node	Asymptotically stable	Not straight lines
$\lambda_1 = 0,\ \lambda_2 < 0$	Real and distinct	Crt. Point(s) on a line	Asymptotically stable	Straight lines
$\lambda_1 = 0,\ \lambda_2 > 0$	Real and distinct	Crt. Point(s) on a line	Unstable	Straight lines
$\lambda_1 = \lambda_2 = 0$	Real, but not distinct	Crt. Point(s) on a line	Unstable	Straight lines
$\lambda_1,\ \lambda_2 = \alpha \pm j\beta,\ \alpha > 0$	Complex conjugate pair	Spiral point	Unstable	Not straight lines
$\lambda_1,\ \lambda_2 = \alpha \pm j\beta,\ \alpha < 0$	Complex conjugate pair	Spiral point	Asymptotically stable	Not straight lines
$\lambda_1,\ \lambda_2 = \alpha \pm j\beta,\ \alpha = 0$	Complex conjugate pair	Center	Stable	Ellipses

Solving eqn. (5.40) gives

$$\begin{aligned} c_1 &= -\frac{3}{5} \\ c_2 &= \frac{3}{5} \end{aligned} . \tag{5.41}$$

Equation (5.39) can now be rewritten as

$$\begin{aligned} x_1(t) &= e^{-t}\left(\frac{2e^{5t}+3}{5}\right) \\ x_2(t) &= 3e^{-t}\left(\frac{e^{5t}-1}{5}\right) \end{aligned} . \tag{5.42}$$

The two coupled equations can be converted to a single second order differential equation in $x_1(t)$ as

$$x_1''(t) - 3x_1'(t) - 4x_1(t) = 0 \tag{5.43}$$

Equation (5.43) can be solved using the methods described in Chapter 3. Note that the roots of the characteristic equation associated with the second order differential equation in eqn. (5.43) will be -1 and 4, matching the eigenvalues obtained directly from the coefficient matrix in eqn. (5.35). The solution for $x_1(t)$ can be expressed in terms of two unknown constants b_1 and b_2 as

$$x_1(t) = b_1 e^{-t} + b_2 e^{4t} . \tag{5.44}$$

Using one of the differential equations in eqn. (5.36), the solution for $x_2(t)$ becomes

$$x_2(t) = -b_1 e^{-t} + \frac{3}{2} b_2 e^{4t} . \tag{5.45}$$

Applying initial conditions given in eqn. (5.37), eqn. (5.44) and (5.45) can be solved for b_1 and b_2 as

$$\begin{aligned} b_1 &= \frac{3}{5} \\ b_2 &= \frac{2}{5} \end{aligned} . \tag{5.46}$$

Substituting the values of b_1 and b_2 in eqns. (5.44) and (5.45), the solutions obtained match those given in eqn. (5.42).

In terms of the Laplace transforms, the Laplace transforms of the solutions to set of differential equations in eqn. (5.36) becomes

$$\begin{aligned} X_1(s) &= -\frac{s-2}{-s^2+3s+4} \\ X_2(s) &= -\frac{3}{-s^2+3s+4} \end{aligned} \tag{5.47}$$

Taking inverse Laplace transforms of eqn. (5.47), the solution set becomes

$$\begin{aligned} x_1(t) &= \frac{2}{5}e^{4t} + \frac{3}{5}e^{-t} \\ x_2(t) &= \frac{3}{5}e^{4t} - \frac{3}{5}e^{-t} \end{aligned} \quad . \tag{5.48}$$

Note that the solution set in eqn. (5.48) matches the set in eqn. (5.42). The phase portrait shown in Figure 5.13 illustrates that the system represented by the coefficient matrix in eqn. (5.35) is unstable.

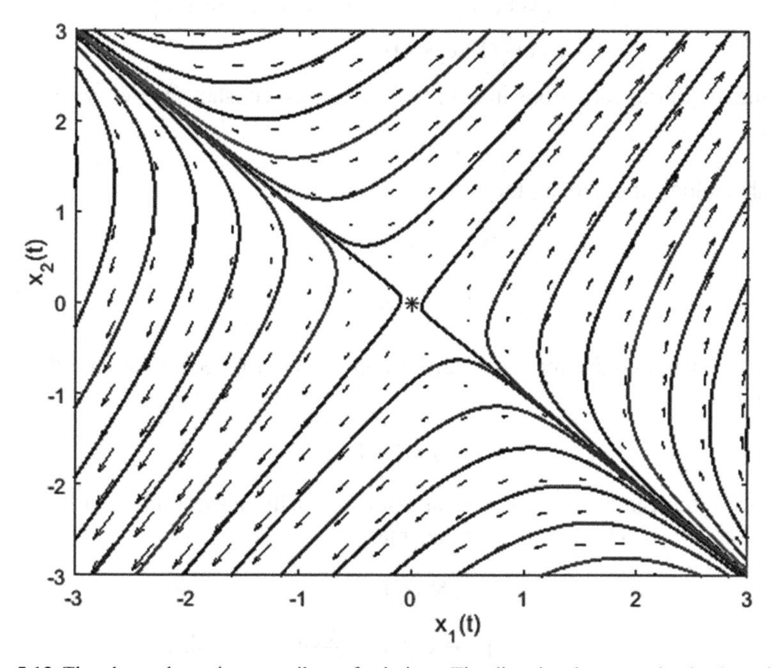

Figure 5.13 The phase plane plots or gallery of solutions. The directional arrows clearly show that the solutions are moving away from the critical point, indicating that the system is unstable.

Example # 5.2 Consider a pair of coupled differential equations represented by the coefficient matrix

$$A = \begin{bmatrix} -1 & 2 \\ -1 & 1 \end{bmatrix} . \tag{5.49}$$

The set of differential equations represented by the coefficient matrix in eqn. (5.49) is

$$\begin{aligned} x_1'(t) &= -x_1(t) + 2x_2(t) \\ x_2'(t) &= -x_1(t) + x_2(t) \end{aligned} \tag{5.50}$$

The initial conditions are

$$\begin{bmatrix} x_1(0) \\ x_2(0) \end{bmatrix} = \begin{bmatrix} 2 \\ 1 \end{bmatrix} . \tag{5.51}$$

The eigenvalues and eigenvectors associated with the coefficient matrix are

$$\begin{aligned} \lambda_1 &= -i \quad \vec{v}_1 = \begin{bmatrix} 1+i \\ 1 \end{bmatrix} \\ \lambda_2 &= i \quad \vec{v}_2 = \begin{bmatrix} 1-i \\ 1 \end{bmatrix} \end{aligned} . \tag{5.52}$$

Notice that the eigenvectors are purely imaginary and therefore, the solution will consist of sines and cosines. The solution set is

$$\begin{aligned} x_1(t) &= c_1\left[\cos(t)+\sin(t)\right]+c_2\left[\cos(t)-\sin(t)\right] \\ x_2(t) &= c_1\cos(t)-c_2\sin(t) \end{aligned} . \tag{5.53}$$

Applying the initial conditions in eqn. (5.51), the two constants are obtained as

$$c_1 = 1 \quad c_2 = 1 . \tag{5.54}$$

Using the values of c_1 and c_2, the solution set becomes

$$\begin{aligned} x_1(t) &= 2\cos(t) \\ x_2(t) &= \sqrt{2}\cos\left(t+\frac{\pi}{4}\right) \end{aligned} . \tag{5.55}$$

The set of differential equations can be converted to a second order differential equations as

$$x_1''(t)+x_1(t)=0 . \tag{5.56}$$

The roots of the characteristic equation associated with the second order differential equation are [I, -i]. The solution set becomes

$$\begin{aligned} x_1(t) &= b_1\cos(t)-b_2\sin(t) \\ x_2(t) &= \frac{b_1}{2}\cos(t)-\frac{b_2}{2}\cos(t)-\frac{b_1}{2}\sin(t)-\frac{b_2}{2}\sin(t) \end{aligned} . \tag{5.57}$$

Applying initial conditions, the two unknown constants become

$$b_1 = 2 \qquad b_2 = 0 . \tag{5.58}$$

Using the values of b_1 and b_2 in eqn. (5.57), one obtains the solution set matching the one given in eqn. (5.55).

The solution to the set of differential equations can also be obtained using Laplace transforms. The Laplace transforms of the two differential equations become (with the initial conditions included)

$$\begin{aligned} X_1(s) &= \frac{2}{s^2+1}+2\frac{s-1}{s^2+1} \\ X_2(s) &= -\frac{2}{s^2+1}+\frac{s+1}{s^2+1} \end{aligned} . \tag{5.59}$$

Taking inverse Laplace transforms, the solution set becomes

$$\begin{aligned} x_1(t) &= 2\cos(t) \\ x_2(t) &= \cos(t) - \sin(t) = \sqrt{2}\cos\left(t + \frac{\pi}{4}\right) \end{aligned} . \tag{5.60}$$

The phase portrait associated with the system described by the coefficient matrix in eqn. (5.49) is shown in Figure 5.14. The oscillatory behavior of the system is clearly seen.

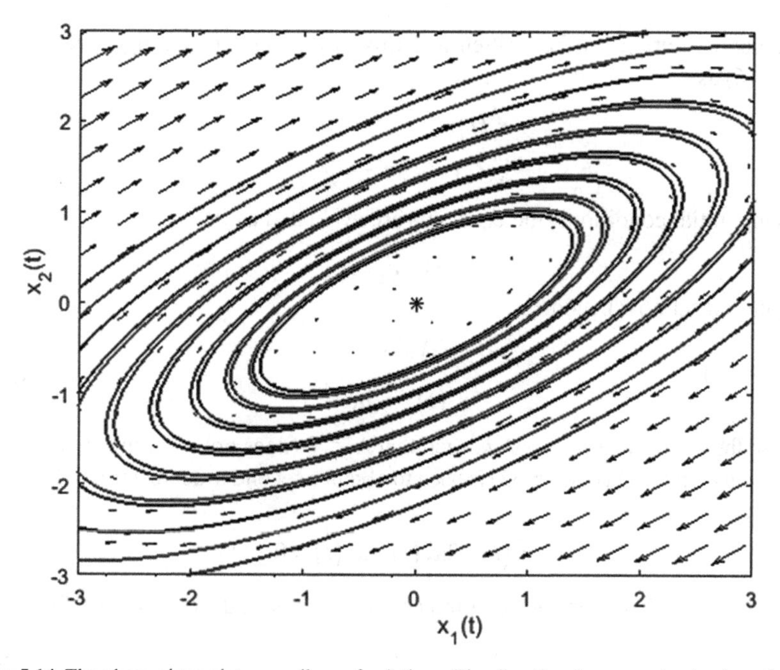

Figure 5.14 The phase plane plots or gallery of solutions. The directional arrows clearly show that the solutions form elliptical orbits indicating stability.

Example # 5.3 Consider a pair of coupled differential equations represented by the coefficient matrix

$$A = \begin{bmatrix} 1 & 0 \\ -2 & 1 \end{bmatrix} . \tag{5.61}$$

The coefficient matrix in eqn. (5.61) represents a pair of differential equations

$$\begin{aligned} x_1'(t) &= x_1(t) \\ x_2'(t) &= -2x_1(t) + x_2(t) \end{aligned} \tag{5.62}$$

The initial conditions are given as

$$\begin{bmatrix} x_1(0) \\ x_2(0) \end{bmatrix} = \begin{bmatrix} 1 \\ 0 \end{bmatrix} \tag{5.63}$$

The two eigenvalues of the matrix are equal, each being 1,

$$\lambda_1 = \lambda_2 = 1. \tag{5.64}$$

The matrix is defective because there is only a single unique eigenvector. This means that a pair of generalized eigenvectors is required. Using the method described in Appendix D, the generalized eigenvectors are (eqn. (5.17))

$$\begin{bmatrix} 1 \\ 0 \end{bmatrix}, \begin{bmatrix} 0 \\ 1 \end{bmatrix}. \tag{5.65}$$

The general solution can be written in terms of the generalized eigenvectors from eqn. (5.16) as

$$\begin{aligned} x_1(t) &= e^t c_1 \\ x_2(t) &= -2te^t c_1 + e^t c_2 \end{aligned}. \tag{5.66}$$

Applying initial conditions, the constants are evaluated as

$$c_1 = 1 \quad c_2 = 0 . \tag{5.67}$$

The solution set now becomes

$$\begin{aligned} x_1(t) &= e^t \\ x_2(t) &= -2te^t \end{aligned}. \tag{5.68}$$

Because the coefficient matrix A is such that $A_{12} = 0$ (does not contain x_2), the second differential equation in eqn. (5.62) is used to obtain a second order differential equation in $x_2(t)$ as

$$x_2''(t) - 2x_2'(t) + x_2(t) = 0 . \tag{5.69}$$

The roots of the characteristic equation are [1, 1]. This leads to a solution for $x_2(t)$ as

$$x_2(t) = e^t(b_1 + b_2 t) . \tag{5.70}$$

Substituting eqn. (5.70) in the second differential equation in eqn. (5.62), the solution for $x_1(t)$ becomes

$$x_1(t) = -\frac{b_2}{2}e^t . \tag{5.71}$$

Applying initial conditions, the two constants are evaluated as

$$b_1 = 0 \quad b_2 = -2 . \tag{5.72}$$

Substituting for b_1 and b_2 in eqn. (5.70) and eqn. (5.71) leads to the solution set obtained from the eigenvector set given in eqn. (5.68).

The Laplace transforms of the set of differential equations (with initial conditions) are

$$\begin{aligned} X_1(s) &= \frac{1}{s-1} \\ X_2(s) &= -2\frac{1}{(s-1)^2} \end{aligned}. \tag{5.73}$$

Taking inverse Laplace transforms, eqn. (5.73) leads to

$$\begin{aligned} x_1(t) &= e^t \\ x_2(t) &= -2te^t \end{aligned} . \tag{5.74}$$

The phase portrait associated with the coefficient matrix in eqn. (5.61) is shown in Figure 5.15. The unstable equilibrium conditions are seen.

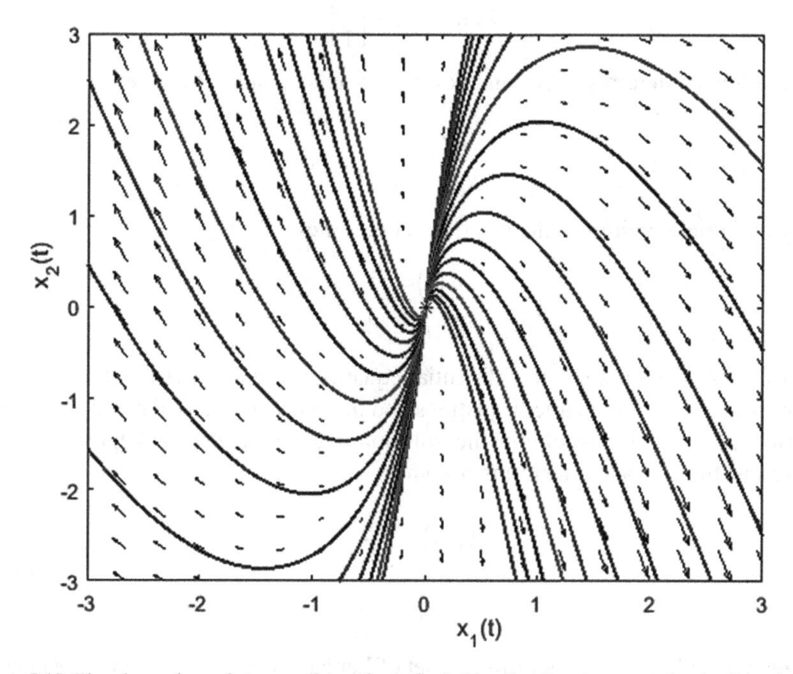

Figure 5.15 The phase plane plots or gallery of solutions. The directional arrows clearly show that the solutions are moving away from the critical point, pointing out that the system is unstable.

Example # 5.4 Consider a pair of coupled differential equations represented by the coefficient matrix

$$A = \begin{bmatrix} 2 & 0 \\ 0 & 2 \end{bmatrix} . \tag{5.75}$$

The set of differential equations associated with the coefficient matrix is

$$\begin{aligned} x_1'(t) &= 2x_1(t) \\ x_2'(t) &= 2x_2(t) \end{aligned} . \tag{5.76}$$

The initial conditions are given as

$$\begin{bmatrix} x_1(0) \\ x_2(0) \end{bmatrix} = \begin{bmatrix} 1 \\ 1 \end{bmatrix} . \tag{5.77}$$

The eigenvalues of the coefficient matrix are equal even though the coefficient matrix is not defective because two distinct eigenvectors are associated with A in eqn. (5.75). The eigenvalues and eigenvectors are

$$\begin{aligned} \lambda_1 &= 2 \quad \vec{v}_1 = \begin{bmatrix} 1 \\ 0 \end{bmatrix} \\ \lambda_2 &= 2 \quad \vec{v}_2 = \begin{bmatrix} 0 \\ 1 \end{bmatrix} \end{aligned} . \tag{5.78}$$

The solution set in terms of two unknown constants c_1 and c_2 becomes

$$\begin{aligned} x_1(t) &= c_1 e^{2t} \\ x_2(t) &= c_2 e^{2t} \end{aligned} . \tag{5.79}$$

Applying initial conditions, the solution set becomes

$$\begin{aligned} x_1(t) &= e^{2t} \\ x_2(t) &= e^{2t} \end{aligned} . \tag{5.80}$$

It is clear from the set of differential equations in eqn. (5.76) that the set does not constitute a pair of coupled equations and therefore, no second order differential equation is necessary to obtain the solution set. Using the concept of Laplace transforms, the Laplace transforms become

$$\begin{aligned} X_1(s) &= \frac{1}{s-2} \\ X_2(s) &= \frac{1}{s-2} \end{aligned} . \tag{5.81}$$

Taking inverse Laplace transforms, the set of Laplace transforms in eqn. (5.81) leads to the solution set in eqn. (5.80).

The phase portrait for the coefficient matrix in eqn. (5.75) is shown in Figure 5.16. The unstable equilibrium conditions are seen.

Example # 5.5 It is interesting to observe how the solutions behave for a system having equal eigenvalues with a degenerate coefficient matrix. Consider a set of differential equations represented by the coefficient matrix

$$A = \begin{bmatrix} 2 & 1 \\ 0 & 2 \end{bmatrix} . \tag{5.82}$$

While the two eigenvalues of the matrix are equal (each being 2), the matrix is defective requiring the use of generalized eigenvectors. The use of the generalized eigenvectors alters the gallery of solutions as seen in Figure 5.17.

The set of differential equations associated with the coefficient matrix in eqn. (5.82) is

$$\begin{aligned} x_1'(t) &= 2x_1(t) + x_2(t) \\ x_2'(t) &= 2x_2(t) \end{aligned} . \tag{5.83}$$

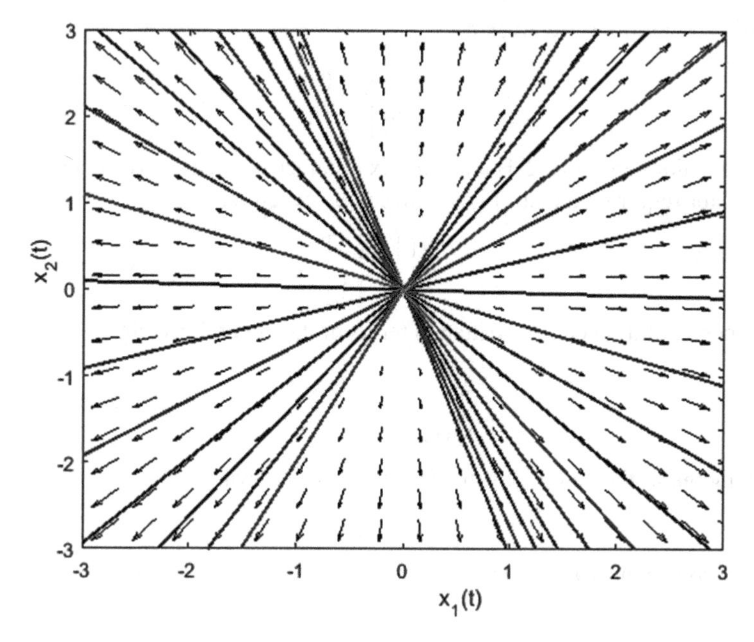

Figure 5.16 The phase plane plots or gallery of solutions. The directional arrows clearly show that the solutions (straight lines) are moving away from the critical point, pointing out that the system is unstable.

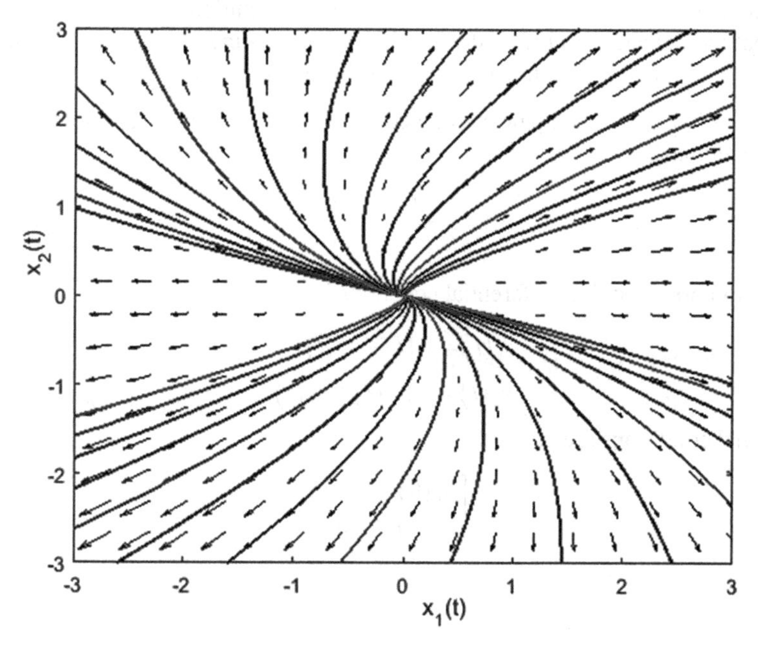

Figure 5.17 The phase plane plots or gallery of solutions. The directional arrows clearly show that the solutions (no longer straight lines) are moving away from the critical point, indicating that the system is unstable. The use of the generalized eigenvector (due to the defective nature of the coefficient matrix) alters the nature of the solutions.

The initial conditions are given as

$$\begin{bmatrix} x_1(0) \\ x_2(0) \end{bmatrix} = \begin{bmatrix} 0 \\ 1 \end{bmatrix}. \tag{5.84}$$

The eigenvalues are 2 and 2. But the matrix is defective because we only have a single eigenvector requiring a pair of generalized eigenvectors,

$$\begin{bmatrix} 1 \\ 0 \end{bmatrix}, \begin{bmatrix} 0 \\ 1 \end{bmatrix}. \tag{5.85}$$

The general solution can be written in terms of the generalized eigenvectors as

$$\begin{aligned} x_1(t) &= e^{2t}c_1 + c_2te^{2t} \\ x_2(t) &= c_2e^{2t} \end{aligned}. \tag{5.86}$$

Applying the initial conditions, the constants can be evaluated as

$$c_1 = 0 \quad c_2 = 1\ . \tag{5.87}$$

The solution set now becomes

$$\begin{aligned} x_1(t) &= te^{2t} \\ x_2(t) &= e^{2t} \end{aligned}. \tag{5.88}$$

If we compare the solution set in eqn. (5.80) and the solution set in eqn. (5.88), it is seen that the defective nature of the matrix leads to scaling by t in eqn. (5.88). This leads to the difference in the trajectory of the solutions as seen in Figures 5.16 and 5.17.

Example # 5.6 Consider a pair of coupled differential equations represented by the coefficient matrix

$$A = \begin{bmatrix} 1 & 1 \\ -2 & 1 \end{bmatrix}. \tag{5.89}$$

The set of corresponding differential equations is

$$\begin{aligned} x_1'(t) &= x_1(t) + x_2(t) \\ x_2'(t) &= -2x_1(t) + x_2(t) \end{aligned}. \tag{5.90}$$

The initial conditions are given as

$$\begin{bmatrix} x_1(0) \\ x_2(0) \end{bmatrix} = \begin{bmatrix} 1 \\ 0 \end{bmatrix}. \tag{5.91}$$

The eigenvalues and eigenvectors of the coefficient matrix are

$$\begin{aligned} \lambda_1 &= 1 - i\sqrt{2} \quad \vec{v}_1 = \begin{bmatrix} \dfrac{i}{\sqrt{2}} \\ 1 \end{bmatrix} \\ \lambda_2 &= 1 + i\sqrt{2} \quad \vec{v}_2 = \begin{bmatrix} -\dfrac{i}{\sqrt{2}} \\ 1 \end{bmatrix} \end{aligned} . \tag{5.92}$$

Using the eigenvalues and eigenvectors, the solution set becomes

$$\begin{aligned} x_1(t) &= \frac{\sqrt{2}}{2} e^t c_1 \sin\left(\sqrt{2}t\right) + \frac{\sqrt{2}}{2} e^t c_2 \cos\left(\sqrt{2}t\right) \\ x_2(t) &= e^t c_1 \cos\left(\sqrt{2}t\right) - e^t c_2 \sin\left(\sqrt{2}t\right) \end{aligned} . \tag{5.93}$$

Applying initial conditions in eqn. (5.91), c_1 and c_2 are evaluated as

$$c_1 = 0 \qquad c_2 = \sqrt{2} \ . \tag{5.94}$$

Using eqn. (5.94), the solution set becomes

$$\begin{aligned} x_1(t) &= e^t \cos\left(\sqrt{2}t\right) \\ x_2(t) &= -\sqrt{2} e^t \sin\left(\sqrt{2}t\right) \end{aligned} . \tag{5.95}$$

A second order differential equation is created as

$$x_1''(t) - 2x_1'(t) + 3x_1(t) = 0 \ . \tag{5.96}$$

The roots of the characteristic equation are

$$\left[1 + i\sqrt{2}\right] \quad \left[1 - i\sqrt{2}\right] . \tag{5.97}$$

The pair of complex roots leads to the solution set as

$$\begin{aligned} x_1(t) &= e^t \left[b_1 \cos\left(\sqrt{2}t\right) - b_2 \sin\left(\sqrt{2}t\right) \right] \\ x_2(t) &= -\sqrt{2} e^t \left[b_1 \sin\left(\sqrt{2}t\right) + b_2 \cos\left(\sqrt{2}t\right) \right] \end{aligned} . \tag{5.98}$$

Applying initial conditions, the two constants b_1 and b_2 are evaluated as

$$b_1 = 1 \quad b_2 = 0 \ . \tag{5.99}$$

Substituting the values of b_1 and b_2 is eqn. (5.98), the solution set matching the set given in eqn. (5.95) is obtained.

The Laplace transform of the differential equations are

$$X_1(s) = \frac{s-1}{s^2-2s+3}$$
$$X_2(s) = -\frac{2}{s^2-2s+3}. \tag{5.100}$$

Taking inverse Laplace transforms of the set in eqn. (5.100), the solution set obtained matches the set in eqn. (5.95).

The phase portrait associated with the coefficient matrix in eqn. (5.89) is shown in Figure 5.18 illustrating that the system is unstable.

Example # 5.7 Consider a pair of coupled differential equations represented by the coefficient matrix

$$A = \begin{bmatrix} 1 & -1 \\ -1 & 1 \end{bmatrix}. \tag{5.101}$$

The set of differential equations associated with the coefficient matrix is

$$x_1'(t) = x_1(t) - x_2(t)$$
$$x_2'(t) = -x_1(t) + x_2(t). \tag{5.102}$$

The initial conditions are given as

$$\begin{bmatrix} x_1(0) \\ x_2(0) \end{bmatrix} = \begin{bmatrix} 1 \\ -1 \end{bmatrix}. \tag{5.103}$$

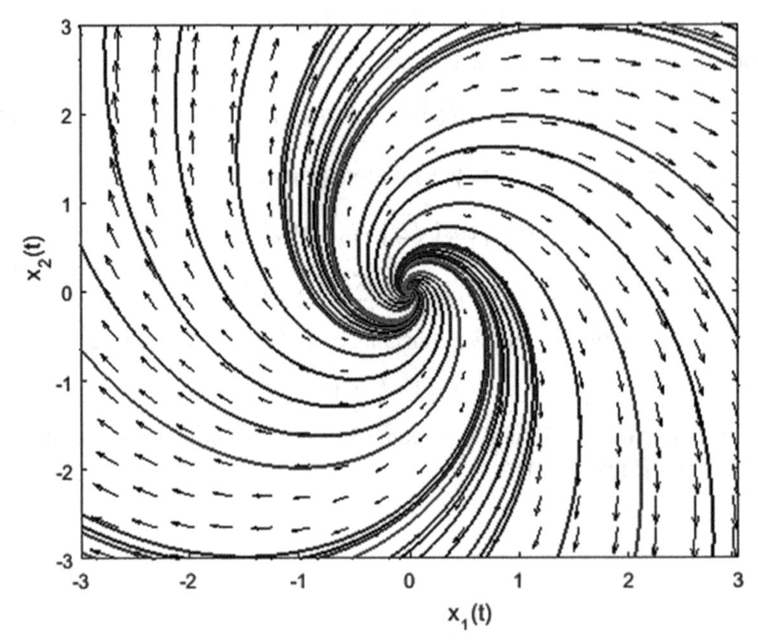

Figure 5.18 The phase plane plots or gallery of solutions. The directional arrows clearly show that the solutions are moving away from the critical point, pointing out that the system is unstable.

The eigenvalues and eigenvectors are

$$\begin{aligned} \lambda_1 &= 0 \quad \vec{v}_1 = \begin{bmatrix} 1 \\ 1 \end{bmatrix} \\ \lambda_2 &= 2 \quad \vec{v}_2 = \begin{bmatrix} -1 \\ 1 \end{bmatrix} \end{aligned} . \tag{5.104}$$

Using the eigenvalues and eigenvectors, the solution set becomes

$$\begin{aligned} x_1(t) &= c_1 - c_2 e^{2t} \\ x_2(t) &= c_1 + c_2 e^{2t} \end{aligned} . \tag{5.105}$$

Applying the initial conditions in eqn. (5.103), the two constants c_1 and c_2 can be evaluated as

$$c_1 = 0 \quad c_2 = -1 . \tag{5.106}$$

The solution set now becomes

$$\begin{aligned} x_1(t) &= e^{2t} \\ x_2(t) &= -e^{2t} \end{aligned} . \tag{5.107}$$

The second order differential equation corresponding to the set of differential equations is

$$x_1''(t) - 2x_1(t) = 0 . \tag{5.108}$$

The roots of the characteristic equation are 0 and 2 which lead to the solution set in eqn. (5.107). The Laplace transforms of the differential equations are

$$\begin{aligned} X_1(s) &= \frac{1}{s-2} \\ X_2(s) &= \frac{1}{2-s} \end{aligned} . \tag{5.109}$$

Taking the inverse Laplace transform of the set in eqn. (5.109), the solution set obtained matches the solution set in eqn. (5.107).

Example # 5.8 It is possible to compare this example to the case of a set of differential equations represented through the coefficient matrix

$$A = \begin{bmatrix} 0 & 1 \\ 0 & -3 \end{bmatrix} . \tag{5.110}$$

The eigenvalues are 0 and -3. The gallery of solutions is displayed in Figure 5.20 which shows that the directional arrows point to the critical point demonstrating the system is asymptotically stable.

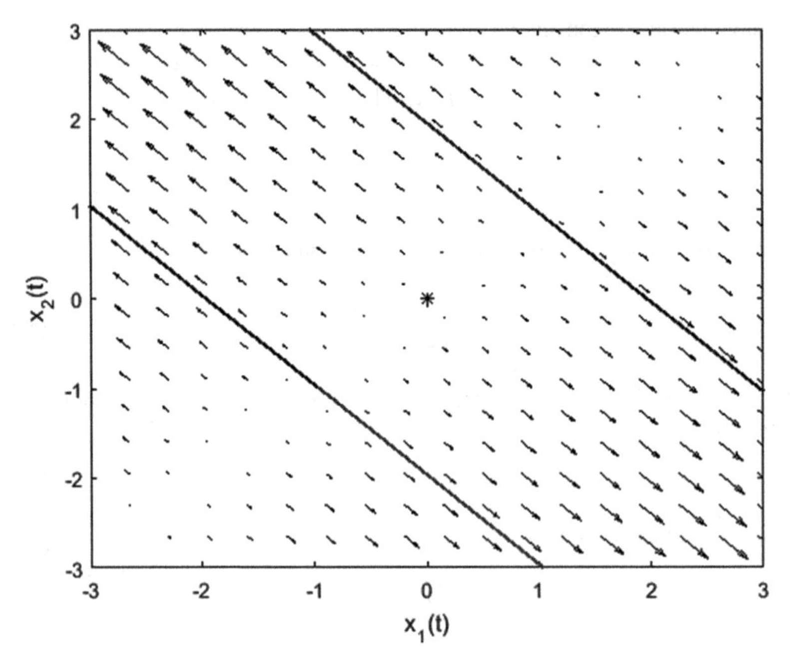

Figure 5.19 The phase plane plots or gallery of solutions. The directional arrows show that the solutions are straight lines and that they are moving away from a line through critical point, indicating that the system is unstable.

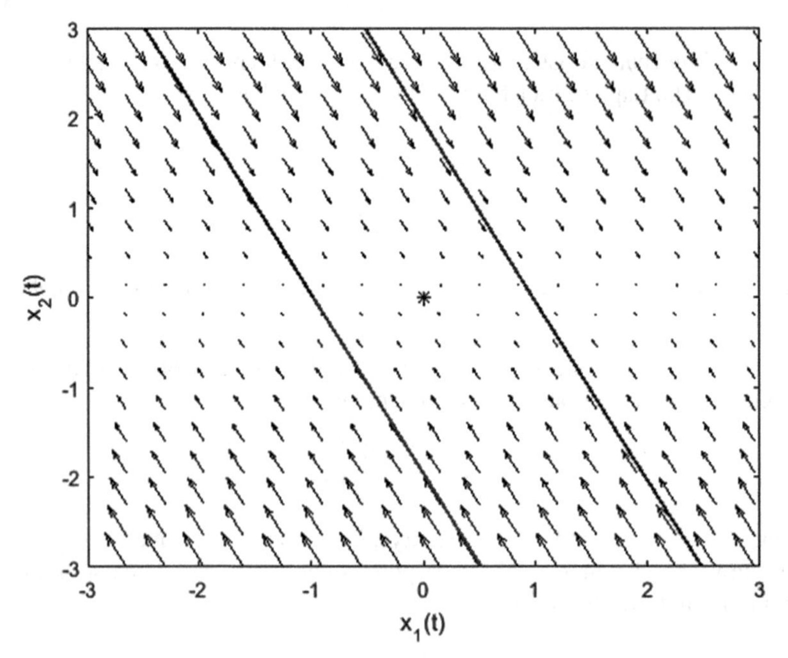

Figure 5.20 The phase plane plots or gallery of solutions. The directional arrows show that the solutions are straight lines and are moving towards a line through the critical point indicating that the system is asymptotically stable.

Example # 5.9 Consider a pair of coupled differential equations represented by the coefficient matrix

$$A = \begin{bmatrix} 2 & 2 \\ -2 & -2 \end{bmatrix}. \tag{5.111}$$

The set of differential equations corresponding to the coefficient matrix is

$$\begin{aligned} x_1'(t) &= 2x_1(t) + 2x_2(t) \\ x_2'(t) &= -2x_1(t) - 2x_2(t) \end{aligned}. \tag{5.112}$$

The initial conditions are

$$\begin{bmatrix} x_1(0) \\ x_2(0) \end{bmatrix} = \begin{bmatrix} -1 \\ -1 \end{bmatrix}. \tag{5.113}$$

The coefficient matrix has a unique set of eigenvalues. They are both equal to 0 and the coefficient matrix is defective. This means that generalized eigenvectors are required. The eigenvalues and generalized eigenvectors are

$$\begin{aligned} \lambda_1 = 0 \quad \vec{v}_1 &= \begin{bmatrix} 1 \\ 0 \end{bmatrix} \\ \lambda_2 = 0 \quad \vec{v}_2 &= \begin{bmatrix} 0 \\ 1 \end{bmatrix} \end{aligned}. \tag{5.114}$$

Using the generalized eigenvectors and the eigenvalue, the solution set in terms of two unknown constants c_1 and c_2 becomes

$$\begin{aligned} x_1(t) &= c_1(2t+1) + 2c_2t \\ x_2(t) &= -2c_1t - c_2(2t-1) \end{aligned}. \tag{5.115}$$

Applying the initial conditions and evaluating the unknown constants, the solution set becomes

$$\begin{aligned} x_1(t) &= -4t - 1 \\ x_2(t) &= 4t - 1 \end{aligned} \tag{5.116}$$

The second order differential equation is unique and it is given by

$$x_1''(t) = 0 . \tag{5.117}$$

The roots of the characteristic equation are equal and each is equal to zero leading to the solution set as

$$\begin{aligned} x_1(t) &= b_1 + b_2t \\ x_2(t) &= -b_1 - b_2t + \frac{b_2}{2} \end{aligned}. \tag{5.118}$$

Applying initial conditions, the values of b_1 and b_2 are obtained as

$$b_1 = -1 \quad b_2 = -4 . \tag{5.119}$$

Substituting for b_1 and b_2, the solution set obtained matches the result in eqn. (5.116).

Taking Laplace transforms of the differential equations, the Laplace transforms are obtained as

$$X_1(s) = \frac{-s-4}{s^2}$$
$$X_2(s) = \frac{-s+4}{s^2}. \qquad (5.120)$$

Taking the inverse Laplace transforms, the solution set obtained matches the solution set in eqn. (5.116).

Example #5.10 Consider a pair of coupled differential equations represented by the coefficient matrix

$$A = \begin{bmatrix} -1 & 0 \\ -2 & -3 \end{bmatrix}. \qquad (5.121)$$

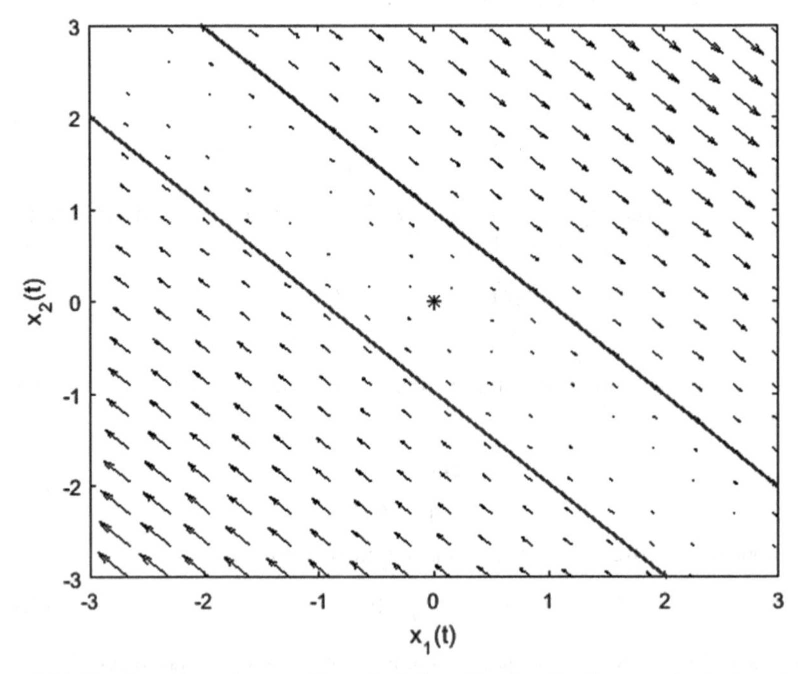

Figure 5.21 The phase plane plots or gallery of solutions. The directional arrows clearly show that the solutions are moving away from a line through the critical point, pointing out that the system is unstable. The case of a single eigenvalue of zero and double zero eigenvalues can be compared and seen that in the latter case arrows only point in two directions while in the former case they are moving away in four directions.

The set of differential equations associated with the coefficient matrix is

$$\begin{aligned} x_1'(t) &= -x_1(t) \\ x_2'(t) &= -2x_1(t) - 3x_2(t) \end{aligned} . \tag{5.122}$$

The initial conditions are given as

$$\begin{bmatrix} x_1(0) \\ x_2(0) \end{bmatrix} = \begin{bmatrix} -1 \\ 1 \end{bmatrix} . \tag{5.123}$$

The eigenvalues and eigenvectors associated with the coefficient matrix are

$$\begin{aligned} \lambda_1 &= -3 \quad \vec{v}_1 = \begin{bmatrix} 0 \\ 1 \end{bmatrix} \\ \lambda_2 &= -1 \quad \vec{v}_2 = \begin{bmatrix} -1 \\ 1 \end{bmatrix} \end{aligned} . \tag{5.124}$$

The solutions set using eigenvalues and eigenvectors becomes

$$\begin{aligned} x_1(t) &= -c_2 e^{-t} \\ x_2(t) &= c_1 e^{-3t} + c_2 e^{-t} \end{aligned} . \tag{5.125}$$

Applying initial conditions, the constants c_1 and c_2 are evaluated as

$$c_1 = 0 \quad c_2 = 1 . \tag{5.126}$$

Using the values of c_1 and c_2, the solution set in eqn. (5.125) becomes

$$\begin{aligned} x_1(t) &= -e^{-t} \\ x_2(t) &= e^{-t} \end{aligned} . \tag{5.127}$$

Because the first differential equation in eqn. (5.122) does not contain x_2, the second order differential equation in x_2 is created starting from the second differential equation in eqn. (5.122) as

$$x_2''(t) + 4x_2'(t) + 3x_2(t) = 0 . \tag{5.128}$$

The roots of the characteristic equation are -3 and -1 and a solution set matching the set in eqn. (5.127) is obtained in this case.

Taking the Laplace transforms of the differential equations, the Laplace transforms become

$$\begin{aligned} X_1(s) &= \frac{-1}{s+1} \\ X_2(s) &= \frac{1}{s+3} + \frac{2}{(s+3)(s+1)} = \frac{1}{s+1} \end{aligned} . \tag{5.129}$$

Taking inverse Laplace transforms, a solution set matching the set in eqn. (5.127) is obtained.

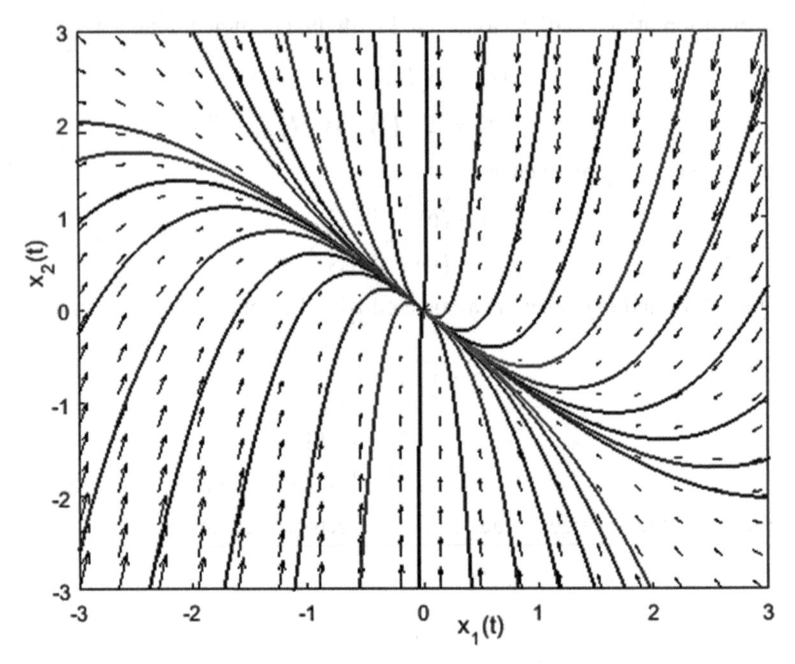

Figure 5.22 The phase plane plots or gallery of solutions. The directional arrows clearly show that the solutions are moving towards to the critical point and converging pointing out that the system is asymptotically stable.

5.2.2 First order non-homogeneous systems

A typical first order coupled equations with constant coefficients with external forcing functions constitute a non-homogeneous system represented in eqn. (5.4). A complete solution to the non-homogeneous system is written as the sum of the solution to the homogeneous part $\vec{x}_h(t)$ and a particular solution $\vec{x}_p(t)$ as

$$\vec{x}_h(t) = c_1\vec{X}_1 + c_2\vec{X}_2 = \sum_{k=1}^{2} c_k\vec{X}_k(t). \tag{5.130}$$

$$\vec{x}_p(t) = \begin{bmatrix} x_{1p}(t) \\ x_{2p}(t) \end{bmatrix}. \tag{5.131}$$

$$\vec{x}(t) = \vec{x}_h(t) + \vec{x}_p(t). \tag{5.132}$$

The particular solutions set can be obtained from the fundamental matrix using the method of variation of parameters described in in Appendix D as (see eqn. (D. 140)) as

$$\vec{x}_p(t) = \begin{bmatrix} x_{1p}(t) \\ x_{2p}(t) \end{bmatrix} = X(t)\int [X(t)]^{-1}\vec{g}(t)\,dt \tag{5.133}$$

Note that X(t) is the fundamental matrix defined in eqn. (5.10). If the initial conditions are given, the unknown coefficients c_1 and c_2 can be obtained. Given the initial conditions, the solution can also be obtained using the concept of Laplace and inverse Laplace transform as

$$\begin{bmatrix} x_1(t) \\ x_2(t) \end{bmatrix} = L^{-1}\left\{ \left[sI_2 - A \right]^{-1} \left[\begin{pmatrix} x_1(0) \\ x_2(0) \end{pmatrix} + L\left(\vec{g}(t)\right) \right] \right\}. \tag{5.134}$$

5.2.2.1 Example of non-homogeneous coupled first order systems

Example # 5.11 Consider a system described through the coefficient matrix A, with initial conditions and forcing functions given by

$$A = \begin{bmatrix} -1 & 2 \\ -1 & 1 \end{bmatrix}. \tag{5.135}$$

$$\vec{g}(t) = \begin{bmatrix} \cos(t) \\ t\sin(t) \end{bmatrix}. \tag{5.136}$$

$$\begin{bmatrix} x_1(0) \\ x_2(0) \end{bmatrix} = \begin{bmatrix} 2 \\ 1 \end{bmatrix}. \tag{5.137}$$

The set of coupled differential equations described through eqns. (5.135)–(5.137) is

$$\begin{aligned} x_1'(t) &= -x_1(t) + 2x_2(t) + \cos(t), \quad x_1(0) = 2 \\ x_2'(t) &= -x_1(t) + x_2(t) + t\sin(t), \quad x_2(0) = 1 \end{aligned} \tag{5.138}$$

The eigenvalues and the eigenvectors of the coefficient matrix A are complex and are given by

$$\begin{aligned} \lambda_1 &= -i \quad \vec{v}_1 = \begin{bmatrix} 1+i \\ 1 \end{bmatrix} \\ \lambda_2 &= i \quad \vec{v}_2 = \begin{bmatrix} 1-i \\ 1 \end{bmatrix} \end{aligned}. \tag{5.139}$$

The fundamental matrix associated with the coefficient matrix (and the homogeneous differential equations set) is

$$X(t) = \left[\vec{X}_1 \quad \vec{X}_2 \right] = \begin{bmatrix} \cos(t)+\sin(t) & \cos(t)-\sin(t) \\ \cos(t) & -\sin(t) \end{bmatrix}. \tag{5.140}$$

The solution to the set of homogeneous differential equations is

$$\vec{x}_h(t) = c_1\vec{X}_1 + c_2\vec{X}_2 = c_1 \begin{bmatrix} \cos(t)+\sin(t) \\ \cos(t) \end{bmatrix} + c_2 \begin{bmatrix} \cos(t)-\sin(t) \\ -\sin(t) \end{bmatrix}. \tag{5.141}$$

The particular solution set is obtained using the expression in eqn. (5.133) as

$$\vec{x}_p(t) = \begin{bmatrix} \dfrac{\sin(t)}{2} - \dfrac{t^2\cos(t)}{2} + \dfrac{t\cos(t)}{2} \\ \dfrac{\sin(t)}{4} - \dfrac{t^2\cos(t)}{4} + \dfrac{t^2\sin(t)}{4} - \dfrac{t\cos(t)}{4} - \dfrac{t\sin(t)}{4} \end{bmatrix}. \tag{5.142}$$

The general solution to the non-homogeneous set of differential equation is the sum of the homogeneous and particular solution sets. Using eqns. (5.141) and (5.142), the general solution becomes

$$\begin{bmatrix} x_1(t) \\ x_2(t) \end{bmatrix} = c_1 \begin{bmatrix} \cos(t) + \sin(t) \\ \cos(t) \end{bmatrix} + c_2 \begin{bmatrix} \cos(t) - \sin(t) \\ -\sin(t) \end{bmatrix} + \begin{bmatrix} \dfrac{\sin(t)}{2} + \dfrac{t\cos(t)}{2}(1-t) \\ \dfrac{\sin(t)}{4}(1-t+t^2) - \dfrac{t\cos(t)}{4}(1+t) \end{bmatrix}. \tag{5.143}$$

Applying initial conditions to eqn. (5.143), the two constants c_1 and c_2 are evaluated as

$$\begin{bmatrix} c_1 \\ c_2 \end{bmatrix} = \begin{bmatrix} 1 \\ 1 \end{bmatrix}. \tag{5.144}$$

Substituting the values of c_1 and c_2 in eqn. (5.143) and simplifying, the solution set becomes

$$\begin{aligned} x_1(t) &= 2\cos(t) + \frac{\sin(t)}{2} + \frac{t\cos(t)}{2}(1-t) \\ x_2(t) &= \cos(t) + \frac{\sin(t)}{4}(t^2 - t - 3) - \frac{t\cos(t)}{4}(1+t) \end{aligned}. \tag{5.145}$$

Incorporating the initial conditions, the Laplace transform of the differential equation can be performed and the Laplace transform set becomes

$$\begin{bmatrix} X_1(s) \\ X_2(s) \end{bmatrix} = \begin{bmatrix} \dfrac{2\left(\dfrac{2s}{(s^2+1)^2} + 1\right) + \left(\dfrac{s}{s^2+1} + 2\right)(s-1)}{s^2+1} \\ \dfrac{\left(\dfrac{2s}{(s^2+1)^2} + 1\right)(s+1) - \dfrac{s}{s^2+1} - 2}{s^2+1} \end{bmatrix}. \tag{5.146}$$

Taking inverse Laplace transforms, the solution set matching the solution set in eqn. (5.145) is obtained.

Example # 5.12 The coefficient matrix, forcing functions, and initial conditions are given as

$$A = \begin{bmatrix} 1 & 2 \\ 2 & 1 \end{bmatrix}. \tag{5.147}$$

$$\vec{g}(t)=\begin{bmatrix} t \\ e^{-t} \end{bmatrix}. \tag{5.148}$$

$$\begin{bmatrix} x_1(0) \\ x_2(0) \end{bmatrix}=\begin{bmatrix} 1 \\ 0 \end{bmatrix}. \tag{5.149}$$

The set of coupled homogeneous set of differential equation becomes

$$\begin{aligned} x_1'(t) &= x_1(t)+2x_2(t)+t, \quad x_1(0)=1 \\ x_2'(t) &= 2x_1(t)+x_2(t)+e^{-t}, \quad x_2(0)=0 \end{aligned}. \tag{5.150}$$

The eigenvalues and eigenvectors of the coefficient matrix are

$$\begin{aligned} \lambda_1 &= -1 \quad \vec{v}_1=\begin{bmatrix} -1 \\ 1 \end{bmatrix} \\ \lambda_2 &= 3 \quad \vec{v}_2=\begin{bmatrix} 1 \\ 1 \end{bmatrix} \end{aligned}. \tag{5.151}$$

The fundamental matrix associated with the coefficient matrix (and the homogeneous differential equations set) is

$$X(t)=\begin{bmatrix} \vec{X}_1 & \vec{X}_2 \end{bmatrix}=\begin{bmatrix} -e^{-t} & e^{3t} \\ e^{-t} & e^{3t} \end{bmatrix}. \tag{5.152}$$

The solution to the set of homogeneous differential equations is

$$\vec{x}_h(t)=c_1\vec{X}_1+c_2\vec{X}_2=c_1\begin{bmatrix} -e^{-t} \\ e^{-t} \end{bmatrix}+c_2\begin{bmatrix} e^{3t} \\ e^{3t} \end{bmatrix}. \tag{5.153}$$

The particular solution set is obtained using the expression in eqn. (5.133) as

$$\vec{x}_p(t)=\begin{bmatrix} \dfrac{t}{3}-\dfrac{e^{-t}}{8}-\dfrac{te^{-t}}{2}-\dfrac{5}{9} \\ \dfrac{te^{-t}}{2}-\dfrac{e^{-t}}{8}-\dfrac{2t}{3}+\dfrac{4}{9} \end{bmatrix}. \tag{5.154}$$

The general solution to the non-homogeneous set of differential equations is the sum of the homogeneous and particular solution sets. Using eqns. (5.153) and (5.154), the general solution becomes

$$\begin{bmatrix} x_1(t) \\ x_2(t) \end{bmatrix}=c_1\begin{bmatrix} -e^{-t} \\ e^{-t} \end{bmatrix}+c_2\begin{bmatrix} e^{3t} \\ e^{3t} \end{bmatrix}+\begin{bmatrix} \dfrac{t}{3}-\dfrac{e^{-t}}{8}-\dfrac{te^{-t}}{2}-\dfrac{5}{9} \\ \dfrac{te^{-t}}{2}-\dfrac{e^{-t}}{8}-\dfrac{2t}{3}+\dfrac{4}{9} \end{bmatrix}. \tag{5.155}$$

Applying initial conditions to eqn. (5.149), the two constants c_1 and c_2 are evaluated as

$$\begin{bmatrix} c_1 \\ c_2 \end{bmatrix} = \begin{bmatrix} -1 \\ \dfrac{49}{72} \end{bmatrix}. \tag{5.156}$$

Using the constants c_1 and c_2 in eqn. (5.156), the solution set in eqn. (5.155) becomes

$$\begin{aligned} x_1(t) &= \frac{t}{3} + \frac{7}{8}e^{-t} + \frac{49}{72}e^{3t} - \frac{t}{2}e^{-t} - \frac{5}{9} \\ x_2(t) &= \frac{49}{72}e^{3t} - \frac{9}{8}e^{-t} - \frac{2t}{3} + \frac{t}{2}e^{-t} + \frac{4}{9} \end{aligned}. \tag{5.157}$$

The Laplace transform relationship is obtained as

$$\begin{bmatrix} X_1(s) \\ X_2(s) \end{bmatrix} = \begin{bmatrix} -\dfrac{2}{(s+1)(-s^2+2s+3)} - \dfrac{(s-1)\left(\dfrac{1}{s^2}+1\right)}{(-s^2+2s+3)} \\ -\dfrac{2\left(\dfrac{1}{s^2}+1\right)}{(-s^2+2s+3)} - \dfrac{s-1}{(s+1)(-s^2+2s+3)} \end{bmatrix}. \tag{5.158}$$

Taking the inverse Laplace transforms, the solution set matching the set in eqn. (5.157) is obtained.

Example # 5.13 The coefficient matrix, forcing functions and initial conditions are given as

$$A = \begin{bmatrix} 1 & 0 \\ -2 & 3 \end{bmatrix}. \tag{5.159}$$

$$\vec{g}(t) = \begin{bmatrix} t^2 \\ te^{-t} \end{bmatrix}. \tag{5.160}$$

$$\begin{bmatrix} x_1(0) \\ x_2(0) \end{bmatrix} = \begin{bmatrix} 1 \\ 1 \end{bmatrix}. \tag{5.161}$$

The set of coupled homogeneous set of differential equation becomes

$$\begin{aligned} x_1'(t) &= x_1(t) + t^2, \quad x_1(0) = 1 \\ x_2'(t) &= -2x_1(t) + 3x_2(t) + te^{-t}, \quad x_2(0) = 1 \end{aligned}. \tag{5.162}$$

The eigenvalues and eigenvectors of the coefficient matrix are

$$\begin{aligned} \lambda_1 &= 1 \quad \vec{v}_1 = \begin{bmatrix} 1 \\ 1 \end{bmatrix} \\ \lambda_2 &= 3 \quad \vec{v}_2 = \begin{bmatrix} 0 \\ 1 \end{bmatrix} \end{aligned}. \tag{5.163}$$

The fundamental matrix associated with the coefficient matrix (and the homogeneous differential equations set) is

$$X(t)=\left[\vec{X}_1 \quad \vec{X}_2\right]=\begin{bmatrix} e^t & 0 \\ e^t & e^{3t} \end{bmatrix}. \tag{5.164}$$

The solution to the set of homogeneous differential equations is

$$\vec{x}_h(t)=c_1\vec{X}_1+c_2\vec{X}_2=c_1\begin{bmatrix} e^t \\ e^t \end{bmatrix}+c_2\begin{bmatrix} 0 \\ e^{3t} \end{bmatrix}. \tag{5.165}$$

The particular solution set is obtained using the expression in eqn. (5.133) as

$$\vec{x}_p(t)=\begin{bmatrix} -t^2-2t-2 \\ -\frac{16}{9}t-\frac{1}{16}e^{-t}-\frac{t}{4}e^{-t}-\frac{2}{3}t^2-\frac{52}{27} \end{bmatrix}. \tag{5.166}$$

The solution set can now be expressed as the sum of the homogeneous and particular solutions as

$$\begin{bmatrix} x_1(t) \\ x_2(t) \end{bmatrix}=c_1\begin{bmatrix} e^t \\ e^t \end{bmatrix}+c_2\begin{bmatrix} 0 \\ e^{3t} \end{bmatrix}+\begin{bmatrix} -t^2-2t-2 \\ -\frac{16}{9}t-\frac{1}{16}e^{-t}-\frac{t}{4}e^{-t}-\frac{2}{3}t^2-\frac{52}{27} \end{bmatrix}. \tag{5.167}$$

Applying initial conditions in eqn. (5.161) to the solution set in eqn. (5.167), the two constants are obtained as

$$\begin{bmatrix} c_1 \\ c_2 \end{bmatrix}=\begin{bmatrix} 3 \\ -\frac{5}{432} \end{bmatrix}. \tag{5.168}$$

Substituting eqn. (5.168) in eqn. (5.167), the solution set becomes

$$\begin{aligned} x_1(t)&=3e^t-2t-t^2-2 \\ x_2(t)&=3e^t-\frac{1}{16}e^{-t}-\frac{5}{432}e^{3t}-\frac{16}{9}t-\frac{t}{4}e^{-t}-\frac{2}{3}t^2-\frac{52}{27} \end{aligned}. \tag{5.169}$$

The Laplace transform relationship is obtained as

$$\begin{bmatrix} X_1(s) \\ X_2(s) \end{bmatrix}=\begin{bmatrix} \dfrac{\frac{2}{s^3}+1}{s-1} \\ \dfrac{\frac{1}{(s+1)^2}+1}{s-3}-2\dfrac{\left(\frac{2}{s^3}+1\right)}{(s-1)(s-3)} \end{bmatrix}. \tag{5.170}$$

Taking the inverse Laplace transform, the solution set matching the set in eqn. (5.169) is obtained.

5.2.2.2 *Non-homogeneous coupled differential equations with constant (no time dependence) forcing functions*

While the method described above can be used even when the forcing functions are numbers instead of time dependent entities, in the absence of time dependent forcing functions, another simple approach exists for solving a pair of coupled non-homogeneous first order differential equations with constant coefficients. The typical non-homogeneous differential equation given in eqn. (5.4) is now expressed as

$$\frac{d\vec{x}}{dt} = \begin{bmatrix} \frac{dx_1}{dt} \\ \frac{dx_2}{dt} \end{bmatrix} = A\vec{x} + \vec{q} \ . \tag{5.171}$$

In eqn. (5.171), the forcing function expressed in vectorial form is

$$\vec{q} = \begin{bmatrix} q_1 \\ q_2 \end{bmatrix} . \tag{5.172}$$

The initial conditions are

$$\vec{x}(0) = \begin{bmatrix} x_1(0) \\ x_2(0) \end{bmatrix} \tag{5.173}$$

While the critical point associated with a pair of coupled first order homogeneous differential equations with constant coefficient is [0, 0], the critical point associated with set of the non-homogeneous differential equations in eqn. (5.171) is the solution of

$$\frac{d\vec{x}}{dt} = A\vec{x} + \vec{q} = \begin{bmatrix} 0 \\ 0 \end{bmatrix} . \tag{5.174}$$

In other words, the critical point is the solution of

$$A\vec{x} + \vec{q} = \begin{bmatrix} 0 \\ 0 \end{bmatrix} \quad \Rightarrow A\vec{x} = -\vec{q} \tag{5.175}$$

Using simple principles of linear algebra, the critical point becomes

$$\begin{bmatrix} x_{10} \\ x_{20} \end{bmatrix} = A^{-1}\left[-\vec{q}\right] = A^{-1}\begin{bmatrix} -q_1 \\ -q_2 \end{bmatrix} . \tag{5.176}$$

In eqn. (5.176), A^{-1} is the inverse of the matrix A. Note that if $\vec{q}$ is a null vector, the critical point will the origin, [0, 0]. But, when $\vec{q}$ is not a null vector, it is obtained as $[x_{10}, x_{20}]$. If $\alpha(t)$ and $\beta(t)$ denote the departure of $x_1(t)$ and $x_2(t)$ from their respective equilibrium values,

$$\begin{aligned} \alpha(t) &= x_1(t) - x_{10} \\ \beta(t) &= x_2(t) - x_{20} \end{aligned} . \tag{5.177}$$

Using eqn. (5.177), the non-homogeneous differential equation set in eqn. (5.171) can be expressed as

$$\begin{aligned}\frac{d}{dt}\left[\alpha(t)+x_{10}\right]&=A_{11}\left[\alpha(t)+x_{10}\right]+A_{12}\left[\beta(t)+x_{20}\right]+q_1\\ \frac{d}{dt}\left[\beta(t)+x_{20}\right]&=A_{21}\left[\alpha(t)+x_{10}\right]+A_{22}\left[\beta(t)+x_{20}\right]+q_2\end{aligned}. \tag{5.178}$$

Carrying out the differentiation leads to

$$\begin{aligned}\frac{d}{dt}\left[\alpha(t)\right]&=A_{11}\alpha(t)+A_{12}\beta(t)+A_{11}x_{10}+A_{12}x_{20}+q_1\\ \frac{d}{dt}\left[\beta(t)\right]&=A_{21}\alpha(t)+A_{22}\beta(t)+A_{21}x_{10}+A_{22}x_{20}+q_2\end{aligned}. \tag{5.179}$$

Because $[x_{10}, x_{20}]$ is the solution to eqn. (5.175),

$$\begin{aligned}A_{11}x_{10}+A_{12}x_{20}&=-q_1\\ A_{21}x_{10}+A_{22}x_{20}&=-q_2\end{aligned}. \tag{5.180}$$

Using eqn. (5.180), eqn. (5.179) becomes

$$\begin{aligned}\frac{d}{dt}\left[\alpha(t)\right]&=A_{11}\alpha(t)+A_{12}\beta(t)\\ \frac{d}{dt}\left[\beta(t)\right]&=A_{21}\alpha(t)+A_{22}\beta(t)\end{aligned}. \tag{5.181}$$

It can be seen that eqn. (5.181) is a homogeneous coupled first order system defined by the same coefficient matrix A. In other words, eqn. (5.181) is the homogeneous equivalent of the non-homogeneous set given in eqn. (5.171) with a new set of initial conditions

$$\begin{aligned}\alpha(0)&=x_1(0)-x_{10}\\ \beta(0)&=x_2(0)-x_{20}\end{aligned}. \tag{5.182}$$

Once eqn. (5.181) is solved for $\alpha(t)$ and $\beta(t)$ with the initial conditions in eqn. (5.182), the solution set of the non-homogeneous differential equation set can be written using eqn. (5.177) as

$$\begin{aligned}x_1(t)&=\alpha(t)+x_{10}\\ x_2(t)&=\beta(t)+x_{20}\end{aligned}. \tag{5.183}$$

It is clear from the discussion above that the approach described here works only when the coefficient matrix is invertible (determinant of the matrix cannot be zero). An example is shown now comparing the two approaches to illustrate that the method described above is rather simple.

Example # 5.14 Consider a non-homogeneous coupled system with the coefficient matrix, forcing functions and initial conditions are given as

$$A=\begin{bmatrix}1 & 1\\ 4 & 1\end{bmatrix}. \tag{5.184}$$

$$\vec{g}(t)=\begin{bmatrix}3\\4\end{bmatrix}. \tag{5.185}$$

$$\begin{bmatrix}x_1(0)\\x_2(0)\end{bmatrix}=\begin{bmatrix}1\\-1\end{bmatrix}. \tag{5.186}$$

The set of differential equations are

$$\begin{aligned}x_1'(t)&=x_1(t)+x_2(t)+3, \quad x_1(0)=1\\x_2'(t)&=4x_1(t)+x_2(t)+4, \quad x_2(0)=-1\end{aligned} \tag{5.187}$$

The equilibrium or the critical point of the non-homogeneous system is obtained using eqn. (5.176)

$$\begin{bmatrix}x_{10}\\x_{20}\end{bmatrix}=\begin{bmatrix}1&1\\4&1\end{bmatrix}^{-1}\begin{bmatrix}-3\\-4\end{bmatrix}=\begin{bmatrix}-\frac{1}{3}\\-\frac{8}{3}\end{bmatrix}. \tag{5.188}$$

The phase plane plot for the non-homogeneous system and the homogeneous system are given. They show that the critical point matches the values in eqn. (5.188) for the non-homogeneous system described by the differential equations in eqn. (5.187).

Using the new critical point, the new initial conditions are

$$\begin{aligned}\alpha(0)&=x_1(0)-x_{10}=\frac{4}{3}\\\beta(0)&=x_2(0)-x_{20}=\frac{5}{3}\end{aligned}. \tag{5.189}$$

The equivalent set of homogeneous differential equations is

$$\begin{aligned}\alpha'(t)&=\alpha(t)+\beta(t), \quad \alpha(0)=\frac{4}{3}\\\beta'(t)&=4\alpha(t)+\beta(t), \quad \beta(0)=\frac{5}{3}\end{aligned} \tag{5.190}$$

The eigenvalues and eigenvectors of the coefficient matrix are

$$\begin{aligned}\lambda_1&=-1 \quad \vec{v}_1=\begin{bmatrix}-\frac{1}{2}\\1\end{bmatrix}\\\lambda_2&=3 \quad \vec{v}_2=\begin{bmatrix}\frac{1}{2}\\1\end{bmatrix}\end{aligned}. \tag{5.191}$$

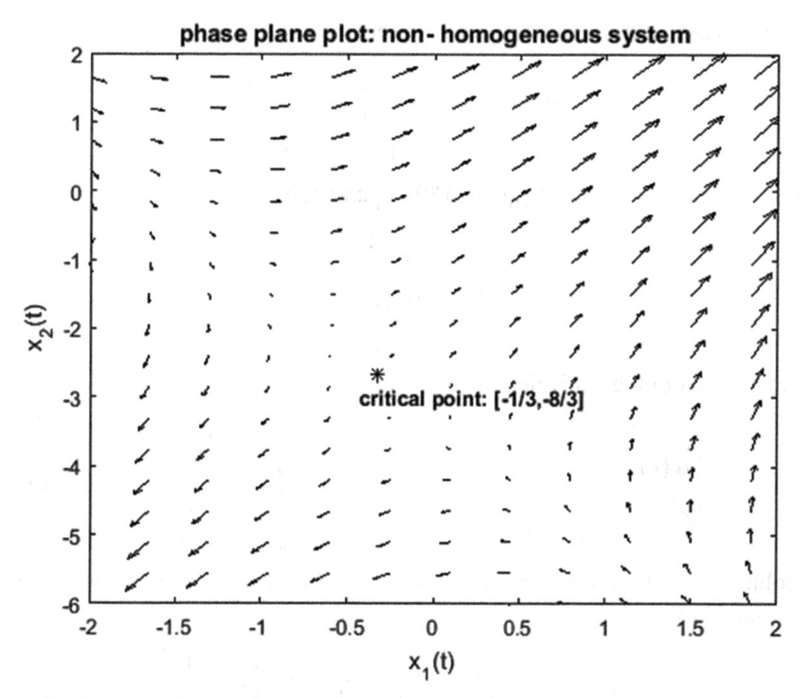

Figure 5.23 Phase portrait associated with the non-homogeneous system in Example # 5.14. The critical point is not located at the origin.

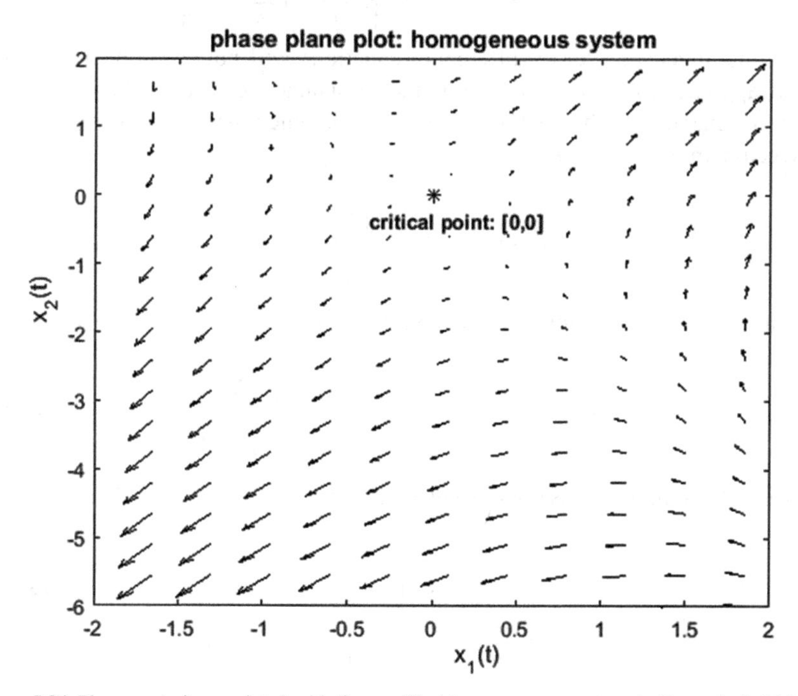

Figure 5.24 Phase portrait associated with the modified homogeneous system in Example # 5.14. The critical point is now at the origin.

The solution set for the new homogeneous set is

$$\begin{bmatrix} \alpha(t) \\ \beta(t) \end{bmatrix} = c_1 e^{-t} \begin{bmatrix} -\frac{1}{2} \\ 1 \end{bmatrix} + c_2 e^{3t} \begin{bmatrix} \frac{1}{2} \\ 1 \end{bmatrix} \tag{5.192}$$

Applying initial conditions in eqn. (5.189), c_1 and c_2 become

$$\begin{bmatrix} c_1 \\ c_2 \end{bmatrix} = \begin{bmatrix} -\frac{1}{2} \\ \frac{13}{6} \end{bmatrix}. \tag{5.193}$$

The solution set of the equivalent homogeneous set becomes

$$\begin{bmatrix} \alpha(t) \\ \beta(t) \end{bmatrix} = -\frac{1}{2} e^{-t} \begin{bmatrix} -\frac{1}{2} \\ 1 \end{bmatrix} + \frac{13}{6} e^{3t} \begin{bmatrix} \frac{1}{2} \\ 1 \end{bmatrix} = \begin{bmatrix} \frac{1}{4} e^{-t} + \frac{13}{12} e^{3t} \\ -\frac{1}{2} e^{-t} + + \frac{13}{6} e^{3t} \end{bmatrix}. \tag{5.194}$$

The solution set of the non-homogeneous set of differential equations are obtained using eqn. (5.183) as

$$\begin{bmatrix} x_1(t) \\ x_2(t) \end{bmatrix} = \begin{bmatrix} \alpha(t) \\ \beta(t) \end{bmatrix} + \begin{bmatrix} x_{10} \\ x_{20} \end{bmatrix} = \begin{bmatrix} \frac{1}{4} e^{-t} + \frac{13}{12} e^{3t} - \frac{1}{3} \\ -\frac{1}{2} e^{-t} + \frac{13}{6} e^{3t} - \frac{8}{3} \end{bmatrix} \tag{5.195}$$

The non-homogeneous set of differential equations will now be solved using the standard approach where the particular solution is obtained separately. The fundamental matrix associated with the coefficient matrix is obtained using the eigenvalues and eigenvectors in eqn. (5.191)

$$X(t) = \begin{bmatrix} \vec{X}_1 & \vec{X}_2 \end{bmatrix} = \begin{bmatrix} \frac{e^{-t}}{2} & \frac{e^{3t}}{2} \\ e^{-t} & e^{3t} \end{bmatrix}. \tag{5.196}$$

The solution to the set of homogeneous differential equations is

$$\vec{x}_h(t) = b_1 \vec{X}_1 + b_2 \vec{X}_2 = b_1 \begin{bmatrix} -\frac{e^{-t}}{2} \\ e^{-t} \end{bmatrix} + b_2 \begin{bmatrix} \frac{e^{3t}}{2} \\ e^{3t} \end{bmatrix}. \tag{5.197}$$

The particular solution set is obtained using the expression in eqn. (5.133) as

$$\vec{x}_p(t) = \begin{bmatrix} -\frac{1}{3} \\ -\frac{8}{3} \end{bmatrix}. \tag{5.198}$$

The solution set can now be expressed as the sum of the homogeneous and particular solutions as

$$\begin{bmatrix} x_1(t) \\ x_2(t) \end{bmatrix} = b_1 \begin{bmatrix} -\dfrac{e^{-t}}{2} \\ e^{-t} \end{bmatrix} + b_2 \begin{bmatrix} \dfrac{e^{3t}}{2} \\ e^{3t} \end{bmatrix} + \begin{bmatrix} -\dfrac{1}{3} \\ -\dfrac{8}{3} \end{bmatrix}. \tag{5.199}$$

Applying the original initial conditions given in eqn. (5.186), the two constants b_1 and b_2 become

$$\begin{bmatrix} b_1 \\ b_2 \end{bmatrix} = \begin{bmatrix} -\dfrac{1}{2} \\ \dfrac{13}{6} \end{bmatrix}. \tag{5.200}$$

Substituting b_1 and b_2 in eqn. (5.199), the solution set of the non-homogeneous set becomes

$$\begin{bmatrix} x_1(t) \\ x_2(t) \end{bmatrix} = \begin{bmatrix} \dfrac{1}{4}e^{-t} + \dfrac{13}{12}e^{3t} - \dfrac{1}{3} \\ -\dfrac{1}{2}e^{-t} + \dfrac{13}{6}e^{3t} - \dfrac{8}{3} \end{bmatrix}. \tag{5.201}$$

Notice that the solution sets in eqn. (5.195) and eqn. (5.201). Furthermore, it can be seen that the particular solution in eqn. (5.198) matches the equilibrium or the critical point in eqn. (5.188). It should be noted that the approach based on converting the non-homogeneous system to an equivalent homogeneous system with a new set of initial conditions works only when the forcing functions are numbers, with the coefficient matrix being invertible.

5.3 Multiple First Order Coupled Differential Equations of Constant Coefficients

The concept of the fundamental matrix can be extended to find solutions to a set of coupled differential equations with constant coefficients. If there are n variables, the set of differential equations are completely defined by the [n x n] coefficient matrix A and the [n x 1] vector $\vec{g}(t)$ g(t) consisting of the forcing functions. Equation (5.4) becomes

$$\frac{d\vec{x}}{dt} = \begin{bmatrix} \dfrac{dx_1}{dt} \\ \dfrac{dx_2}{dt} \\ \cdot \\ \cdot \\ \dfrac{dx_{n-1}}{dt} \\ \dfrac{dx_n}{dt} \end{bmatrix} = A\vec{x} + \vec{g}(t). \tag{5.202}$$

The initial conditions will consist of another [n x 1] vector

$$\vec{x}(0)=\begin{bmatrix} x_1(0) \\ x_2(0) \\ x_3(0) \\ . \\ x_{n-1}(0) \\ x_n(0) \end{bmatrix} . \tag{5.203}$$

The forcing function also constitute another [n x 1] vector,

$$\vec{g}(t)=\begin{bmatrix} g_1(t) \\ g_2(t) \\ g_3(t) \\ . \\ g_{n-1}(t) \\ g_n(t) \end{bmatrix} . \tag{5.204}$$

Even though the forcing functions are represented as functions of t, if the forcing functions constitute a null vector, the set of equations in eqn. (5.202) becomes a set of homogeneous first order differential equations. Another important point to be noted is that it is not necessary for all the elements of the vector in eqn. (5.204) be time dependent, they may even be constants represented by real numbers.

The procedure for obtaining the solution to a set of non-homogeneous first order coupled differential equations of constant coefficients is identical to the procedure used for obtaining the solution in the case of a pair of coupled equations. The solution is once again written as the sum of the solution to the homogeneous part and the particular solutions set.

5.3.1 Solution using eigenvalues and eigenvectors

The first step in this procedure is the determination of whether the coefficient matrix is defective or not. If the matrix is not defective as demonstrated in terms of having n distinct eigenvectors, the solution to the homogeneous part can be written as

$$\vec{x}_h(t)=\sum_{k=1}^{n} c_k \vec{x}_k(t) . \tag{5.205}$$

In eqn. (5.205),

$$\vec{x}_k(t)=\vec{v}_k e^{\lambda_k t}=\vec{X}_k,\quad k=1,2,...,n . \tag{5.206}$$

The distinct eigenvectors are represented by $\vec{v}_k$ and the corresponding eigenvalues are λ_k. The fundamental matrix associated with the set of differential equations is

$$X(t)=\left[\vec{X}_1 \quad \vec{X}_2 \quad \cdots \quad \vec{X}_n\right] . \tag{5.207}$$

Depending on the algebraic and geometric multiplicities of the eigenvalues, if the matrix is deemed to be defective, the solution to the homogeneous part can still be expressed by eqn. (5.205) with the individual solution vectors represented in terms of the generalized eigenvectors. This aspect is discussed in detail in Appendix D.

The particular solution is obtained as described in eqn. (5.133). The general solution can be written as

$$\vec{x}(t) = X(t)\vec{c} + \vec{x}_p(t). \tag{5.208}$$

In eqn. (5.208), $\vec{x}_p(t)$ is an [n x 1] vector and $\vec{c}$ is a vector of size [n x 1] consisting of unknown constants to be determined by applying the initial conditions. The particular solution $\vec{x}_p(t)$ is obtained using eqn. (5.133) noting that we now have an [n x n] fundamental matrix and the solution vectors and the forcing functions are of size [n x 1]. The homogeneous solution set as a vector can be written as

$$\vec{x}_h(t) = X(t)\vec{c} \ . \tag{5.209}$$

While obtaining the unknown constants is relatively simple in the case of a pair of coupled first order differential equations, the evaluation of the constants becomes cumbersome as n goes up. The vector $\vec{c}$ can be evaluated using the concepts of inverse of a matrix. Setting t = 0 eqn. (5.208) becomes

$$X(0)\vec{c} = \vec{x}(0) - \vec{x}_p(0) \ . \tag{5.210}$$

In eqn. (5.210),

$$\begin{aligned} X(0) &= X(t)\big|_{t=0} \\ \vec{x}_p(0) &= \vec{x}_p(t)\big|_{t=0} \end{aligned} \tag{5.211}$$

Equation (5.210) can be rewritten as a simple matrix equation (matrix-vector equation) in terms a new [n x 1] vector $\vec{x}_d(0)$,

$$\vec{x}_d(0) = \vec{x}(0) - \vec{x}_p(0) \ . \tag{5.212}$$

The solution for vector $\vec{c}$ is given by

$$\vec{c} = \left[X(0)\right]^{-1} \vec{x}_d(0) \ . \tag{5.213}$$

The power of (-1) on the right hand side of eqn. (5.213) implies the inverse of the matrix. Once $\vec{c}$ is substituted in eqn. (5.208), the complete solution is obtained.

5.3.2 *Solution using Laplace transforms*

With the availability of initial conditions, the complete solution can be obtained in a way similar to the method adopted for a pair of coupled first order differential equations in eqn. (5.134) which can be rewritten as

$$\vec{x}(t) = L^{-1}\left\{\left[sI_n - A\right]^{-1} \vec{x}(0) + L\left(\vec{g}(t)\right)\right\} \ . \tag{5.214}$$

5.3.3 Examples

Example # 5.15 Consider the coefficient matrix A, initial conditions $\vec{x}(0)$ and the forcing function $\vec{g}(t)$ given by

$$A = \begin{bmatrix} 0 & -1 & 1 \\ 0 & 2 & 0 \\ -2 & -1 & 3 \end{bmatrix}. \tag{5.215}$$

$$\vec{x}(0) = \begin{bmatrix} 0 \\ 0 \\ 1 \end{bmatrix}. \tag{5.216}$$

$$\vec{g}(t) = \begin{bmatrix} 1 \\ t \\ e^{-t} \end{bmatrix}. \tag{5.217}$$

The coupled differential equations can be expressed as

$$\begin{aligned} x_1'(t) &= -x_2(t) + x_3(t) + 1 \\ x_2'(t) &= 2x_2(t) + t \\ x_3'(t) &= -2x_1(t) - x_2(t) + 3x_3(t) + e^{-t} \end{aligned} \quad . \tag{5.218}$$

The eigenvalues and eigenvectors of the coefficient matrix are

$$\begin{aligned} \lambda_1 &= 1 \quad \vec{v}_1 = \begin{bmatrix} 1 \\ 0 \\ 1 \end{bmatrix} \\ \lambda_2 &= 2 \quad \vec{v}_2 = \begin{bmatrix} -\frac{1}{2} \\ 1 \\ 0 \end{bmatrix} \\ \lambda_3 &= 2 \quad \vec{v}_3 = \begin{bmatrix} \frac{1}{2} \\ 1 \\ 0 \end{bmatrix} \end{aligned} . \tag{5.219}$$

Even though two of the eigenvalues are equal, eigenvectors associated with these eigenvalues are distinct and therefore, the coefficient t matrix is not defective. The fundamental matrix in eqn. (5.207) becomes

$$X(t) = \begin{bmatrix} e^{t} & -\frac{e^{2t}}{2} & \frac{e^{2t}}{2} \\ 0 & e^{2t} & 0 \\ e^{t} & 0 & e^{2t} \end{bmatrix}. \tag{5.220}$$

Using a set of unknown constants to be determined, the solutions to the associated homogeneous set of differential equations become

$$\vec{x}_h(t) = X(t)\vec{c} = \begin{bmatrix} \frac{e^t}{2}\left(2c_1 - c_2 e^t + c_3 e^t\right) \\ c_2 e^{2t} \\ e^t\left(c_1 + c_3 e^t\right) \end{bmatrix} \tag{5.221}$$

Using the method of variation parameters expressed in eqn. (5.133), the particular solution set is given by

$$\vec{x}_p(t) = \begin{bmatrix} \frac{e^{-t}}{6} - \frac{t}{2} - \frac{9}{4} \\ -\frac{t}{2} - \frac{1}{4} \\ -\frac{e^{-t}}{6} - \frac{t}{2} - \frac{7}{4} \end{bmatrix}. \tag{5.222}$$

The complete solution is the sum of the homogeneous and particular solutions and the solutions set is given as

$$\vec{x}(t) = \begin{bmatrix} \frac{e^t}{2}\left(2c_1 - c_2 e^t + c_3 e^t\right) + \frac{e^{-t}}{6} - \frac{t}{2} - \frac{9}{4} \\ c_2 e^{2t} - \frac{t}{2} - \frac{1}{4} \\ e^t\left(c_1 + c_3 e^t\right) - \frac{e^{-t}}{6} - \frac{t}{2} - \frac{7}{4} \end{bmatrix}. \tag{5.223}$$

Applying the initial conditions in eqn. (5.216), the unknown constants are

$$\vec{c} = \begin{bmatrix} c_1 \\ c_2 \\ c_3 \end{bmatrix} = \begin{bmatrix} \frac{3}{2} \\ \frac{1}{4} \\ \frac{17}{12} \end{bmatrix}. \tag{5.224}$$

Substituting the constants from eqn. (5.224), the solutions set for the differential equations in eqn. (5.218) becomes

$$\vec{x}(t) = \begin{bmatrix} \frac{7}{12}e^{2t} + \frac{3}{2}e^t + \frac{e^{-t}}{6} - \frac{t}{2} - \frac{9}{4} \\ \frac{e^{2t}}{4} - \frac{t}{2} - \frac{1}{4} \\ \frac{17}{12}e^{2t} + \frac{3}{2}e^t - \frac{e^{-t}}{6} - \frac{t}{2} - \frac{7}{4} \end{bmatrix}. \tag{5.225}$$

Based on the relationship of the Laplace transforms and differential equations given in eqn. (5.214), the Laplace transform of the solution set becomes

$$\vec{x}_L(s) = \begin{bmatrix} -\dfrac{-2s^3+4s+1}{s^2(s-1)(s+1)(s-2)} \\ \dfrac{1}{s^2(s-2)} \\ -\dfrac{-s^4-2s^3+2s^2+3s+1}{s^2(s-1)(s+1)(s-2)} \end{bmatrix}. \tag{5.226}$$

Taking the inverse Laplace transform, the solution set obtained matches the set in eqn. (5.225).

Example # 5.16 Consider the coefficient matrix A, initial conditions $\vec{x}(0)$ and the forcing function $\vec{g}(t)$ given by

$$A = \begin{bmatrix} -1 & -1 & 1 \\ -1 & 2 & -1 \\ -1 & 1 & 0 \end{bmatrix}. \tag{5.227}$$

$$\vec{x}(0) = \begin{bmatrix} 0 \\ 1 \\ 1 \end{bmatrix}. \tag{5.228}$$

$$\vec{g}(t) = \begin{bmatrix} te^{-t} \\ e^{-t} \\ e^{t} \end{bmatrix}. \tag{5.229}$$

The coupled differential equations can be expressed as

$$\begin{aligned} x_1'(t) &= -x_1(t) - x_2(t) + x_3(t) + te^{-t} \\ x_2'(t) &= -x_1(t) + 2x_2(t) - x_3(t) + e^{-t} \\ x_3'(t) &= -x_1(t) + x_2(t) + e^{t} \end{aligned}. \tag{5.230}$$

The eigenvalues and eigenvectors of the coefficient matrix are

$$\begin{aligned} \lambda_1 &= -1 \quad \vec{v}_1 = \begin{bmatrix} 2 \\ 1 \\ 1 \end{bmatrix} \\ \lambda_{2,3} &= 1 \quad \vec{v}_2 = \begin{bmatrix} 0 \\ 1 \\ 1 \end{bmatrix} \end{aligned}. \tag{5.231}$$

Equation (5.231) shows that the coefficient matrix is defective because distinct eigenvectors do not exist for the pair of equal eigenvalues. This requires the generation of generalized eigenvectors ($\vec{v}_{g2}$ and $\vec{v}_{g3}$) for the eigenvalue pair. These can be obtained using the method described in Appendix D and they are

$$\lambda_{2,3} = 1 \quad \vec{v}_2 = \begin{bmatrix} -\dfrac{1}{2} \\ 1 \\ 0 \end{bmatrix} \quad \vec{v}_3 = \begin{bmatrix} \dfrac{1}{2} \\ 0 \\ 1 \end{bmatrix}. \tag{5.232}$$

The fundamental matrix becomes

$$X(t) = \begin{bmatrix} 2e^{-t} & -\dfrac{e^t}{2} & \dfrac{e^t}{2} \\ e^{-t} & e^t\left(\dfrac{3t}{2}+1\right) & -\dfrac{3e^t t}{2} \\ e^{-t} & \dfrac{3e^t t}{2} & -e^t\left(\dfrac{3t}{2}-1\right) \end{bmatrix}. \tag{5.233}$$

Using a set of unknown constants to be determined, the solutions to the associated homogeneous set of differential equations become

$$\vec{x}_h(t) = X(t)\vec{c} = \begin{bmatrix} 2c_1e^{-t} - \dfrac{c_2}{2}e^t + \dfrac{c_3}{2}e^t \\ c_1e^{-t} + \dfrac{c_2}{2}e^t(3t+2) - \dfrac{3c_3te^t}{2} \\ c_1e^{-t} + \dfrac{3c_2te^t}{2} - \dfrac{c_3}{2}e^t(3t-2) \end{bmatrix} \tag{5.234}$$

Using the method of variation parameters expressed in eqn. (5.133), the particular solution set is given by

$$\vec{x}_p(t) = \begin{bmatrix} e^{-t}\dfrac{\left(2t - e^{2t} + 2te^{2t} + 2t^2 + 1\right)}{4} \\ e^{-t}\dfrac{\left(4t - e^{2t} + 2te^{2t} - 6t^2e^{2t} + 2t^2 + 1\right)}{8} \\ e^{-t}\dfrac{\left(4t - e^{2t} + 10te^{2t} - 6t^2e^{2t} + 2t^2 + 5\right)}{8} \end{bmatrix}. \tag{5.235}$$

The complete solution is the sum of the homogeneous and particular solutions and the solutions set is given as

$$\vec{x}(t)=\begin{bmatrix} 2c_1e^{-t}-\frac{c_2}{2}e^{t}+\frac{c_3}{2}e^{t}+e^{-t}\frac{\left(2t-e^{2t}+2te^{2t}+2t^2+1\right)}{4} \\ c_1e^{-t}+\frac{c_2}{2}e^{t}\left(3t+2\right)-\frac{3c_3te^{t}}{2}+e^{-t}\frac{\left(4t-e^{2t}+2te^{2t}-6t^2e^{2t}+2t^2+1\right)}{8} \\ c_1e^{-t}+\frac{3c_2te^{t}}{2}-\frac{c_3}{2}e^{t}\left(3t-2\right)+e^{-t}\frac{\left(4t-e^{2t}+10te^{2t}-6t^2e^{2t}+2t^2+5\right)}{8} \end{bmatrix}. \tag{5.236}$$

Applying the initial conditions in eqn. (5.216), the unknown constants are

$$\vec{c}=\begin{bmatrix} c_1 \\ c_2 \\ c_3 \end{bmatrix}=\begin{bmatrix} \frac{1}{8} \\ \frac{7}{8} \\ \frac{3}{8} \end{bmatrix}. \tag{5.237}$$

Substituting the constants from eqn. (5.224), the solutions set for the differential equations in eqn. (5.218) becomes

$$\vec{x}(t)=\begin{bmatrix} e^{-t}\frac{\left(t-e^{2t}+2te^{2t}+2t^2+1\right)}{2} \\ e^{-t}\frac{\left(2t+3e^{2t}+4te^{2t}-3t^2e^{2t}+t^2+1\right)}{4} \\ e^{-t}\frac{\left(2t+e^{2t}+8te^{2t}-3t^2e^{2t}+t^2+3\right)}{4} \end{bmatrix}. \tag{5.238}$$

Based on the relationship of the Laplace transforms and differential equations given in eqn. (5.214), the Laplace transform of the solution set becomes

$$\vec{x}_L(s)=\begin{bmatrix} \frac{s^3+3}{\left(s-1\right)^3\left(s+1\right)^3} \\ -\frac{-s^5-2s^4+2s^3+7s^2+3s+3}{\left(s^2-1\right)^3} \\ -\frac{-s^5-2s^4+5s^2+5s+5}{\left(s^2-1\right)^3} \end{bmatrix}. \tag{5.239}$$

Taking the inverse Laplace transforms, the solution set obtained matches the set in eqn. (5.238).

Example # 5.17 Consider the coefficient matrix A of size [4 x 4], initial conditions ( ) and the forcing function $\vec{g}(t)$ given by

$$A=\begin{bmatrix} -1 & 0 & 0 & 0 \\ 0 & -1 & 0 & 1 \\ 0 & 0 & 1 & -1 \\ 0 & 0 & 0 & 1 \end{bmatrix}. \tag{5.240}$$

$$\vec{x}(0)=\begin{bmatrix} 0 \\ 1 \\ 1 \\ -1 \end{bmatrix}. \tag{5.241}$$

$$\vec{g}(t)=\begin{bmatrix} e^{t} \\ te^{-t} \\ e^{-t} \\ t \end{bmatrix}. \tag{5.242}$$

The set of differential equations associated with the coefficient matrix is

$$\begin{bmatrix} x_1^{'}(t)=-x_1(t)+e^{t} \\ x_2^{'}(t)=-x_2(t)+x_4(t)+te^{-t} \\ x_3^{'}(t)=x_3(t)-x_4(t)+e^{-t} \\ x_4^{'}(t)=x_4(t)+t \end{bmatrix}. \tag{5.243}$$

The eigenvalues of the matrix are

$$\begin{aligned} \lambda_{1,2} &= 1 \\ \lambda_{3,4} &= -1 \end{aligned}. \tag{5.244}$$

The difficulties arise because the coefficient matrix only has three eigenvectors,

$$\begin{bmatrix} 0 \\ 0 \\ 1 \\ 0 \end{bmatrix}, \begin{bmatrix} 1 \\ 0 \\ 0 \\ 0 \end{bmatrix}, \begin{bmatrix} 0 \\ 1 \\ 0 \\ 0 \end{bmatrix}. \tag{5.245}$$

There is a need to determine which of the eigenvalues are degenerate so that appropriate generalized eigenvectors can be created. To find this out, the first step is to determine the eigenvectors associated with the eigenvalue of 1. These are obtained as the solution of

$$(A-\lambda I_4)=0 . \tag{5.246}$$

There is only a single solution to eqn. (5.246),

$$\vec{v} = \begin{bmatrix} 0 \\ 0 \\ 1 \\ 0 \end{bmatrix}. \tag{5.247}$$

Repeating the procedure for the second eigenvalue of -1, the solution consists of two eigenvectors

$$\vec{v} = \begin{bmatrix} 1 \\ 0 \\ 0 \\ 0 \end{bmatrix}, \begin{bmatrix} 0 \\ 1 \\ 0 \\ 0 \end{bmatrix}. \tag{5.248}$$

This means that the eigenvalue of 1 is degenerate and there is a need to find a pair of generalized eigenvectors associated with this eigenvalue. Using the technique described in the Appendix D, the generalized eigenvectors are

$$\vec{v}_g = \begin{bmatrix} 0 \\ 0 \\ 1 \\ 0 \end{bmatrix}, \begin{bmatrix} 0 \\ \frac{1}{2} \\ 0 \\ 1 \end{bmatrix}. \tag{5.249}$$

Using the eigenvalues and paired eigenvectors, the fundamental matrix becomes

$$X(t) = \begin{bmatrix} e^{-t} & 0 & 0 & 0 \\ 0 & e^{-t} & 0 & \frac{1}{2}e^{t} \\ 0 & 0 & e^{t} & -te^{t} \\ 0 & 0 & 0 & e^{t} \end{bmatrix} \tag{5.250}$$

The solution for the homogeneous set is

$$\vec{x}_h(t) = \begin{bmatrix} c_1 e^{-t} \\ c_2 e^{-t} + \frac{c_4}{2} e^{t} \\ (c_3 - c_4 t) e^{t} \\ c_4 e^{t} \end{bmatrix}. \tag{5.251}$$

The particular solutions set is

$$\vec{x}_p(t) = \begin{bmatrix} \frac{e^t}{2} \\ \frac{t^2 e^{-t}}{2} - t \\ -t - \frac{e^{-t}}{2} - 2 \\ -t - 1 \end{bmatrix}. \tag{5.252}$$

Combining the homogeneous and particular solutions, the general solution becomes

$$\vec{x}(t) = \begin{bmatrix} c_1 e^{-t} + \frac{e^t}{2} \\ c_2 e^{-t} + \frac{c_4}{2} e^t + \frac{t^2 e^{-t}}{2} - t \\ (c_3 - c_4 t) e^t - t - \frac{e^{-t}}{2} - 2 \\ c_4 e^t - t - 1 \end{bmatrix}. \tag{5.253}$$

Applying initial conditions, the constants become

$$\vec{c} = \begin{bmatrix} c_1 \\ c_2 \\ c_3 \\ c_4 \end{bmatrix} = \begin{bmatrix} -\frac{1}{2} \\ 1 \\ \frac{7}{2} \\ 0 \end{bmatrix}. \tag{5.254}$$

The solutions set now becomes

$$\vec{x}(t) = \begin{bmatrix} \sinh(t) \\ e^{-t} + \frac{t^2 e^{-t}}{2} - t \\ \frac{7}{2} e^t - t - \frac{e^{-t}}{2} - 2 \\ -t - 1 \end{bmatrix}. \tag{5.255}$$

The Laplace transforms become

$$X(s)=\begin{bmatrix} \frac{1}{s^2-1} \\ \frac{1}{s+1}+\frac{1}{(s+1)^3}-\frac{1}{s^2} \\ \frac{s^3+3s^2+2s+1}{s^2(s^2-1)} \\ -\frac{s+1}{s^2} \end{bmatrix}. \tag{5.256}$$

Taking the inverse Laplace transforms, the solution set matching to the one in eqn. (5.255) is obtained.

Example # 5.18 Consider the coefficient matrix A of size [4 x 4], initial conditions $\vec{x}(0)$ and the forcing function $\vec{g}(t)$ given by

$$A=\begin{bmatrix} 1 & 0 & 0 & 0 \\ 0 & 1 & 0 & 0 \\ 1 & 4 & -3 & 0 \\ -1 & -2 & 0 & -3 \end{bmatrix}. \tag{5.257}$$

$$\vec{x}(0)=\begin{bmatrix} 1 \\ 1 \\ 1 \\ -1 \end{bmatrix}. \tag{5.258}$$

$$\vec{g}(t)=\begin{bmatrix} e^{-3t} \\ te^{3t} \\ e^{t} \\ t+e^{t} \end{bmatrix}. \tag{5.259}$$

The set of differential equations associated with the coefficient matrix is

$$\begin{bmatrix} x_1'(t)=x_1(t)+e^{-3t} \\ x_2'(t)=x_2(t)+te^{3t} \\ x_3'(t)=x_1(t)+4x_2(t)-3x_3(t)+e^{t} \\ x_4'(t)=-x_1(t)-2x_2(t)-3x_4(t)+t+e^{t} \end{bmatrix}. \tag{5.260}$$

The eigenvalues of the matrix are

$$\begin{aligned} \lambda_{1,2} &= -3 \\ \lambda_{3,4} &= -1 \end{aligned}. \tag{5.261}$$

The matrix is not defective because four distinct eigenvectors exist. These are

$$\begin{bmatrix}0\\0\\1\\0\end{bmatrix},\begin{bmatrix}0\\0\\0\\1\end{bmatrix},\begin{bmatrix}-4\\2\\1\\0\end{bmatrix},\begin{bmatrix}-8\\2\\0\\1\end{bmatrix}. \tag{5.262}$$

The first two eigenvectors are associated with the eigenvalue of -3 and the other two are associated with the eigenvalue of 1. Using the eigenvalues and paired eigenvectors, the fundamental matrix becomes

$$X(t)=\begin{bmatrix}0 & 0 & -4e^{t} & -8e^{t}\\ 0 & 0 & 2e^{t} & 2e^{t}\\ e^{-3t} & 0 & e^{t} & 0\\ 0 & e^{-3t} & 0 & e^{t}\end{bmatrix} \tag{5.263}$$

The solution for the homogeneous set is

$$\vec{x}_h(t)=\begin{bmatrix}-4e^{t}\left(c_3+2c_4\right)\\ 2e^{t}\left(c_3+c_4\right)\\ c_3e^{t}+c_1e^{-3t}\\ c_4e^{t}+c_2e^{-3t}\end{bmatrix}. \tag{5.264}$$

The particular solutions set is

$$\vec{x}_p(t)=\begin{bmatrix}-\frac{e^{-3t}}{4}\\ \frac{e^{3t}}{4}(2t-1)\\ -\frac{e^{-3t}}{144}\left(36t-36e^{4t}+32e^{6t}-48te^{6t}+9\right)\\ \frac{t}{3}+\frac{e^{-3t}}{16}+\frac{e^{3t}}{9}+\frac{e^{t}}{4}+\frac{te^{-3t}}{4}-\frac{te^{3t}}{6}-\frac{1}{9}\end{bmatrix}. \tag{5.265}$$

Combining the homogeneous and particular solutions, the general solution becomes

$$\vec{x}(t)=\begin{bmatrix}-4e^{t}\left(c_3+2c_4\right)-\frac{e^{-3t}}{4}\\ 2e^{t}\left(c_3+c_4\right)+\frac{e^{3t}}{4}(2t-1)\\ c_3e^{t}+c_1e^{-3t}-\frac{e^{-3t}}{144}\left(36t-36e^{4t}+32e^{6t}-48te^{6t}+9\right)\\ c_4e^{t}+c_2e^{-3t}+\frac{t}{3}+\frac{e^{-3t}}{16}+\frac{e^{3t}}{9}+\frac{e^{t}}{4}+\frac{te^{-3t}}{4}-\frac{te^{3t}}{6}-\frac{1}{9}\end{bmatrix}. \tag{5.266}$$

Applying initial conditions, the constants become

$$\vec{c} = \begin{bmatrix} c_1 \\ c_2 \\ c_3 \\ c_4 \end{bmatrix} = \begin{bmatrix} -\frac{19}{36} \\ -\frac{3}{8} \\ \frac{25}{16} \\ -\frac{15}{16} \end{bmatrix}. \tag{5.267}$$

The solutions set now becomes

$$\vec{x}(t) = \begin{bmatrix} \frac{5}{4}e^{t} - \frac{1}{4}e^{-3t} \\ \frac{5}{4}e^{t} + \frac{e^{3t}}{4}(2t-1) \\ -\frac{e^{-3t}}{144}\left(36t - 261e^{4t} + 32e^{6t} - 48te^{6t} + 85\right) \\ \frac{t}{3} - \frac{5e^{-3t}}{16} + \frac{e^{3t}}{9} - \frac{11e^{t}}{16} + \frac{te^{-3t}}{4} - \frac{te^{3t}}{6} - \frac{1}{9} \end{bmatrix}. \tag{5.268}$$

The Laplace transforms become

$$X(s) = \begin{bmatrix} \frac{5}{4(s-1)} - \frac{1}{4(s+3)} \\ \frac{(s-3)^2 + 1}{(s-1)(s-3)^2} \\ \frac{s^4 + 2s^3 - 23s^2 - 20s + 156}{(s-1)(s^2-9)^2} \\ -\frac{s^6 - 2s^5 - 12s^4 + 18s^3 + 48s^2 - 36s + 27}{s^2(s-1)(s^2-9)^2} \end{bmatrix}. \tag{5.269}$$

Taking the inverse Laplace transforms, the solution set matching the one in eqn. (5.268) is obtained.

Example # 5.19 Consider the coefficient matrix A of size [4 x 4], initial conditions $\vec{x}(0)$ and the forcing function $\vec{g}(t)$ given by

$$A = \begin{bmatrix} -1 & -4 & 0 & 0 \\ 1 & 3 & 0 & 0 \\ 1 & 2 & 1 & 0 \\ 0 & 1 & 0 & 1 \end{bmatrix}. \tag{5.270}$$

$$\vec{x}(0) = \begin{bmatrix} 1 \\ -1 \\ 1 \\ -1 \end{bmatrix} . \tag{5.271}$$

$$\vec{g}(t) = \begin{bmatrix} e^t \\ te^t \\ t^2 e^{-t} \\ 4 \end{bmatrix} . \tag{5.272}$$

The set of differential equations associated with the coefficient matrix is

$$\begin{bmatrix} x_1'(t) = -x_1(t) - 4x_2(t) + e^t \\ x_2'(t) = x_1(t) + 3x_2(t) + te^t \\ x_3'(t) = x_1(t) + 2x_2(t) + x_3(t) + t^2 e^{-t} \\ x_4'(t) = x_2(t) + x_4(t) + 4 \end{bmatrix} . \tag{5.273}$$

The eigenvalues of the matrix are

$$\lambda_{1,2,3,4} = 1 . \tag{5.274}$$

The matrix is defective because only two distinct eigenvectors exist,

$$\begin{bmatrix} 0 \\ 0 \\ 1 \\ 0 \end{bmatrix}, \begin{bmatrix} 0 \\ 0 \\ 0 \\ 1 \end{bmatrix} \tag{5.275}$$

Because all the eigenvalues are equal and the matrix is defective, this case can be treated as an example of a defective matrix with algebraic multiplicity of 4. The four generalized eigenvectors in this case constitute an identity matrix of size 4 and these eigenvectors are

$$\begin{bmatrix} 1 \\ 0 \\ 0 \\ 0 \end{bmatrix}, \begin{bmatrix} 0 \\ 1 \\ 0 \\ 0 \end{bmatrix}, \begin{bmatrix} 0 \\ 0 \\ 1 \\ 0 \end{bmatrix}, \begin{bmatrix} 0 \\ 0 \\ 0 \\ 1 \end{bmatrix} . \tag{5.276}$$

Using the eigenvalues and generalized eigenvectors, the fundamental matrix becomes

$$X(t) = \begin{bmatrix} -e^t(2t-1) & -4te^t & 0 & 0 \\ te^t & e^t(2t+1) & 0 & 0 \\ te^t & 2te^t & e^t & 0 \\ \dfrac{t^2 e^t}{2} & e^t\left(t^2+t\right) & 0 & e^t \end{bmatrix} \tag{5.277}$$

The solution for the homogeneous set is

$$\vec{x}_h(t)=\begin{bmatrix} -e^t\left(2c_1t-c_1+4c_2t\right) \\ e^t\left(c_2+c_1t+2c_2t\right) \\ e^t\left(c_3+c_1t+2c_2t\right) \\ \dfrac{e^t\left(2c_4+2c_2t+c_1t^2+2c_2t^2\right)}{2} \end{bmatrix}. \tag{5.278}$$

The particular solutions set is

$$\vec{x}_p(t)=\begin{bmatrix} -\dfrac{te^t\left(2t^2+3t-3\right)}{3} \\ \dfrac{t^2e^t}{3}(t+3) \\ -\dfrac{e^{-t}}{12}\left(6t-6t^2e^{2t}-4t^3e^{2t}+6t^2+3\right) \\ \dfrac{t^3e^t}{3}+\dfrac{t^4e^t}{12}-4 \end{bmatrix}. \tag{5.279}$$

Combining the homogeneous and particular solutions, the general solution becomes

$$\vec{x}(t)=\begin{bmatrix} -e^t\left(2c_1t-c_1+4c_2t\right)-\dfrac{te^t\left(2t^2+3t-3\right)}{3} \\ e^t\left(c_2+c_1t+2c_2t\right)+\dfrac{t^2e^t}{3}(t+1) \\ e^t\left(c_3+c_1t+2c_2t\right)-\dfrac{e^{-t}}{12}\left(6t-6t^2e^{2t}-4t^3e^{2t}+6t^2+3\right) \\ \dfrac{e^t\left(2c_4+2c_2t+c_1t^2+2c_2t^2\right)}{2}+\dfrac{t^3e^t}{3}+\dfrac{t^4e^t}{12}-4 \end{bmatrix}. \tag{5.280}$$

Applying initial conditions, the constants become

$$\vec{c}=\begin{bmatrix} c_1 \\ c_2 \\ c_3 \\ c_4 \end{bmatrix}=\begin{bmatrix} 1 \\ -1 \\ \dfrac{5}{4} \\ 3 \end{bmatrix}. \tag{5.281}$$

The solutions set now becomes

$$\vec{x}(t) = \begin{bmatrix} \dfrac{e^t\left(-2t^3 - 3t^2 + 9t + 3\right)}{3} \\ -\dfrac{e^t\left(-t^3 - 3t^2 + 3t + 3\right)}{3} \\ -\dfrac{e^{-t}}{12}\left(6t - 15e^{2t} + 12te^{2t} - 6t^2e^{2t} - 4t^3e^{2t} + 6t^2 + 3\right) \\ 3e^t - \dfrac{t^2e^t}{2} + \dfrac{t^3e^t}{3} + \dfrac{t^4e^t}{12} - te^t - 4 \end{bmatrix}. \tag{5.282}$$

The Laplace transforms become

$$X(s) = \begin{bmatrix} \dfrac{s\left(s^2 - 5\right)}{(s-1)^4} \\ \dfrac{s\left(-s^2 + 2s + 1\right)}{(s-1)^4} \\ \dfrac{s^6 - s^5 - 3s^4 + 8s^3 + 5s^2 + 9s - 3}{(s-1)^4(s+1)^3} \\ -\dfrac{s^5 - 7s^4 + 20s^3 - 29s^2 + 17s - 4}{s(s-1)^5} \end{bmatrix}. \tag{5.283}$$

Taking the inverse Laplace transforms, the solution set matching to the one in eqn. (5.282) is obtained.

5.4 Numerical Solutions

It is clear that as the size of the coefficient matrix increases, the complexity involved in obtaining the solutions using eigenvalues and eigenvectors increases because defective matrices require detailed exploration to identify the degenerate eigenvalues and obtain the appropriate set of generalized eigenvectors. On the other hand, the approach based on Laplace transforms is clear, straightforward and formulaic and therefore, it can be implemented easily. In addition to these analytical approaches, it is also possible to use the Runge-Kutta methods for the numerical evaluation of the solutions. The use of numerical approaches also permits another level of verification of the results.

Example # 5.20 An example of a 3 x 3 coefficient matrix and the set of coupled first order differential equations is now presented. Consider the coefficient matrix A, initial conditions $\vec{x}(0)$ and the forcing function $\vec{g}(t)$ given by

$$A = \begin{bmatrix} 2 & -1 & -1 \\ -1 & 2 & -1 \\ -1 & -1 & 2 \end{bmatrix}. \tag{5.284}$$

$$\vec{x}(0) = \begin{bmatrix} 0 \\ -1 \\ 1 \end{bmatrix}. \tag{5.285}$$

$$\vec{g}(t) = \begin{bmatrix} t \\ t^2 \\ e^t \end{bmatrix}. \tag{5.286}$$

The coupled differential equations can be expressed as

$$\begin{aligned} x_1'(t) &= 2x_1(t) - x_2(t) - x_3(t) + t \\ x_2'(t) &= -x_1(t) + 2x_2(t) - x_3(t) + t^2 \\ x_3'(t) &= -x_1(t) - x_2(t) + 2x_3(t) + e^t \end{aligned}. \tag{5.287}$$

The approach based on Laplace transform is used to get the analytical solutions and the numerical solutions are obtained using ode45 in MATLAB. The MATLAB script and results are given.

```
function example_numerical_firstorder_multiple
% June 2017 p m shankar solve a coupled system using Laplace transforms and
% compare the results to those obtained numerically
close all
 A=[2,-1,-1;-1,2,-1;-1,-1,2]; % coefficient matrix
 xx0=[0; -1; 1];% initial conditions
 syms t s;
 ft={t;t^2;exp(t)}; % forcing functions
 B=(s*eye(3)-sym(A)); % this sI-A
F=laplace(ft,t,s); % Laplace transforms of forcing functions
W=inv(B)*(F+xx0); % Laplace transforms
w=simplify(ilaplace(W,s,t),'steps',100); % inverse Laplac: solution set
% create inline MATLAB functions for the analytical solutions
fy1=MATLABFunction(w(1,1));
fy2=MATLABFunction(w(2,1));
fy3=MATLABFunction(w(3,1));
tspan=[0 5]; % time span for numerical integration
[tt,yy]=ode45(@examplef,tspan,xx0);
plot(tt,fy1(tt),'r',tt,yy(:,1),'r*',tt,fy2(tt),'k',tt,yy(:,2),'ko',...
 tt,fy3(tt),'b',tt,yy(:,3),'bs','linewidth',1.5)
```

```
xlabel('time t'),ylabel('solutions')
legend('x_1(t)-analyt','x_1(t)-ode','x_2(t)-analyt','x_2(t)-ode',...
 'x_3(t)-analyt','x_3(t)-ode','location','southwest')
syms x_1(t) x_2(t) x_3(t)
% for solution display on the plot
yt1=[x_1(t)==w(1,1)];
yt2=[x_2(t)==w(2,1)];
yt3=[x_3(t)==w(3,1)];
text(0.5,4e6,['$' latex(yt1) '$'],'interpreter','latex',...
 'fontsize',16,'color','b','fontweight','bold')
text(0.5,3e6,['$' latex(yt2) '$'],'interpreter','latex',...
 'fontsize',16,'color','b','fontweight','bold')
text(0.5,2e6,['$' latex(yt3) '$'],'interpreter','latex',...
 'fontsize',16,'color','b','fontweight','bold')
end

function dydt = examplef(t,y)
dydt=zeros(3,1);
dydt(1,1)=t + 2*y(1,1) - y(2,1) - y(3,1);
dydt(2,1)= -y(1,1)+2*y(2,1) - y(3,1)+t^2;
dydt(3,1)=exp(t) -y(1,1) - y(2,1)+2*y(3,1);
end
```

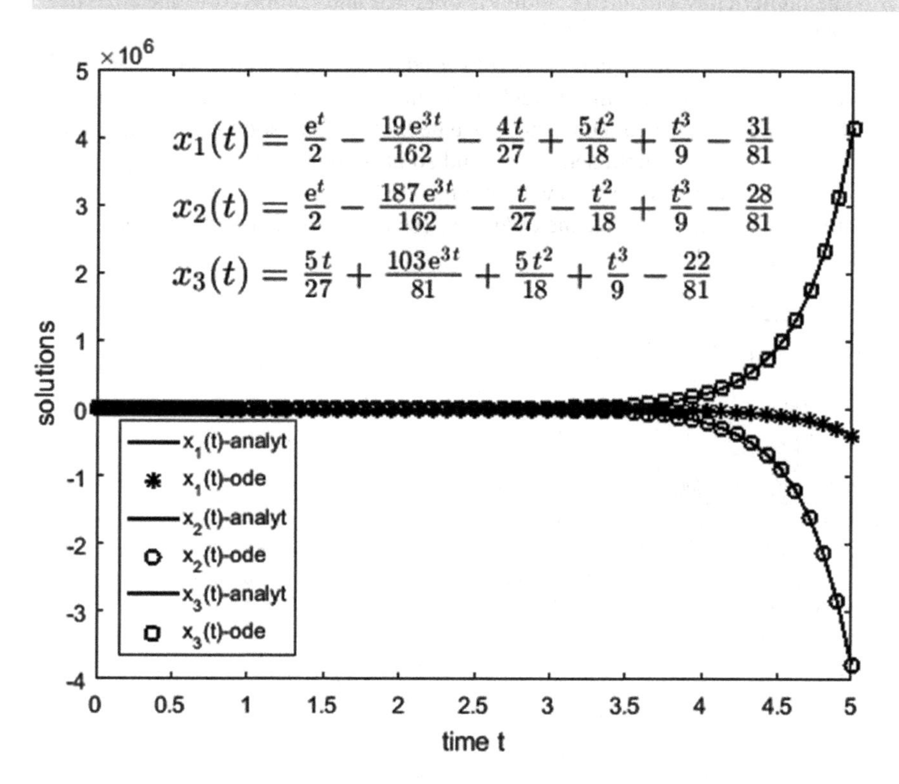

5.5 Examples

Part 1 A pair of first order equations

Starting with the theoretical descriptions of the methods for solving the pair of equations. The MATLAB script displays the following:

1. The summary of all the theoretical methods used are shown first. The methods are based on the use of eigenvectors and eigenvalues (including the case when the coefficient matrix is defective), conversion into a second order homogeneous equation of constant coefficients and the Laplace transforms.
2. The phase plane plot is shown with the eigenvalues and the state of equilibrium.
3. The mechanics of forming the solution using the eigenvalues and eigenvectors are illustrated next. The results contain the eigenvalues (indications when they form a complex conjugate pair) and, eigenvectors (the need for generalized eigenvectors if the matrix is defective).
4. The general solution is shown and the unknown constants are evaluated based on the initial conditions.
5. The approach based on Laplace transforms is presented next along with the transforms of the two solutions and the solution.
6. The solution obtained directly using MATLAB command dsolve(.) is also shown.
7. The numerically obtained solution is obtained and compared to the analytical solution in plots.
8. The set of first order equations is converted to the second order equation and is shown with the roots of the characteristic equation (same as the eigenvalues) and the solution is derived, constants evaluated, one of the solutions is determined first and using the input set, the second solution is extracted. The display also indicates whether x(t) or y(t) was obtained first. If the two differential equations are independent, dx/dt=ax and dy/dt=by, a statement indicates that the equations are not coupled.

Example # 5.21

```
A=[1,2,3,2,1,0];
```

Solution of a first order coupled system: Three methods (Theory)

$$\frac{d\vec{x}}{dt} = A\vec{x} \Rightarrow \begin{bmatrix} x\prime(t) \\ y\prime(t) \end{bmatrix} = \begin{bmatrix} A_{11}\ A_{12} \\ A_{21}\ A_{22} \end{bmatrix} \begin{bmatrix} x(t) \\ y(t) \end{bmatrix} \Leftarrow INPUT$$

General solution using eigenvalues and eigenvectors

Eigenvalues $\lambda_{1,2} \Rightarrow det(A - \lambda I_2) = 0$; **Eigenvectors** $\vec{v}_{1,2} \Rightarrow (A - \lambda_{1,2} I_2)\vec{v} = 0$

Complex Eigenvalues and Eigenvectors $\Rightarrow \lambda = \alpha \pm j\beta;\ \vec{v} = \vec{u} \pm j\vec{w}$

$$[\vec{X}_1, \vec{X}_2] = \left[e^{\alpha t}\left[\vec{u}\,cos(\beta t) - \vec{w}\,sin(\beta t)\right], e^{\alpha t}\left[\vec{u}\,sin(\beta t) + \vec{w}\,cos(\beta t)\right]\right]$$

Real Eigenvalues, distinct Eigenvectors $\Rightarrow \lambda_{1,2};\ \vec{v}_{1,2}$

$$[\vec{X}_1, \vec{X}_2] = \left[\vec{v}(:,1)e^{\lambda_1 t}, \vec{v}(:,2)e^{\lambda_2 t}\right]$$

[A] is defective: $\lambda_{1,2} = \lambda;\ \vec{v}_{1,2} = \vec{v}_0)$; **solve** $[A - \lambda I]^2\vec{v} = 0$ **get generalized eigenvectors**

$$[\vec{X}_1, \vec{X}_2] = \left[e^{\lambda t}\left[\vec{v}(:,1) + t(A - \lambda I)\vec{v}(:,1)\right], e^{\lambda t}\left[\vec{v}(:,2) + t(A - \lambda I)\vec{v}(:,2)\right]\right]$$

$$\text{General Solution} \Rightarrow \begin{bmatrix} x(t) \\ y(t) \end{bmatrix} = c_1\vec{X}_1 + c_2\vec{X}_2$$

General Solution after conversion to a 2nd order Differential Equation

$x''(t) - [A_{11} + A_{22}]x'(t) + [A_{11}A_{22} - A_{12}A_{21}]x(t) = 0$ Solve for x(t) and get y(t) from input

Characteristic Eqn. $r^2 - [A_{11} + A_{22}]r + [A_{11}A_{22} - A_{12}A_{21}] = 0 \Rightarrow$ Roots $r_{1,2}$ or $r_\pm$

Real Roots $r_1 \neq r_2$, $x(t) = b_1 e^{r_1 t} + b_2 e^{r_2 t}$; $r_{1,2} = r$, $x(t) = b_1 e^{rt} + b_2 t e^{rt}$

Complex Roots $r_\pm = \alpha \pm j\beta$; $x(t) = e^{\alpha t}[b_1 cos(\beta t) + b_2 sin(\beta t)]$

If A_{12}=0, set differential EQn. in y(t) & get x(t); If A_{12}=A_{21}=0, solve for x(t) & y(t) directly

---

Complete Solution using inverse Laplace transforms with initial conditions [x(0) y(0)]

$$\text{Complete Solution} \Rightarrow \begin{bmatrix} x(t) \\ y(t) \end{bmatrix} = L^{-1}\left[[sI_2 - A]^{-1}\begin{pmatrix} x(0) \\ y(0) \end{pmatrix}\right]$$

$$\begin{pmatrix} x'(t) = x(t) + 2\,y(t) \\ y'(t) = 3\,x(t) + 2\,y(t) \end{pmatrix}$$

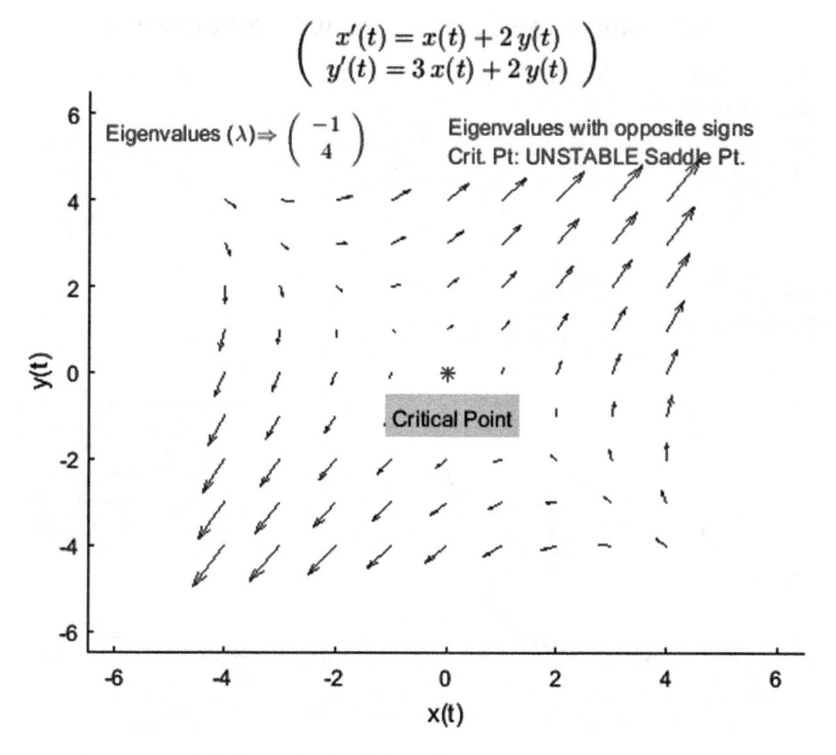

Complete Solutions: Pair of First Order Linear Differential Equations

Coefficient Matrix [A] $\begin{pmatrix} 1 & 2 \\ 3 & 2 \end{pmatrix}$ Differential Equations $\begin{pmatrix} x'(t) = x(t) + 2\,y(t) \\ y'(t) = 3\,x(t) + 2\,y(t) \end{pmatrix}$ IC $\begin{pmatrix} 1 \\ 0 \end{pmatrix}$

Real and distinct eigenvalues & eigenvectors

Eigenvalues (λ) $\begin{pmatrix} -1 \\ 4 \end{pmatrix}$ Eigenvectors (Columns) $(\vec{v})$ $\begin{pmatrix} -1 & \frac{2}{3} \\ 1 & 1 \end{pmatrix}$

Solution basis $[\vec{X}_1\ \vec{X}_2]$ (columns) $\begin{pmatrix} -e^{-t} & \frac{2e^{4t}}{3} \\ e^{-t} & e^{4t} \end{pmatrix}$

General Solution set

$$x_g(t) = \frac{2\,c_2\,e^{4t}}{3} - c_1\,e^{-t}$$

$$y_g(t) = c_1\,e^{-t} + c_2\,e^{4t}$$

Apply initial conditions ⇒

$$\begin{pmatrix} \frac{2c_2}{3} - c_1 = 1 \\ c_1 + c_2 = 0 \end{pmatrix} \Rightarrow \begin{bmatrix} c_1 \\ c_2 \end{bmatrix} = \begin{pmatrix} -\frac{3}{5} \\ \frac{3}{5} \end{pmatrix}$$

Solution set Eigenvalues/vectors

$$x_e(t) = \frac{e^{-t}\left(2e^{5t}+3\right)}{5}$$

$$y_e(t) = \frac{3\,e^{-t}\left(e^{5t}-1\right)}{5}$$

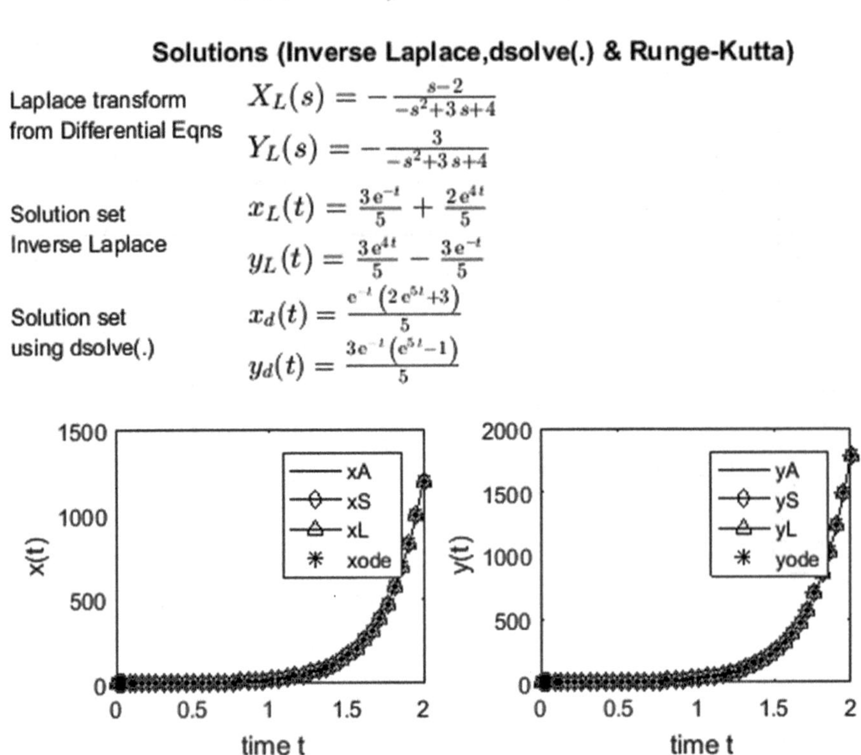

2nd order Differential Equation based solution set

Coefficient Matrix [A] $\begin{pmatrix} 1 & 2 \\ 3 & 2 \end{pmatrix}$ Differential Equations $\begin{pmatrix} x'(t) = x(t) + 2\,y(t) \\ y'(t) = 3\,x(t) + 2\,y(t) \end{pmatrix}$ IC $\begin{pmatrix} 1 \\ 0 \end{pmatrix}$

$A_{12} \neq 0 \Rightarrow$ Create Equation in x $\Rightarrow a x''(t) + b x'(t) + c x(t) = 0$

Equivalent 2nd order DE in x

$x''(t) - 3\,x'(t) - 4\,x(t) = 0$

Roots of the Characteristic Equation

$(\,-1 \quad 4\,)$

General Solution in terms of unknown constants b_1, b_2

$x_2(t) = b_1\,e^{-t} + b_2\,e^{4\,t}$

$y_2(t) = \frac{3 b_2 e^{4t}}{2} - b_1\,e^{-t}$

Apply Initial Conditions $[b_1, b_2] \Rightarrow \quad (\,\frac{3}{5} \quad \frac{2}{5}\,)$

Solution set using roots

$x_1(t) = \frac{e^{-t}\left(2e^{5t}+3\right)}{5}$

$y_1(t) = \frac{3e^{-t}\left(e^{5t}-1\right)}{5}$

Example # 5.22

```
A=[-1,0,-2,-3,-1,1];
```

Solution of a first order coupled system: Three methods (Theory)

$\frac{d\vec{x}}{dt} = A\vec{x} \Rightarrow \begin{bmatrix} x\prime(t) \\ y\prime(t) \end{bmatrix} = \begin{bmatrix} A_{11}\,A_{12} \\ A_{21}\,A_{22} \end{bmatrix} \begin{bmatrix} x(t) \\ y(t) \end{bmatrix} \Leftarrow INPUT$

General solution using eigenvalues and eigenvectors

Eigenvalues $\lambda_{1,2} \Rightarrow det(A - \lambda I_2) = 0$; Eigenvectors $\vec{v}_{1,2} \Rightarrow (A - \lambda_{1,2} I_2)\vec{v} = 0$

Complex Eigenvalues and Eigenvectors $\Rightarrow \lambda = \alpha \pm j\beta$; $\vec{v} = \vec{u} \pm j\vec{w}$

$[\vec{X}_1, \vec{X}_2] = \left[e^{\alpha t}\left[\vec{u}\,cos(\beta t) - \vec{w}\,sin(\beta t)\right], e^{\alpha t}\left[\vec{u}\,sin(\beta t) + \vec{w}\,cos(\beta t)\right]\right]$

Real Eigenvalues, distinct Eigenvectors $\Rightarrow \lambda_{1,2}$; $\vec{v}_{1,2}$

$[\vec{X}_1, \vec{X}_2] = \left[\vec{v}(:,1)e^{\lambda_1 t}, \vec{v}(:,2)e^{\lambda_2 t}\right]$

[A] is defective: $\lambda_{1,2} = \lambda$; $\vec{v}_{1,2} = \vec{v}_0$); solve $[A - \lambda I]^2\vec{v} = 0$ get generalized eigenvectors

$[\vec{X}_1, \vec{X}_2] = \left[e^{\lambda t}\left[\vec{v}(:,1) + t(A - \lambda I)\vec{v}(:,1)\right], e^{\lambda t}\left[\vec{v}(:,2) + t(A - \lambda I)\vec{v}(:,2)\right]\right]$

General Solution $\Rightarrow \begin{bmatrix} x(t) \\ y(t) \end{bmatrix} = c_1\vec{X}_1 + c_2\vec{X}_2$

General Solution after conversion to a 2nd order Differential Equation

$x''(t) - [A_{11} + A_{22}]x'(t) + [A_{11}A_{22} - A_{12}A_{21}]x(t) = 0$ Solve for x(t) and get y(t) from input

Characteristic Eqn. $r^2 - [A_{11} + A_{22}]r + [A_{11}A_{22} - A_{12}A_{21}] = 0 \Rightarrow$ Roots $r_{1,2}$ *or* $r_{\pm}$

Real Roots $r_1 \neq r_2$, $x(t) = b_1 e^{r_1 t} + b_2 e^{r_2 t}$; $r_{1,2} = r$, $x(t) = b_1 e^{rt} + b_2 t e^{rt}$

Complex Roots $r_{\pm} = \alpha \pm j\beta$; $x(t) = e^{\alpha t}[b_1 cos(\beta t) + b_2 sin(\beta t)]$

If A_{12}=0, set differential EQn. in y(t) & get x(t); If A_{12}=A_{21}=0, solve for x(t) & y(t) directly

Complete Solution using inverse Laplace transforms with initial conditions [x(0) y(0)]

Complete Solution $\Rightarrow \begin{bmatrix} x(t) \\ y(t) \end{bmatrix} = L^{-1}\left[[sI_2 - A]^{-1}\begin{pmatrix} x(0) \\ y(0) \end{pmatrix}\right]$

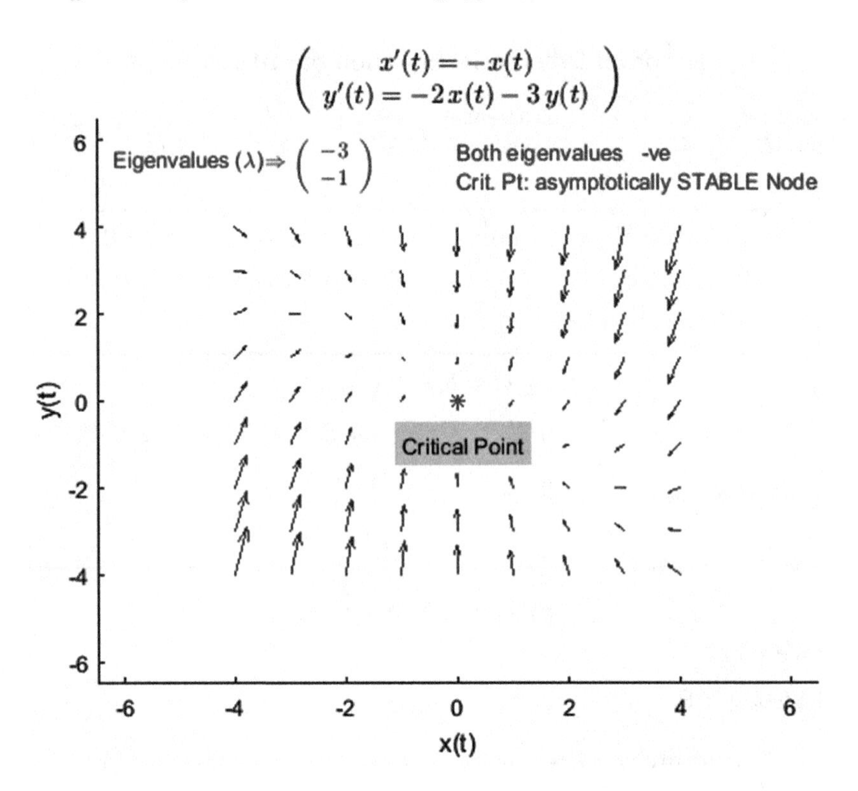

Complete Solutions: Pair of First Order Linear Differential Equations

Coefficient Matrix [A] $\begin{pmatrix} -1 & 0 \\ -2 & -3 \end{pmatrix}$ Differential Equations $\begin{pmatrix} x'(t) = -x(t) \\ y'(t) = -2\,x(t) - 3\,y(t) \end{pmatrix}$ IC $\begin{pmatrix} -1 \\ 1 \end{pmatrix}$

Real and distinct eigenvalues & eigenvectors

Eigenvalues (λ) $\begin{pmatrix} -3 \\ -1 \end{pmatrix}$ Eigenvectors (Columns) ($\vec{v}$) $\begin{pmatrix} 0 & -1 \\ 1 & 1 \end{pmatrix}$

Solution basis $\left[\vec{X}_1\ \vec{X}_2\right]$ (columns) $\begin{pmatrix} 0 & -e^{-t} \\ e^{-3t} & e^{-t} \end{pmatrix}$

General Solution set

$$x_g(t) = -c_2\,e^{-t}$$
$$y_g(t) = c_2\,e^{-t} + c_1\,e^{-3t}$$

Apply initial conditions ⇒ $\begin{pmatrix} -c_2 = -1 \\ c_1 + c_2 = 1 \end{pmatrix}$ $\Rightarrow \begin{bmatrix} c_1 \\ c_2 \end{bmatrix} = \begin{pmatrix} 0 \\ 1 \end{pmatrix}$

Solution set Eigenvalues/vectors

$$x_e(t) = -e^{-t}$$
$$y_e(t) = e^{-t}$$

Solutions (Inverse Laplace,dsolve(.) & Runge-Kutta)

Laplace transform from Differential Eqns
$$X_L(s) = -\frac{1}{s+1}$$
$$Y_L(s) = \frac{1}{s+3} + \frac{2}{(s+1)\,(s+3)}$$

Solution set Inverse Laplace
$$x_L(t) = -e^{-t}$$
$$y_L(t) = e^{-t}$$

Solution set using dsolve(.)
$$x_d(t) = -e^{-t}$$
$$y_d(t) = e^{-t}$$

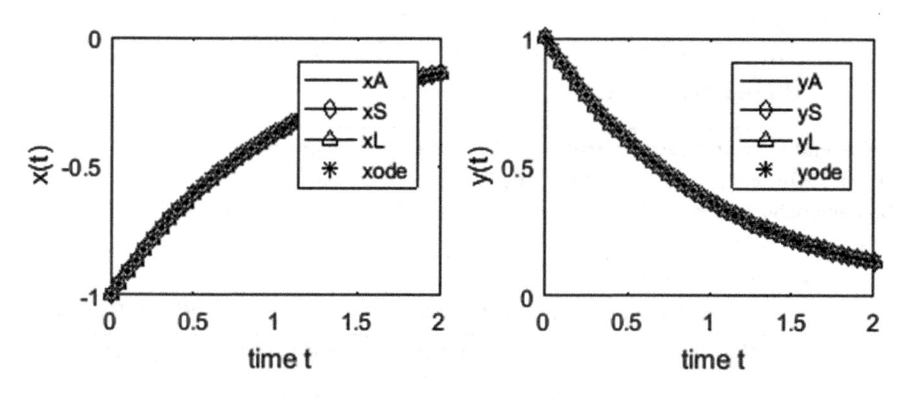

2nd order Differential Equation based solution set

Coefficient Matrix [A] $\begin{pmatrix} -1 & 0 \\ -2 & -3 \end{pmatrix}$ Differential Equations $\begin{pmatrix} x'(t) = -x(t) \\ y'(t) = -2\,x(t) - 3\,y(t) \end{pmatrix}$ IC $\begin{pmatrix} -1 \\ 1 \end{pmatrix}$

$$A_{12} = 0 \Rightarrow \text{Create Equation in y} \Rightarrow ay''(t) + by'(t) + cy(t) = 0$$

Equivalent 2nd order DE in y

$$y''(t) + 4\,y'(t) + 3\,y(t) = 0$$

Roots of the Characteristic Equation

$$(\; -3 \quad -1 \;)$$

General Solution in terms of unknown constants b_1, b_2
$$x_2(t) = -b_2\,e^{-t}$$
$$y_2(t) = b_2\,e^{-t} + b_1\,e^{-3t}$$

Apply Initial Conditions $[b_1, b_2] \Rightarrow \quad (\; 0 \quad 1 \;)$

Solution set using roots
$$x_1(t) = -e^{-t}$$
$$y_1(t) = e^{-t}$$

Example # 5.23

```
A=[0,1,0,-2,0,1];
```

Solution of a first order coupled system: Three methods (Theory)

$$\frac{d\vec{x}}{dt} = A\vec{x} \Rightarrow \begin{bmatrix} x'(t) \\ y'(t) \end{bmatrix} = \begin{bmatrix} A_{11} & A_{12} \\ A_{21} & A_{22} \end{bmatrix} \begin{bmatrix} x(t) \\ y(t) \end{bmatrix} \Leftarrow INPUT$$

General solution using eigenvalues and eigenvectors

Eigenvalues $\lambda_{1,2} \Rightarrow det(A - \lambda I_2) = 0$; Eigenvectors $\vec{v}_{1,2} \Rightarrow (A - \lambda_{1,2} I_2)\vec{v} = 0$

Complex Eigenvalues and Eigenvectors $\Rightarrow \lambda = \alpha \pm j\beta$; $\vec{v} = \vec{u} \pm j\vec{w}$

$$[\vec{X}_1, \vec{X}_2] = \left[e^{\alpha t}\left[\vec{u}\,cos(\beta t) - \vec{w}\,sin(\beta t)\right], e^{\alpha t}\left[\vec{u}\,sin(\beta t) + \vec{w}\,cos(\beta t)\right]\right]$$

Real Eigenvalues, distinct Eigenvectors $\Rightarrow \lambda_{1,2}$; $\vec{v}_{1,2}$

$$[\vec{X}_1, \vec{X}_2] = \left[\vec{v}(:,1)e^{\lambda_1 t}, \vec{v}(:,2)e^{\lambda_2 t}\right]$$

[A] is defective: $\lambda_{1,2} = \lambda$; $\vec{v}_{1,2} = \vec{v}_0$); solve $[A - \lambda I]^2\vec{v} = 0$ get generalized eigenvectors

$$[\vec{X}_1, \vec{X}_2] = \left[e^{\lambda t}\left[\vec{v}(:,1) + t(A - \lambda I)\vec{v}(:,1)\right], e^{\lambda t}\left[\vec{v}(:,2) + t(A - \lambda I)\vec{v}(:,2)\right]\right]$$

General Solution $\Rightarrow \begin{bmatrix} x(t) \\ y(t) \end{bmatrix} = c_1\vec{X}_1 + c_2\vec{X}_2$

General Solution after conversion to a 2nd order Differential Equation

$x''(t) - [A_{11} + A_{22}]x'(t) + [A_{11}A_{22} - A_{12}A_{21}]x(t) = 0$ Solve for x(t) and get y(t) from input

Characteristic Eqn. $r^2 - [A_{11} + A_{22}]r + [A_{11}A_{22} - A_{12}A_{21}] = 0 \Rightarrow$ Roots $r_{1,2}$ *or* $r_\pm$

Real Roots $r_1 \neq r_2$, $x(t) = b_1e^{r_1 t} + b_2e^{r_2 t}$; $r_{1,2} = r$, $x(t) = b_1e^{rt} + b_2te^{rt}$

Complex Roots $r_\pm = \alpha \pm j\beta$; $x(t) = e^{\alpha t}[b_1cos(\beta t) + b_2sin(\beta t)]$

If A_{12}=0, set differential EQn. in y(t) & get x(t); If A_{12}=A_{21}=0, solve for x(t) & y(t) directly

Complete Solution using inverse Laplace transforms with initial conditions [x(0) y(0)]

Complete Solution $\Rightarrow \begin{bmatrix} x(t) \\ y(t) \end{bmatrix} = L^{-1}\left[[sI_2 - A]^{-1}\begin{pmatrix} x(0) \\ y(0) \end{pmatrix}\right]$

$$\begin{pmatrix} x'(t) = y(t) \\ y'(t) = -2\,y(t) \end{pmatrix}$$

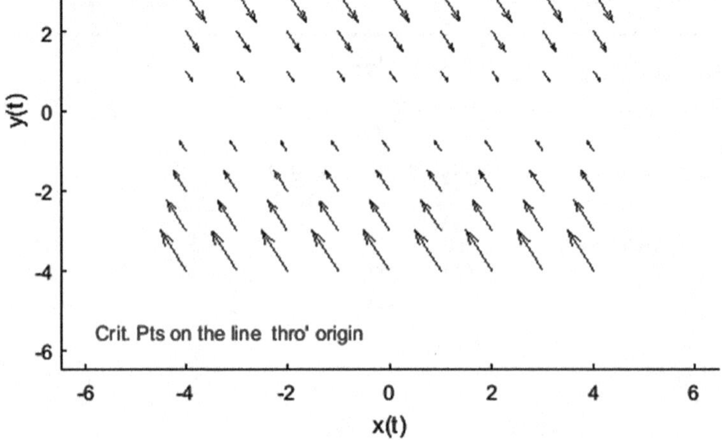

Complete Solutions: Pair of First Order Linear Differential Equations

Coefficient Matrix [A] $\begin{pmatrix} 0 & 1 \\ 0 & -2 \end{pmatrix}$ Differential Equations $\begin{pmatrix} x'(t) = y(t) \\ y'(t) = -2\,y(t) \end{pmatrix}$ IC $\begin{pmatrix} 0 \\ 1 \end{pmatrix}$

Real and distinct eigenvalues & eigenvectors

Eigenvalues (λ) $\begin{pmatrix} 0 \\ -2 \end{pmatrix}$ Eigenvectors (Columns) $(\vec{v})$ $\begin{pmatrix} 1 & -\frac{1}{2} \\ 0 & 1 \end{pmatrix}$

Solution basis $\left[\vec{X}_1\ \vec{X}_2\right]$ (columns) $\begin{pmatrix} 1 & -\frac{e^{-2t}}{2} \\ 0 & e^{-2t} \end{pmatrix}$

General Solution set

$$x_g(t) = c_1 - \frac{c_2\,e^{-2t}}{2}$$

$$y_g(t) = c_2\,e^{-2t}$$

Apply initial conditions $\Rightarrow$ $\begin{pmatrix} c_1 - \frac{c_2}{2} = 0 \\ c_2 = 1 \end{pmatrix}$ $\Rightarrow \begin{bmatrix} c_1 \\ c_2 \end{bmatrix} = \begin{pmatrix} \frac{1}{2} \\ 1 \end{pmatrix}$

Solution set Eigenvalues/vectors

$$x_e(t) = \frac{1}{2} - \frac{e^{-2t}}{2}$$

$$y_e(t) = e^{-2t}$$

Solutions (Inverse Laplace,dsolve(.) & Runge-Kutta)

Laplace transform from Differential Eqns

$$X_L(s) = \frac{1}{s\,(s+2)}$$

$$Y_L(s) = \frac{1}{s+2}$$

Solution set Inverse Laplace

$$x_L(t) = \frac{1}{2} - \frac{e^{-2t}}{2}$$

$$y_L(t) = e^{-2t}$$

Solution set using dsolve(.)

$$x_d(t) = \frac{1}{2} - \frac{e^{-2t}}{2}$$

$$y_d(t) = e^{-2t}$$

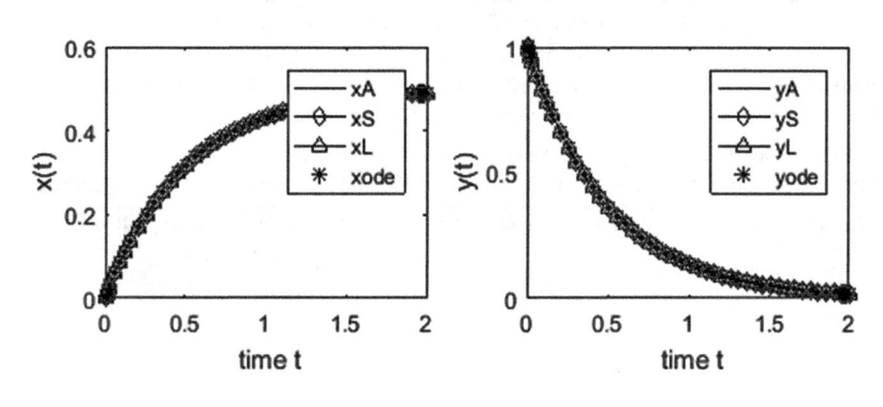

2nd order Differential Equation based solution set

Coefficient Matrix [A] $\begin{pmatrix} 0 & 1 \\ 0 & -2 \end{pmatrix}$ Differential Equations $\begin{pmatrix} x'(t) = y(t) \\ y'(t) = -2\,y(t) \end{pmatrix}$ IC $\begin{pmatrix} 0 \\ 1 \end{pmatrix}$

$A_{12} \neq 0 \Rightarrow$ Create Equation in x $\Rightarrow ax''(t) + bx'(t) + cx(t) = 0$

Equivalent 2nd order DE in x

$x''(t) + 2\,x'(t) = 0$

Roots of the Characteristic Equation

$(0 \quad -2)$

General Solution in terms of unknown constants b_1, b_2

$x_2(t) = b_1 + b_2\,e^{-2t}$

$y_2(t) = -2\,b_2\,e^{-2t}$

Apply Initial Conditions $[b_1,b_2] \Rightarrow \quad (\frac{1}{2} \quad -\frac{1}{2})$

Solution set using roots

$x_1(t) = \frac{1}{2} - \frac{e^{-2t}}{2}$

$y_1(t) = e^{-2\,t}$

Example # 5.24

```
A=[1,-1,-1,1,1,-1];
```

Solution of a first order coupled system: Three methods (Theory)

$$\frac{d\vec{x}}{dt} = A\vec{x} \Rightarrow \begin{bmatrix} x'(t) \\ y'(t) \end{bmatrix} = \begin{bmatrix} A_{11}\,A_{12} \\ A_{21}\,A_{22} \end{bmatrix} \begin{bmatrix} x(t) \\ y(t) \end{bmatrix} \Leftarrow INPUT$$

General solution using eigenvalues and eigenvectors

Eigenvalues $\lambda_{1,2} \Rightarrow det(A - \lambda I_2) = 0$; Eigenvectors $\vec{v}_{1,2} \Rightarrow (A - \lambda_{1,2} I_2)\vec{v} = 0$

Complex Eigenvalues and Eigenvectors $\Rightarrow \lambda = \alpha \pm j\beta$; $\vec{v} = \vec{u} \pm j\vec{w}$

$$[\vec{X}_1, \vec{X}_2] = \left[e^{\alpha t}\left[\vec{u}\,cos(\beta t) - \vec{w}\,sin(\beta t)\right], e^{\alpha t}\left[\vec{u}\,sin(\beta t) + \vec{w}\,cos(\beta t)\right]\right]$$

Real Eigenvalues, distinct Eigenvectors $\Rightarrow \lambda_{1,2}$; $\vec{v}_{1,2}$

$$[\vec{X}_1, \vec{X}_2] = \left[\vec{v}(:,1)e^{\lambda_1 t}, \vec{v}(:,2)e^{\lambda_2 t}\right]$$

[A] is defective: $\lambda_{1,2} = \lambda$; $\vec{v}_{1,2} = \vec{v}_0$); solve $[A - \lambda I]^2\vec{v} = 0$ get generalized eigenvectors

$$[\vec{X}_1, \vec{X}_2] = \left[e^{\lambda t}\left[\vec{v}(:,1) + t(A - \lambda I)\vec{v}(:,1)\right], e^{\lambda t}\left[\vec{v}(:,2) + t(A - \lambda I)\vec{v}(:,2)\right]\right]$$

$$\text{General Solution} \Rightarrow \begin{bmatrix} x(t) \\ y(t) \end{bmatrix} = c_1\vec{X}_1 + c_2\vec{X}_2$$

General Solution after conversion to a 2nd order Differential Equation

$x''(t) - [A_{11} + A_{22}]x'(t) + [A_{11}A_{22} - A_{12}A_{21}]x(t) = 0$ Solve for x(t) and get y(t) from input

Characteristic Eqn. $r^2 - [A_{11} + A_{22}]r + [A_{11}A_{22} - A_{12}A_{21}] = 0 \Rightarrow$ Roots $r_{1,2}$ or $r_{\pm}$

Real Roots $r_1 \neq r_2$, $x(t) = b_1 e^{r_1 t} + b_2 e^{r_2 t}$; $r_{1,2} = r$, $x(t) = b_1 e^{rt} + b_2 t e^{rt}$

Complex Roots $r_{\pm} = \alpha \pm j\beta$; $x(t) = e^{\alpha t}[b_1 cos(\beta t) + b_2 sin(\beta t)]$

If A_{12}=0, set differential EQn. in y(t) & get x(t); If A_{12}=A_{21}=0, solve for x(t) & y(t) directly

Complete Solution using inverse Laplace transforms with initial conditions [x(0) y(0)]

$$\text{Complete Solution} \Rightarrow \begin{bmatrix} x(t) \\ y(t) \end{bmatrix} = L^{-1}\left[[sI_2 - A]^{-1}\begin{pmatrix} x(0) \\ y(0) \end{pmatrix}\right]$$

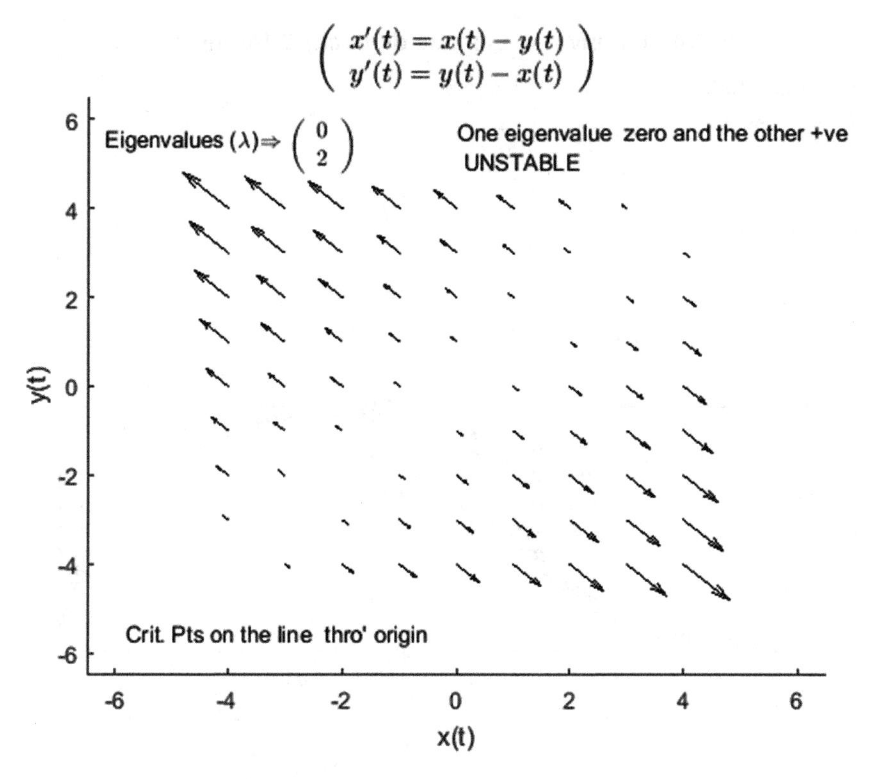

Complete Solutions: Pair of First Order Linear Differential Equations

Coefficient Matrix [A] $\begin{pmatrix} 1 & -1 \\ -1 & 1 \end{pmatrix}$ Differential Equations $\begin{pmatrix} x'(t) = x(t) - y(t) \\ y'(t) = y(t) - x(t) \end{pmatrix}$ IC $\begin{pmatrix} 1 \\ -1 \end{pmatrix}$

Real and distinct eigenvalues & eigenvectors

Eigenvalues (λ) $\begin{pmatrix} 0 \\ 2 \end{pmatrix}$ Eigenvectors (Columns) $(\vec{v})$ $\begin{pmatrix} 1 & -1 \\ 1 & 1 \end{pmatrix}$

Solution basis $\left[\vec{X}_1\ \vec{X}_2\right]$ (columns) $\begin{pmatrix} 1 & -e^{2t} \\ 1 & e^{2t} \end{pmatrix}$

General Solution set

$$x_g(t) = c_1 - c_2\,e^{2t}$$

$$y_g(t) = c_1 + c_2\,e^{2t}$$

Apply initial conditions ⇒ $\begin{pmatrix} c_1 - c_2 = 1 \\ c_1 + c_2 = -1 \end{pmatrix}$ $\Rightarrow \begin{bmatrix} c_1 \\ c_2 \end{bmatrix} = \begin{pmatrix} 0 \\ -1 \end{pmatrix}$

Solution set Eigenvalues/vectors

$$x_e(t) = e^{2t}$$

$$y_e(t) = -e^{2t}$$

Solutions (Inverse Laplace,dsolve(.) & Runge-Kutta)

Laplace transform from Differential Eqns
$$X_L(s) = -\frac{1}{2s-s^2} - \frac{s-1}{2s-s^2}$$
$$Y_L(s) = \frac{1}{2s-s^2} + \frac{s-1}{2s-s^2}$$

Solution set Inverse Laplace
$$x_L(t) = e^{2t}$$
$$y_L(t) = -e^{2t}$$

Solution set using dsolve(.)
$$x_d(t) = e^{2t}$$
$$y_d(t) = -e^{2t}$$

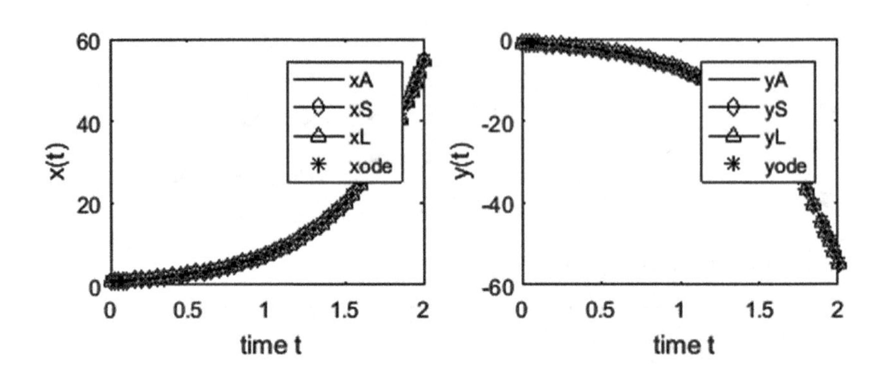

2nd order Differential Equation based solution set

Coefficient Matrix [A] $\begin{pmatrix} 1 & -1 \\ -1 & 1 \end{pmatrix}$ Differential Equations $\begin{pmatrix} x'(t) = x(t) - y(t) \\ y'(t) = y(t) - x(t) \end{pmatrix}$ IC $\begin{pmatrix} 1 \\ -1 \end{pmatrix}$

$A_{12} \neq 0 \Rightarrow$ Create Equation in x $\Rightarrow a x''(t) + b x'(t) + c x(t) = 0$

Equivalent 2nd order DE in x
$$x''(t) - 2x'(t) = 0$$

Roots of the Characteristic Equation
$$(0 \quad 2)$$

General Solution in terms of unknown constants b_1, b_2
$$x_2(t) = b_1 + b_2 e^{2t}$$
$$y_2(t) = b_1 - b_2 e^{2t}$$

Apply Initial Conditions $[b_1, b_2] \Rightarrow \quad (0 \quad 1)$

Solution set using roots
$$x_1(t) = e^{2t}$$
$$y_1(t) = -e^{2t}$$

For the remaining examples, the first display containing the theoretical aspects is not shown.

Example # 5.25

```
A=[1,-1,-1,-1,0,1];
```

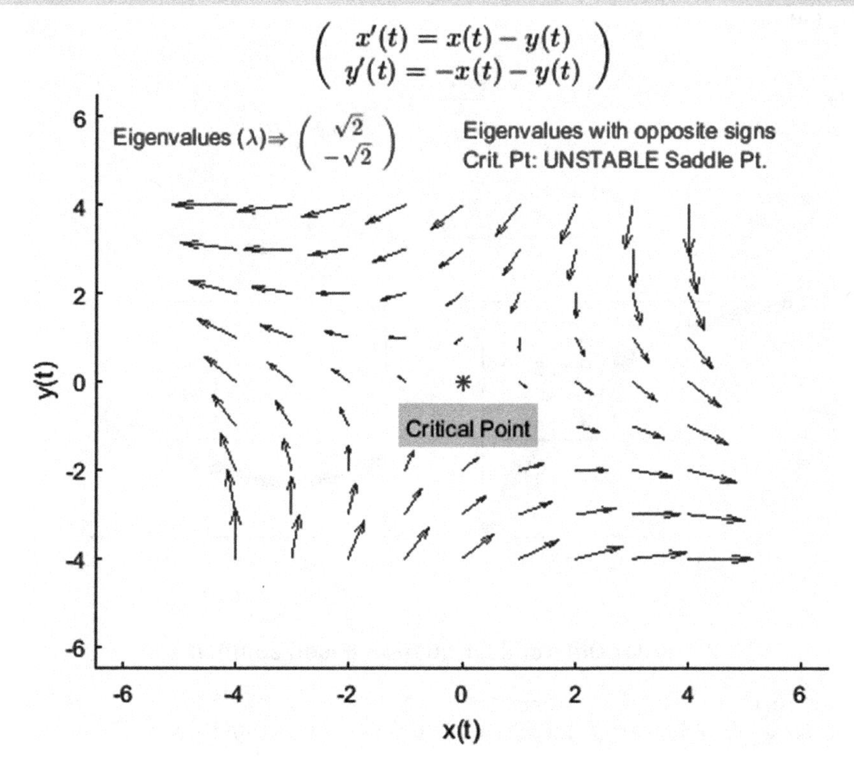

Complete Solutions: Pair of First Order Linear Differential Equations

Coefficient Matrix [A] $\begin{pmatrix} 1 & -1 \\ -1 & -1 \end{pmatrix}$ Differential Equations $\begin{pmatrix} x'(t) = x(t) - y(t) \\ y'(t) = -x(t) - y(t) \end{pmatrix}$ IC $\begin{pmatrix} 0 \\ 1 \end{pmatrix}$

Real and distinct eigenvalues & eigenvectors

Eigenvalues (λ) $\begin{pmatrix} \sqrt{2} \\ -\sqrt{2} \end{pmatrix}$ Eigenvectors (Columns) $(\vec{v})$ $\begin{pmatrix} -\sqrt{2}-1 & \sqrt{2}-1 \\ 1 & 1 \end{pmatrix}$

Solution basis $[\vec{X}_1\ \vec{X}_2]$ (columns)

$$\begin{pmatrix} -e^{\sqrt{2}t}\left(\sqrt{2}+1\right) & e^{-\sqrt{2}t}\left(\sqrt{2}-1\right) \\ e^{\sqrt{2}t} & e^{-\sqrt{2}t} \end{pmatrix}$$

General Solution set

$$x_g(t) = c_2\, e^{-\sqrt{2}t}\left(\sqrt{2}-1\right) - c_1\, e^{\sqrt{2}t}\left(\sqrt{2}+1\right)$$

$$y_g(t) = c_1\, e^{\sqrt{2}t} + c_2\, e^{-\sqrt{2}t}$$

Apply initial conditions ⇒

$$\begin{pmatrix} c_2\left(\sqrt{2}-1\right) - c_1\left(\sqrt{2}+1\right) = 0 \\ c_1 + c_2 = 1 \end{pmatrix} \Rightarrow \begin{bmatrix} c_1 \\ c_2 \end{bmatrix} = \begin{pmatrix} \frac{1}{2} - \frac{\sqrt{2}}{4} \\ \frac{\sqrt{2}\left(\sqrt{2}+1\right)}{4} \end{pmatrix}$$

Solution set Eigenvalues/vectors

$$x_e(t) = -\frac{\sqrt{2}\sinh\left(\sqrt{2}t\right)}{2}$$

$$y_e(t) = \frac{e^{-\sqrt{2}t}\left(\sqrt{2}+2\right)}{4} - e^{\sqrt{2}t}\left(\frac{\sqrt{2}}{4} - \frac{1}{2}\right)$$

Solutions (Inverse Laplace,dsolve(.) & Runge-Kutta)

Laplace transform from Differential Eqns

$$X_L(s) = -\frac{1}{s^2-2}$$

$$Y_L(s) = \frac{s-1}{s^2-2}$$

Solution set Inverse Laplace

$$x_L(t) = -\frac{\sqrt{2}\sinh(\sqrt{2}\,t)}{2}$$

$$y_L(t) = \cosh(\sqrt{2}\,t) - \frac{\sqrt{2}\sinh(\sqrt{2}\,t)}{2}$$

Solution set using dsolve(.)

$$x_d(t) = -\frac{\sqrt{2}\sinh(\sqrt{2}\,t)}{2}$$

$$y_d(t) = \frac{e^{-\sqrt{2}t}\,(\sqrt{2}+2)}{4} - e^{\sqrt{2}\,t}\left(\frac{\sqrt{2}}{4} - \frac{1}{2}\right)$$

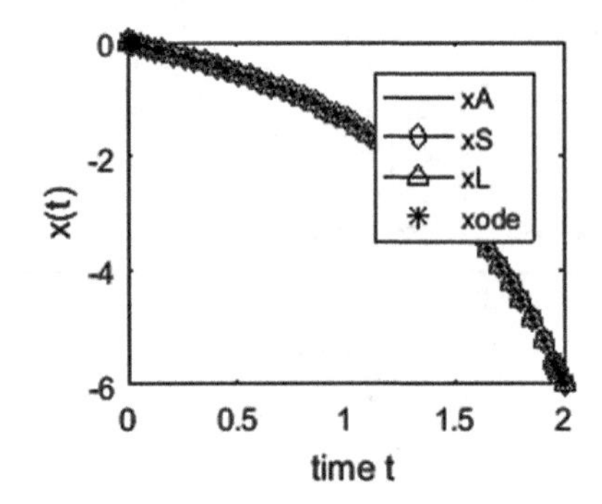

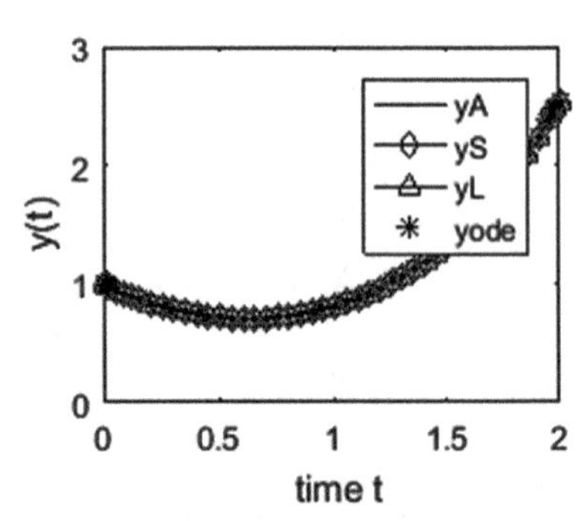

2nd order Differential Equation based solution set

Coefficient Matrix [A] $\begin{pmatrix} 1 & -1 \\ -1 & -1 \end{pmatrix}$ Differential Equations $\begin{pmatrix} x'(t) = x(t) - y(t) \\ y'(t) = -x(t) - y(t) \end{pmatrix}$ IC $\begin{pmatrix} 0 \\ 1 \end{pmatrix}$

$A_{12} \neq 0 \Rightarrow$ Create Equation in x $\Rightarrow a x''(t) + b x'(t) + c x(t) = 0$

Equivalent 2nd order DE in x

$$x''(t) - 2\,x(t) = 0$$

Roots of the Characteristic Equation

$$(\ \sqrt{2} \quad -\sqrt{2}\ )$$

General Solution in terms of unknown constants b_1, b_2

$$x_2(t) = b_1\,e^{\sqrt{2}\,t} + b_2\,e^{-\sqrt{2}\,t}$$

$$y_2(t) = e^{-\sqrt{2}t}\left(b_2 + \sqrt{2}\,b_2 + b_1\,e^{2\sqrt{2}t} - \sqrt{2}\,b_1\,e^{2\sqrt{2}\,t}\right)$$

Apply Initial Conditions $[b_1, b_2] \Rightarrow \left(\ -\frac{\sqrt{2}}{4} \quad \frac{\sqrt{2}}{4}\ \right)$

Solution set using roots

$$x_1(t) = -\frac{\sqrt{2}\sinh(\sqrt{2}\,t)}{2}$$

$$y_1(t) = \frac{e^{-\sqrt{2}t}\,(2e^{2\sqrt{2}t} - \sqrt{2}\,e^{2\sqrt{2}t} + \sqrt{2} + 2)}{4}$$

Example # 5.26

```
A=[-1,2,-1,1,2,1];
```

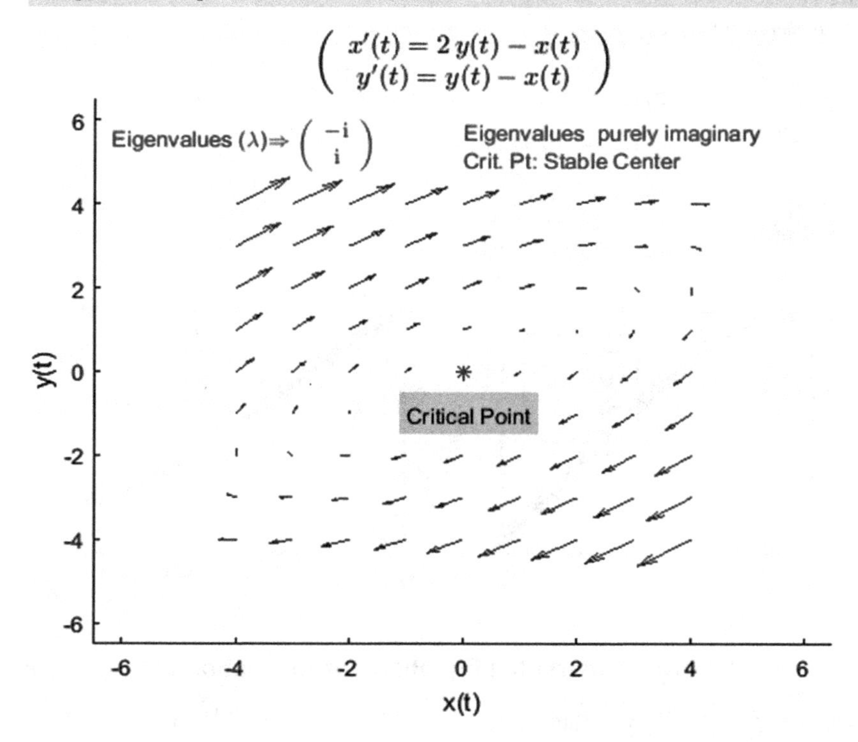

Complete Solutions: Pair of First Order Linear Differential Equations

Coefficient Matrix [A] $\begin{pmatrix} -1 & 2 \\ -1 & 1 \end{pmatrix}$ Differential Equations $\begin{pmatrix} x'(t) = 2\,y(t) - x(t) \\ y'(t) = y(t) - x(t) \end{pmatrix}$ IC $\begin{pmatrix} 2 \\ 1 \end{pmatrix}$

Complex eigenvalues $\lambda = \alpha \pm j\beta$; complex eigenvectors $\vec{v} = \vec{u} \pm j\vec{w}$

Eigenvalues (λ) $\begin{pmatrix} -i \\ i \end{pmatrix}$ Eigenvectors (Columns) $(\vec{v})$ $\begin{pmatrix} 1+i & 1-i \\ 1 & 1 \end{pmatrix}$

Solution basis $\left[\vec{X}_1\ \vec{X}_2\right]$ (columns) $\begin{pmatrix} \cos(t)+\sin(t) & \cos(t)-\sin(t) \\ \cos(t) & -\sin(t) \end{pmatrix}$

General Solution set

$$x_g(t) = c_2\ (\cos(t) - \sin(t)) + c_1\ (\cos(t) + \sin(t))$$

$$y_g(t) = c_1\ \cos(t) - c_2\ \sin(t)$$

Apply initial conditions ⇒ $\begin{pmatrix} c_1 + c_2 = 2 \\ c_1 = 1 \end{pmatrix}$ $\Rightarrow \begin{bmatrix} c_1 \\ c_2 \end{bmatrix} = \begin{pmatrix} 1 \\ 1 \end{pmatrix}$

Solution set Eigenvalues/vectors

$$x_e(t) = 2\ \cos(t)$$

$$y_e(t) = \sqrt{2}\ \cos\left(t + \tfrac{\pi}{4}\right)$$

Solutions (Inverse Laplace,dsolve(.) & Runge-Kutta)

Laplace transform from Differential Eqns

$$X_L(s) = \frac{2}{s^2+1} + \frac{2(s-1)}{s^2+1}$$

$$Y_L(s) = \frac{s+1}{s^2+1} - \frac{2}{s^2+1}$$

Solution set Inverse Laplace

$$x_L(t) = 2\cos(t)$$

$$y_L(t) = \cos(t) - \sin(t)$$

Solution set using dsolve(.)

$$x_d(t) = 2\cos(t)$$

$$y_d(t) = \sqrt{2}\cos\left(t + \frac{\pi}{4}\right)$$

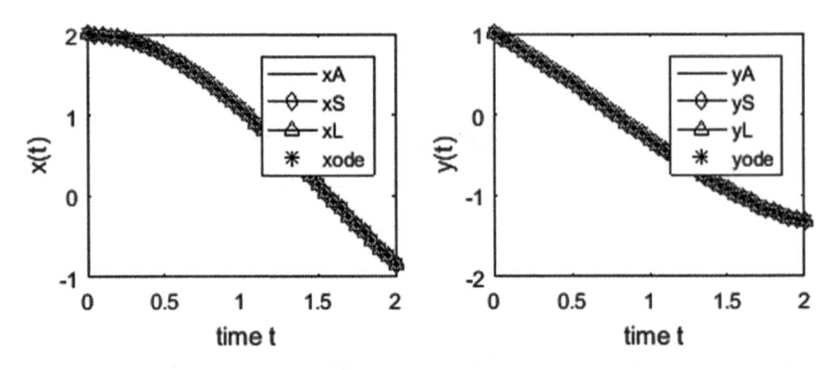

2nd order Differential Equation based solution set

Coefficient Matrix [A] $\begin{pmatrix} -1 & 2 \\ -1 & 1 \end{pmatrix}$ Differential Equations $\begin{pmatrix} x'(t) = 2\,y(t) - x(t) \\ y'(t) = y(t) - x(t) \end{pmatrix}$ IC $\begin{pmatrix} 2 \\ 1 \end{pmatrix}$

$A_{12} \neq 0 \Rightarrow$ Create Equation in x $\Rightarrow ax''(t) + bx'(t) + cx(t) = 0$

Equivalent 2nd order DE in x

$$x''(t) + x(t) = 0$$

Roots of the Characteristic Equation

$$(\, i \quad -i \,)$$

General Solution in terms of unknown constants b_1, b_2

$$x_2(t) = b_1\cos(t) - b_2\sin(t)$$

$$y_2(t) = \frac{b_1\cos(t)}{2} - \frac{b_2\cos(t)}{2} - \frac{b_1\sin(t)}{2} - \frac{b_2\sin(t)}{2}$$

Apply Initial Conditions $[b_1, b_2] \Rightarrow \quad (\, 2 \quad 0 \,)$

Solution set using roots

$$x_1(t) = 2\cos(t)$$

$$y_1(t) = \sqrt{2}\cos\left(t + \frac{\pi}{4}\right)$$

Example # 5.27

```
A=[-3,1,-4,2,-1,1];
```

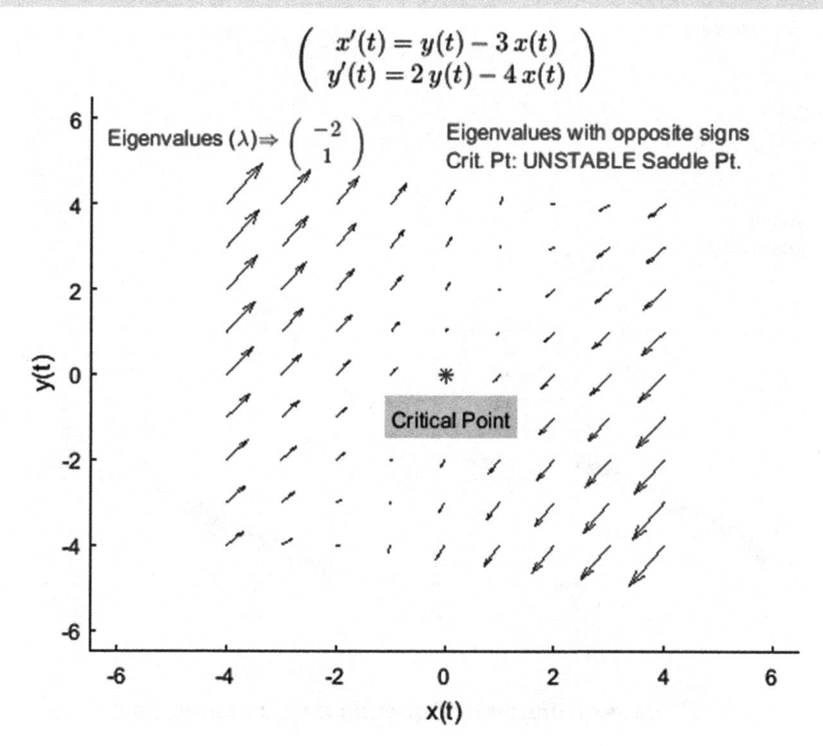

Complete Solutions: Pair of First Order Linear Differential Equations

Coefficient Matrix [A] $\begin{pmatrix} -3 & 1 \\ -4 & 2 \end{pmatrix}$ Differential Equations $\begin{pmatrix} x'(t) = y(t) - 3x(t) \\ y'(t) = 2y(t) - 4x(t) \end{pmatrix}$ IC $\begin{pmatrix} -1 \\ 1 \end{pmatrix}$

Real and distinct eigenvalues & eigenvectors

Eigenvalues (λ) $\begin{pmatrix} -2 \\ 1 \end{pmatrix}$ Eigenvectors (Columns) ($\vec{v}$) $\begin{pmatrix} 1 & \frac{1}{4} \\ 1 & 1 \end{pmatrix}$

Solution basis $\left[\vec{X}_1\ \vec{X}_2\right]$ (columns) $\begin{pmatrix} e^{-2t} & \frac{e^t}{4} \\ e^{-2t} & e^t \end{pmatrix}$

General Solution set
$$x_g(t) = \frac{c_2 e^t}{4} + c_1 e^{-2t}$$
$$y_g(t) = c_2 e^t + c_1 e^{-2t}$$

Apply initial conditions ⇒ $\begin{pmatrix} c_1 + \frac{c_2}{4} = -1 \\ c_1 + c_2 = 1 \end{pmatrix}$ ⇒ $\begin{bmatrix} c_1 \\ c_2 \end{bmatrix} = \begin{pmatrix} -\frac{5}{3} \\ \frac{8}{3} \end{pmatrix}$

Solution set Eigenvalues/vectors
$$x_e(t) = \frac{2e^t}{3} - \frac{5e^{-2t}}{3}$$
$$y_e(t) = \frac{8e^t}{3} - \frac{5e^{-2t}}{3}$$

Solutions (Inverse Laplace,dsolve(.) & Runge-Kutta)

Laplace transform from Differential Eqns

$$X_L(s) = \frac{1}{s^2+s-2} - \frac{s-2}{s^2+s-2}$$

$$Y_L(s) = \frac{4}{s^2+s-2} + \frac{s+3}{s^2+s-2}$$

Solution set Inverse Laplace

$$x_L(t) = \frac{2e^t}{3} - \frac{5e^{-2t}}{3}$$

$$y_L(t) = \frac{8e^t}{3} - \frac{5e^{-2t}}{3}$$

Solution set using dsolve(.)

$$x_d(t) = \frac{2e^t}{3} - \frac{5e^{-2t}}{3}$$

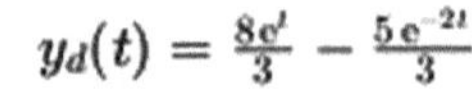

$$y_d(t) = \frac{8e^t}{3} - \frac{5e^{-2t}}{3}$$

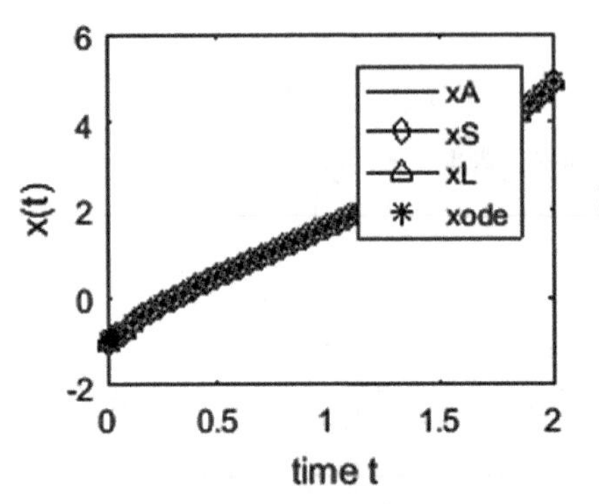

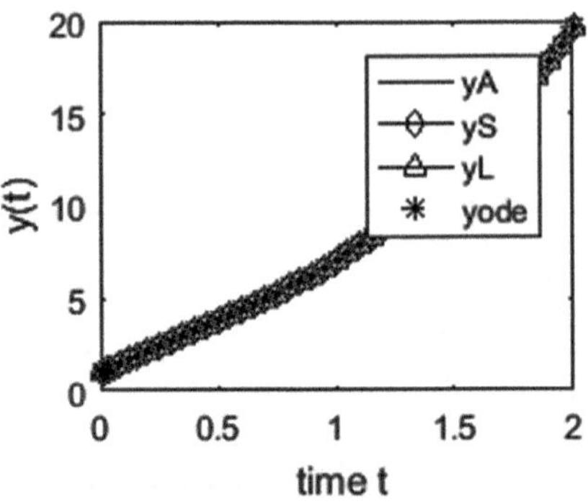

2nd order Differential Equation based solution set

Coefficient Matrix [A] $\begin{pmatrix} -3 & 1 \\ -4 & 2 \end{pmatrix}$ Differential Equations $\begin{pmatrix} x'(t) = y(t) - 3x(t) \\ y'(t) = 2y(t) - 4x(t) \end{pmatrix}$ IC $\begin{pmatrix} -1 \\ 1 \end{pmatrix}$

$A_{12} \neq 0 \Rightarrow$ Create Equation in x $\Rightarrow ax''(t) + bx'(t) + cx(t) = 0$

Equivalent 2nd order DE in x

$x''(t) + x'(t) - 2x(t) = 0$

Roots of the Characteristic Equation

$(\ -2 \quad 1\ )$

General Solution in terms of unknown constants b_1, b_2

$$x_2(t) = b_2\,e^t + b_1\,e^{-2t}$$

$$y_2(t) = 4\,b_2\,e^t + b_1\,e^{-2t}$$

Apply Initial Conditions $[b_1, b_2] \Rightarrow \quad (\ -\frac{5}{3} \quad \frac{2}{3}\ )$

Solution set using roots

$$x_1(t) = \frac{2e^t}{3} - \frac{5e^{-2t}}{3}$$

$$y_1(t) = \frac{8e^t}{3} - \frac{5e^{-2t}}{3}$$

Example # 5.28

```
A=[1,0,-2,1,1,0];
```

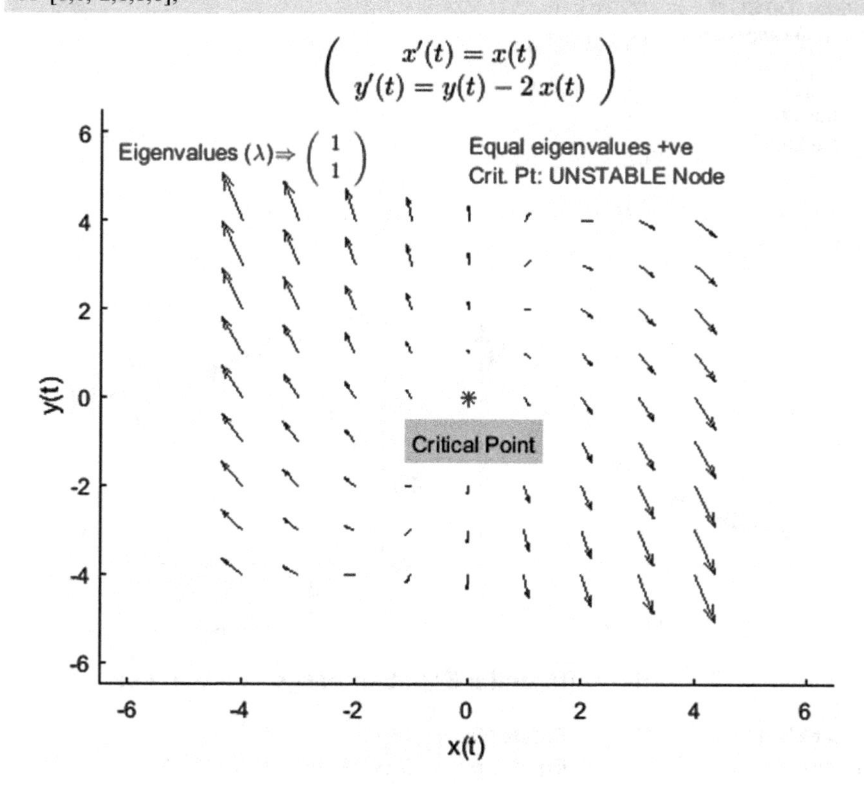

Complete Solutions: Pair of First Order Linear Differential Equations

Coefficient Matrix [A] $\begin{pmatrix} 1 & 0 \\ -2 & 1 \end{pmatrix}$ Differential Equations $\begin{pmatrix} x'(t) = x(t) \\ y'(t) = y(t) - 2\,x(t) \end{pmatrix}$ IC $\begin{pmatrix} 1 \\ 0 \end{pmatrix}$

$\Rightarrow$ [A] is defective: solve $[A - \lambda I_2]^2 \vec{v} = 0$ to get generalized eigenvectors

Eigenvalues (λ) $\begin{pmatrix} 1 \\ 1 \end{pmatrix}$ Generalized Eigenvectors (Columns) $(\vec{v})$ $\begin{pmatrix} 1 & 0 \\ 0 & 1 \end{pmatrix}$

Solution basis $[\vec{X}_1\ \vec{X}_2]$ (columns) $\begin{pmatrix} e^t & 0 \\ -2\,t\,e^t & e^t \end{pmatrix}$

General Solution set

$$x_g(t) = c_1\,e^t$$

$$y_g(t) = c_2\,e^t - 2\,c_1\,t\,e^t$$

Apply initial conditions $\Rightarrow$ $\begin{pmatrix} c_1 = 1 \\ c_2 = 0 \end{pmatrix}$ $\Rightarrow \begin{bmatrix} c_1 \\ c_2 \end{bmatrix} = \begin{pmatrix} 1 \\ 0 \end{pmatrix}$

Solution set Eigenvalues/vectors

$$x_e(t) = e^t$$

$$y_e(t) = -2\,t\,e^t$$

Solutions (Inverse Laplace,dsolve(.) & Runge-Kutta)

Laplace transform from Differential Eqns $X_L(s) = \frac{1}{s-1}$

$Y_L(s) = -\frac{2}{(s-1)^2}$

Solution set Inverse Laplace $x_L(t) = e^t$

$y_L(t) = -2\,t\,e^t$

Solution set using dsolve(.) $x_d(t) = e^t$

$y_d(t) = -2\,t\,e^t$

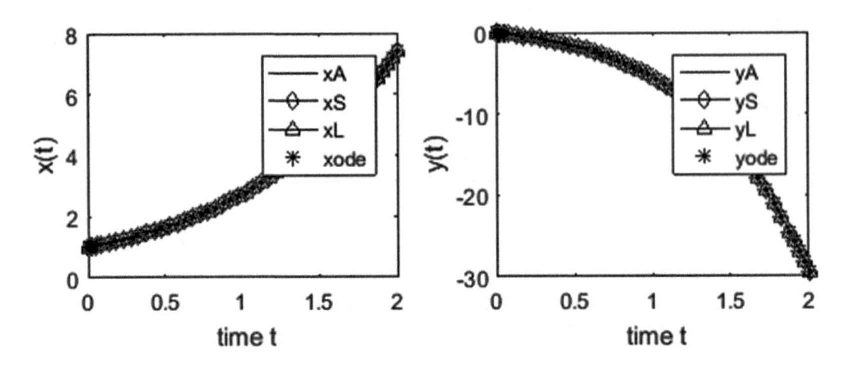

2nd order Differential Equation based solution set

Coefficient Matrix [A] $\begin{pmatrix} 1 & 0 \\ -2 & 1 \end{pmatrix}$ Differential Equations $\begin{pmatrix} x'(t) = x(t) \\ y'(t) = y(t) - 2\,x(t) \end{pmatrix}$ IC $\begin{pmatrix} 1 \\ 0 \end{pmatrix}$

$A_{12} = 0 \Rightarrow$ Create Equation in y $\Rightarrow ay''(t) + by'(t) + cy(t) = 0$

Equivalent 2nd order DE in y

$y''(t) - 2\,y'(t) + y(t) = 0$

Roots of the Characteristic Equation

$(1 \quad 1)$

General Solution in terms of unknown constants b_1, b_2 $x_2(t) = -\frac{b_2\,e^t}{2}$

$y_2(t) = e^t\,(b_1 + b_2\,t)$

Apply Initial Conditions $[b_1,b_2] \Rightarrow$ $(0 \quad -2)$

Solution set using roots $x_1(t) = e^t$

$y_1(t) = -2\,t\,e^t$

Example # 5.29

```
A=[0,1,-1,-2,1,-1];
```

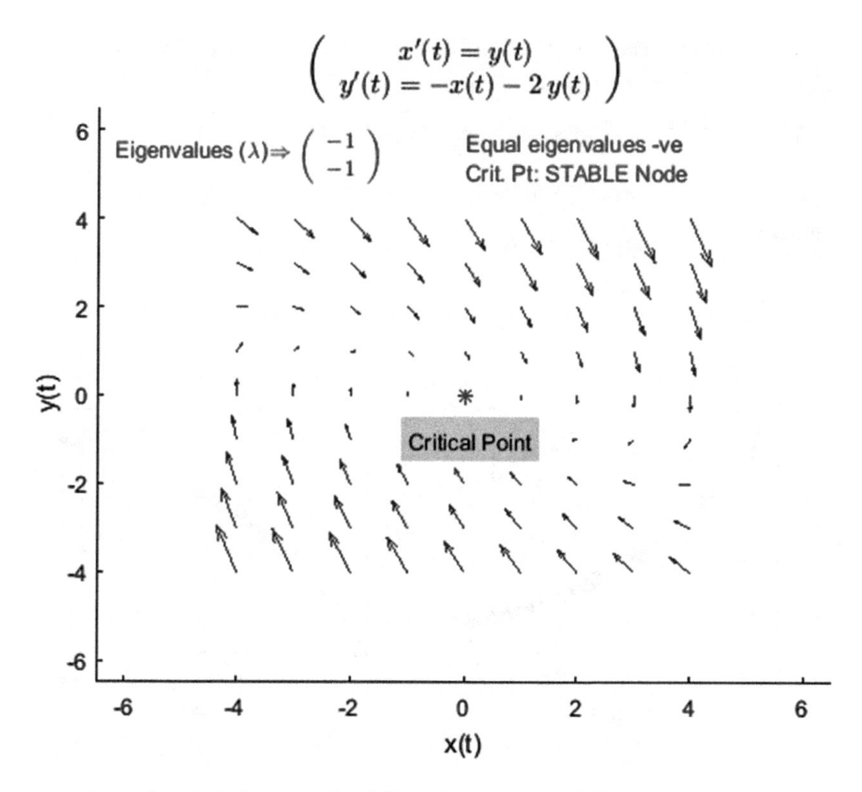

Complete Solutions: Pair of First Order Linear Differential Equations

Coefficient Matrix [A] $\begin{pmatrix} 0 & 1 \\ -1 & -2 \end{pmatrix}$ Differential Equations $\begin{pmatrix} x'(t) = y(t) \\ y'(t) = -x(t) - 2\,y(t) \end{pmatrix}$ IC $\begin{pmatrix} 1 \\ -1 \end{pmatrix}$

$\Rightarrow$ [A] is defective: solve $[A - \lambda I_2]^2 \vec{v} = 0$ to get generalized eigenvectors

Eigenvalues (λ) $\begin{pmatrix} -1 \\ -1 \end{pmatrix}$ Generalized Eigenvectors (Columns) $(\vec{v})$ $\begin{pmatrix} 1 & 0 \\ 0 & 1 \end{pmatrix}$

Solution basis $[\vec{X}_1\ \vec{X}_2]$ (columns) $\begin{pmatrix} e^{-t}\,(t+1) & t\,e^{-t} \\ -t\,e^{-t} & -e^{-t}\,(t-1) \end{pmatrix}$

General Solution set

$$x_g(t) = c_2\,t\,e^{-t} + c_1\,e^{-t}\,(t+1)$$

$$y_g(t) = -c_1\,t\,e^{-t} - c_2\,e^{-t}\,(t-1)$$

Apply initial conditions $\Rightarrow$ $\begin{pmatrix} c_1 = 1 \\ c_2 = -1 \end{pmatrix}$ $\Rightarrow \begin{bmatrix} c_1 \\ c_2 \end{bmatrix} = \begin{pmatrix} 1 \\ -1 \end{pmatrix}$

Solution set Eigenvalues/vectors

$$x_e(t) = e^{-t}$$

$$y_e(t) = -e^{-t}$$

Solutions (Inverse Laplace,dsolve(.) & Runge-Kutta)

Laplace transform from Differential Eqns

$$X_L(s) = \frac{s+2}{s^2+2s+1} - \frac{1}{s^2+2s+1}$$

$$Y_L(s) = -\frac{s}{s^2+2s+1} - \frac{1}{s^2+2s+1}$$

Solution set Inverse Laplace

$$x_L(t) = e^{-t}$$

$$y_L(t) = -e^{-t}$$

Solution set using dsolve(.)

$$x_d(t) = e^{-t}$$

$$y_d(t) = -e^{-t}$$

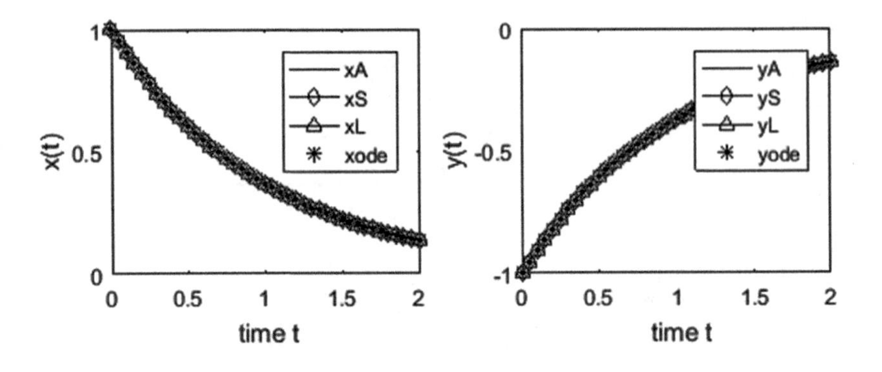

2nd order Differential Equation based solution set

Coefficient Matrix [A] $\begin{pmatrix} 0 & 1 \\ -1 & -2 \end{pmatrix}$ Differential Equations $\begin{pmatrix} x'(t) = y(t) \\ y'(t) = -x(t) - 2\,y(t) \end{pmatrix}$ IC $\begin{pmatrix} 1 \\ -1 \end{pmatrix}$

$A_{12} \neq 0 \Rightarrow$ Create Equation in x $\Rightarrow ax''(t) + bx'(t) + cx(t) = 0$

Equivalent 2nd order DE in x

$$x''(t) + 2\,x'(t) + x(t) = 0$$

Roots of the Characteristic Equation

$$\begin{pmatrix} -1 & -1 \end{pmatrix}$$

General Solution in terms of unknown constants b_1, b_2

$$x_2(t) = e^{-t}\,(b_1 + b_2\,t)$$

$$y_2(t) = -e^{-t}\,(b_1 - b_2 + b_2\,t)$$

Apply Initial Conditions $[b_1, b_2] \Rightarrow \begin{pmatrix} 1 & 0 \end{pmatrix}$

Solution set using roots

$$x_1(t) = e^{-t}$$

$$y_1(t) = -e^{-t}$$

Example # 5.30

```
A=[2,0,0,2,1,1];
```

$$\begin{pmatrix} x'(t) = 2\,x(t) \\ y'(t) = 2\,y(t) \end{pmatrix}$$

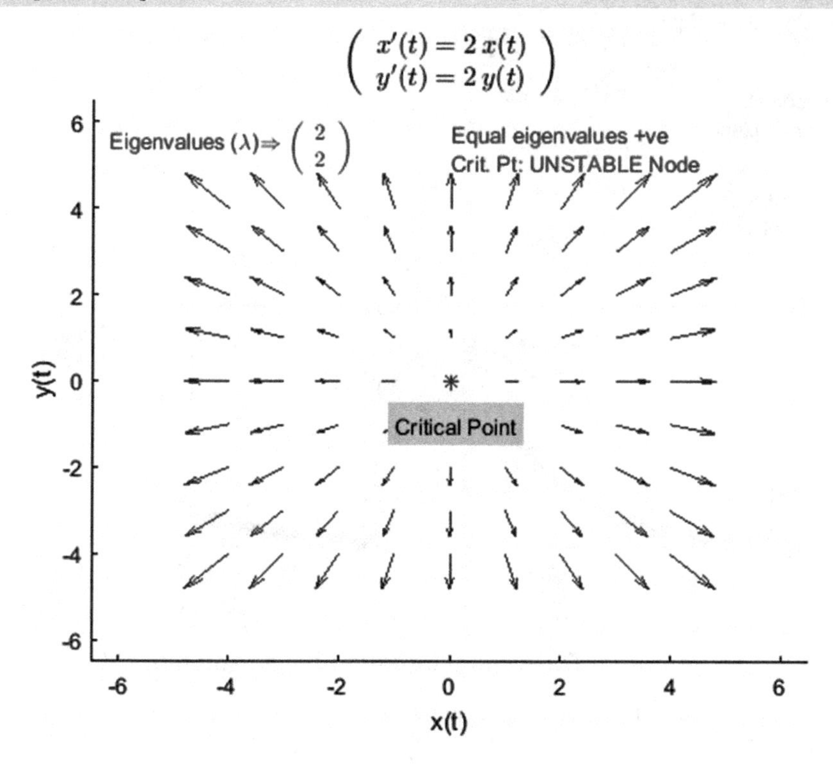

Complete Solutions: Pair of First Order Linear Differential Equations

Coefficient Matrix [A] $\begin{pmatrix} 2 & 0 \\ 0 & 2 \end{pmatrix}$ Differential Equations $\begin{pmatrix} x'(t) = 2\,x(t) \\ y'(t) = 2\,y(t) \end{pmatrix}$ IC $\begin{pmatrix} 1 \\ 1 \end{pmatrix}$

⇒ [A] is NOT defective; equal eigenvalues with distinct eigenvectors

Eigenvalues (λ) $\begin{pmatrix} 2 \\ 2 \end{pmatrix}$ Eigenvectors (Columns) $(\vec{v})$ $\begin{pmatrix} 1 & 0 \\ 0 & 1 \end{pmatrix}$

Solution basis $[\vec{X}_1\ \vec{X}_2]$ (columns) $\begin{pmatrix} e^{2t} & 0 \\ 0 & e^{2t} \end{pmatrix}$

General Solution set

$$x_g(t) = c_1\, e^{2t}$$

$$y_g(t) = c_2\, e^{2t}$$

Apply initial conditions ⇒ $\begin{pmatrix} c_1 = 1 \\ c_2 = 1 \end{pmatrix}$ $\Rightarrow \begin{bmatrix} c_1 \\ c_2 \end{bmatrix} = \begin{pmatrix} 1 \\ 1 \end{pmatrix}$

Solution set Eigenvalues/vectors

$$x_e(t) = e^{2t}$$

$$y_e(t) = e^{2t}$$

Solutions (Inverse Laplace,dsolve(.) & Runge-Kutta)

Laplace transform from Differential Eqns $X_L(s) = \frac{1}{s-2}$

$Y_L(s) = \frac{1}{s-2}$

Solution set Inverse Laplace $x_L(t) = e^{2t}$

$y_L(t) = e^{2t}$

Solution set using dsolve(.) $x_d(t) = e^{2t}$

$y_d(t) = e^{2t}$

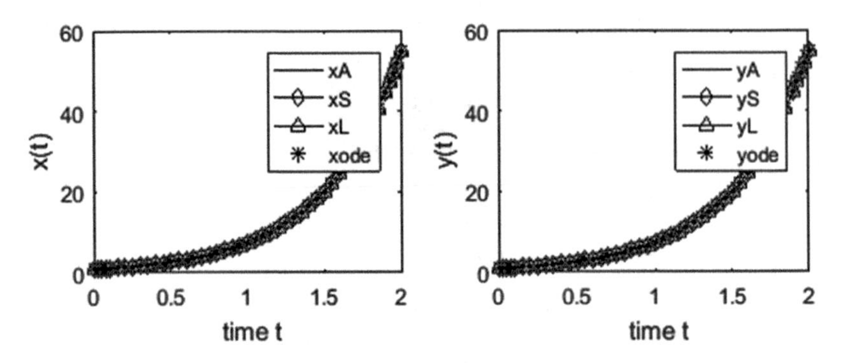

2nd order Differential Equation based solution set

Coefficient Matrix [A] $\begin{pmatrix} 2 & 0 \\ 0 & 2 \end{pmatrix}$ Differential Equations $\begin{pmatrix} x'(t) = 2\,x(t) \\ y'(t) = 2\,y(t) \end{pmatrix}$ IC $\begin{pmatrix} 1 \\ 1 \end{pmatrix}$

$A_{12}=A_{21}=0 \Rightarrow$ No Equivalent 2nd order differential EQn; Equations are not coupled Independent First Order Differential Equations

---

General Solution in terms of unknown constants b_1, b_2 $x_2(t) = b_1\, e^{2t}$

$y_2(t) = b_2\, e^{2t}$

Apply Initial Conditions $[b_1, b_2] \Rightarrow \begin{pmatrix} 1 & 1 \end{pmatrix}$

Solution set using roots $x_1(t) = e^{2t}$

$y_1(t) = e^{2t}$

Example # 5.31

```
A=[2,2,-2,-2,-1,-1];
```

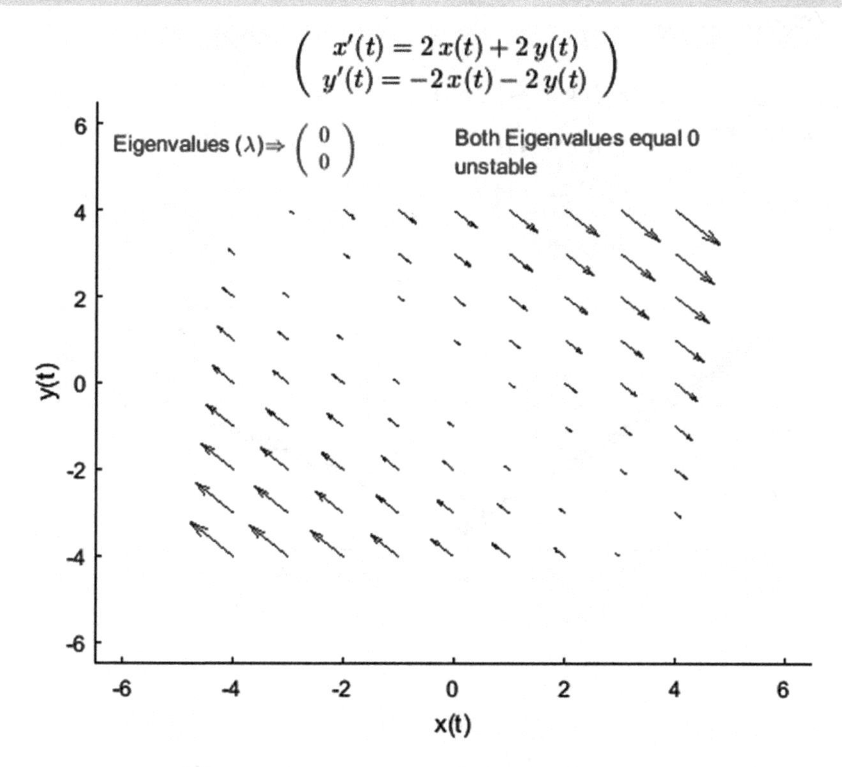

Complete Solutions: Pair of First Order Linear Differential Equations

Coefficient Matrix [A] $\begin{pmatrix} 2 & 2 \\ -2 & -2 \end{pmatrix}$ Differential Equations $\begin{pmatrix} x'(t) = 2\,x(t) + 2y(t) \\ y'(t) = -2\,x(t) - 2y(t) \end{pmatrix}$ IC $\begin{pmatrix} -1 \\ -1 \end{pmatrix}$

$\Rightarrow$ [A] is defective: solve $[A - \lambda I_2]^2 \vec{v} = 0$ to get generalized eigenvectors

Eigenvalues (λ) $\begin{pmatrix} 0 \\ 0 \end{pmatrix}$ Generalized Eigenvectors (Columns) $(\vec{v})$ $\begin{pmatrix} 1 & 0 \\ 0 & 1 \end{pmatrix}$

Solution basis $\left[\vec{X}_1\ \vec{X}_2\right]$ (columns) $\begin{pmatrix} 2t+1 & 2t \\ -2t & 1-2t \end{pmatrix}$

General Solution set

$$x_g(t) = 2\,c_2\,t + c_1\,(2\,t + 1)$$

$$y_g(t) = -2\,c_1\,t - c_2\,(2\,t - 1)$$

Apply initial conditions ⇒ $\begin{pmatrix} c_1 = -1 \\ c_2 = -1 \end{pmatrix}$ $\Rightarrow \begin{bmatrix} c_1 \\ c_2 \end{bmatrix} = \begin{pmatrix} -1 \\ -1 \end{pmatrix}$

Solution set Eigenvalues/vectors

$$x_e(t) = -4\,t - 1$$

$$y_e(t) = 4\,t - 1$$

Solutions (Inverse Laplace,dsolve(.) & Runge-Kutta)

Laplace transform from Differential Eqns

$$X_L(s) = -\frac{s+2}{s^2} - \frac{2}{s^2}$$

$$Y_L(s) = \frac{2}{s^2} - \frac{s-2}{s^2}$$

Solution set Inverse Laplace

$$x_L(t) = -4\,t - 1$$

$$y_L(t) = 4\,t - 1$$

Solution set using dsolve(.)

$$x_d(t) = -4t - 1$$

$$y_d(t) = 4t - 1$$

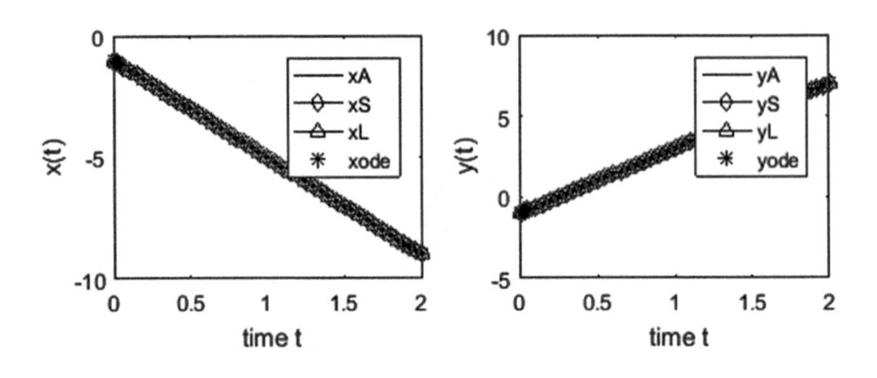

2nd order Differential Equation based solution set

Coefficient Matrix [A] $\begin{pmatrix} 2 & 2 \\ -2 & -2 \end{pmatrix}$ Differential Equations $\begin{pmatrix} x'(t) = 2\,x(t) + 2\,y(t) \\ y'(t) = -2\,x(t) - 2\,y(t) \end{pmatrix}$ IC $\begin{pmatrix} -1 \\ -1 \end{pmatrix}$

$A_{12} \neq 0 \Rightarrow$ Create Equation in x $\Rightarrow ax''(t) + bx'(t) + cx(t) = 0$

Equivalent 2nd order DE in x

$x''(t) = 0$

Roots of the Characteristic Equation

$(\;0 \quad 0\;)$

General Solution in terms of unknown constants b_1, b_2

$$x_2(t) = b_1 + b_2\,t$$

$$y_2(t) = \frac{b_2}{2} - b_1 - b_2\,t$$

Apply Initial Conditions $[b_1, b_2] \Rightarrow$ $(\;-1 \quad -4\;)$

Solution set using roots

$$x_1(t) = -4\,t - 1$$

$$y_1(t) = 4\,t - 1$$

Example # 5.32

```
A=[1,1,-2,1,1,0];
```

$$\begin{pmatrix} x'(t) = x(t) + y(t) \\ y'(t) = y(t) - 2\,x(t) \end{pmatrix}$$

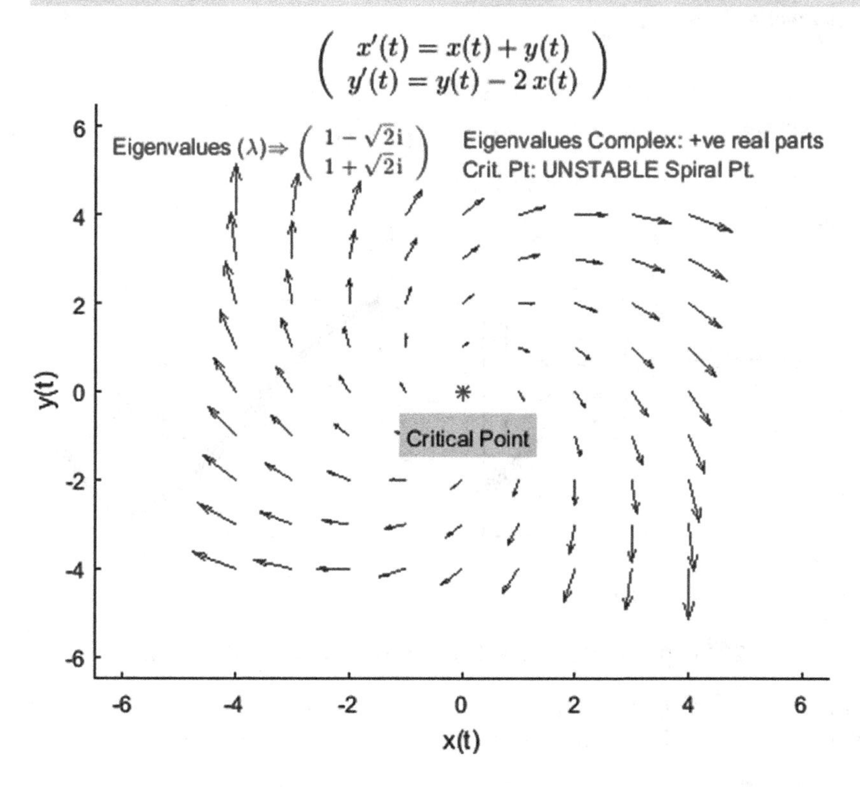

Complete Solutions: Pair of First Order Linear Differential Equations

Coefficient Matrix [A] $\begin{pmatrix} 1 & 1 \\ -2 & 1 \end{pmatrix}$ Differential Equations $\begin{pmatrix} x'(t) = x(t) + y(t) \\ y'(t) = y(t) - 2\,x(t) \end{pmatrix}$ IC $\begin{pmatrix} 1 \\ 0 \end{pmatrix}$

Complex eigenvalues $\lambda = \alpha \pm j\beta$; complex eigenvectors $\vec{v} = \vec{u} \pm j\vec{w}$

Eigenvalues (λ) $\begin{pmatrix} 1-\sqrt{2}\,i \\ 1+\sqrt{2}\,i \end{pmatrix}$ Eigenvectors (Columns) $(\vec{v})$ $\begin{pmatrix} \frac{\sqrt{2}i}{2} & -\frac{\sqrt{2}i}{2} \\ 1 & 1 \end{pmatrix}$

Solution basis $\left[\vec{X}_1\ \vec{X}_2\right]$ (columns) $\begin{pmatrix} \frac{\sqrt{2}\,e^t \sin(\sqrt{2}\,t)}{2} & \frac{\sqrt{2}\,e^t \cos(\sqrt{2}\,t)}{2} \\ e^t \cos(\sqrt{2}\,t) & -e^t \sin(\sqrt{2}\,t) \end{pmatrix}$

General Solution set
$$x_g(t) = \frac{\sqrt{2}\,c_2 e^t \cos(\sqrt{2}\,t)}{2} + \frac{\sqrt{2}\,c_1 e^t \sin(\sqrt{2}\,t)}{2}$$
$$y_g(t) = c_1\,e^t \cos(\sqrt{2}\,t) - c_2\,e^t \sin(\sqrt{2}\,t)$$

Apply initial conditions ⇒ $\begin{pmatrix} \frac{\sqrt{2}c_2}{2} = 1 \\ c_1 = 0 \end{pmatrix}$ $\Rightarrow \begin{bmatrix} c_1 \\ c_2 \end{bmatrix} = \begin{pmatrix} 0 \\ \sqrt{2} \end{pmatrix}$

Solution set Eigenvalues/vectors
$$x_e(t) = e^t \cos(\sqrt{2}\,t)$$
$$y_e(t) = -\sqrt{2}\,e^t \sin(\sqrt{2}\,t)$$

Solutions (Inverse Laplace,dsolve(.) & Runge-Kutta)

Laplace transform from Differential Eqns

$$X_L(s) = \frac{s-1}{s^2-2s+3}$$

$$Y_L(s) = -\frac{2}{s^2-2s+3}$$

Solution set Inverse Laplace

$$x_L(t) = e^t \cos(\sqrt{2}\,t)$$

$$y_L(t) = -\sqrt{2}\,e^t \sin(\sqrt{2}\,t)$$

Solution set using dsolve(.)

$$x_d(t) = e^t \cos(\sqrt{2}t)$$

$$y_d(t) = -\sqrt{2}e^t \sin(\sqrt{2}\,t)$$

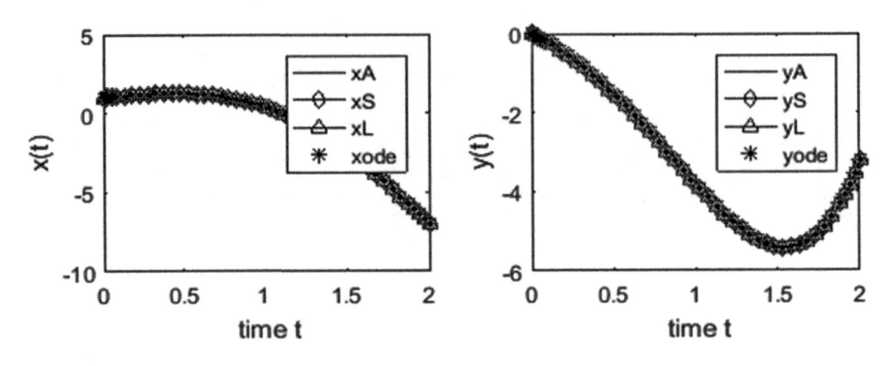

2nd order Differential Equation based solution set

Coefficient Matrix [A] $\begin{pmatrix} 1 & 1 \\ -2 & 1 \end{pmatrix}$ Differential Equations $\begin{pmatrix} x'(t) = x(t) + y(t) \\ y'(t) = y(t) - 2\,x(t) \end{pmatrix}$ IC $\begin{pmatrix} 1 \\ 0 \end{pmatrix}$

$A_{12} \neq 0 \Rightarrow$ Create Equation in x $\Rightarrow ax''(t) + bx'(t) + cx(t) = 0$

Equivalent 2nd order DE in x

$$x''(t) - 2\,x'(t) + 3\,x(t) = 0$$

Roots of the Characteristic Equation

$$\begin{pmatrix} 1+\sqrt{2}\mathrm{i} & 1-\sqrt{2}\mathrm{i} \end{pmatrix}$$

General Solution in terms of unknown constants b_1, b_2

$$x_2(t) = e^t \left(b_1 \cos(\sqrt{2}\,t) - b_2 \sin(\sqrt{2}t)\right)$$

$$y_2(t) = -\sqrt{2}e^t \left(b_2 \cos(\sqrt{2}\,t) + b_1 \sin(\sqrt{2}t)\right)$$

Apply Initial Conditions $[b_1, b_2] \Rightarrow \begin{pmatrix} 1 & 0 \end{pmatrix}$

Solution set using roots

$$x_1(t) = e^t \cos(\sqrt{2}\,t)$$

$$y_1(t) = -\sqrt{2}\,e^t \sin(\sqrt{2}\,t)$$

Part 2 A pair of coupled first order non-homogeneous equations

Example # 5.33

```
abc=[1,1,4,1,1,-1];gt1='3';gt2='4';
```

Particular and Complete Solutions
Pair of First Order Linear Differential Equations

Coefficient Matrix [A] $\begin{pmatrix} 1 & 1 \\ 4 & 1 \end{pmatrix}$ Forcing functions $\vec{g}(t)$ $\begin{pmatrix} 3 \\ 4 \end{pmatrix}$ IC $\begin{pmatrix} 1 \\ -1 \end{pmatrix}$

Coupled Differential Equations $\begin{pmatrix} x'(t) = x(t) + y(t) + 3 \\ y'(t) = 4x(t) + y(t) + 4 \end{pmatrix}$

Eigenvalues (λ) $\begin{pmatrix} -1 \\ 3 \end{pmatrix}$ Eigenvectors ($\vec{v}$) $\begin{pmatrix} -\frac{1}{2} & \frac{1}{2} \\ 1 & 1 \end{pmatrix}$

Fundamental Matrix $X_f(t) = \left[\vec{X}_1, \vec{X}_2\right]$ of the Homogeneous Differential Equation $A\vec{x} = 0$

$\left[\vec{X}_1, \vec{X}_2\right] = \left[v(:,1)e^{\lambda_1 t}, v(:,2)e^{\lambda_2 t}\right]$

$$X_f(t) = \left[\vec{X}_1, \vec{X}_2\right] \Rightarrow \begin{pmatrix} -\frac{e^{-t}}{2} & \frac{e^{3t}}{2} \\ e^{-t} & e^{3t} \end{pmatrix}$$

Particular Solution set

$$\begin{bmatrix} x_p(t) \\ y_p(t) \end{bmatrix} = X_f(t) \int [X_f(t)]^{-1} g(t)\, dt \Rightarrow \begin{pmatrix} -\frac{1}{3} \\ -\frac{8}{3} \end{pmatrix}$$

General Solution set $\Rightarrow$ $\begin{bmatrix} x(t) \\ y(t) \end{bmatrix} = \begin{bmatrix} x_p(t) \\ y_p(t) \end{bmatrix} + c_1\vec{X}_1 + c_2\vec{X}_2$

Complete Solutions

Apply initial conditions $\Rightarrow \begin{pmatrix} \frac{c_2}{2} - \frac{c_1}{2} - \frac{1}{3} = 1 \\ c_1 + c_2 - \frac{8}{3} = -1 \end{pmatrix} \Rightarrow \begin{bmatrix} c_1 \\ c_2 \end{bmatrix} = \begin{pmatrix} -\frac{1}{2} \\ \frac{13}{6} \end{pmatrix}$

Solution Eigenvalues

$$x_e(t) = \frac{e^{-t}}{4} + \frac{13e^{3t}}{12} - \frac{1}{3}$$

$$y_e(t) = \frac{13e^{3t}}{6} - \frac{e^{-t}}{2} - \frac{8}{3}$$

Laplace Transform Theory $\begin{bmatrix} X(s) \\ Y(s) \end{bmatrix} = \left[[sI_2 - A]^{-1}\left(\begin{pmatrix} x(0) \\ y(0) \end{pmatrix} + L[g(t)]\right)\right]$

Laplace Transform $\begin{bmatrix} X(s) \\ Y(s) \end{bmatrix} = \begin{pmatrix} -\frac{\frac{4}{s}-1}{-s^2+2s+3} - \frac{\left(\frac{3}{s}+1\right)(s-1)}{-s^2+2s+3} \\ -\frac{4\left(\frac{3}{s}+1\right)}{-s^2+2s+3} - \frac{\left(\frac{4}{s}-1\right)(s-1)}{-s^2+2s+3} \end{pmatrix}$

Laplace Transform based solution

$$x_L(t) = \frac{e^{-t}}{4} + \frac{13e^{3t}}{12} - \frac{1}{3}$$

$$y_L(t) = \frac{13e^{3t}}{6} - \frac{e^{-t}}{2} - \frac{8}{3}$$

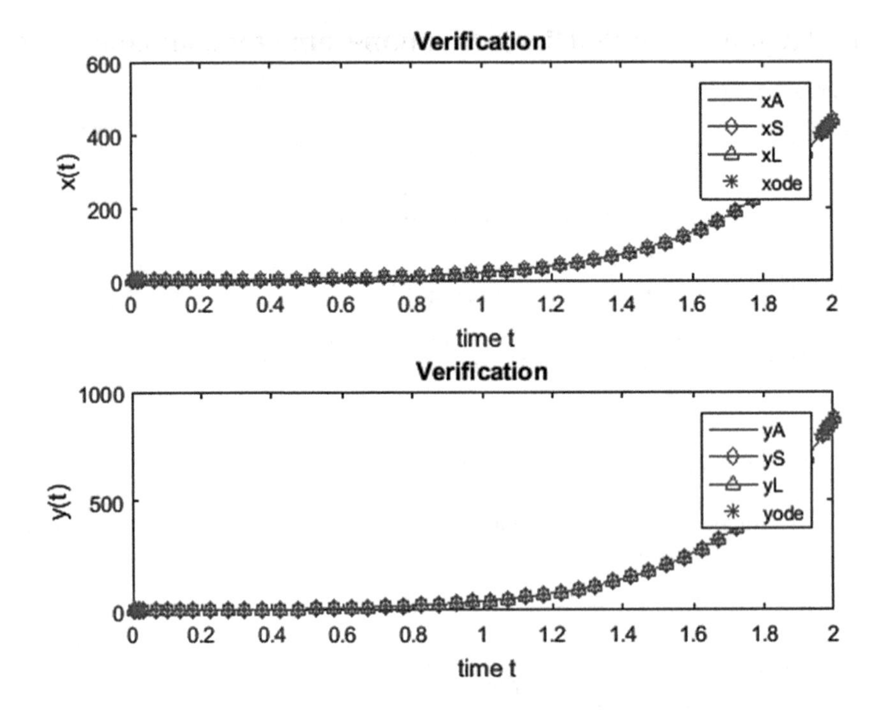

overview

Differential Equations $\begin{pmatrix} x'(t) = x(t) + y(t) + 3 \\ y'(t) = 4\,x(t) + y(t) + 4 \end{pmatrix}$ Initial Conditions $\begin{pmatrix} 1 \\ -1 \end{pmatrix}$

Particular Solution set $\begin{bmatrix} x_p(t) \\ y_p(t) \end{bmatrix} = \begin{pmatrix} -\frac{1}{3} \\ -\frac{8}{3} \end{pmatrix}$

Verification by substitution in the differential Equations: Must be a NULL Vector $\begin{pmatrix} 0 \\ 0 \end{pmatrix}$

Solution set using eigenvalues & eigenvectors

$$x_e(t) = \frac{e^{-t}}{4} + \frac{13\,e^{3t}}{12} - \frac{1}{3}$$

$$y_e(t) = \frac{13\,e^{3t}}{6} - \frac{e^{-t}}{2} - \frac{8}{3}$$

Solution set using dsolve(.)

$$x_d(t) = \frac{e^{-t}}{4} + \frac{13\,e^{3t}}{12} - \frac{1}{3}$$

$$y_d(t) = \frac{13\,e^{3t}}{6} - \frac{e^{-t}}{2} - \frac{8}{3}$$

Example # 5.34

```
abc=[1,1,0,2,1,0];gt1='t*exp(2*t)';gt2='exp(t)';
```

Particular and Complete Solutions
Pair of First Order Linear Differential Equations

Coefficient Matrix [A] $\begin{pmatrix} 1 & 1 \\ 0 & 2 \end{pmatrix}$ Forcing functions $\vec{g}(t)$ $\begin{pmatrix} t e^{2t} \\ e^t \end{pmatrix}$ IC $\begin{pmatrix} 1 \\ 0 \end{pmatrix}$

Coupled Differential Equations $\begin{pmatrix} x'(t) = x(t) + y(t) + t e^{2t} \\ y'(t) = e^t + 2y(t) \end{pmatrix}$

Eigenvalues (λ) $\begin{pmatrix} 1 \\ 2 \end{pmatrix}$ Eigenvectors ($\vec{v}$) $\begin{pmatrix} 1 & 1 \\ 0 & 1 \end{pmatrix}$

Fundamental Matrix $X_f(t) = \left[\vec{X}_1, \vec{X}_2\right]$ of the Homogeneous Differential Equation $A\vec{x} = 0$

$\left[\vec{X}_1, \vec{X}_2\right] = \left[v(:,1)e^{\lambda_1 t}, v(:,2)e^{\lambda_2 t}\right]$

$$X_f(t) = \left[\vec{X}_1, \vec{X}_2\right] \Rightarrow \begin{pmatrix} e^t & e^{2t} \\ 0 & e^{2t} \end{pmatrix}$$

Particular Solution set

$$\begin{bmatrix} x_p(t) \\ y_p(t) \end{bmatrix} = X_f(t) \int [X_f(t)]^{-1} g(t)\, dt \Rightarrow \begin{pmatrix} -e^t\,(t + e^t - t e^t + 1) \\ -e^t \end{pmatrix}$$

General Solution set $\Rightarrow$ $\begin{bmatrix} x(t) \\ y(t) \end{bmatrix} = \begin{bmatrix} x_p(t) \\ y_p(t) \end{bmatrix} + c_1\vec{X}_1 + c_2\vec{X}_2$

Complete Solutions

Apply initial conditions $\Rightarrow \begin{pmatrix} c_1 + c_2 - 2 = 1 \\ c_2 - 1 = 0 \end{pmatrix} \Rightarrow \begin{bmatrix} c_1 \\ c_2 \end{bmatrix} = \begin{pmatrix} 2 \\ 1 \end{pmatrix}$

Solution Eigenvalues

$$x_e(t) = e^t\,(t\,e^t - t + 1)$$

$$y_e(t) = e^t\,(e^t - 1)$$

Laplace Transform Theory $\begin{bmatrix} X(s) \\ Y(s) \end{bmatrix} = \left[[sI_2 - A]^{-1}\left(\begin{pmatrix} x(0) \\ y(0) \end{pmatrix} + L[g(t)]\right)\right]$

Laplace Transform $\begin{bmatrix} X(s) \\ Y(s) \end{bmatrix} = \begin{pmatrix} \frac{\frac{1}{(s-2)^2}+1}{s-1} + \frac{1}{(s-1)^2(s-2)} \\ \frac{1}{(s-1)(s-2)} \end{pmatrix}$

Laplace Transform based solution

$$x_L(t) = e^t + t e^{2t} - t e^t$$

$$y_L(t) = e^{2t} - e^t$$

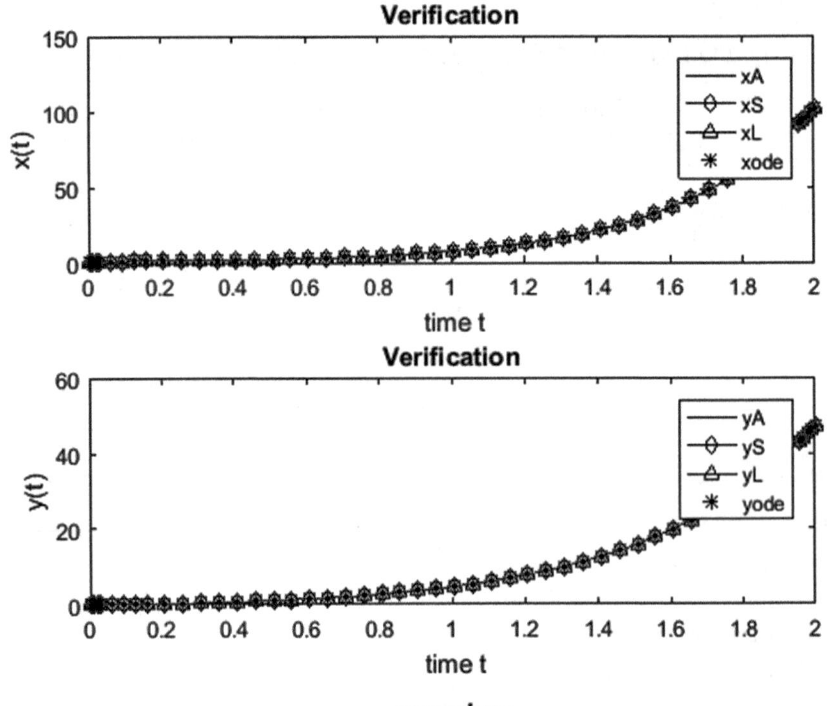

overview

Differential Equations $\begin{pmatrix} x'(t) = x(t) + y(t) + t\,e^{2t} \\ y'(t) = e^{t} + 2\,y(t) \end{pmatrix}$ Initial Conditions $\begin{pmatrix} 1 \\ 0 \end{pmatrix}$

Particular Solution set $\begin{bmatrix} x_p(t) \\ y_p(t) \end{bmatrix} = \begin{pmatrix} -e^{t}\;(t + e^{t} - t\,e^{t} + 1) \\ -e^{t} \end{pmatrix}$

Verification by substitution in the differential Equations: Must be a NULL Vector $\begin{pmatrix} 0 \\ 0 \end{pmatrix}$

Solution set using eigenvalues & eigenvectors

$$x_e(t) = e^{t}\;(t\,e^{t} - t + 1)$$

$$y_e(t) = e^{t}\;(e^{t} - 1)$$

Solution set using dsolve(.)

$$x_d(t) = e^{t}\;(t\,e^{t} - t + 1)$$

$$y_d(t) = e^{t}\;(e^{t} - 1)$$

Example # 5.35

```
abc=[-1,-2,1, -1,-1,1];gt1='t';gt2='exp(t)';
```

Particular and Complete Solutions
Pair of First Order Linear Differential Equations

Coefficient Matrix [A] $\begin{pmatrix} -1 & -2 \\ 1 & -1 \end{pmatrix}$ Forcing functions $\vec{g}(t)\begin{pmatrix} t \\ e^t \end{pmatrix}$ IC $\begin{pmatrix} -1 \\ 1 \end{pmatrix}$

Coupled Differential Equations $\begin{pmatrix} x'(t) = t - x(t) - 2y(t) \\ y'(t) = e^t + x(t) - y(t) \end{pmatrix}$

Complex eigenvalues and eigenvectors!

Eigenvalues (λ) $\begin{pmatrix} -1-\sqrt{2}\,i \\ -1+\sqrt{2}\,i \end{pmatrix}$ Eigenvectors ($\vec{v}$) $\begin{pmatrix} -\sqrt{2}\,i & \sqrt{2}\,i \\ 1 & 1 \end{pmatrix}$

$\Rightarrow \lambda = \alpha \pm j\beta$ $\Rightarrow \vec{v} = \vec{u} \pm j\vec{w}$

Fundamental Matrix $X_f(t) = \left[\vec{X}_1, \vec{X}_2\right]$ of the Homogeneous Differential Equation $A\vec{x} = 0$

$$\left[\vec{X}_1, \vec{X}_2\right] = \left[e^{\alpha t}\left[\vec{u}\,cos(\beta t) - \vec{w}\,sin(\beta t)\right], e^{\alpha t}\left[\vec{u}\,sin(\beta t) + \vec{w}\,cos(\beta t)\right]\right]$$

$$X_f(t) = \left[\vec{X}_1, \vec{X}_2\right] \Rightarrow \begin{pmatrix} -\sqrt{2}\,e^{-t}\sin(\sqrt{2}\,t) & -\sqrt{2}\,e^{-t}\cos(\sqrt{2}\,t) \\ e^{-t}\cos(\sqrt{2}\,t) & -e^{-t}\sin(\sqrt{2}\,t) \end{pmatrix}$$

Particular Solution set

$$\begin{bmatrix} x_p(t) \\ y_p(t) \end{bmatrix} = X_f(t)\int [X_f(t)]^{-1} g(t)\,dt \Rightarrow \begin{pmatrix} \frac{t}{3} - \frac{e^t}{3} + \frac{1}{9} \\ \frac{t}{3} + \frac{e^t}{3} - \frac{2}{9} \end{pmatrix}$$

General Solution set $\Rightarrow$ $\begin{bmatrix} x(t) \\ y(t) \end{bmatrix} = \begin{bmatrix} x_p(t) \\ y_p(t) \end{bmatrix} + c_1\vec{X}_1 + c_2\vec{X}_2$

Complete Solutions

Apply initial conditions $\Rightarrow \begin{pmatrix} -\sqrt{2}\,c_2 - \frac{2}{9} = -1 \\ c_1 + \frac{1}{9} = 1 \end{pmatrix} \Rightarrow \begin{bmatrix} c_1 \\ c_2 \end{bmatrix} = \begin{pmatrix} \frac{8}{9} \\ \frac{7\sqrt{2}}{18} \end{pmatrix}$

Solution Eigenvalues

$$x_e(t) = \frac{t}{3} - \frac{e^t}{3} - \frac{7e^{-t}\cos(\sqrt{2}\,t)}{9} - \frac{8\sqrt{2}\,e^{-t}\sin(\sqrt{2}\,t)}{9} + \frac{1}{9}$$

$$y_e(t) = \frac{t}{3} + \frac{e^t}{3} + \frac{8e^{-t}\cos(\sqrt{2}\,t)}{9} - \frac{7\sqrt{2}\,e^{-t}\sin(\sqrt{2}\,t)}{18} - \frac{2}{9}$$

Laplace Transform Theory $\begin{bmatrix} X(s) \\ Y(s) \end{bmatrix} = \left[[sI_2 - A]^{-1}\left(\begin{pmatrix} x(0) \\ y(0) \end{pmatrix} + L[g(t)]\right)\right]$

Laplace Transform $\begin{bmatrix} X(s) \\ Y(s) \end{bmatrix} = \begin{pmatrix} \frac{\left(\frac{1}{s^2}-1\right)(s+1)}{s^2+2s+3} - \frac{2\left(\frac{1}{s-1}+1\right)}{s^2+2s+3} \\ \frac{\frac{1}{s^2}-1}{s^2+2s+3} + \frac{\left(\frac{1}{s-1}+1\right)(s+1)}{s^2+2s+3} \end{pmatrix}$

Laplace Transform based solution

$$x_L(t) = \frac{t}{3} - \frac{e^t}{3} - \frac{7e^{-t}\left(\cos(\sqrt{2}\,t) + \frac{8\sqrt{2}\sin(\sqrt{2}t)}{7}\right)}{9} + \frac{1}{9}$$

$$y_L(t) = \frac{t}{3} + \frac{e^t}{3} + \frac{8e^{-t}\left(\cos(\sqrt{2}\,t) - \frac{7\sqrt{2}\sin(\sqrt{2}t)}{16}\right)}{9} - \frac{2}{9}$$

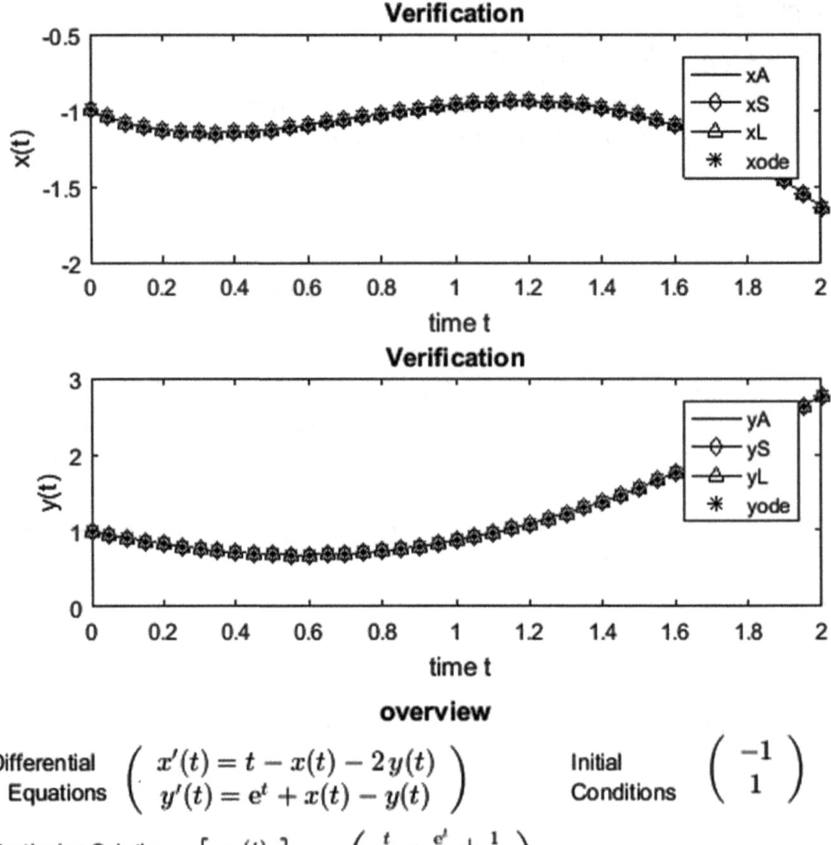

overview

Differential Equations $\begin{pmatrix} x'(t) = t - x(t) - 2\,y(t) \\ y'(t) = e^t + x(t) - y(t) \end{pmatrix}$ Initial Conditions $\begin{pmatrix} -1 \\ 1 \end{pmatrix}$

Particular Solution set $\begin{bmatrix} x_p(t) \\ y_p(t) \end{bmatrix} = \begin{pmatrix} \frac{t}{3} - \frac{e^t}{3} + \frac{1}{9} \\ \frac{t}{3} + \frac{e^t}{3} - \frac{2}{9} \end{pmatrix}$

Verification by substitution in the differential Equations: Must be a NULL Vector $\begin{pmatrix} 0 \\ 0 \end{pmatrix}$

Solution set using eigenvalues & eigenvectors

$$x_e(t) = \frac{t}{3} - \frac{e^t}{3} - \frac{7\,e^{-t}\cos\left(\sqrt{2}\,t\right)}{9} - \frac{8\sqrt{2}\,e^{-t}\sin\left(\sqrt{2}\,t\right)}{9} + \frac{1}{9}$$

$$y_e(t) = \frac{t}{3} + \frac{e^t}{3} + \frac{8\,e^{-t}\cos\left(\sqrt{2}\,t\right)}{9} - \frac{7\sqrt{2}\,e^{-t}\sin\left(\sqrt{2}\,t\right)}{18} - \frac{2}{9}$$

Solution set using dsolve(.)

$$x_d(t) = \frac{t}{3} - \frac{e^t}{3} - \frac{7\,e^{-t}\cos\left(\sqrt{2}\,t\right)}{9} - \frac{8\sqrt{2}\,e^{-t}\sin\left(\sqrt{2}\,t\right)}{9} + \frac{1}{9}$$

$$y_d(t) = \frac{t}{3} + \frac{e^t}{3} + \frac{8\,e^{-t}\cos\left(\sqrt{2}\,t\right)}{9} - \frac{7\sqrt{2}\,e^{-t}\sin\left(\sqrt{2}\,t\right)}{18} - \frac{2}{9}$$

Part 3 Multiple first order homogeneous systems

Multiple coupled first order homogeneous solutions are solved and the results are obtained by (1) using the concepts of eigenvectors (2) inverse Laplace transforms and (3) ODE. The results generate the following displays.

1. The description of the method based on eigenvalues and eigenvectors is shown.
2. The method to generate generalized eigenvectors if needed is shown.
3. Eigenvalues and eigenvectors are shown along with the statement identifying the need for generalized eigenvectors.
4. If generalized eigenvectors are required, they are generated with statement identifying the specific eigenvalue that leads to the generalized eigenvector along with the specific value of the algebraic and geometric multiplicity.
5. The fundamental matrix is shown along with the general solution. The result obtained from dsolve(.) is also shown with the caveat that the unknown constants are different in both solutions.
6. Solutions obtained after the application of the initial conditions are shown for both dsolve(.) and the solution is generated using eigenvalues and eigenvectors.
7. The equation that forms the basis for the Laplace transform based approach is shown along with the Laplace transform of the solution.
8. The Laplace place based solution is compared to the one from dsolve(.) and eigenvector/eigenvalues and the symbolic differences are shown.
9. The solutions are compared to the one obtained from ODE.

Example # 5.36

```
A=[-1,-2;1,-1];x0=[-1 1];gt={'t';'exp(t)'};
```

General Solution associated with [n x n] coefficient matrix A

⇒ Matrix not defective

Eigenvalues/vectors: $\lambda_1,\ \vec{v}_1;\ \lambda_2,\ \vec{v}_2;\ \cdots;\ \lambda_n, \vec{v}_n$

Solution set: $\vec{X}_k = e^{\lambda_k t}\vec{v}_k,\ k = 1, 2, \cdots, n$

⇒ Matrix defective: An example with two distinct eigenvectors

Eigenvalues & distinct eigenvectors (two): $\lambda_1, \vec{v}_1$ & $\lambda_2, \vec{v}_2$

Eigenvalue (λ_3, algebraic multiplicity of $m = n - 2,\ n \geq 3$)

generalized eigenvectors: $\vec{v}_3, \vec{v}_4, \cdots, \vec{v}_n$

Solution set with distinct eigenvectors: $\vec{X}_k = \vec{v}_k e^{\lambda_k t},\ k = 1, 2$

Solution set with eigenvalue λ_3 :

$$\vec{X}_j = [I_n + t(A - \lambda_3 I_n) + \cdots + [\Gamma(m)]^{-1} t^{m-1}(A - \lambda_3 I_n)^{m-1}]\vec{v}_j e^{\lambda_3 t}$$
$$j = 3, \cdots, n;\ m = n - 2$$

⇒ Fundamental Matrix of A: $X(t) = [\vec{X}_1(t), \vec{X}_2(t), \cdots, \vec{X}_n(t)]$

Homogeneous Solution ⇒ $\vec{x}_h(t) = X(t)\vec{c}$

$\vec{c}$ (from initial conditions)

Particular and Complete Solutions

Input ⇒ $\frac{d\vec{x}}{dt} = A\vec{x}(t) + \vec{g}(t)$

Fundamental Matrix of A ⇒ $X(t) = [\vec{X}_1(t), \vec{X}_2(t), \cdots, \vec{X}_n(t)]$

Initial Conditions ⇒ $\vec{x}(0) = \begin{pmatrix} x_1(0) \\ x_2(0) \\ \cdots \\ x_n(0) \end{pmatrix}$

Particular Solution ⇒ $\vec{x}_p(t) = X(t) \int [X(t)]^{-1} \vec{g}(t) dt$

Complete Solution ⇒ $\vec{x}(t) = X(t)\vec{c} + \vec{x}_p(t)$

Constants ⇒ $\vec{c} = [X(0)]^{-1} [\vec{x}(0) - \vec{x}_p(0)]$

$$\vec{x}(t) = \begin{pmatrix} x_1(t) \\ x_2(t) \\ \cdots \\ x_n(t) \end{pmatrix}$$

Coupled 1st order System: Problem Statement

$$\frac{d\vec{x}}{dt} = A\vec{x}(t) + \vec{g}(t)$$

Coefficient Matrix: A $\begin{pmatrix} -1 & -2 \\ 1 & -1 \end{pmatrix}$

Forcing functions $\vec{g}(t)$ $\begin{pmatrix} t \\ e^t \end{pmatrix}$

Initial Conditions: $\vec{x}(0)$ $\begin{pmatrix} -1 \\ 1 \end{pmatrix}$

Differential Equations: $\frac{d\vec{x}}{dt}$ $\begin{pmatrix} x_1'(t) = t - x_1(t) - 2\,x_2(t) \\ x_2'(t) = e^t + x_1(t) - x_2(t) \end{pmatrix}$

Eigenvalues and Eigenvectors

Input Matrix A

-1	-2
1	-1

Eigenvalues

-1-1.41i
-1+1.41i

Eigenvectors

0-1.41i	0+1.41i
1+0i	1+0i

Eigenvalues (2) & Eigenvectors (2): Matrix NOT defective

General solutions and unknown constants

Homogeneous solutions set

$$\begin{pmatrix} -\sqrt{2}\,\mathrm{e}^{-t}\left(c_2\cos(\sqrt{2}\,t)+c_1\sin(\sqrt{2}\,t)\right) \\ \mathrm{e}^{-t}\left(c_1\cos(\sqrt{2}\,t)-c_2\sin(\sqrt{2}\,t)\right) \end{pmatrix}$$

Complete solutions set (homogeneous + particular)

$$\begin{pmatrix} \frac{t}{3}-\frac{\mathrm{e}^t}{3}-\sqrt{2}\,\mathrm{e}^{-t}\left(c_2\cos(\sqrt{2}\,t)+c_1\sin(\sqrt{2}\,t)\right)+\frac{1}{9} \\ \frac{t}{3}+\frac{\mathrm{e}^t}{3}+\mathrm{e}^{-t}\left(c_1\cos(\sqrt{2}\,t)-c_2\sin(\sqrt{2}\,t)\right)-\frac{2}{9} \end{pmatrix}$$

constants c_1 and c_2

$$\begin{pmatrix} \frac{8}{9} & \frac{7\sqrt{2}}{18} \end{pmatrix}$$

Fundamental Matrix, Particular & Complete Solutions

Fundamental Matrix X(t):

$$\begin{pmatrix} -\sqrt{2}e^{-t}\sin(\sqrt{2}t) & -\sqrt{2}e^{-t}\cos(\sqrt{2}t) \\ e^{-t}\cos(\sqrt{2}t) & -e^{-t}\sin(\sqrt{2}t) \end{pmatrix}$$

Particular Solution: $\vec{Y}_p(t)$

$$\begin{matrix} \frac{t}{3} - \frac{e^t}{3} + \frac{1}{9} \\ \frac{t}{3} + \frac{e^t}{3} - \frac{2}{9} \end{matrix}$$

Complete Solution: $\vec{Y}$(t)

$$\begin{matrix} \frac{t}{3} - \frac{e^t}{3} - \frac{e^{-t}\left(7\cos(\sqrt{2}t) + 8\sqrt{2}\sin(\sqrt{2}t)\right)}{9} + \frac{1}{9} \\ \frac{t}{3} + \frac{e^t}{3} + \frac{8e^{-t}\cos(\sqrt{2}t)}{9} - \frac{7\sqrt{2}e^{-t}\sin(\sqrt{2}t)}{18} - \frac{2}{9} \end{matrix}$$

Coupled 1st Order Differential Equations: Laplace Transforms

Input⇒

$$\begin{bmatrix} x'_1(t) \\ x'_2(t) \\ \cdots \\ x'_n(t) \end{bmatrix} = A \begin{pmatrix} x_1(t) \\ x_2(t) \\ \cdots \\ x_n(t) \end{pmatrix} + \begin{pmatrix} g_1(t) \\ g_2(t) \\ \cdots \\ g_n(t) \end{pmatrix}$$

Initial⇒ Conditions

$$\begin{pmatrix} x_1(0) \\ x_2(0) \\ \cdots \\ x_n(0) \end{pmatrix}$$

Laplace⇒ Transform

$$\begin{bmatrix} X_1(s) \\ X_2(s) \\ \cdots \\ X_n(s) \end{bmatrix} = [sI_n - A]^{-1} \begin{pmatrix} x_1(0) \\ x_2(0) \\ \cdots \\ x_n(0) \end{pmatrix} + [sI_n - A]^{-1} \begin{pmatrix} G_1(s) \\ G_2(s) \\ \cdots \\ G_n(0) \end{pmatrix}$$

Solution⇒

$$\begin{pmatrix} x_1(t) \\ x_2(t) \\ \cdots \\ x_n(t) \end{pmatrix} = L^{-1} \begin{bmatrix} X_1(s) \\ X_2(s) \\ \cdots \\ X_n(s) \end{bmatrix}$$

Laplace Transforms and Solutions using inverse Laplace Transforms

Laplace Transform Y(s)

$$\begin{pmatrix} -\dfrac{s^4+2s^3-2s^2+1}{s^2(s-1)(s^2+2s+3)} \\ \dfrac{s^4+s^2+s-1}{s^2(s-1)(s^2+2s+3)} \end{pmatrix}$$

Solution: Inverse Laplace Transform Y(t)

$$\begin{pmatrix} \dfrac{t}{3}-\dfrac{e^t}{3}-\dfrac{7e^{-t}\left(\cos(\sqrt{2}t)+\frac{8\sqrt{2}\sin(\sqrt{2}t)}{7}\right)}{9}+\dfrac{1}{9} \\ \dfrac{t}{3}+\dfrac{e^t}{3}+\dfrac{8e^{-t}\left(\cos(\sqrt{2}t)-\frac{7\sqrt{2}\sin(\sqrt{2}t)}{16}\right)}{9}-\dfrac{2}{9} \end{pmatrix}$$

Comparison of Analytical Solutions

Solution: Eigenvalues & Eigenvectors: $Y_e(t)$

$$\begin{pmatrix} \dfrac{t}{3}-\dfrac{e^t}{3}-\dfrac{e^{-t}(7\cos(\sqrt{2}t)+8\sqrt{2}\sin(\sqrt{2}t))}{9}+\dfrac{1}{9} \\ \dfrac{t}{3}+\dfrac{e^t}{3}+\dfrac{8e^{-t}\cos(\sqrt{2}t)}{9}-\dfrac{7\sqrt{2}e^{-t}\sin(\sqrt{2}t)}{18}-\dfrac{2}{9} \end{pmatrix}$$

Solution: dsolve(.): $Y_d(t)$

$$\begin{pmatrix} \dfrac{t}{3}-\dfrac{e^t}{3}-\dfrac{7e^{-t}\cos(\sqrt{2}t)}{9}-\dfrac{8\sqrt{2}e^{-t}\sin(\sqrt{2}t)}{9}+\dfrac{1}{9} \\ \dfrac{t}{3}+\dfrac{e^t}{3}+\dfrac{8e^{-t}\cos(\sqrt{2}t)}{9}-\dfrac{7\sqrt{2}e^{-t}\sin(\sqrt{2}t)}{18}-\dfrac{2}{9} \end{pmatrix}$$

$Y_d(t)-Y_e(t) \Rightarrow \begin{pmatrix} 0 \\ 0 \end{pmatrix}$

Solution: Laplace Transform $Y_L(t)$

$$\begin{pmatrix} \dfrac{t}{3}-\dfrac{e^t}{3}-\dfrac{7e^{-t}\left(\cos(\sqrt{2}t)+\frac{8\sqrt{2}\sin(\sqrt{2}t)}{7}\right)}{9}+\dfrac{1}{9} \\ \dfrac{t}{3}+\dfrac{e^t}{3}+\dfrac{8e^{-t}\left(\cos(\sqrt{2}t)-\frac{7\sqrt{2}\sin(\sqrt{2}t)}{16}\right)}{9}-\dfrac{2}{9} \end{pmatrix}$$

$Y_L(t)-Y_e(t) \Rightarrow \begin{pmatrix} 0 \\ 0 \end{pmatrix}$

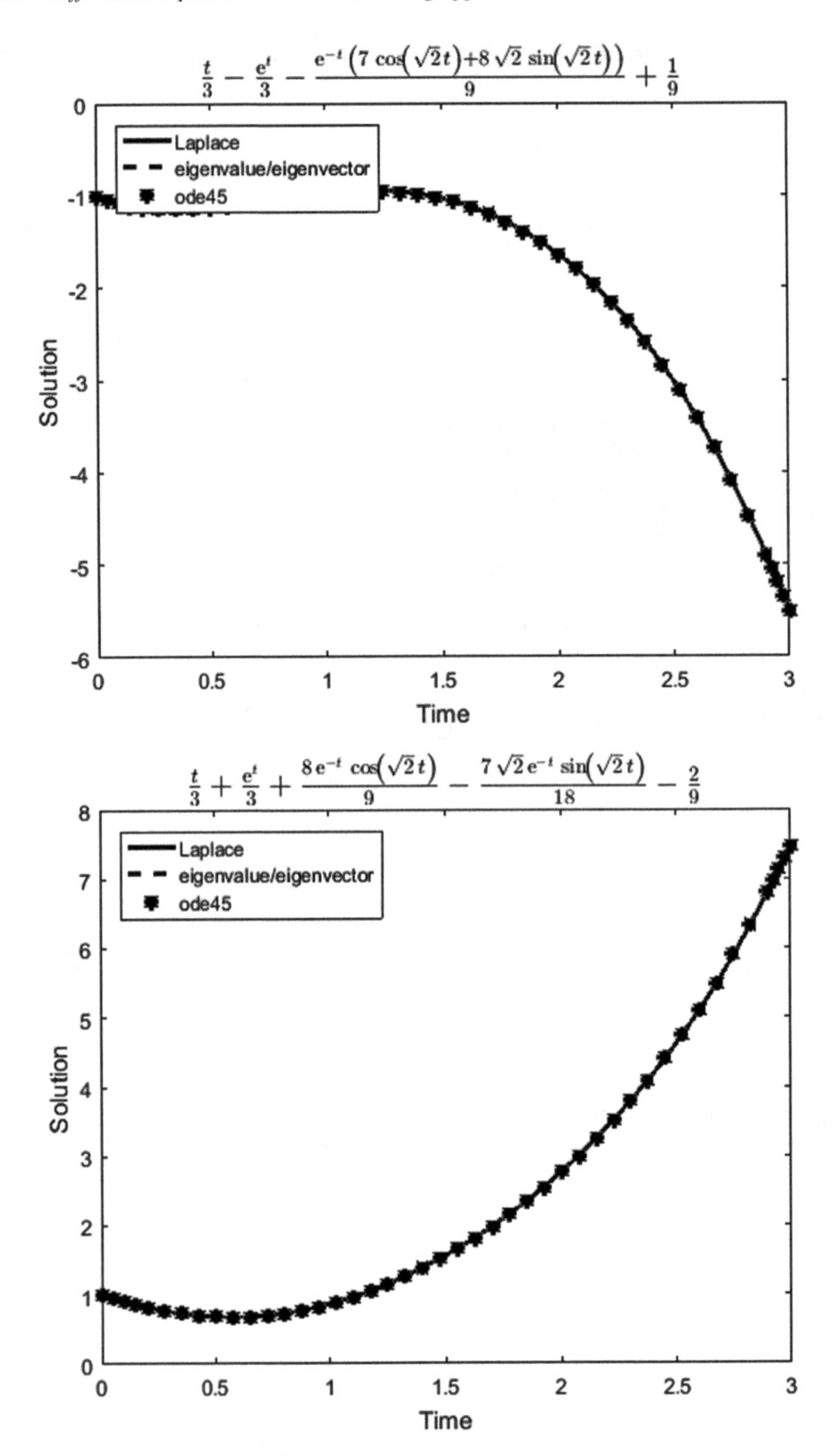
$\frac{t}{3} - \frac{e^t}{3} - \frac{e^{-t}\left(7\cos\left(\sqrt{2}t\right)+8\sqrt{2}\sin\left(\sqrt{2}t\right)\right)}{9} + \frac{1}{9}$
Laplace
eigenvalue/eigenvector
ode45
Solution
Time
0
-1
-2
-3
-4
-5
-6
0
0.5
1
1.5
2
2.5
3
$\frac{t}{3} + \frac{e^t}{3} + \frac{8e^{-t}\cos\left(\sqrt{2}t\right)}{9} - \frac{7\sqrt{2}e^{-t}\sin\left(\sqrt{2}t\right)}{18} - \frac{2}{9}$
Laplace
eigenvalue/eigenvector
ode45
Solution
Time
8
7
6
5
4
3
2
1
0
0
0.5
1
1.5
2
2.5
3

Example # 5.37

```
abc=[2,-5,1,-2,0,1];gt1='-cos(t)';gt2='sin(t)';
```

Particular and Complete Solutions
Pair of First Order Linear Differential Equations

Coefficient Matrix [A] $\begin{pmatrix} 2 & -5 \\ 1 & -2 \end{pmatrix}$ Forcing functions $\vec{g}(t)$ $\begin{pmatrix} -\cos(t) \\ \sin(t) \end{pmatrix}$ IC $\begin{pmatrix} 0 \\ 1 \end{pmatrix}$

Coupled Differential Equations $\begin{pmatrix} x'(t) = 2\,x(t) - \cos(t) - 5\,y(t) \\ y'(t) = \sin(t) + x(t) - 2\,y(t) \end{pmatrix}$

Complex eigenvalues and eigenvectors!

Eigenvalues (λ) $\begin{pmatrix} -i \\ i \end{pmatrix}$ $\Rightarrow \lambda = \alpha \pm j\beta$ Eigenvectors ($\vec{v}$) $\begin{pmatrix} 2-i & 2+i \\ 1 & 1 \end{pmatrix}$ $\Rightarrow \vec{v} = \vec{u} \pm j\vec{w}$

Fundamental Matrix $X_f(t) = \left[\vec{X}_1, \vec{X}_2\right]$ of the Homogeneous Differential Equation $A\vec{x} = 0$

$$\left[\vec{X}_1, \vec{X}_2\right] = \left[e^{\alpha t}\left[\vec{u}\,cos(\beta t) - \vec{w}\,sin(\beta t)\right], e^{\alpha t}\left[\vec{u}\,sin(\beta t) + \vec{w}\,cos(\beta t)\right]\right]$$

$$X_f(t) = \left[\vec{X}_1, \vec{X}_2\right] \Rightarrow \begin{pmatrix} 2\cos(t) - \sin(t) & -\cos(t) - 2\sin(t) \\ \cos(t) & -\sin(t) \end{pmatrix}$$

Particular Solution set

$$\begin{bmatrix} x_p(t) \\ y_p(t) \end{bmatrix} = X_f(t)\int [X_f(t)]^{-1} g(t)\,dt \Rightarrow \begin{pmatrix} 2t\cos(t) - \frac{3\sin(t)}{2} - \frac{\cos(t)}{2} - t\sin(t) \\ t\cos(t) - \frac{\sin(t)}{2} - \frac{\cos(t)}{2} \end{pmatrix}$$

General Solution set $\Rightarrow$ $\begin{bmatrix} x(t) \\ y(t) \end{bmatrix} = \begin{bmatrix} x_p(t) \\ y_p(t) \end{bmatrix} + c_1\vec{X}_1 + c_2\vec{X}_2$

Complete Solutions

Apply initial conditions $\Rightarrow \begin{pmatrix} 2c_1 - c_2 - \frac{1}{2} = 0 \\ c_1 - \frac{1}{2} = 1 \end{pmatrix} \Rightarrow \begin{bmatrix} c_1 \\ c_2 \end{bmatrix} = \begin{pmatrix} \frac{3}{2} \\ \frac{5}{2} \end{pmatrix}$

Solution Eigenvalues

$$x_e(t) = 2t\cos(t) - 8\sin(t) - t\sin(t)$$

$$y_e(t) = \cos(t) - 3\sin(t) + t\cos(t)$$

Laplace Transform Theory $\begin{bmatrix} X(s) \\ Y(s) \end{bmatrix} = \left[[sI_2 - A]^{-1}\left(\begin{pmatrix} x(0) \\ y(0) \end{pmatrix} + L[g(t)]\right)\right]$

Laplace Transform $\begin{bmatrix} X(s) \\ Y(s) \end{bmatrix} = \begin{pmatrix} -\frac{5\left(\frac{1}{s^2+1}+1\right)}{s^2+1} - \frac{s(s+2)}{(s^2+1)^2} \\ \frac{(s-2)\left(\frac{1}{s^2+1}+1\right)}{s^2+1} - \frac{s}{(s^2+1)^2} \end{pmatrix}$

Laplace Transform based solution

$$x_L(t) = 2t\cos(t) - 8\sin(t) - t\sin(t)$$

$$y_L(t) = \cos(t) - 3\sin(t) + t\cos(t)$$

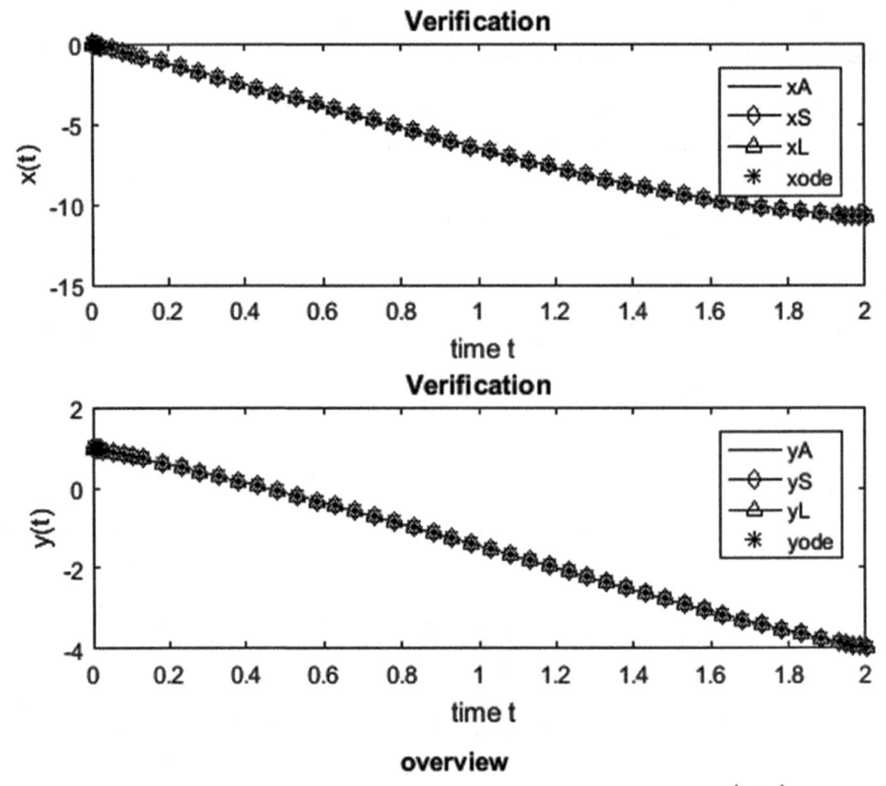

overview

Differential Equations $\begin{pmatrix} x'(t) = 2\,x(t) - \cos(t) - 5\,y(t) \\ y'(t) = \sin(t) + x(t) - 2\,y(t) \end{pmatrix}$ Initial Conditions $\begin{pmatrix} 0 \\ 1 \end{pmatrix}$

Particular Solution set $\begin{bmatrix} x_p(t) \\ y_p(t) \end{bmatrix} = \begin{pmatrix} 2t\,\cos(t) - \frac{3\,\sin(t)}{2} - \frac{\cos(t)}{2} - t\,\sin(t) \\ t\,\cos(t) - \frac{\sin(t)}{2} - \frac{\cos(t)}{2} \end{pmatrix}$

Verification by substitution in the differential Equations: Must be a NULL Vector $\begin{pmatrix} 0 \\ 0 \end{pmatrix}$

Solution set using eigenvalues & eigenvectors

$$x_e(t) = 2t\,\cos(t) - 8\,\sin(t) - t\,\sin(t)$$

$$y_e(t) = \cos(t) - 3\,\sin(t) + t\,\cos(t)$$

Solution set using dsolve(.)

$$x_d(t) = 2\,t\,\cos(t) - 8\,\sin(t) - t\,\sin(t)$$

$$y_d(t) = \cos(t) - 3\,\sin(t) + t\,\cos(t)$$

Example # 5.38

```
A=[0 -1 1;0 2 0;-2 -1 3];x0=[0 0 1];gt={'1';'t';'exp(-t)'};
```

General Solution associated with [n x n] coefficient matrix A

⇒ Matrix not defective

Eigenvalues/vectors: $\lambda_1,\ \vec{v}_1;\ \lambda_2,\ \vec{v}_2;\ \cdots;\lambda_n,\vec{v}_n$

Solution set: $\vec{X}_k = e^{\lambda_k t}\vec{v}_k,\ k = 1, 2, \cdots, n$

⇒ Matrix defective: An example with two distinct eigenvectors

Eigenvalues & distinct eigenvectors (two): $\lambda_1, \vec{v}_1$ & $\lambda_2, \vec{v}_2$

Eigenvalue (λ_3, algebraic multiplicity of $m = n - 2,\ n \geq 3$)

generalized eigenvectors: $\vec{v}_3, \vec{v}_4, \cdots, \vec{v}_n$

Solution set with distinct eigenvectors: $\vec{X}_k = \vec{v}_k e^{\lambda_k t},\ k = 1, 2$

Solution set with eigenvalue λ_3 :

$$\vec{X}_j = [I_n + t(A - \lambda_3 I_n) + \cdots + [\Gamma(m)]^{-1} t^{m-1} (A - \lambda_3 I_n)^{m-1}]\vec{v}_j e^{\lambda_3 t}$$
$$j = 3, \cdots, n;\ m = n - 2$$

⇒ Fundamental Matrix of A: $X(t) = [\vec{X}_1(t), \vec{X}_2(t), \cdots, \vec{X}_n(t)]$

Homogeneous Solution ⇒ $\vec{x}_h(t) = X(t)\vec{c}$

$\vec{c}$ (from initial conditions)

Particular and Complete Solutions

Input ⇒ $$\frac{d\vec{x}}{dt} = A\vec{x}(t) + \vec{g}(t)$$

Fundamental Matrix of A ⇒ $$X(t) = [\vec{X}_1(t), \vec{X}_2(t), \cdots, \vec{X}_n(t)]$$

Initial Conditions ⇒ $$\vec{x}(0) = \begin{pmatrix} x_1(0) \\ x_2(0) \\ \cdots \\ x_n(0) \end{pmatrix}$$

Particular Solution ⇒ $$\vec{x}_p(t) = X(t) \int [X(t)]^{-1} \vec{g}(t) dt$$

Complete Solution ⇒ $$\vec{x}(t) = X(t)\vec{c} + \vec{x}_p(t)$$

Constants ⇒ $$\vec{c} = [X(0)]^{-1} [\vec{x}(0) - \vec{x}_p(0)]$$

$$\vec{x}(t) = \begin{pmatrix} x_1(t) \\ x_2(t) \\ \cdots \\ x_n(t) \end{pmatrix}$$

Coupled 1st order System: Problem Statement

$$\frac{d\vec{x}}{dt} = A\vec{x}(t) + \vec{g}(t)$$

Coefficient Matrix: A $\begin{pmatrix} 0 & -1 & 1 \\ 0 & 2 & 0 \\ -2 & -1 & 3 \end{pmatrix}$

Forcing functions $\vec{g}(t)$ $\begin{pmatrix} 1 \\ t \\ e^{-t} \end{pmatrix}$ Initial Conditions: $\vec{x}(0)$ $\begin{pmatrix} 0 \\ 0 \\ 1 \end{pmatrix}$

Differential Equations: $\frac{d\vec{x}}{dt}$ $\begin{pmatrix} x_1'(t) = x_3(t) - x_2(t) + 1 \\ x_2'(t) = t + 2\,x_2(t) \\ x_3'(t) = e^{-t} - 2\,x_1(t) - x_2(t) + 3\,x_3(t) \end{pmatrix}$

Eigenvalues and Eigenvectors

Input Matrix A

0 -1 1
0 2 0
-2 -1 3

	1		1	-0.5	0.5
Eigenvalues	2	Eigenvectors	0	1	0
	2		1	0	1

Eigenvalues (3) & Eigenvectors (3): Matrix NOT defective

General solutions and unknown constants

Homogeneous solutions set

$$\begin{pmatrix} \frac{e^t (2c_1 - c_2 e^t + c_3 e^t)}{2} \\ c_2 e^{2t} \\ e^t (c_1 + c_3 e^t) \end{pmatrix}$$

Complete solutions set (homogeneous + particular)

$$\begin{pmatrix} \frac{e^{-t}}{6} - \frac{t}{2} + \frac{e^t (2c_1 - c_2 e^t + c_3 e^t)}{2} - \frac{9}{4} \\ c_2 e^{2t} - \frac{t}{2} - \frac{1}{4} \\ e^t (c_1 + c_3 e^t) - \frac{e^{-t}}{6} - \frac{t}{2} - \frac{7}{4} \end{pmatrix}$$

constants c_1, c_2 and c_3

$$\begin{pmatrix} \frac{3}{2} & \frac{1}{4} & \frac{17}{12} \end{pmatrix}$$

Fundamental Matrix, Particular & Complete Solutions

Fundamental Matrix X(t):

$$\begin{pmatrix} e^t & -\frac{e^{2t}}{2} & \frac{e^{2t}}{2} \\ 0 & e^{2t} & 0 \\ e^t & 0 & e^{2t} \end{pmatrix}$$

Particular Solution: $\vec{Y}_p(t)$

$$\begin{matrix} \frac{e^{-t}}{6} - \frac{t}{2} - \frac{9}{4} \\ -\frac{t}{2} - \frac{1}{4} \\ -\frac{t}{2} - \frac{e^{-t}}{6} - \frac{7}{4} \end{matrix}$$

Complete Solution: $\vec{Y}$(t)

$$\begin{matrix} \frac{e^{-t}}{6} - \frac{t}{2} + \frac{7e^{2t}}{12} + \frac{3e^t}{2} - \frac{9}{4} \\ \frac{e^{2t}}{4} - \frac{t}{2} - \frac{1}{4} \\ \frac{17e^{2t}}{12} - \frac{e^{-t}}{6} - \frac{t}{2} + \frac{3e^t}{2} - \frac{7}{4} \end{matrix}$$

Coupled 1st Order Differential Equations: Laplace Transforms

Input⇒

$$\begin{bmatrix} x\prime_1(t) \\ x\prime_2(t) \\ \dots \\ x\prime_n(t) \end{bmatrix} = A\begin{pmatrix} x_1(t) \\ x_2(t) \\ \dots \\ x_n(t) \end{pmatrix} + \begin{pmatrix} g_1(t) \\ g_2(t) \\ \dots \\ g_n(t) \end{pmatrix}$$

Initial⇒
Conditions

$$\begin{pmatrix} x_1(0) \\ x_2(0) \\ \dots \\ x_n(0) \end{pmatrix}$$

Laplace⇒
Transform

$$\begin{bmatrix} X_1(s) \\ X_2(s) \\ \dots \\ X_n(s) \end{bmatrix} = [sI_n - A]^{-1}\begin{pmatrix} x_1(0) \\ x_2(0) \\ \dots \\ x_n(0) \end{pmatrix} + [sI_n - A]^{-1}\begin{pmatrix} G_1(s) \\ G_2(s) \\ \dots \\ G_n(0) \end{pmatrix}$$

Solution⇒

$$\begin{pmatrix} x_1(t) \\ x_2(t) \\ \dots \\ x_n(t) \end{pmatrix} = L^{-1}\begin{bmatrix} X_1(s) \\ X_2(s) \\ \dots \\ X_n(s) \end{bmatrix}$$

Laplace Transforms and Solutions using inverse Laplace Transforms

Laplace
Transform
Y(s)

$$-\frac{-2\,s^3+4\,s+1}{s^2\,(s-1)\,(s+1)\,(s-2)}$$

$$\frac{1}{s^2\,(s-2)}$$

$$-\frac{-s^4-2\,s^3+2\,s^2+3\,s+1}{s^2\,(s-1)\,(s+1)\,(s-2)}$$

Solution:
Inverse
Laplace
Transform
Y(t)

$$\frac{e^{-t}}{6} - \frac{t}{2} + \frac{7e^{2t}}{12} + \frac{3\,e^{t}}{2} - \frac{9}{4}$$

$$\frac{e^{2t}}{4} - \frac{t}{2} - \frac{1}{4}$$

$$\frac{17e^{2t}}{12} - \frac{e^{-t}}{6} - \frac{t}{2} + \frac{3e^{t}}{2} - \frac{7}{4}$$

Comparison of Analytical Solutions

Solution:
Eigenvalues
& Eigenvectors: $Y_e(t)$

$$\begin{array}{l} \frac{e^{-t}}{6} - \frac{t}{2} + \frac{7e^{2t}}{12} + \frac{3e^{t}}{2} - \frac{9}{4} \\ \frac{e^{2t}}{4} - \frac{t}{2} - \frac{1}{4} \\ \frac{17e^{2t}}{12} - \frac{e^{-t}}{6} - \frac{t}{2} + \frac{3e^{t}}{2} - \frac{7}{4} \end{array}$$

Solution:
dsolve(.): $Y_d(t)$

$$\begin{array}{l} \frac{e^{-t}}{6} - \frac{t}{2} + \frac{7e^{2t}}{12} + \frac{3e^{t}}{2} - \frac{9}{4} \\ \frac{e^{2t}}{4} - \frac{t}{2} - \frac{1}{4} \\ \frac{17e^{2t}}{12} - \frac{e^{-t}}{6} - \frac{t}{2} + \frac{3e^{t}}{2} - \frac{7}{4} \end{array}$$

$Y_d(t)-Y_e(t) \Rightarrow \begin{pmatrix} 0 \\ 0 \\ 0 \end{pmatrix}$

Solution:
Laplace Transform
$Y_L(t)$

$$\begin{array}{l} \frac{e^{-t}}{6} - \frac{t}{2} + \frac{7e^{2t}}{12} + \frac{3e^{t}}{2} - \frac{9}{4} \\ \frac{e^{2t}}{4} - \frac{t}{2} - \frac{1}{4} \\ \frac{17e^{2t}}{12} - \frac{e^{-t}}{6} - \frac{t}{2} + \frac{3e^{t}}{2} - \frac{7}{4} \end{array}$$

$Y_L(t)-Y_e(t) \Rightarrow \begin{pmatrix} 0 \\ 0 \\ 0 \end{pmatrix}$

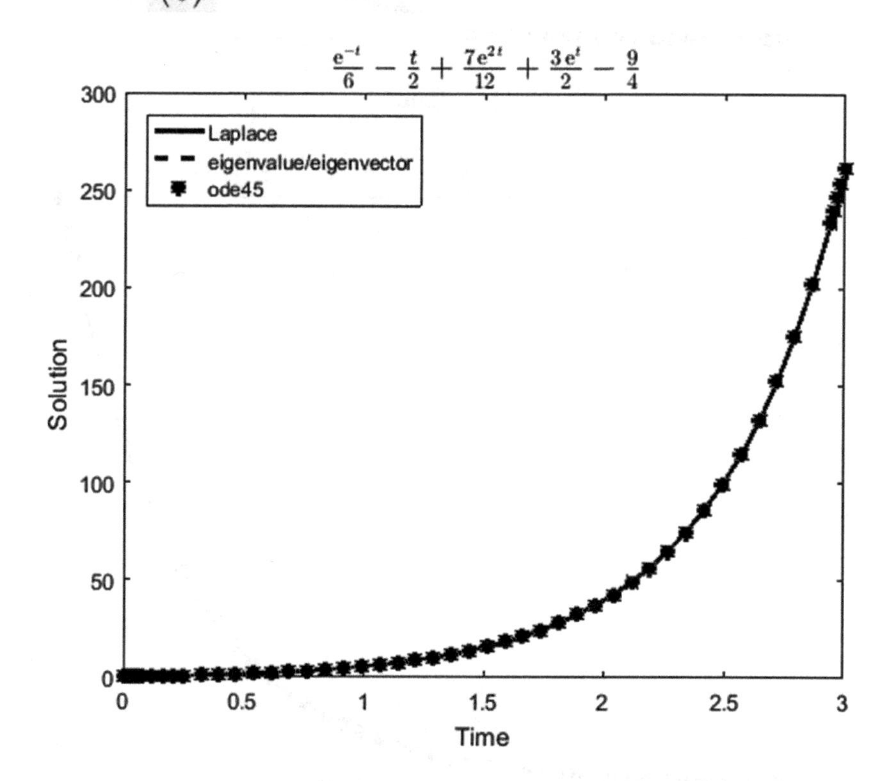

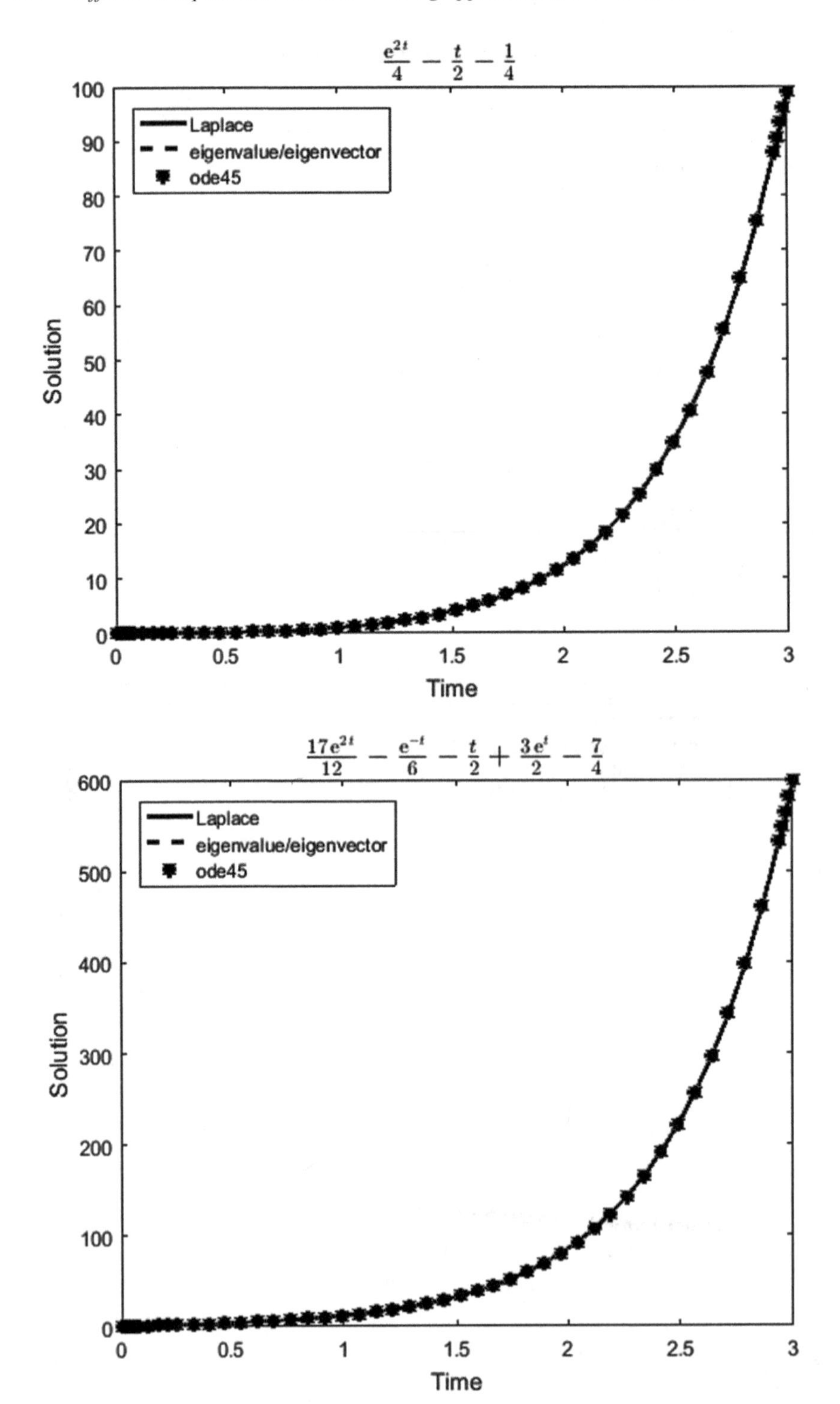

$\frac{e^{2t}}{4} - \frac{t}{2} - \frac{1}{4}$
Laplace
eigenvalue/eigenvector
ode45
Solution
Time
0
10
20
30
40
50
60
70
80
90
100
0.5
1
1.5
2
2.5
3
$\frac{17e^{2t}}{12} - \frac{e^{-t}}{6} - \frac{t}{2} + \frac{3e^{t}}{2} - \frac{7}{4}$
Laplace
eigenvalue/eigenvector
ode45
Solution
Time
0
100
200
300
400
500
600
0.5
1
1.5
2
2.5
3

Example # 5.39

```
A=[2,-1,-1;-1,2,-1;-1,-1,2];x0=[0 -1 1];gt={'t';'t^2';'exp(t)'};
```

General Solution associated with [n x n] coefficient matrix A

⇒ Matrix not defective

Eigenvalues/vectors: $\lambda_1,\ \vec{v}_1;\ \lambda_2,\ \vec{v}_2;\ \cdots;\lambda_n,\vec{v}_n$

Solution set: $\vec{X}_k = e^{\lambda_k t}\vec{v}_k,\ \ k = 1, 2, \cdots, n$

⇒ Matrix defective: An example with two distinct eigenvectors

Eigenvalues & distinct eigenvectors (two): $\lambda_1, \vec{v}_1$ & $\lambda_2, \vec{v}_2$

Eigenvalue (λ_3, algebraic multiplicity of $m = n - 2,\ n \geq 3$)

generalized eigenvectors: $\vec{v}_3, \vec{v}_4, \cdots, \vec{v}_n$

Solution set with distinct eigenvectors: $\vec{X}_k = \vec{v}_k e^{\lambda_k t},\ \ k = 1, 2$

Solution set with eigenvalue λ_3 :

$$\vec{X}_j = [I_n + t(A - \lambda_3 I_n) + \cdots + [\Gamma(m)]^{-1} t^{m-1}(A - \lambda_3 I_n)^{m-1}]\vec{v}_j e^{\lambda_3 t}$$
$$j = 3, \cdots, n;\ \ m = n - 2$$

⇒ Fundamental Matrix of A: $X(t) = [\vec{X}_1(t), \vec{X}_2(t), \cdots, \vec{X}_n(t)]$

Homogeneous Solution ⇒ $\vec{x}_h(t) = X(t)\vec{c}$

$\vec{c}$ (from initial conditions)

Particular and Complete Solutions

Input ⇒ $$\frac{d\vec{x}}{dt} = A\vec{x}(t) + \vec{g}(t)$$

Fundamental Matrix of A ⇒ $$X(t) = [\vec{X}_1(t), \vec{X}_2(t), \cdots, \vec{X}_n(t)]$$

Initial Conditions ⇒ $$\vec{x}(0) = \begin{pmatrix} x_1(0) \\ x_2(0) \\ \cdots \\ x_n(0) \end{pmatrix}$$

Particular Solution ⇒ $$\vec{x}_p(t) = X(t)\int [X(t)]^{-1}\vec{g}(t)dt$$

Complete Solution ⇒ $$\vec{x}(t) = X(t)\vec{c} + \vec{x}_p(t)$$

Constants ⇒ $$\vec{c} = [X(0)]^{-1}\,[\vec{x}(0) - \vec{x}_p(0)]$$

$$\vec{x}(t) = \begin{pmatrix} x_1(t) \\ x_2(t) \\ \cdots \\ x_n(t) \end{pmatrix}$$

Coupled 1st order System: Problem Statement

$$\frac{d\vec{x}}{dt} = A\vec{x}(t) + \vec{g}(t)$$

Coefficient Matrix: A $\begin{pmatrix} 2 & -1 & -1 \\ -1 & 2 & -1 \\ -1 & -1 & 2 \end{pmatrix}$

Forcing functions $\vec{g}(t)$ $\begin{pmatrix} t \\ t^2 \\ e^t \end{pmatrix}$ Initial Conditions: $\vec{x}(0)$ $\begin{pmatrix} 0 \\ -1 \\ 1 \end{pmatrix}$

Differential Equations: $\frac{d\vec{x}}{dt}$ $\begin{pmatrix} x_1'(t) = t + 2x_1(t) - x_2(t) - x_3(t) \\ x_2'(t) = 2x_2(t) - x_1(t) - x_3(t) + t^2 \\ x_3'(t) = e^t - x_1(t) - x_2(t) + 2x_3(t) \end{pmatrix}$

Eigenvalues and Eigenvectors

Input Matrix A

2 -1 -1
-1 2 -1
-1 -1 2

	0		1	-1	-1
Eigenvalues	3	Eigenvectors	1	1	0
	3		1	0	1

Eigenvalues (3) & Eigenvectors (3): Matrix NOT defective

General solutions and unknown constants

Homogeneous solutions set

$$\begin{pmatrix} c_1 - c_2\, e^{3t} - c_3\, e^{3t} \\ c_1 + c_2\, e^{3t} \\ c_1 + c_3\, e^{3t} \end{pmatrix}$$

Complete solutions set (homogeneous + particular)

$$\begin{pmatrix} c_1 - \frac{4t}{27} + \frac{e^t}{2} + \frac{5t^2}{18} + \frac{t^3}{9} - c_2\, e^{3t} - c_3\, e^{3t} - \frac{4}{81} \\ c_1 - \frac{t}{27} + \frac{e^t}{2} - \frac{t^2}{18} + \frac{t^3}{9} + c_2\, e^{3t} - \frac{1}{81} \\ c_1 + \frac{5t}{27} + \frac{5t^2}{18} + \frac{t^3}{9} + c_3\, e^{3t} + \frac{5}{81} \end{pmatrix}$$

constants c_1, c_2 and c_3

$$\begin{pmatrix} -\frac{1}{3} & -\frac{187}{162} & \frac{103}{81} \end{pmatrix}$$

Fundamental Matrix, Particular & Complete Solutions

Fundamental Matrix X(t):

$$\begin{pmatrix} 1 & -e^{3t} & -e^{3t} \\ 1 & e^{3t} & 0 \\ 1 & 0 & e^{3t} \end{pmatrix}$$

Particular Solution: $\vec{Y}_p(t)$

$$\begin{matrix} \frac{e^t}{2} - \frac{4t}{27} + \frac{5t^2}{18} + \frac{t^3}{9} - \frac{4}{81} \\ \frac{e^t}{2} - \frac{t}{27} - \frac{t^2}{18} + \frac{t^3}{9} - \frac{1}{81} \\ \frac{t^3}{9} + \frac{5t^2}{18} + \frac{5t}{27} + \frac{5}{81} \end{matrix}$$

Complete Solution: $\vec{Y}$(t)

$$\begin{matrix} \frac{e^t}{2} - \frac{19\, e^{3t}}{162} - \frac{4t}{27} + \frac{5t^2}{18} + \frac{t^3}{9} - \frac{31}{81} \\ \frac{e^t}{2} - \frac{187\, e^{3t}}{162} - \frac{t}{27} - \frac{t^2}{18} + \frac{t^3}{9} - \frac{28}{81} \\ \frac{5t}{27} + \frac{103\, e^{3t}}{81} + \frac{5t^2}{18} + \frac{t^3}{9} - \frac{22}{81} \end{matrix}$$

Coupled 1st Order Differential Equations: Laplace Transforms

Input⇒
$$\begin{bmatrix} x'_1(t) \\ x'_2(t) \\ \dots \\ x'_n(t) \end{bmatrix} = A\begin{pmatrix} x_1(t) \\ x_2(t) \\ \dots \\ x_n(t) \end{pmatrix} + \begin{pmatrix} g_1(t) \\ g_2(t) \\ \dots \\ g_n(t) \end{pmatrix}$$

Initial⇒ Conditions
$$\begin{pmatrix} x_1(0) \\ x_2(0) \\ \dots \\ x_n(0) \end{pmatrix}$$

Laplace⇒ Transform
$$\begin{bmatrix} X_1(s) \\ X_2(s) \\ \dots \\ X_n(s) \end{bmatrix} = [sI_n - A]^{-1}\begin{pmatrix} x_1(0) \\ x_2(0) \\ \dots \\ x_n(0) \end{pmatrix} + [sI_n - A]^{-1}\begin{pmatrix} G_1(s) \\ G_2(s) \\ \dots \\ G_n(0) \end{pmatrix}$$

Solution⇒
$$\begin{pmatrix} x_1(t) \\ x_2(t) \\ \dots \\ x_n(t) \end{pmatrix} = L^{-1}\begin{bmatrix} X_1(s) \\ X_2(s) \\ \dots \\ X_n(s) \end{bmatrix}$$

Laplace Transforms and Solutions using inverse Laplace Transforms

Laplace Transform Y(s)

$$-\frac{2s^2+s-2}{s^4(s-1)(s-3)}$$

$$-\frac{s^5-s^4+s^3-s^2+3s-2}{s^4(s-1)(s-3)}$$

$$-\frac{-s^4-s^3+s+2}{s^4(s-3)}$$

Solution: Inverse Laplace Transform Y(t)

$$\frac{e^t}{2} - \frac{19e^{3t}}{162} - \frac{4t}{27} + \frac{5t^2}{18} + \frac{t^3}{9} - \frac{31}{81}$$

$$\frac{e^t}{2} - \frac{187e^{3t}}{162} - \frac{t}{27} - \frac{t^2}{18} + \frac{t^3}{9} - \frac{28}{81}$$

$$\frac{5t}{27} + \frac{103e^{3t}}{81} + \frac{5t^2}{18} + \frac{t^3}{9} - \frac{22}{81}$$

Comparison of Analytical Solutions

Solution:
Eigenvalues
& Eigenvectors: $Y_e(t)$

$$\frac{e^t}{2} - \frac{19e^{3t}}{162} - \frac{4t}{27} + \frac{5t^2}{18} + \frac{t^3}{9} - \frac{31}{81}$$
$$\frac{e^t}{2} - \frac{187e^{3t}}{162} - \frac{t}{27} - \frac{t^2}{18} + \frac{t^3}{9} - \frac{28}{81}$$
$$\frac{5t}{27} + \frac{103e^{3t}}{81} + \frac{5t^2}{18} + \frac{t^3}{9} - \frac{22}{81}$$

Solution:
dsolve(.): $Y_d(t)$

$$\frac{e^t}{2} - \frac{19e^{3t}}{162} - \frac{4t}{27} + \frac{5t^2}{18} + \frac{t^3}{9} - \frac{31}{81}$$
$$\frac{e^t}{2} - \frac{187e^{3t}}{162} - \frac{t}{27} - \frac{t^2}{18} + \frac{t^3}{9} - \frac{28}{81}$$
$$\frac{5t}{27} + \frac{103e^{3t}}{81} + \frac{5t^2}{18} + \frac{t^3}{9} - \frac{22}{81}$$

$$Y_d(t)-Y_e(t) \Rightarrow \begin{pmatrix} 0 \\ 0 \\ 0 \end{pmatrix}$$

Solution:
Laplace Transform
$Y_L(t)$

$$\frac{e^t}{2} - \frac{19e^{3t}}{162} - \frac{4t}{27} + \frac{5t^2}{18} + \frac{t^3}{9} - \frac{31}{81}$$
$$\frac{e^t}{2} - \frac{187e^{3t}}{162} - \frac{t}{27} - \frac{t^2}{18} + \frac{t^3}{9} - \frac{28}{81}$$
$$\frac{5t}{27} + \frac{103e^{3t}}{81} + \frac{5t^2}{18} + \frac{t^3}{9} - \frac{22}{81}$$

$$Y_L(t)-Y_e(t) \Rightarrow \begin{pmatrix} 0 \\ 0 \\ 0 \end{pmatrix}$$

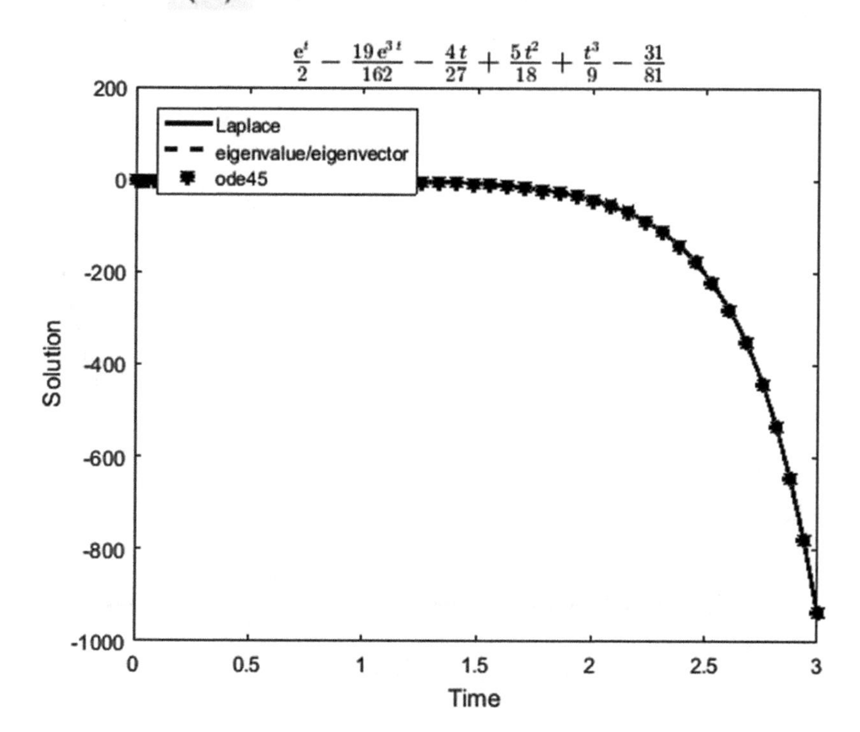

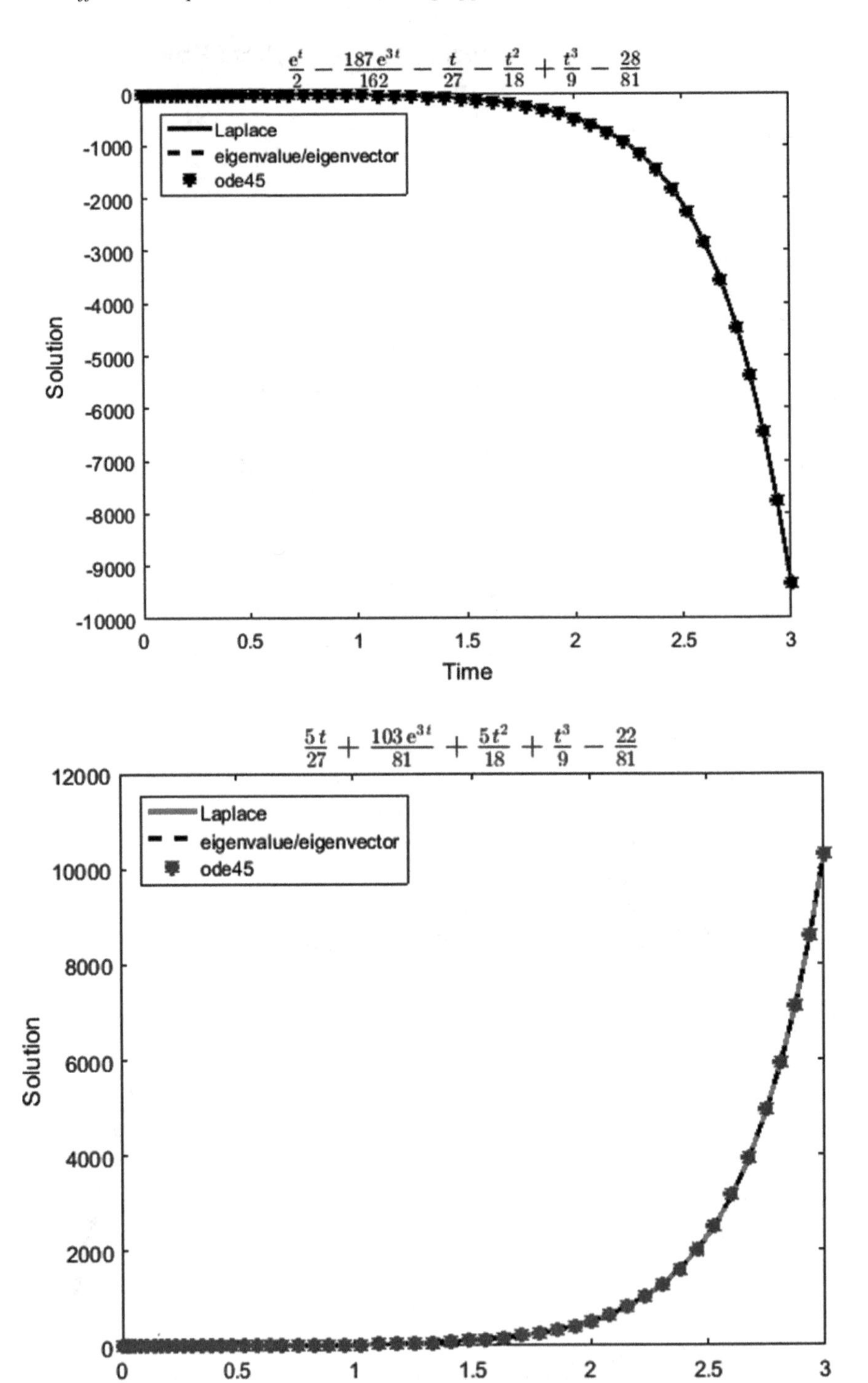

$\frac{e^t}{2} - \frac{187\,e^{3t}}{162} - \frac{t}{27} - \frac{t^2}{18} + \frac{t^3}{9} - \frac{28}{81}$
Laplace
eigenvalue/eigenvector
ode45
Solution
Time
0
-1000
-2000
-3000
-4000
-5000
-6000
-7000
-8000
-9000
-10000
0.5
1
1.5
2
2.5
3
$\frac{5t}{27} + \frac{103\,e^{3t}}{81} + \frac{5t^2}{18} + \frac{t^3}{9} - \frac{22}{81}$
Laplace
eigenvalue/eigenvector
ode45
Solution
Time
12000
10000
8000
6000
4000
2000
0
0.5
1
1.5
2
2.5
3

For the remaining examples, the first two displays containing the theoretical aspects are not shown.

Example # 5.40

```
A=[-1,-1,1;-1,2,-1;-1,1,0];x0=[0 1 1];gt={'t*exp(-t)';'exp(-t)';'exp(t)'};
```

Coupled 1st order System: Problem Statement

$$\frac{d\vec{x}}{dt} = A\vec{x}(t) + \vec{g}(t)$$

Coefficient Matrix: A $\begin{pmatrix} -1 & -1 & 1 \\ -1 & 2 & -1 \\ -1 & 1 & 0 \end{pmatrix}$

Forcing functions $\vec{g}(t)$ $\begin{pmatrix} t\,e^{-t} \\ e^{-t} \\ e^{t} \end{pmatrix}$ Initial Conditions: $\vec{x}(0)$ $\begin{pmatrix} 0 \\ 1 \\ 1 \end{pmatrix}$

Differential Equations: $\frac{d\vec{x}}{dt}$ $\begin{pmatrix} x_1'(t) = x_3(t) - x_2(t) - x_1(t) + t\,e^{-t} \\ x_2'(t) = e^{-t} - x_1(t) + 2\,x_2(t) - x_3(t) \\ x_3'(t) = e^{t} - x_1(t) + x_2(t) \end{pmatrix}$

Eigenvalues and Eigenvectors

Input Matrix A

-1 -1 1
-1 2 -1
-1 1 0

	-1		2	0
Eigenvalues	1	Eigenvectors	1	1
	1		1	1

Eigenvalues (3) & Eigenvectors (2) ⇒ Matrix DEFECTIVE:
Generalized eigenvectors required!

	2	-0.5	0.5
Eigenvectors	1	1	0
incl. Generalized	1	0	1

Algebraic multiplicity (p) of the eigenvalue λ > geometric multiplicity (m)

⇒ Generalized Eigenvectors are solutions of $[\,(A - \lambda I_n)^p\,]v = 0$

λ = -1

λ=1: p = 2

General solutions and unknown constants

Homogeneous solutions set

$$\begin{pmatrix} \frac{c_3 e^t}{2} - \frac{c_2 e^t}{2} + 2\,c_1\,e^{-t} \\ c_1\,e^{-t} - \frac{3\,c_3\,t\,e^t}{2} + \frac{c_2\,e^t\,(3t+2)}{2} \\ c_1\,e^{-t} + \frac{3\,c_2\,t\,e^t}{2} - \frac{c_3\,e^t\,(3t-2)}{2} \end{pmatrix}$$

Complete solutions set (homogeneous + particular)

$$\begin{pmatrix} \frac{e^{-t}\left(8\,c_1+2\,t-e^{2\,t}+2\,t\,e^{2\,t}+2\,t^2-2\,c_2\,e^{2\,t}+2\,c_3\,e^{2\,t}+1\right)}{4} \\ \frac{e^{-t}\left(4\,t-e^{2\,t}+2\,t\,e^{2\,t}-6\,t^2\,e^{2\,t}+2\,t^2+1\right)}{8} + c_1\,e^{-t} - \frac{3\,c_3\,t\,e^t}{2} + \frac{c_2\,e^t\,(3\,t+2)}{2} \\ \frac{e^{-t}\left(4\,t-e^{2\,t}+10\,t\,e^{2\,t}-6\,t^2\,e^{2\,t}+2\,t^2+5\right)}{8} + c_1\,e^{-t} + \frac{3\,c_2\,t\,e^t}{2} - \frac{c_3\,e^t\,(3\,t-2)}{2} \end{pmatrix}$$

constants c_1, c_2 and c_3

$$\begin{pmatrix} \frac{1}{8} & \frac{7}{8} & \frac{3}{8} \end{pmatrix}$$

Fundamental Matrix, Particular & Complete Solutions

Fundamental Matrix X(t):

$$\begin{pmatrix} 2\,e^{-t} & -\frac{e^t}{2} & \frac{e^t}{2} \\ e^{-t} & e^t\left(\frac{3t}{2}+1\right) & -\frac{3\,t\,e^t}{2} \\ e^{-t} & \frac{3\,t\,e^t}{2} & -e^t\left(\frac{3\,t}{2}-1\right) \end{pmatrix}$$

Particular Solution: $\vec{Y}_p(t)$

$$\begin{matrix} \frac{e^{-t}\left(2\,t-e^{2\,t}+2\,t\,e^{2\,t}+2\,t^2+1\right)}{4} \\ \frac{e^{-t}\left(4\,t-e^{2\,t}+2\,t\,e^{2\,t}-6\,t^2\,e^{2\,t}+2\,t^2+1\right)}{8} \\ \frac{e^{-t}\left(4\,t-e^{2\,t}+10\,t\,e^{2\,t}-6\,t^2\,e^{2\,t}+2\,t^2+5\right)}{8} \end{matrix}$$

Complete Solution: $\vec{Y}$(t)

$$\begin{matrix} \frac{e^{-t}\left(t-e^{2\,t}+t\,e^{2\,t}+t^2+1\right)}{2} \\ \frac{e^{-t}\left(2\,t+3\,e^{2\,t}+4\,t\,e^{2\,t}-3\,t^2\,e^{2\,t}+t^2+1\right)}{4} \\ \frac{e^{-t}\left(2\,t+e^{2\,t}+8\,t\,e^{2\,t}-3\,t^2\,e^{2\,t}+t^2+3\right)}{4} \end{matrix}$$

Coupled 1st Order Differential Equations: Laplace Transforms

Input⇒

$$\begin{bmatrix} x'_1(t) \\ x'_2(t) \\ \ldots \\ x'_n(t) \end{bmatrix} = A \begin{pmatrix} x_1(t) \\ x_2(t) \\ \ldots \\ x_n(t) \end{pmatrix} + \begin{pmatrix} g_1(t) \\ g_2(t) \\ \ldots \\ g_n(t) \end{pmatrix}$$

Initial⇒
Conditions

$$\begin{pmatrix} x_1(0) \\ x_2(0) \\ \ldots \\ x_n(0) \end{pmatrix}$$

Laplace⇒
Transform

$$\begin{bmatrix} X_1(s) \\ X_2(s) \\ \ldots \\ X_n(s) \end{bmatrix} = [sI_n - A]^{-1} \begin{pmatrix} x_1(0) \\ x_2(0) \\ \ldots \\ x_n(0) \end{pmatrix} + [sI_n - A]^{-1} \begin{pmatrix} G_1(s) \\ G_2(s) \\ \ldots \\ G_n(0) \end{pmatrix}$$

Solution⇒

$$\begin{pmatrix} x_1(t) \\ x_2(t) \\ \ldots \\ x_n(t) \end{pmatrix} = L^{-1} \begin{bmatrix} X_1(s) \\ X_2(s) \\ \ldots \\ X_n(s) \end{bmatrix}$$

Laplace Transforms and Solutions using inverse Laplace Transforms

Laplace
Transform
Y(s)

$$\frac{s^2+3}{(s-1)^2\,(s+1)^3}$$

$$-\frac{-s^5-2\,s^4+2\,s^3+7\,s^2+3\,s+3}{(s^2-1)^3}$$

$$-\frac{-s^5-2\,s^4+5\,s^2+5\,s+5}{(s^2-1)^3}$$

Solution:
Inverse
Laplace
Transform
Y(t)

$$\frac{e^{-t}\left(t-e^{2t}+t\,e^{2t}+t^2+1\right)}{2}$$

$$\frac{e^{-t}\left(2\,t+3e^{2t}+4\,te^{2t}-3\,t^2\,e^{2t}+t^2+1\right)}{4}$$

$$\frac{e^{-t}\left(2\,t+e^{2t}+8\,t\,e^{2t}-3\,t^2\,e^{2t}+t^2+3\right)}{4}$$

Comparison of Analytical Solutions

Solution:
Eigenvalues
& Eigenvectors: $Y_e(t)$

$$\begin{pmatrix} \frac{e^{-t}\left(t-e^{2t}+t\,e^{2t}+t^2+1\right)}{2} \\ \frac{e^{-t}\left(2t+3\,e^{2t}+4\,t\,e^{2t}-3\,t^2\,e^{2t}+t^2+1\right)}{4} \\ \frac{e^{-t}\left(2t+e^{2t}+8\,t\,e^{2t}-3t^2\,e^{2t}+t^2+3\right)}{4} \end{pmatrix}$$

Solution:
dsolve(.): $Y_d(t)$

$$\begin{pmatrix} \frac{e^{-t}\left(t-e^{2t}+t\,e^{2t}+t^2+1\right)}{2} \\ \frac{e^{-t}\left(2t+3\,e^{2t}+4\,t\,e^{2t}-3\,t^2\,e^{2t}+t^2+1\right)}{4} \\ \frac{e^{-t}\left(2t+e^{2t}+8\,t\,e^{2t}-3t^2\,e^{2t}+t^2+3\right)}{4} \end{pmatrix}$$

$Y_d(t)-Y_e(t) \Rightarrow \begin{pmatrix} 0 \\ 0 \\ 0 \end{pmatrix}$

Solution:
Laplace Transform
$Y_L(t)$

$$\begin{pmatrix} \frac{e^{-t}\left(t-e^{2t}+t\,e^{2t}+t^2+1\right)}{2} \\ \frac{e^{-t}\left(2t+3\,e^{2t}+4\,t\,e^{2t}-3\,t^2\,e^{2t}+t^2+1\right)}{4} \\ \frac{e^{-t}\left(2t+e^{2t}+8\,t\,e^{2t}-3t^2\,e^{2t}+t^2+3\right)}{4} \end{pmatrix}$$

$Y_L(t)-Y_e(t) \Rightarrow \begin{pmatrix} 0 \\ 0 \\ 0 \end{pmatrix}$

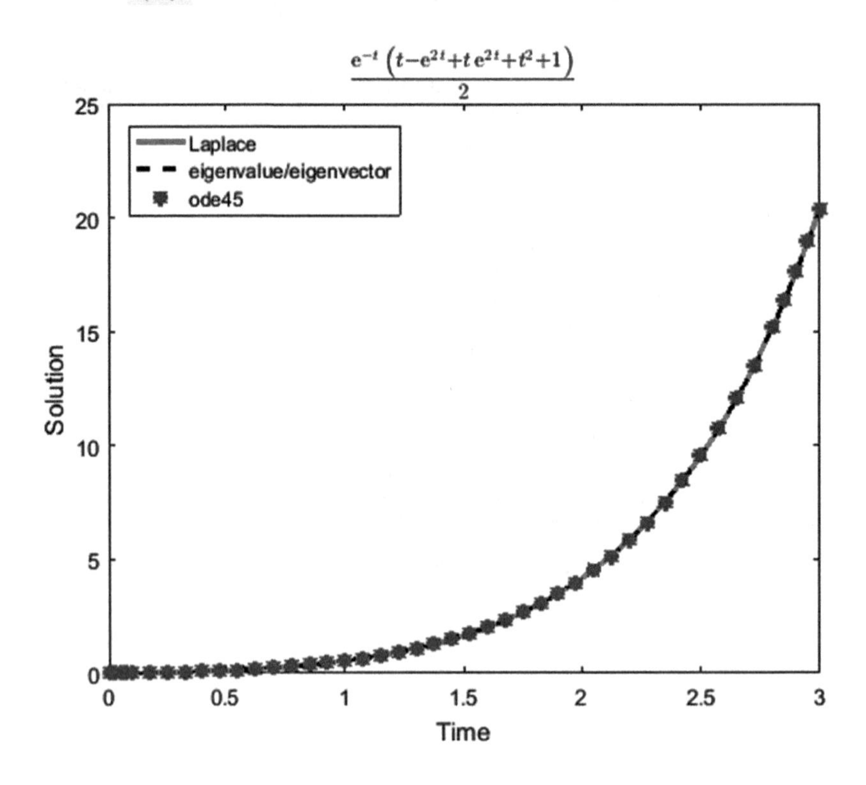

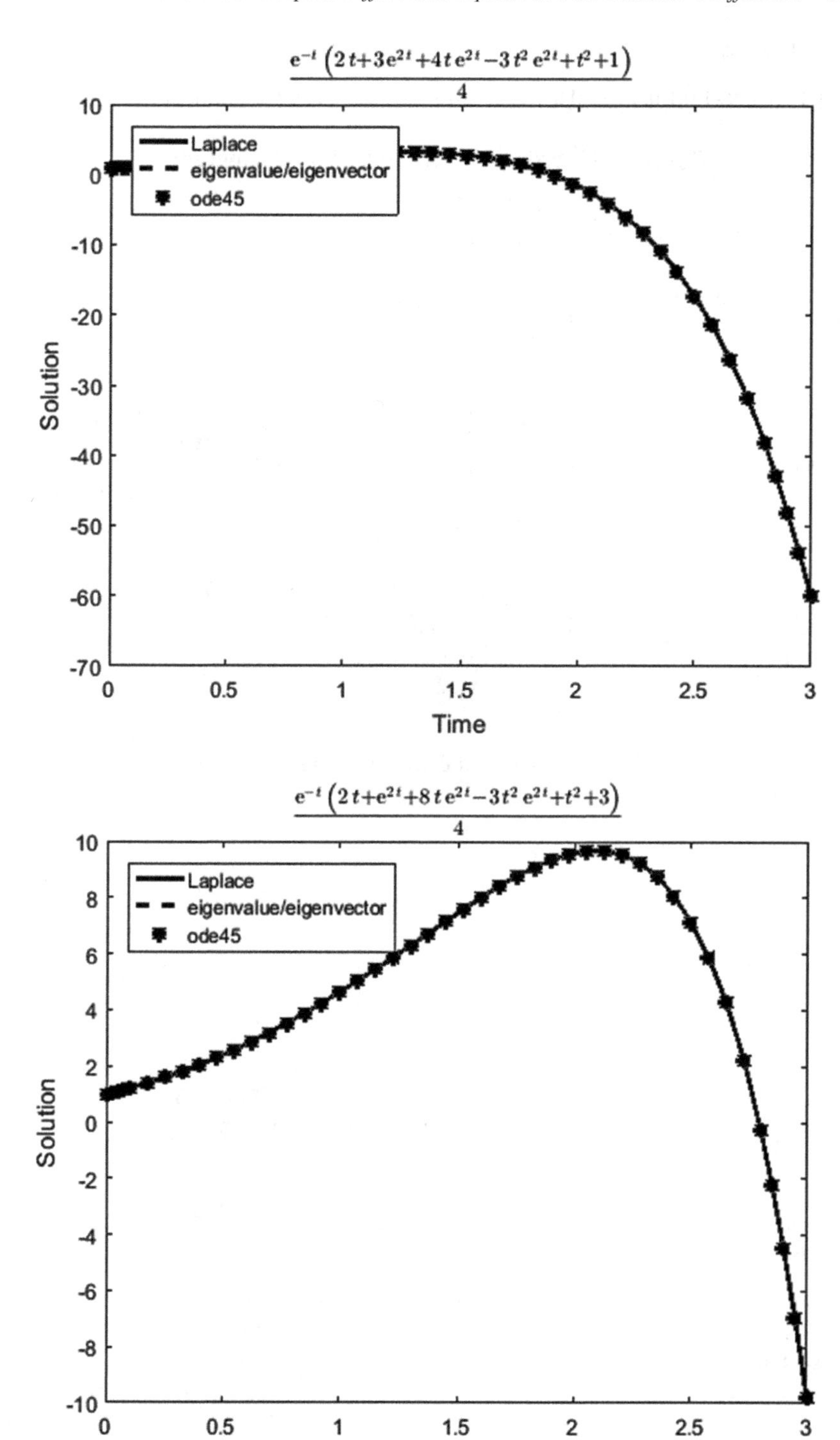
$\frac{e^{-t}\left(2t+3e^{2t}+4te^{2t}-3t^2e^{2t}+t^2+1\right)}{4}$
Laplace
eigenvalue/eigenvector
ode45
Solution
Time
10
0
-10
-20
-30
-40
-50
-60
-70
0
0.5
1
1.5
2
2.5
3
$\frac{e^{-t}\left(2t+e^{2t}+8te^{2t}-3t^2e^{2t}+t^2+3\right)}{4}$
Laplace
eigenvalue/eigenvector
ode45
Solution
Time
10
8
6
4
2
0
-2
-4
-6
-8
-10
0
0.5
1
1.5
2
2.5
3

Example # 5.41

```
A=[-1,0,0,0;0,-1,0,1;0,0,1,-1;0,0,0,1];x0=[0 1 1 -1];gt={'exp(t)';'t*exp(-t)';'exp(-t)';'t'};
```

Coupled 1st order System: Problem Statement

$$\frac{d\vec{x}}{dt} = A\vec{x}(t) + \vec{g}(t)$$

Coefficient Matrix: A $\begin{pmatrix} -1 & 0 & 0 & 0 \\ 0 & -1 & 0 & 1 \\ 0 & 0 & 1 & -1 \\ 0 & 0 & 0 & 1 \end{pmatrix}$

Forcing functions $\vec{g}(t)$ $\begin{pmatrix} e^t \\ te^{-t} \\ e^{-t} \\ t \end{pmatrix}$ Initial Conditions: $\vec{x}(0)$ $\begin{pmatrix} 0 \\ 1 \\ 1 \\ -1 \end{pmatrix}$

Differential Equations: $\frac{d\vec{x}}{dt}$ $\begin{pmatrix} x_1'(t) = e^t - x_1(t) \\ x_2'(t) = x_4(t) - x_2(t) + te^{-t} \\ x_3'(t) = e^{-t} + x_3(t) - x_4(t) \\ x_4'(t) = t + x_4(t) \end{pmatrix}$

Eigenvalues and Eigenvectors

Input Matrix A

-1	0	0	0
0	-1	0	1
0	0	1	-1
0	0	0	1

Eigenvalues	Eigenvectors		
1	0	1	0
1	0	0	1
-1	1	0	0
-1	0	0	0

Eigenvalues (4) & Eigenvectors (3) ⇒ Matrix DEFECTIVE: Generalized eigenvectors required!

Eigenvectors incl. Generalized

1	0	0	0
0	1	0	0.5
0	0	1	0
0	0	0	1

Algebraic multiplicity (p) of the eigenvalue λ > geometric multiplicity (m)

⇒ Generalized Eigenvectors are solutions of $[(A - \lambda I_n)^p]v = 0$

λ=-1: p = 2, m = 2

λ=1: p = 2

General solutions and unknown constants

Homogeneous solutions set

$$\begin{pmatrix} c_1\, e^{-t} \\ \frac{c_4\, e^t}{2} + c_2\, e^{-t} \\ e^t\,(c_3 - c_4\, t) \\ c_4\, e^t \end{pmatrix}$$

Complete solutions set (homogeneous + particular)

$$\begin{pmatrix} \frac{e^t}{2} + c_1\, e^{-t} \\ \frac{t^2\, e^{-t}}{2} - t + \frac{c_4\, e^t}{2} + c_2\, e^{-t} \\ e^t\,(c_3 - c_4\, t) - \frac{e^{-t}}{2} - t - 2 \\ c_4\, e^t - t - 1 \end{pmatrix}$$

constants c_1, c_2, c_3 and c_4

$$\begin{pmatrix} -\frac{1}{2} & 1 & \frac{7}{2} & 0 \end{pmatrix}$$

Fundamental Matrix, Particular & Complete Solutions

Fundamental Matrix X(t):

$$\begin{pmatrix} e^{-t} & 0 & 0 & 0 \\ 0 & e^{-t} & 0 & \frac{e^t}{2} \\ 0 & 0 & e^t & -t\, e^t \\ 0 & 0 & 0 & e^t \end{pmatrix}$$

Particular Solution: $\vec{Y}_p(t)$

$$\begin{matrix} \frac{e^t}{2} \\ \frac{t^2\, e^{-t}}{2} - t \\ -t - \frac{e^{-t}}{2} - 2 \\ -t - 1 \end{matrix}$$

Complete Solution: $\vec{Y}(t)$

$$\begin{matrix} \sinh(t) \\ e^{-t} - t + \frac{t^2\, e^{-t}}{2} \\ \frac{7\, e^t}{2} - \frac{e^{-t}}{2} - t - 2 \\ -t - 1 \end{matrix}$$

Coupled 1st Order Differential Equations: Laplace Transforms

Input⇒

$$\begin{bmatrix} x\prime_1(t) \\ x\prime_2(t) \\ \cdots \\ x\prime_n(t) \end{bmatrix} = A \begin{pmatrix} x_1(t) \\ x_2(t) \\ \cdots \\ x_n(t) \end{pmatrix} + \begin{pmatrix} g_1(t) \\ g_2(t) \\ \cdots \\ g_n(t) \end{pmatrix}$$

Initial⇒ Conditions

$$\begin{pmatrix} x_1(0) \\ x_2(0) \\ \cdots \\ x_n(0) \end{pmatrix}$$

Laplace⇒ Transform

$$\begin{bmatrix} X_1(s) \\ X_2(s) \\ \cdots \\ X_n(s) \end{bmatrix} = [sI_n - A]^{-1} \begin{pmatrix} x_1(0) \\ x_2(0) \\ \cdots \\ x_n(0) \end{pmatrix} + [sI_n - A]^{-1} \begin{pmatrix} G_1(s) \\ G_2(s) \\ \cdots \\ G_n(0) \end{pmatrix}$$

Solution⇒

$$\begin{pmatrix} x_1(t) \\ x_2(t) \\ \cdots \\ x_n(t) \end{pmatrix} = L^{-1} \begin{bmatrix} X_1(s) \\ X_2(s) \\ \cdots \\ X_n(s) \end{bmatrix}$$

Laplace Transforms and Solutions using inverse Laplace Transforms

Laplace Transform Y(s)

$$\frac{1}{s^2-1}$$

$$\frac{1}{s+1} + \frac{1}{(s+1)^3} - \frac{1}{s^2}$$

$$\frac{s^3+3\,s^2+2\,s+1}{s^2\,(s^2-1)}$$

$$-\frac{s+1}{s^2}$$

Solution: Inverse Laplace Transform Y(t)

$$\sinh(t)$$

$$\mathrm{e}^{-t} - t + \frac{t^2\,\mathrm{e}^{-t}}{2}$$

$$\frac{7\,\mathrm{e}^{t}}{2} - \frac{\mathrm{e}^{-t}}{2} - t - 2$$

$$-t - 1$$

Comparison of Analytical Solutions

Solution: Eigenvalues & Eigenvectors: $Y_e(t)$

$$\begin{pmatrix} \sinh(t) \\ e^{-t} - t + \frac{t^2 e^{-t}}{2} \\ \frac{7e^{t}}{2} - \frac{e^{-t}}{2} - t - 2 \\ -t - 1 \end{pmatrix}$$

Solution: dsolve(.): $Y_d(t)$

$$\begin{pmatrix} \sinh(t) \\ e^{-t} - t + \frac{t^2 e^{-t}}{2} \\ \frac{7e^{t}}{2} - \frac{e^{-t}}{2} - t - 2 \\ -t - 1 \end{pmatrix}$$

$Y_d(t)$-$Y_e(t)$⇒ $\begin{pmatrix} 0 \\ 0 \\ 0 \\ 0 \end{pmatrix}$

Solution: Laplace Transform $Y_L(t)$

$$\begin{pmatrix} \sinh(t) \\ e^{-t} - t + \frac{t^2 e^{-t}}{2} \\ \frac{7e^{t}}{2} - \frac{e^{-t}}{2} - t - 2 \\ -t - 1 \end{pmatrix}$$

$Y_L(t)$-$Y_e(t)$⇒ $\begin{pmatrix} 0 \\ 0 \\ 0 \\ 0 \end{pmatrix}$

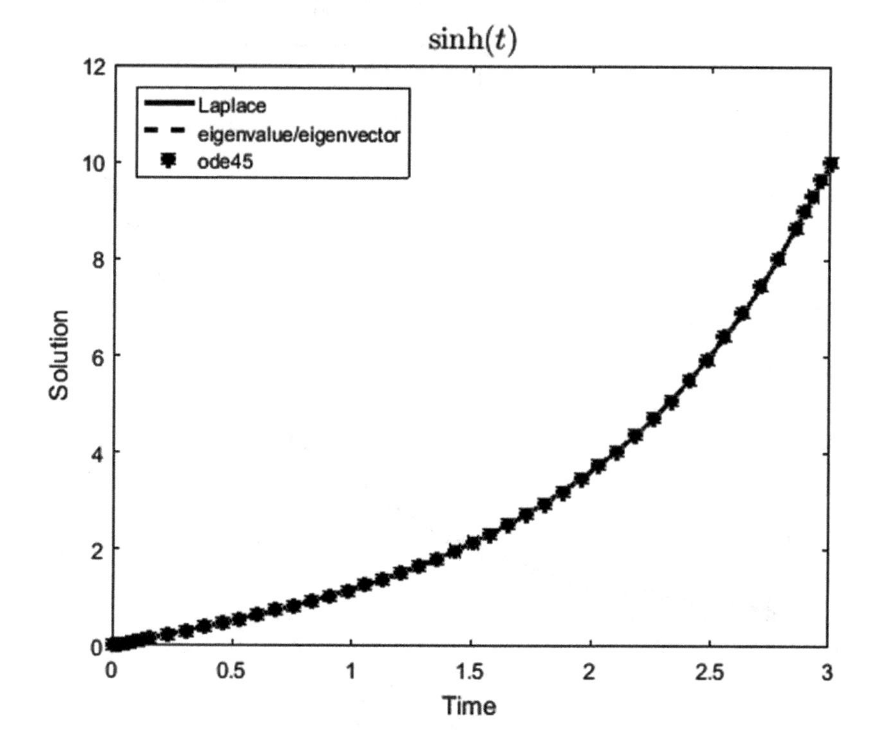

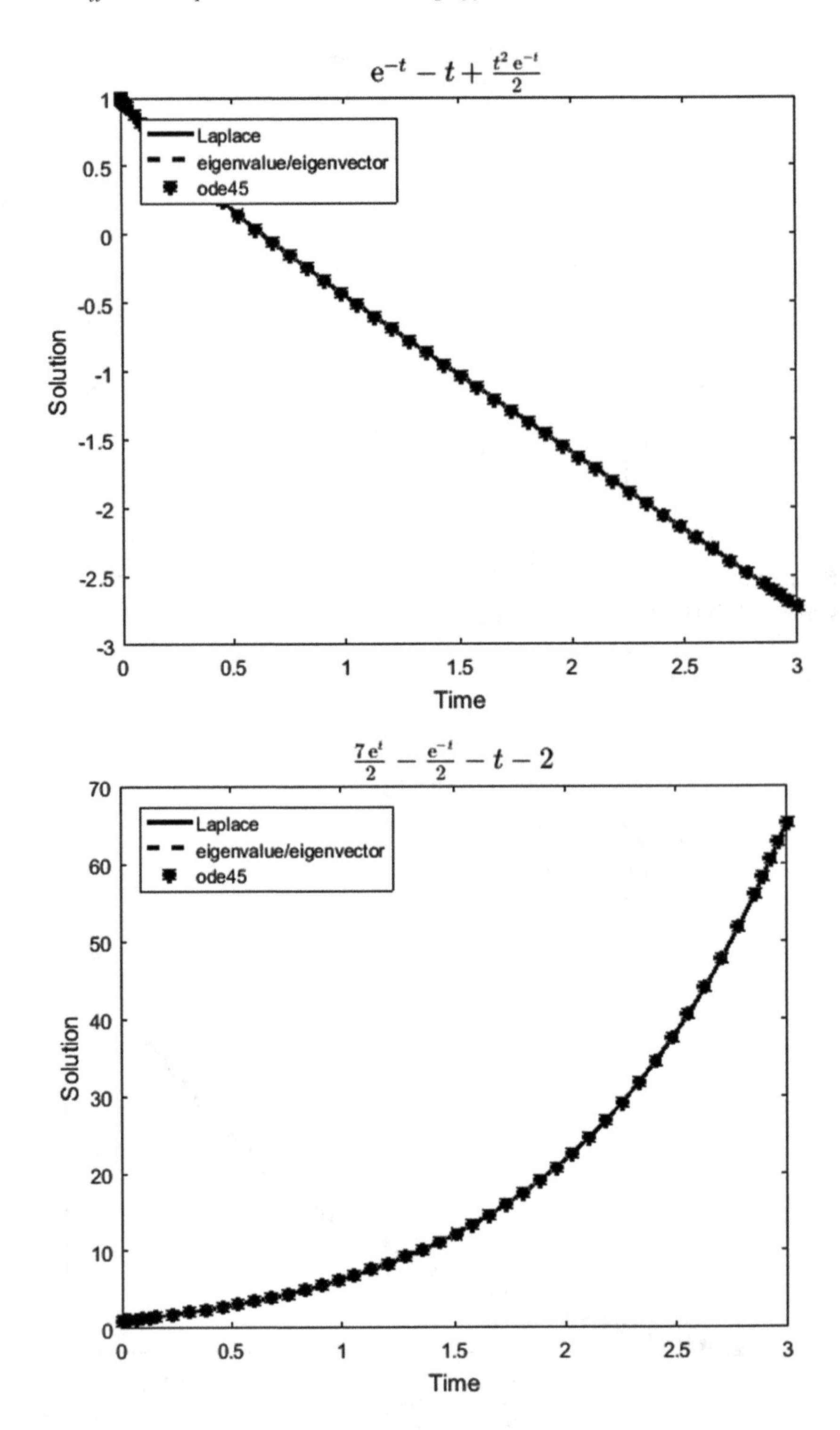

$e^{-t} - t + \frac{t^2 e^{-t}}{2}$
Laplace
eigenvalue/eigenvector
ode45
Solution
Time
$\frac{7e^t}{2} - \frac{e^{-t}}{2} - t - 2$
Laplace
eigenvalue/eigenvector
ode45
Solution
Time

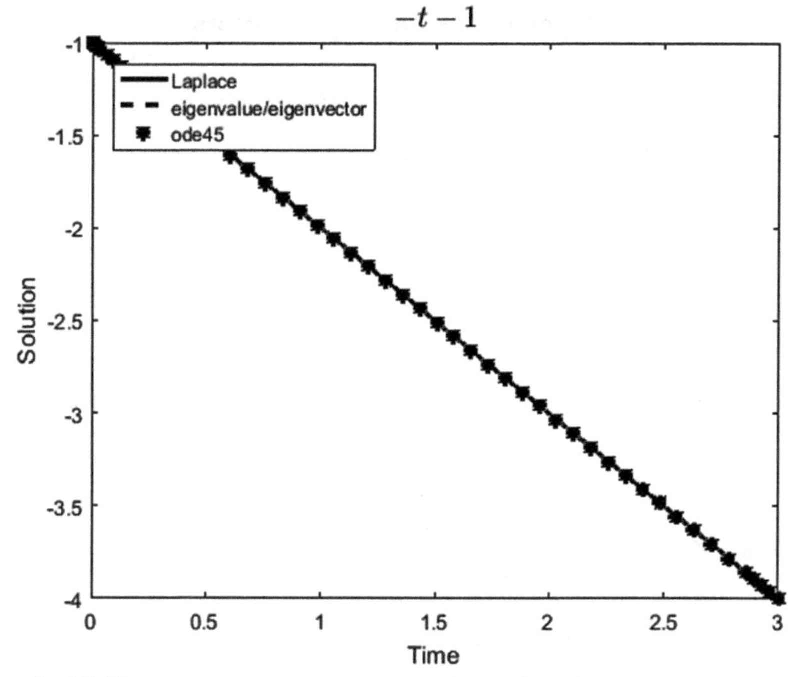

Example # 5.42

```
A=[-1 -4 0 0; 1 3 0 0; 1 2 1 0;0 1 0 1];x0=[1 -1 1 -1];gt={'exp(t)';'t*exp(t)';'t^2*exp(-t)';'4'};
```

Coupled 1st order System: Problem Statement

$$\frac{d\vec{x}}{dt} = A\vec{x}(t) + \vec{g}(t)$$

Coefficient Matrix: A

$$\begin{pmatrix} -1 & -4 & 0 & 0 \\ 1 & 3 & 0 & 0 \\ 1 & 2 & 1 & 0 \\ 0 & 1 & 0 & 1 \end{pmatrix}$$

Forcing functions $\vec{g}(t)$

$$\begin{pmatrix} e^{t} \\ t\,e^{t} \\ t^2\,e^{-t} \\ 4 \end{pmatrix}$$

Initial Conditions: $\vec{x}(0)$

$$\begin{pmatrix} 1 \\ -1 \\ 1 \\ -1 \end{pmatrix}$$

Differential Equations: $\frac{d\vec{x}}{dt}$

$$\begin{pmatrix} x_1'(t) = e^{t} - x_1(t) - 4\,x_2(t) \\ x_2'(t) = x_1(t) + 3\,x_2(t) + t\,e^{t} \\ x_3'(t) = x_1(t) + 2\,x_2(t) + x_3(t) + t^2\,e^{-t} \\ x_4'(t) = x_2(t) + x_4(t) + 4 \end{pmatrix}$$

Eigenvalues and Eigenvectors

Input Matrix A

$$\begin{matrix} -1 & -4 & 0 & 0 \\ 1 & 3 & 0 & 0 \\ 1 & 2 & 1 & 0 \\ 0 & 1 & 0 & 1 \end{matrix}$$

Eigenvalues

$$\begin{matrix} 1 \\ 1 \\ 1 \\ 1 \end{matrix}$$

Eigenvectors

$$\begin{matrix} 0 & 0 \\ 0 & 0 \\ 1 & 0 \\ 0 & 1 \end{matrix}$$

Eigenvalues (4) & Eigenvectors (2) ⇒ Matrix DEFECTIVE:
Generalized eigenvectors required!

Eigenvectors incl. Generalized

$$\begin{matrix} 1 & 0 & 0 & 0 \\ 0 & 1 & 0 & 0 \\ 0 & 0 & 1 & 0 \\ 0 & 0 & 0 & 1 \end{matrix}$$

Algebraic multiplicity (p) of the eigenvalue λ > geometric multiplicity (m)
⇒ Generalized Eigenvectors are solutions of $[(A-\lambda I_n)^p]v=0$

$\lambda=1$: p = 4

General solutions and unknown constants

Homogeneous solutions set

$$\begin{pmatrix} -e^t\,(2c_1t-c_1+4c_2t) \\ e^t\,(c_2+c_1t+2c_2t) \\ e^t\,(c_3+c_1t+2c_2t) \\ \frac{e^t\left(2c_4+2c_2t+c_1t^2+2c_2t^2\right)}{2} \end{pmatrix}$$

Complete solutions set (homogeneous + particular)

$$\begin{pmatrix} -e^t\,(2c_1t-c_1+4c_2t)-\frac{t\,e^t\left(2t^2+3t-3\right)}{3} \\ e^t\,(c_2+c_1t+2c_2t)+\frac{t^2\,e^t\,(t+3)}{3} \\ e^t\,(c_3+c_1t+2c_2t)-\frac{e^{-t}\left(6t-6t^2e^{2t}-4t^3e^{2t}+6t^2+3\right)}{12} \\ \frac{t^3e^t}{3}+\frac{t^4e^t}{12}+\frac{e^t\left(2c_4+2c_2t+c_1t^2+2c_2t^2\right)}{2}-4 \end{pmatrix}$$

constants c_1, c_2, c_3 and c_4

$$\begin{pmatrix} 1 & -1 & \frac{5}{4} & 3 \end{pmatrix}$$

Fundamental Matrix, Particular & Complete Solutions

Fundamental Matrix X(t):

$$\begin{pmatrix} -e^t\,(2t-1) & -4t\,e^t & 0 & 0 \\ t\,e^t & e^t\,(2t+1) & 0 & 0 \\ t\,e^t & 2t\,e^t & e^t & 0 \\ \frac{t^2 e^t}{2} & e^t\,(t^2+t) & 0 & e^t \end{pmatrix}$$

Particular Solution: $\vec{Y}_p(t)$

$$\begin{matrix} -\frac{t\,e^t\,(2t^2+3t-3)}{3} \\ \frac{t^2\,e^t\,(t+3)}{3} \\ -\frac{e^{-t}\,(6t-6t^2 e^{2t}-4t^3 e^{2t}+6t^2+3)}{12} \\ \frac{t^3 e^t}{3}+\frac{t^4 e^t}{12}-4 \end{matrix}$$

Complete Solution: $\vec{Y}$(t)

$$\begin{matrix} \frac{e^t\,(-2t^3-3t^2+9t+3)}{3} \\ -\frac{e^t\,(-t^3-3t^2+3t+3)}{3} \\ -\frac{e^{-t}\,(6t-15e^{2t}+12t\,e^{2t}-6t^2 e^{2t}-4t^3 e^{2t}+6t^2+3)}{12} \\ 3e^t-\frac{t^2 e^t}{2}+\frac{t^3 e^t}{3}+\frac{t^4 e^t}{12}-t\,e^t-4 \end{matrix}$$

Coupled 1st Order Differential Equations: Laplace Transforms

Input⇒

$$\begin{bmatrix} x'_1(t) \\ x'_2(t) \\ \cdots \\ x'_n(t) \end{bmatrix} = A\begin{pmatrix} x_1(t) \\ x_2(t) \\ \cdots \\ x_n(t) \end{pmatrix} + \begin{pmatrix} g_1(t) \\ g_2(t) \\ \cdots \\ g_n(t) \end{pmatrix}$$

Initial⇒ Conditions

$$\begin{pmatrix} x_1(0) \\ x_2(0) \\ \cdots \\ x_n(0) \end{pmatrix}$$

Laplace⇒ Transform

$$\begin{bmatrix} X_1(s) \\ X_2(s) \\ \cdots \\ X_n(s) \end{bmatrix} = [sI_n - A]^{-1}\begin{pmatrix} x_1(0) \\ x_2(0) \\ \cdots \\ x_n(0) \end{pmatrix} + [sI_n - A]^{-1}\begin{pmatrix} G_1(s) \\ G_2(s) \\ \cdots \\ G_n(0) \end{pmatrix}$$

Solution⇒

$$\begin{pmatrix} x_1(t) \\ x_2(t) \\ \cdots \\ x_n(t) \end{pmatrix} = L^{-1}\begin{bmatrix} X_1(s) \\ X_2(s) \\ \cdots \\ X_n(s) \end{bmatrix}$$

Laplace Transforms and Solutions using inverse Laplace Transforms

Laplace Transform Y(s)

$$\frac{s\left(s^2-5\right)}{(s-1)^4}$$

$$\frac{s\left(-s^2+2s+1\right)}{(s-1)^4}$$

$$\frac{s^6-s^5-3s^4+8s^3+5s^2+9s-3}{(s-1)^4(s+1)^3}$$

$$-\frac{s^5-7s^4+20s^3-29s^2+17s-4}{s(s-1)^5}$$

Solution: Inverse Laplace Transform Y(t)

$$\frac{e^t\left(-2t^3-3t^2+9t+3\right)}{3}$$

$$-\frac{e^t\left(-t^3-3t^2+3t+3\right)}{3}$$

$$-\frac{e^{-t}\left(6t-15e^{2t}+12te^{2t}-6t^2e^{2t}-4t^3e^{2t}+6t^2+3\right)}{12}$$

$$3e^t-\frac{t^2e^t}{2}+\frac{t^3e^t}{3}+\frac{t^4e^t}{12}-te^t-4$$

Comparison of Analytical Solutions

Solution: Eigenvalues & Eigenvectors: $Y_e(t)$

$$\frac{e^t\left(-2t^3-3t^2+9t+3\right)}{3}$$

$$-\frac{e^t\left(-t^3-3t^2+3t+3\right)}{3}$$

$$-\frac{e^{-t}\left(6t-15e^{2t}+12te^{2t}-6t^2e^{2t}-4t^3e^{2t}+6t^2+3\right)}{12}$$

$$3e^t-\frac{t^2e^t}{2}+\frac{t^3e^t}{3}+\frac{t^4e^t}{12}-te^t-4$$

Solution: dsolve(.): $Y_d(t)$

$Y_d(t)-Y_e(t) \Rightarrow \begin{pmatrix}0\\0\\0\\0\end{pmatrix}$

$$\frac{e^t\left(-2t^3-3t^2+9t+3\right)}{3}$$

$$-\frac{e^t\left(-t^3-3t^2+3t+3\right)}{3}$$

$$-\frac{e^{-t}\left(6t-15e^{2t}+12te^{2t}-6t^2e^{2t}-4t^3e^{2t}+6t^2+3\right)}{12}$$

$$3e^t-\frac{t^2e^t}{2}+\frac{t^3e^t}{3}+\frac{t^4e^t}{12}-te^t-4$$

Solution: Laplace Transform $Y_L(t)$

$Y_L(t)-Y_e(t) \Rightarrow \begin{pmatrix}0\\0\\0\\0\end{pmatrix}$

$$\frac{e^t\left(-2t^3-3t^2+9t+3\right)}{3}$$

$$-\frac{e^t\left(-t^3-3t^2+3t+3\right)}{3}$$

$$-\frac{e^{-t}\left(6t-15e^{2t}+12te^{2t}-6t^2e^{2t}-4t^3e^{2t}+6t^2+3\right)}{12}$$

$$3e^t-\frac{t^2e^t}{2}+\frac{t^3e^t}{3}+\frac{t^4e^t}{12}-te^t-4$$

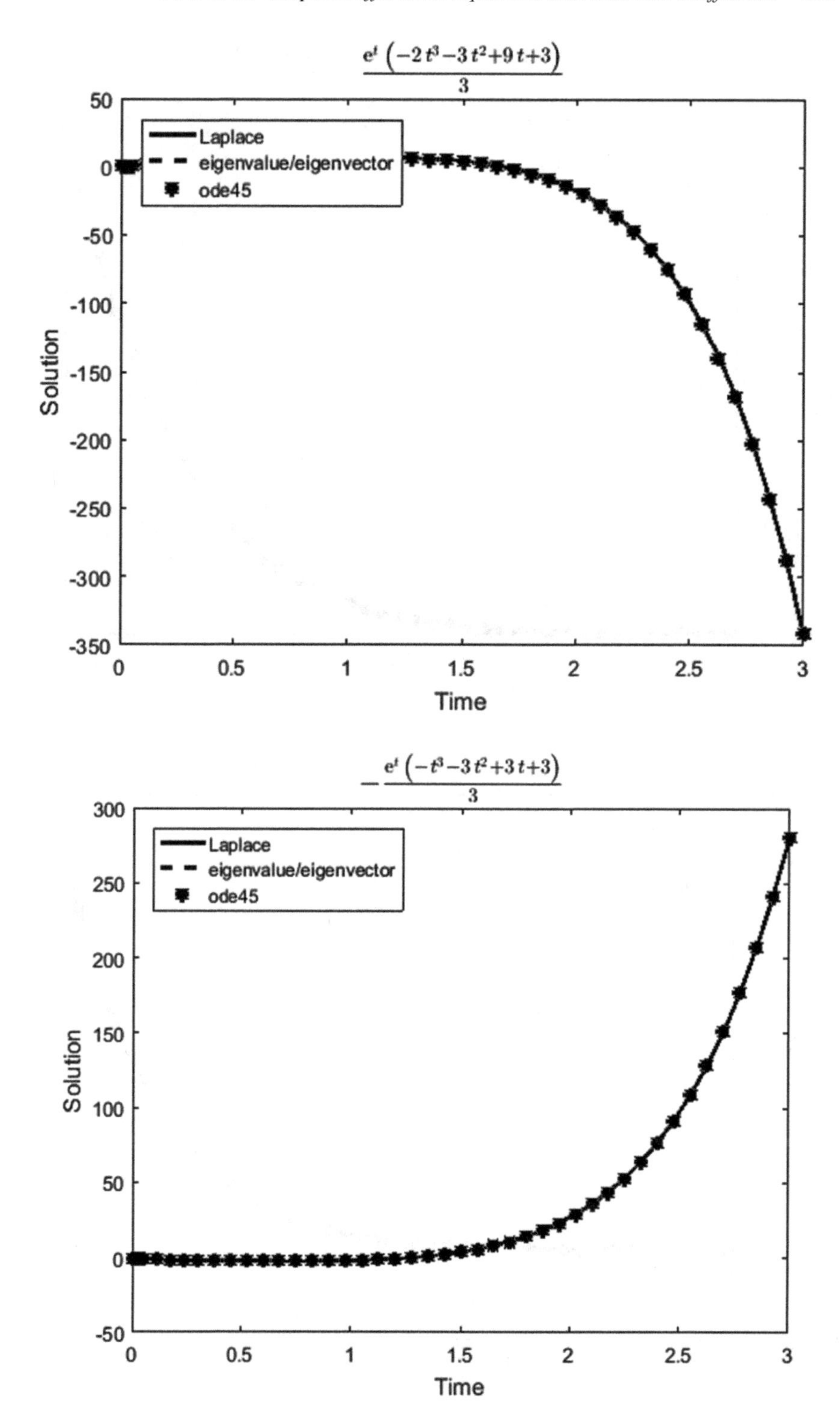

$\frac{e^t\left(-2t^3-3t^2+9t+3\right)}{3}$
Laplace
eigenvalue/eigenvector
ode45
Solution
Time
$-\frac{e^t\left(-t^3-3t^2+3t+3\right)}{3}$
Laplace
eigenvalue/eigenvector
ode45
Solution
Time

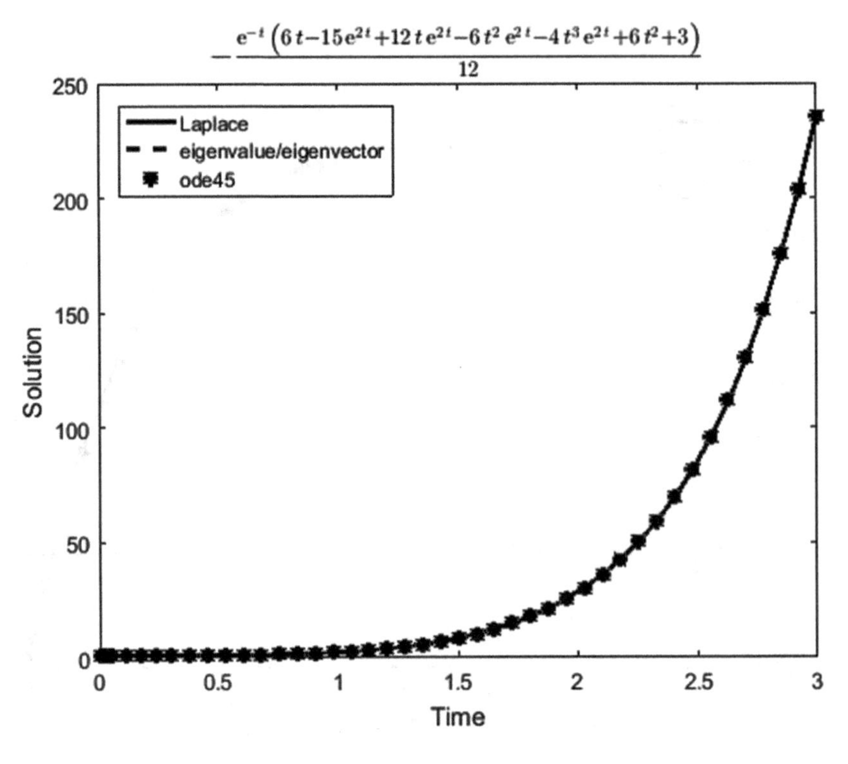
$-\frac{e^{-t}\left(6t-15e^{2t}+12te^{2t}-6t^2e^{2t}-4t^3e^{2t}+6t^2+3\right)}{12}$
Laplace
eigenvalue/eigenvector
ode45
Solution
Time
250
200
150
100
50
0
0
0.5
1
1.5
2
2.5
3

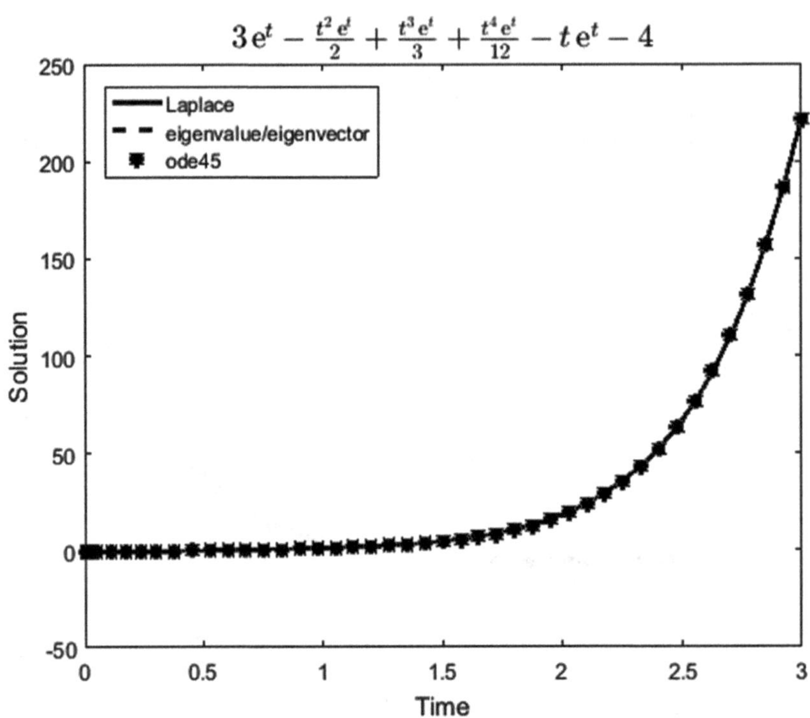
$3e^t-\frac{t^2e^t}{2}+\frac{t^3e^t}{3}+\frac{t^4e^t}{12}-te^t-4$
Laplace
eigenvalue/eigenvector
ode45
Solution
Time
250
200
150
100
50
0
-50
0
0.5
1
1.5
2
2.5
3

Example # 5.43

```
A=[1,0,0,0;0,1,0,0;1,4,-3,0;-1,-2,0,-3];x0=[1 1 1 -1];gt={'exp(-3*t)';'t*exp(3*t)';'exp(t)';'t+exp(t)'};
```

Coupled 1st order System: Problem Statement

$$\frac{d\vec{x}}{dt} = A\vec{x}(t) + \vec{g}(t)$$

Coefficient Matrix: A
$$\begin{pmatrix} 1 & 0 & 0 & 0 \\ 0 & 1 & 0 & 0 \\ 1 & 4 & -3 & 0 \\ -1 & -2 & 0 & -3 \end{pmatrix}$$

Forcing functions $\vec{g}(t)$
$$\begin{pmatrix} e^{-3t} \\ t\,e^{3t} \\ e^{t} \\ t + e^{t} \end{pmatrix}$$

Initial Conditions: $\vec{x}(0)$
$$\begin{pmatrix} 1 \\ 1 \\ 1 \\ -1 \end{pmatrix}$$

Differential Equations: $\frac{d\vec{x}}{dt}$
$$\begin{pmatrix} x_1'(t) = e^{-3t} + x_1(t) \\ x_2'(t) = x_2(t) + t\,e^{3t} \\ x_3'(t) = e^{t} + x_1(t) + 4\,x_2(t) - 3\,x_3(t) \\ x_4'(t) = t + e^{t} - x_1(t) - 2\,x_2(t) - 3\,x_4(t) \end{pmatrix}$$

Eigenvalues and Eigenvectors

Input Matrix A

1	0	0	0
0	1	0	0
1	4	-3	0
-1	-2	0	-3

Eigenvalues	Eigenvectors			
-3	0	0	-4	-8
-3	0	0	2	2
1	1	0	1	0
1	0	1	0	1

Eigenvalues (4) & Eigenvectors (4): Matrix NOT defective

General solutions and unknown constants

Homogeneous solutions set

$$\begin{pmatrix} -4\,e^{t}\,(c_3+2\,c_4) \\ 2\,e^{t}\,(c_3+c_4) \\ c_3\,e^{t}+c_1\,e^{-3t} \\ c_4\,e^{t}+c_2\,e^{-3t} \end{pmatrix}$$

Complete solutions set (homogeneous + particular)

$$\begin{pmatrix} -\frac{e^{-3t}}{4}-4\,e^{t}\,(c_3+2\,c_4) \\ 2\,e^{t}\,(c_3+c_4)+\frac{e^{3t}\,(2\,t-1)}{4} \\ \frac{e^{-3t}\left(144\,c_1-36\,t+36\,e^{4t}-32\,e^{6t}+48\,t\,e^{6t}+144\,c_3\,e^{4t}-9\right)}{144} \\ \frac{t}{3}+\frac{e^{-3t}}{16}+\frac{e^{3t}}{9}+\frac{e^{t}}{4}+\frac{t\,e^{-3t}}{4}-\frac{t\,e^{3t}}{6}+c_4\,e^{t}+c_2\,e^{-3t}-\frac{1}{9} \end{pmatrix}$$

constants c_1, c_2, c_3 and c_4

$$\left(-\frac{19}{36}\quad -\frac{3}{8}\quad \frac{25}{16}\quad -\frac{15}{16}\right)$$

Fundamental Matrix, Particular & Complete Solutions

Fundamental Matrix X(t):

$$\begin{pmatrix} 0 & 0 & -4e^{t} & -8e^{t} \\ 0 & 0 & 2\,e^{t} & 2e^{t} \\ e^{-3t} & 0 & e^{t} & 0 \\ 0 & e^{-3t} & 0 & e^{t} \end{pmatrix}$$

Particular Solution: $\vec{Y}_p(t)$

$$-\frac{e^{-3t}}{4}$$
$$\frac{e^{3t}\,(2\,t-1)}{4}$$
$$-\frac{e^{-3t}\left(36\,t-36\,e^{4t}+32\,e^{6t}-48\,t\,e^{6t}+9\right)}{144}$$
$$\frac{t}{3}+\frac{e^{-3t}}{16}+\frac{e^{3t}}{9}+\frac{e^{t}}{4}+\frac{t\,e^{-3t}}{4}-\frac{t\,e^{3t}}{6}-\frac{1}{9}$$

Complete Solution: $\vec{Y}$(t)

$$\frac{5\,e^{t}}{4}-\frac{e^{-3t}}{4}$$
$$\frac{5\,e^{t}}{4}+\frac{e^{3t}\,(2\,t-1)}{4}$$
$$-\frac{e^{-3t}\left(36\,t-261\,e^{4t}+32e^{6t}-48\,t\,e^{6t}+85\right)}{144}$$
$$\frac{t}{3}-\frac{5\,e^{-3t}}{16}+\frac{e^{3t}}{9}-\frac{11\,e^{t}}{16}+\frac{t\,e^{-3t}}{4}-\frac{t\,e^{3t}}{6}-\frac{1}{9}$$

Laplace Transforms and Solutions using inverse Laplace Transforms

Laplace Transform Y(s)

$$\frac{5}{4(s-1)} - \frac{1}{4(s+3)}$$

$$\frac{(s-3)^2+1}{(s-1)(s-3)^2}$$

$$\frac{s^4+2s^3-23s^2-20s+156}{(s^2-9)^2(s-1)}$$

$$-\frac{s^6-2s^5-12s^4+18s^3+48s^2-36s+27}{s^2(s^2-9)^2(s-1)}$$

Solution: Inverse Laplace Transform Y(t)

$$\frac{5e^t}{4} - \frac{e^{-3t}}{4}$$

$$\frac{e^t\left(2te^{2t}-e^{2t}+5\right)}{4}$$

$$-\frac{e^{-3t}\left(36t-261e^{4t}+32e^{6t}-48te^{6t}+85\right)}{144}$$

$$\frac{t}{3} - \frac{5e^{-3t}}{16} + \frac{e^{3t}}{9} - \frac{11e^t}{16} + \frac{te^{-3t}}{4} - \frac{te^{3t}}{6} - \frac{1}{9}$$

Comparison of Analytical Solutions

Solution: Eigenvalues & Eigenvectors: $Y_e(t)$

$$\frac{5e^t}{4} - \frac{e^{-3t}}{4}$$

$$\frac{5e^t}{4} + \frac{e^{3t}(2t-1)}{4}$$

$$-\frac{e^{-3t}\left(36t-261e^{4t}+32e^{6t}-48te^{6t}+85\right)}{144}$$

$$\frac{t}{3} - \frac{5e^{-3t}}{16} + \frac{e^{3t}}{9} - \frac{11e^t}{16} + \frac{te^{-3t}}{4} - \frac{te^{3t}}{6} - \frac{1}{9}$$

Solution: dsolve(.): $Y_d(t)$

$$\frac{5e^t}{4} - \frac{e^{-3t}}{4}$$

$$\frac{e^t\left(2te^{2t}-e^{2t}+5\right)}{4}$$

$$-\frac{e^{-3t}\left(36t-261e^{4t}+32e^{6t}-48te^{6t}+85\right)}{144}$$

$$\frac{t}{3} - \frac{5e^{-3t}}{16} + \frac{e^{3t}}{9} - \frac{11e^t}{16} + \frac{te^{-3t}}{4} - \frac{te^{3t}}{6} - \frac{1}{9}$$

$Y_d(t)-Y_e(t) \Rightarrow \begin{pmatrix} 0 \\ 0 \\ 0 \\ 0 \end{pmatrix}$

Solution: Laplace Transform $Y_L(t)$

$$\frac{5e^t}{4} - \frac{e^{-3t}}{4}$$

$$\frac{e^t\left(2te^{2t}-e^{2t}+5\right)}{4}$$

$$-\frac{e^{-3t}\left(36t-261e^{4t}+32e^{6t}-48te^{6t}+85\right)}{144}$$

$$\frac{t}{3} - \frac{5e^{-3t}}{16} + \frac{e^{3t}}{9} - \frac{11e^t}{16} + \frac{te^{-3t}}{4} - \frac{te^{3t}}{6} - \frac{1}{9}$$

$Y_L(t)-Y_e(t) \Rightarrow \begin{pmatrix} 0 \\ 0 \\ 0 \\ 0 \end{pmatrix}$

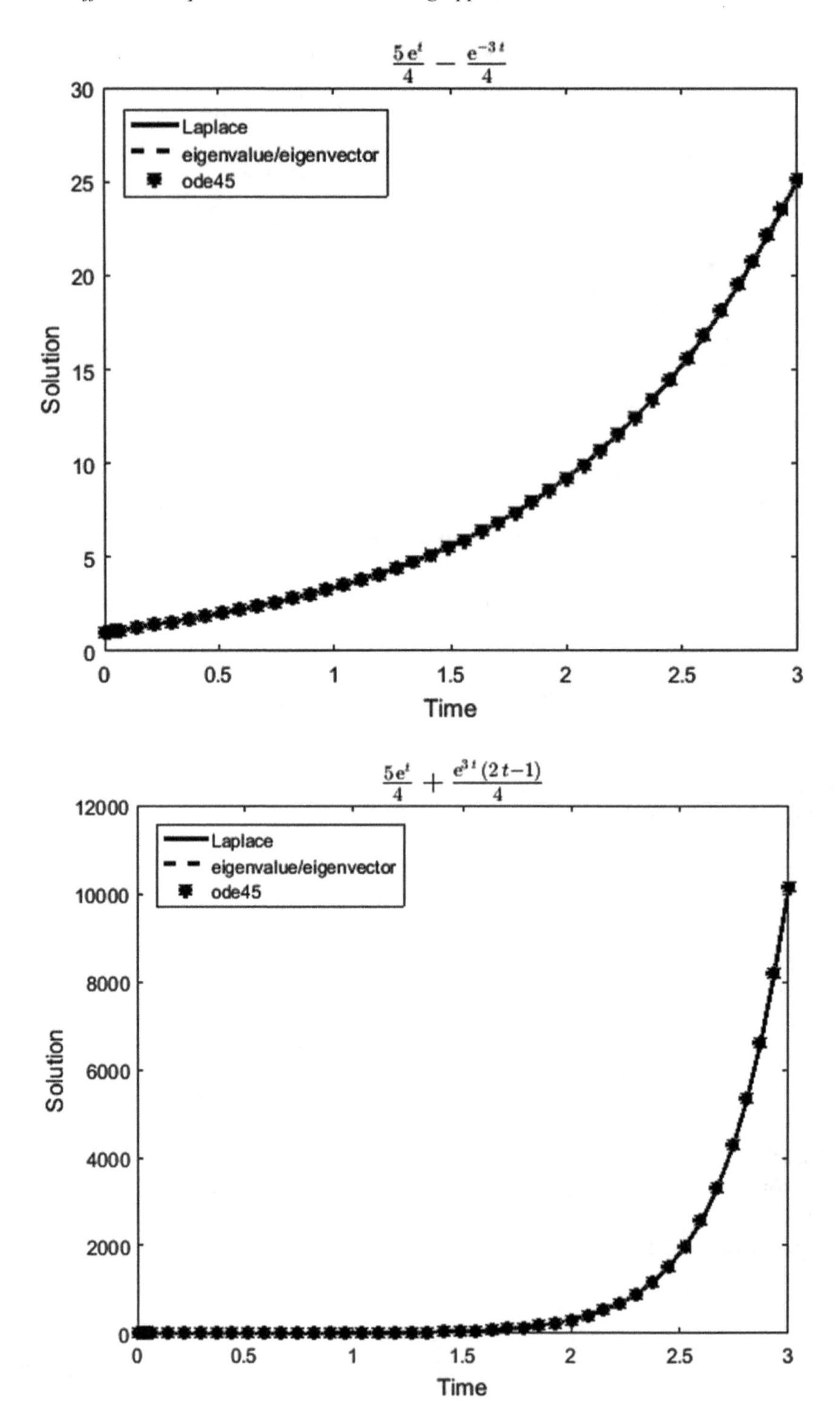

$\frac{5e^t}{4} - \frac{e^{-3t}}{4}$
Laplace
eigenvalue/eigenvector
ode45
Solution
Time
0
5
10
15
20
25
30
0.5
1
1.5
2
2.5
3
$\frac{5e^t}{4} + \frac{e^{3t}(2t-1)}{4}$
Laplace
eigenvalue/eigenvector
ode45
Solution
Time
0
2000
4000
6000
8000
10000
12000
0.5
1
1.5
2
2.5
3

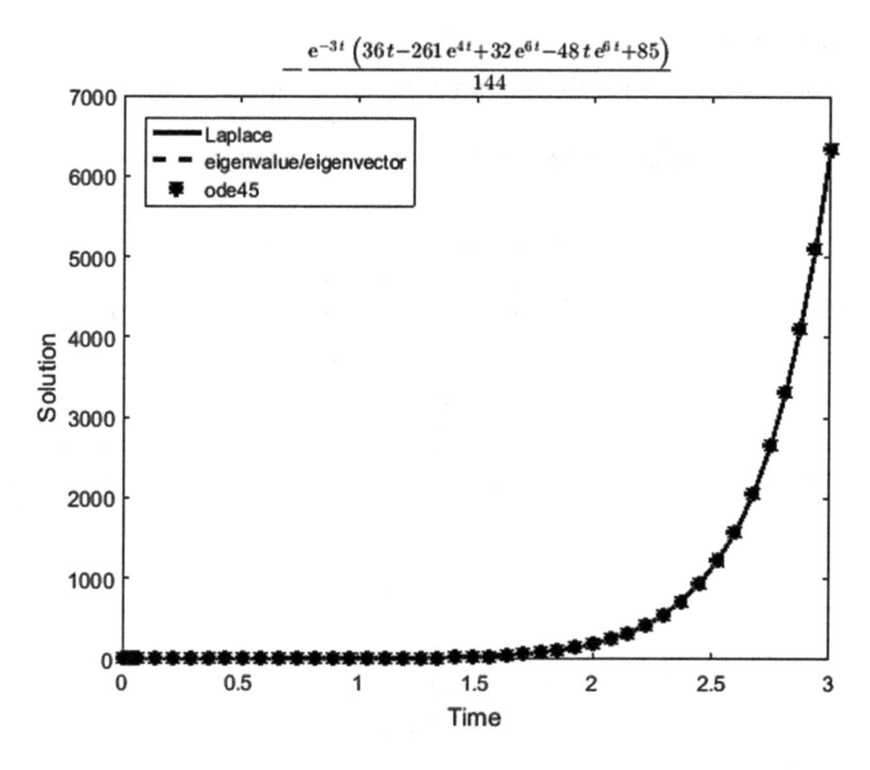
$-\frac{e^{-3t}\left(36t-261\,e^{4t}+32\,e^{6t}-48\,t\,e^{6t}+85\right)}{144}$
7000
6000
5000
4000
3000
2000
1000
0
Solution
0
0.5
1
1.5
2
2.5
3
Time
Laplace
eigenvalue/eigenvector
ode45

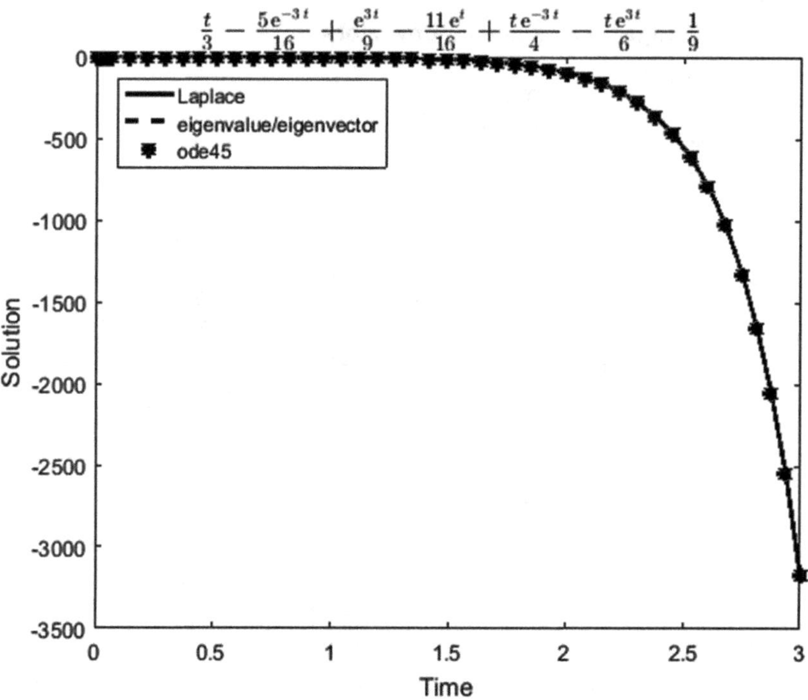
$\frac{t}{3}-\frac{5e^{-3t}}{16}+\frac{e^{3t}}{9}-\frac{11\,e^{t}}{16}+\frac{t\,e^{-3t}}{4}-\frac{t\,e^{3t}}{6}-\frac{1}{9}$
0
-500
-1000
-1500
-2000
-2500
-3000
-3500
Solution
0
0.5
1
1.5
2
2.5
3
Time
Laplace
eigenvalue/eigenvector
ode45

Example # 5.44

```
A=[0,0,1,0;0,0,0,1;2,1,1,1;-5,2,5,-1];x0=[0 0 0 -1];gt={'t';'exp(-t)';'5';'cos(t)'};
```

Coupled 1st order System: Problem Statement

$$\frac{d\vec{x}}{dt} = A\vec{x}(t) + \vec{g}(t)$$

Coefficient Matrix: A $\begin{pmatrix} 0 & 0 & 1 & 0 \\ 0 & 0 & 0 & 1 \\ 2 & 1 & 1 & 1 \\ -5 & 2 & 5 & -1 \end{pmatrix}$

Forcing functions $\vec{g}(t)$ $\begin{pmatrix} t \\ e^{-t} \\ 5 \\ \cos(t) \end{pmatrix}$ Initial Conditions: $\vec{x}(0)$ $\begin{pmatrix} 0 \\ 0 \\ 0 \\ -1 \end{pmatrix}$

Differential Equations: $\frac{d\vec{x}}{dt}$ $\begin{pmatrix} x_1'(t) = t + x_3(t) \\ x_2'(t) = e^{-t} + x_4(t) \\ x_3'(t) = 2\,x_1(t) + x_2(t) + x_3(t) + x_4(t) + 5 \\ x_4'(t) = \cos(t) - 5\,x_1(t) + 2\,x_2(t) + 5\,x_3(t) - x_4(t) \end{pmatrix}$

Eigenvalues and Eigenvectors

Input Matrix A

0	0	1	0
0	0	0	1
2	1	1	1
-5	2	5	-1

Eigenvalues	Eigenvectors			
1	-1	0.07	0.33	-0.2
-3	1	-0.33	0.33	-1
3	-1	-0.2	1	0.2
-1	1	1	1	1

Eigenvalues (4) & Eigenvectors (4): Matrix NOT defective

General solutions and unknown constants

Homogeneous solutions set

$$\begin{pmatrix} \frac{c_2 e^{-3t}}{15} - c_1 e^t - \frac{c_4 e^{-t}}{5} + \frac{c_3 e^{3t}}{3} \\ c_1 e^t - \frac{c_2 e^{-3t}}{3} - c_4 e^{-t} + \frac{c_3 e^{3t}}{3} \\ \frac{c_4 e^{-t}}{5} - \frac{c_2 e^{-3t}}{5} - c_1 e^t + c_3 e^{3t} \\ c_1 e^t + c_2 e^{-3t} + c_4 e^{-t} + c_3 e^{3t} \end{pmatrix}$$

Complete solutions set (homogeneous + particular)

$$\begin{pmatrix} \frac{\cos(t)}{20} - \frac{t}{3} - \frac{\sin(t)}{20} + e^{-t}\left(\frac{t}{8} - \frac{c_4}{5} + \frac{3}{32}\right) - c_1 e^t + \frac{c_2 e^{-3t}}{15} + \frac{c_3 e^{3t}}{3} - 2 \\ \frac{5t}{3} - \frac{3\cos(t)}{20} + \frac{\sin(t)}{20} + c_1 e^t + \frac{5 e^{-t}(4t-1)}{32} - \frac{c_2 e^{-3t}}{3} - c_4 e^{-t} + \frac{c_3 e^{3t}}{3} - \frac{10}{3} \\ \frac{e^{-t}}{32} - t - \frac{t e^{-t}}{8} - c_1 e^t - \frac{\sqrt{2}\sin\left(t+\frac{\pi}{4}\right)}{20} - \frac{c_2 e^{-3t}}{5} + \frac{c_4 e^{-t}}{5} + c_3 e^{3t} - \frac{1}{3} \\ \frac{\cos(t)}{20} + \frac{3\sin(t)}{20} + c_1 e^t - \frac{e^{-t}(20t+7)}{32} + c_2 e^{-3t} + c_4 e^{-t} + c_3 e^{3t} + \frac{5}{3} \end{pmatrix}$$

constants c_1, c_2, c_3 and c_4

$$\begin{pmatrix} -\frac{9}{16} & \frac{71}{48} & \frac{461}{480} & -\frac{35}{8} \end{pmatrix}$$

Fundamental Matrix, Particular & Complete Solutions

Fundamental Matrix X(t):

$$\begin{pmatrix} -e^t & \frac{e^{-3t}}{15} & \frac{e^{3t}}{3} & -\frac{e^{-t}}{5} \\ e^t & -\frac{e^{-3t}}{3} & \frac{e^{3t}}{3} & -e^{-t} \\ -e^t & -\frac{e^{-3t}}{5} & e^{3t} & \frac{e^{-t}}{5} \\ e^t & e^{-3t} & e^{3t} & e^{-t} \end{pmatrix}$$

Particular Solution: $\vec{Y}_p(t)$

$$\begin{matrix} \frac{3e^{-t}}{32} - \frac{t}{3} + \frac{t e^{-t}}{8} + \frac{\sqrt{2}\cos\left(t+\frac{\pi}{4}\right)}{20} - 2 \\ \frac{5t}{3} - \frac{3\cos(t)}{20} + \frac{\sin(t)}{20} + \frac{5e^{-t}(4t-1)}{32} - \frac{10}{3} \\ \frac{e^{-t}}{32} - t - \frac{t e^{-t}}{8} - \frac{\sqrt{2}\sin\left(t+\frac{\pi}{4}\right)}{20} - \frac{1}{3} \\ \frac{\cos(t)}{20} + \frac{3\sin(t)}{20} - \frac{e^{-t}(20t+7)}{32} + \frac{5}{3} \end{matrix}$$

Complete Solution: $\vec{Y}(t)$

$$\begin{matrix} \frac{71 e^{-3t}}{720} - \frac{t}{3} + \frac{461 e^{3t}}{1440} + \frac{\cos(t)}{20} + \frac{9e^t}{16} - \frac{\sin(t)}{20} + \frac{e^{-t}(4t+31)}{32} - 2 \\ \frac{5t}{3} - \frac{71 e^{-3t}}{144} + \frac{461 e^{3t}}{1440} - \frac{3\cos(t)}{20} - \frac{9e^t}{16} + \frac{\sin(t)}{20} + \frac{5e^{-t}(4t+27)}{32} - \frac{10}{3} \\ \frac{461 e^{3t}}{480} - \frac{71 e^{-3t}}{240} - t - \frac{\cos(t)}{20} + \frac{9e^t}{16} - \frac{\sin(t)}{20} - \frac{e^{-t}(4t+27)}{32} - \frac{1}{3} \\ \frac{71 e^{-3t}}{48} + \frac{461 e^{3t}}{480} + \frac{\cos(t)}{20} - \frac{9e^t}{16} + \frac{3\sin(t)}{20} - \frac{e^{-t}(20t+147)}{32} + \frac{5}{3} \end{matrix}$$

Laplace Transforms and Solutions using Inverse Laplace Transforms

Laplace Transform Y(s)

$$-\frac{-5s^6-11s^5+4s^4+10s^3+14s^2+21s+3}{s^2(s^2+1)(s-1)(s+1)^2(s-3)(s+3)}$$

$$\frac{s^6+20s^5-3s^4+3s^3+15s^2-15s+15}{s^2(s^2+1)(s-1)(s+1)^2(s-3)(s+3)}$$

$$-\frac{-4s^7-10s^6-5s^5+s^4+13s^3+20s^2+12s+9}{s^2(s^2+1)(s-1)(s+1)^2(s-3)(s+3)}$$

$$\frac{-s^7+s^6+29s^5-3s^4+4s^3+15s^2-24s+15}{s(s^2+1)(s-1)(s+1)^2(s-3)(s+3)}$$

Solution: Inverse Laplace Transform Y(t)

$$\frac{31e^{-t}}{32}-\frac{t}{3}+\frac{71e^{-3t}}{720}+\frac{461e^{3t}}{1440}+\frac{9e^{t}}{16}+\frac{te^{-t}}{8}+\frac{\sqrt{2}\cos\left(t+\frac{\pi}{4}\right)}{20}-2$$

$$\frac{5t}{3}+\frac{135e^{-t}}{32}-\frac{71e^{-3t}}{144}+\frac{461e^{3t}}{1440}-\frac{3\cos(t)}{20}-\frac{9e^{t}}{16}+\frac{\sin(t)}{20}+\frac{5te^{-t}}{8}-1$$

$$\frac{461e^{3t}}{480}-\frac{27e^{-t}}{32}-\frac{71e^{-3t}}{240}-t+\frac{9e^{t}}{16}-\frac{te^{-t}}{8}-\frac{\sqrt{2}\sin\left(t+\frac{\pi}{4}\right)}{20}-\frac{1}{3}$$

$$\frac{71e^{-3t}}{48}-\frac{147e^{-t}}{32}+\frac{461e^{3t}}{480}+\frac{\cos(t)}{20}-\frac{9e^{t}}{16}+\frac{3\sin(t)}{20}-\frac{5te^{-t}}{8}+\frac{5}{3}$$

Comparison of Analytical Solutions

Solution: Eigenvalues & Eigenvectors: $Y_e(t)$

$$\frac{71e^{-3t}}{720}-\frac{t}{3}+\frac{461e^{3t}}{1440}+\frac{\cos(t)}{20}+\frac{9e^{t}}{16}-\frac{\sin(t)}{20}+\frac{e^{-t}(4t+31)}{32}-2$$

$$\frac{5t}{3}-\frac{71e^{-3t}}{144}+\frac{461e^{3t}}{1440}-\frac{3\cos(t)}{20}-\frac{9e^{t}}{16}+\frac{\sin(t)}{20}+\frac{5e^{-t}(4t+27)}{32}-\frac{10}{3}$$

$$\frac{461e^{3t}}{480}-\frac{71e^{-3t}}{240}-t-\frac{\cos(t)}{20}+\frac{9e^{t}}{16}-\frac{\sin(t)}{20}-\frac{e^{-t}(4t+27)}{32}-\frac{1}{3}$$

$$\frac{71e^{-3t}}{48}+\frac{461e^{3t}}{480}+\frac{\cos(t)}{20}-\frac{9e^{t}}{16}+\frac{3\sin(t)}{20}-\frac{e^{-t}(20t+147)}{32}+\frac{5}{3}$$

Solution: dsolve(.): $Y_d(t)$

$$\frac{71e^{-3t}}{720}-\frac{t}{3}+\frac{461e^{3t}}{1440}+\frac{\cos(t)}{20}+\frac{9e^{t}}{16}-\frac{\sin(t)}{20}+e^{-t}\left(\frac{t}{8}+\frac{31}{32}\right)-2$$

$$\frac{5t}{3}-\frac{71e^{-3t}}{144}+\frac{461e^{3t}}{1440}-\frac{3\cos(t)}{20}-\frac{9e^{t}}{16}+\frac{\sin(t)}{20}+e^{-t}\left(\frac{5t}{8}+\frac{135}{32}\right)-$$

$$\frac{461e^{3t}}{480}-\frac{71e^{-3t}}{240}-t-\frac{\cos(t)}{20}+\frac{9e^{t}}{16}-\frac{\sin(t)}{20}-e^{-t}\left(\frac{t}{8}+\frac{27}{32}\right)-\frac{1}{3}$$

$$\frac{71e^{-3t}}{48}+\frac{461e^{3t}}{480}+\frac{\cos(t)}{20}-\frac{9e^{t}}{16}+\frac{3\sin(t)}{20}-e^{-t}\left(\frac{5t}{8}+\frac{147}{32}\right)+\frac{5}{3}$$

$Y_d(t)-Y_e(t) \Rightarrow \begin{pmatrix}0\\0\\0\\0\end{pmatrix}$

Solution: Laplace Transform $Y_L(t)$

$$\frac{31e^{-t}}{32}-\frac{t}{3}+\frac{71e^{-3t}}{720}+\frac{461e^{3t}}{1440}+\frac{9e^{t}}{16}+\frac{te^{-t}}{8}+\frac{\sqrt{2}\cos\left(t+\frac{\pi}{4}\right)}{20}-2$$

$$\frac{5t}{3}+\frac{135e^{-t}}{32}-\frac{71e^{-3t}}{144}+\frac{461e^{3t}}{1440}-\frac{3\cos(t)}{20}-\frac{9e^{t}}{16}+\frac{\sin(t)}{20}+\frac{5te^{-t}}{8}-$$

$$\frac{461e^{3t}}{480}-\frac{27e^{-t}}{32}-\frac{71e^{-3t}}{240}-t+\frac{9e^{t}}{16}-\frac{te^{-t}}{8}-\frac{\sqrt{2}\sin\left(t+\frac{\pi}{4}\right)}{20}-\frac{1}{3}$$

$$\frac{71e^{-3t}}{48}-\frac{147e^{-t}}{32}+\frac{461e^{3t}}{480}+\frac{\cos(t)}{20}-\frac{9e^{t}}{16}+\frac{3\sin(t)}{20}-\frac{5te^{-t}}{8}+\frac{5}{3}$$

$Y_L(t)-Y_e(t) \Rightarrow \begin{pmatrix}0\\0\\0\\0\end{pmatrix}$

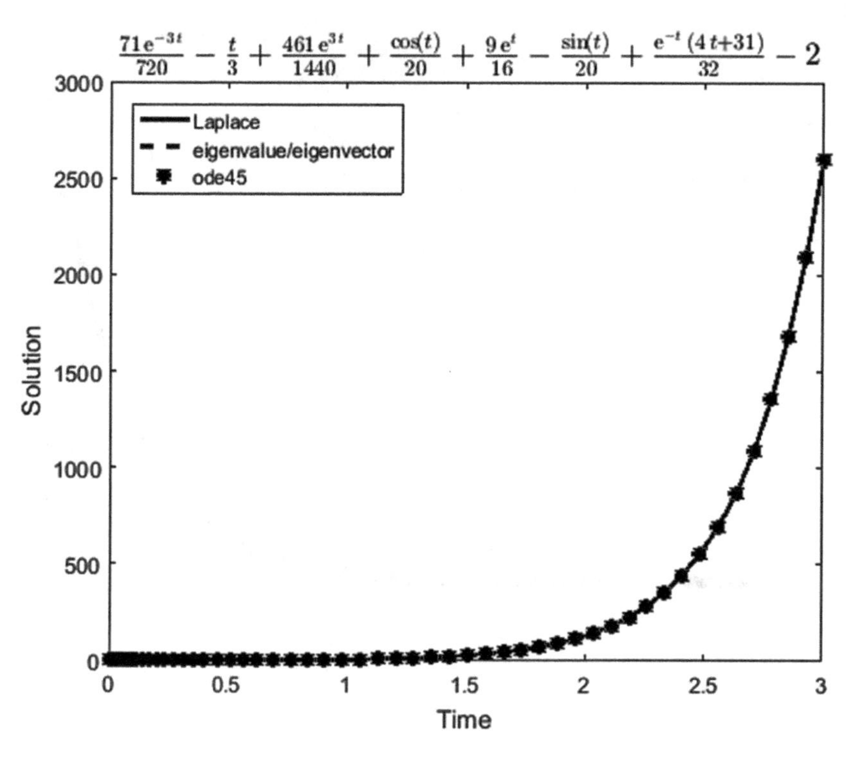
$\frac{71e^{-3t}}{720} - \frac{t}{3} + \frac{461e^{3t}}{1440} + \frac{\cos(t)}{20} + \frac{9e^{t}}{16} - \frac{\sin(t)}{20} + \frac{e^{-t}(4t+31)}{32} - 2$
Laplace
eigenvalue/eigenvector
ode45
Solution
Time
0
500
1000
1500
2000
2500
3000
0.5
1
1.5
2
2.5
3

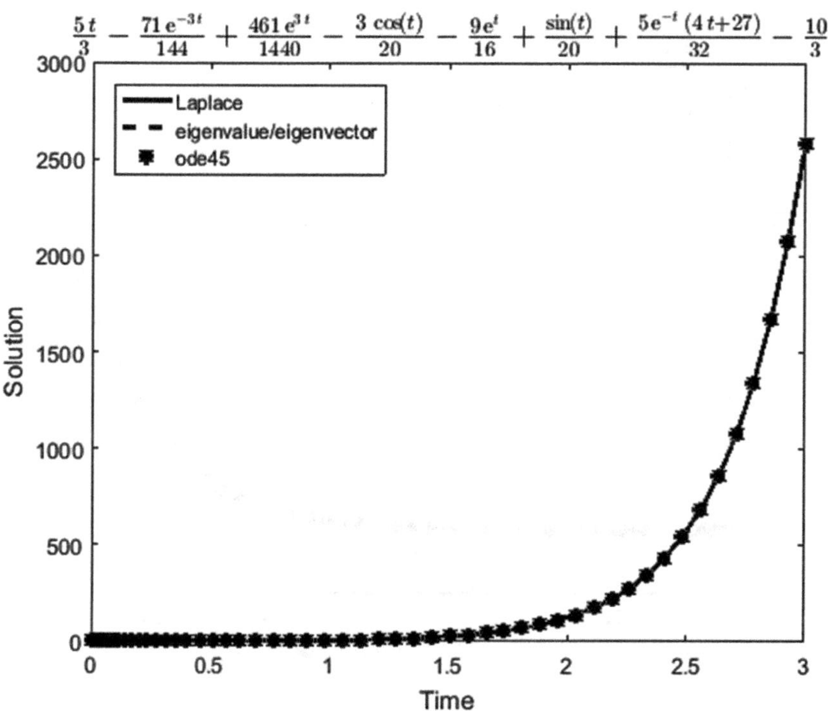
$\frac{5t}{3} - \frac{71e^{-3t}}{144} + \frac{461e^{3t}}{1440} - \frac{3\cos(t)}{20} - \frac{9e^{t}}{16} + \frac{\sin(t)}{20} + \frac{5e^{-t}(4t+27)}{32} - \frac{10}{3}$
Laplace
eigenvalue/eigenvector
ode45
Solution
Time
0
500
1000
1500
2000
2500
3000
0.5
1
1.5
2
2.5
3

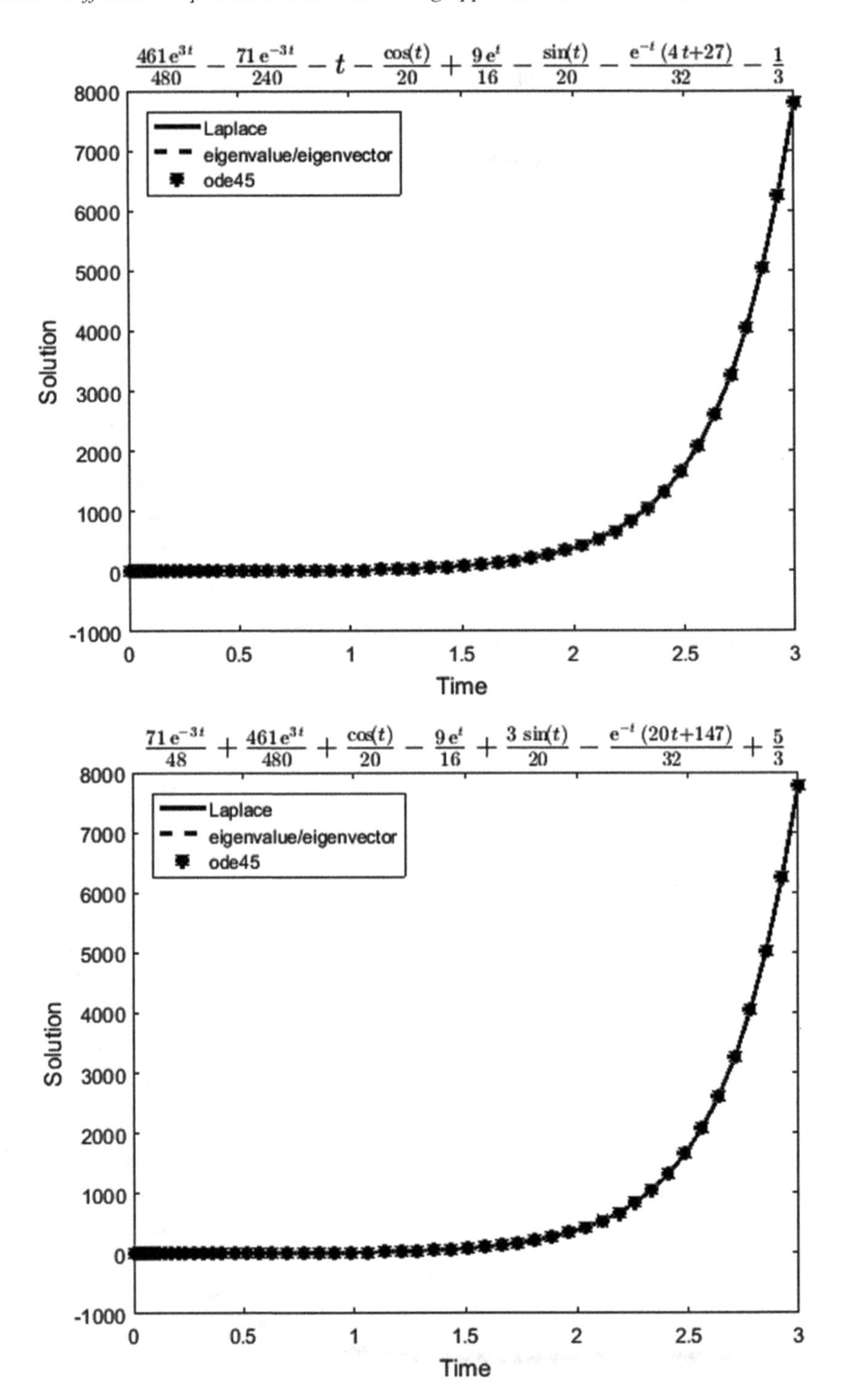
$\frac{461e^{3t}}{480} - \frac{71e^{-3t}}{240} - t - \frac{\cos(t)}{20} + \frac{9e^t}{16} - \frac{\sin(t)}{20} - \frac{e^{-t}(4t+27)}{32} - \frac{1}{3}$
Laplace
eigenvalue/eigenvector
ode45
Solution
Time
$\frac{71e^{-3t}}{48} + \frac{461e^{3t}}{480} + \frac{\cos(t)}{20} - \frac{9e^t}{16} + \frac{3\sin(t)}{20} - \frac{e^{-t}(20t+147)}{32} + \frac{5}{3}$
Laplace
eigenvalue/eigenvector
ode45
Solution
Time

Example # 5.46

```
A=[0,0,0,0,1;0,0,0,1,1;0,0,0,1,1;0,2,0,-1,0;0,0,0,0,1];x0=[0 0 0 1 -
1];gt={'t';'t';'t^2';'cos(t)';'5'};
```

Coupled 1st order System: $\frac{d\vec{x}}{dt} = A\vec{x}(t) + \vec{f}(t)$

Coefficient Matrix: A $\begin{pmatrix} 0 & 0 & 0 & 0 & 1 \\ 0 & 0 & 0 & 1 & 1 \\ 0 & 0 & 0 & 1 & 1 \\ 0 & 2 & 0 & -1 & 0 \\ 0 & 0 & 0 & 0 & 1 \end{pmatrix}$

Forcing functions $\vec{x}(t)$ $\begin{pmatrix} t \\ t \\ t^2 \\ \cos(t) \\ 5 \end{pmatrix}$ Initial Conditions: $\vec{x}(0)$ $\begin{pmatrix} 0 \\ 0 \\ 0 \\ 1 \\ -1 \end{pmatrix}$

Differential Equations: $\frac{d\vec{x}}{dt}$ $\begin{pmatrix} x_1'(t) = t + x_5(t) \\ x_2'(t) = t + x_4(t) + x_5(t) \\ x_3'(t) = x_4(t) + x_5(t) + t^2 \\ x_4'(t) = \cos(t) + 2\,x_2(t) - x_4(t) \\ x_5'(t) = x_5(t) + 5 \end{pmatrix}$

Coupled 1st Order Differential Equations: Laplace Transforms

Input⇒ $\begin{bmatrix} y'_1(t) \\ y'_2(t) \\ \ldots \\ y'_n(t) \end{bmatrix} = A \begin{pmatrix} y_1(t) \\ y_2(t) \\ \ldots \\ y_n(t) \end{pmatrix} + \begin{pmatrix} g_1(t) \\ g_2(t) \\ \ldots \\ g_n(t) \end{pmatrix}$

Initial⇒ Conditions $\begin{pmatrix} y_1(0) \\ y_2(0) \\ \ldots \\ y_n(0) \end{pmatrix}$

Laplace⇒ Transform $\begin{bmatrix} Y_1(s) \\ Y_2(s) \\ \ldots \\ Y_n(s) \end{bmatrix} = [sI_n - A]^{-1} \begin{pmatrix} y_1(0) \\ y_2(0) \\ \ldots \\ y_n(0) \end{pmatrix} + [sI_n - A]^{-1} \begin{pmatrix} G_1(s) \\ G_2(s) \\ \ldots \\ G_n(0) \end{pmatrix}$

Solution⇒ $\begin{pmatrix} y_1(t) \\ y_2(t) \\ \ldots \\ y_n(t) \end{pmatrix} = L^{-1} \begin{bmatrix} Y_1(s) \\ Y_2(s) \\ \ldots \\ Y_n(s) \end{bmatrix}$

Laplace Transforms and Solutions using inverse Laplace Transforms

$$\begin{pmatrix} \frac{5}{s^2\,(s-1)} - \frac{1}{s\,(s-1)} + \frac{1}{s^3} \\ \frac{1}{s^2+s-2} + \frac{s+1}{s^2\,(s^2+s-2)} - \frac{s+1}{(s-1)\,(s^2+s-2)} + \frac{s}{(s^2+1)\,(s^2+s-2)} + \frac{5\,(s+1)}{s\,(s-1)\,(s^2+s-2)} \\ \frac{1}{s^2+s-2} + \frac{2}{s^2\,(s^3+s^2-2\,s)} + \frac{2}{s^4} - \frac{s+1}{(s-1)\,(s^2+s-2)} + \frac{s}{(s^2+1)\,(s^2+s-2)} + \frac{5\,(s+1)}{s\,(s-1)\,(s^2+s-2)} \\ \frac{2}{s^2\,(s^2+s-2)} - \frac{2}{(s-1)\,(s^2+s-2)} + \frac{s}{s^2+s-2} + \frac{10}{s\,(s-1)\,(s^2+s-2)} + \frac{s^2}{(s^2+1)\,(s^2+s-2)} \\ \frac{5}{s\,(s-1)} - \frac{1}{s-1} \end{pmatrix}$$

$$\begin{pmatrix} 4\,e^t - 5\,t + \frac{t^2}{2} - 4 \\ \frac{49\,e^{-2\,t}}{180} - \frac{t}{2} - \frac{3\cos(t)}{10} - \frac{31\,e^t}{18} + \frac{\sin(t)}{10} + \frac{8\,t\,e^t}{3} + \frac{7}{4} \\ \frac{49\,e^{-2\,t}}{180} - \frac{t}{2} - \frac{3\cos(t)}{10} - \frac{31\,e^t}{18} + \frac{\sin(t)}{10} + \frac{8\,t\,e^t}{3} - \frac{t^2}{2} + \frac{t^3}{3} + \frac{7}{4} \\ \frac{\cos(t)}{10} - \frac{49\,e^{-2\,t}}{90} - t - \frac{55\,e^t}{18} + \frac{3\sin(t)}{10} + \frac{8\,t\,e^t}{3} + \frac{9}{2} \\ 4\,e^t - 5 \end{pmatrix}$$

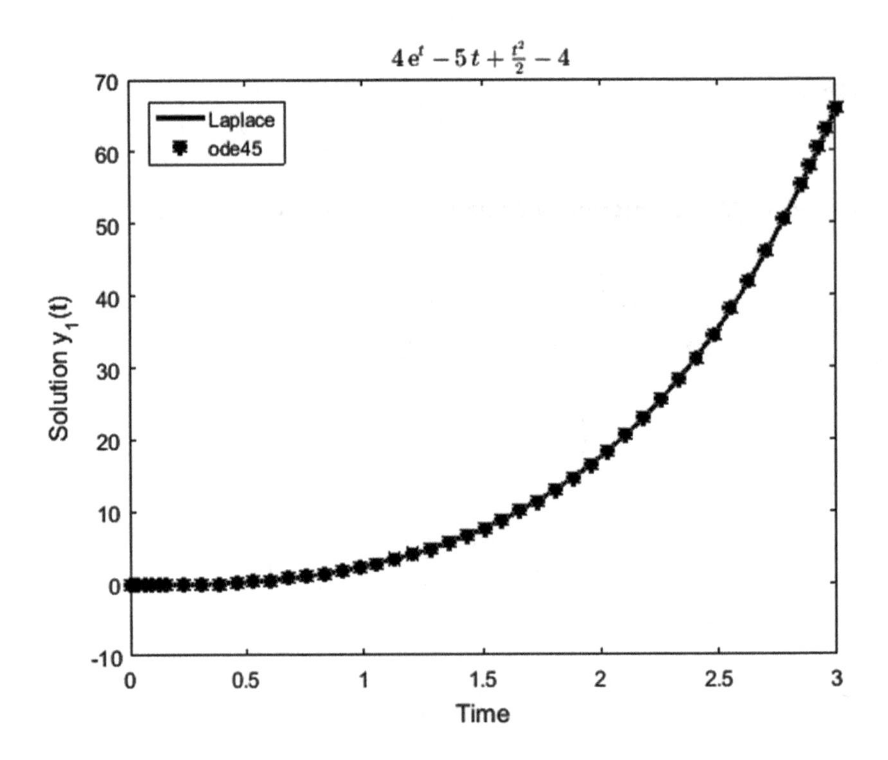

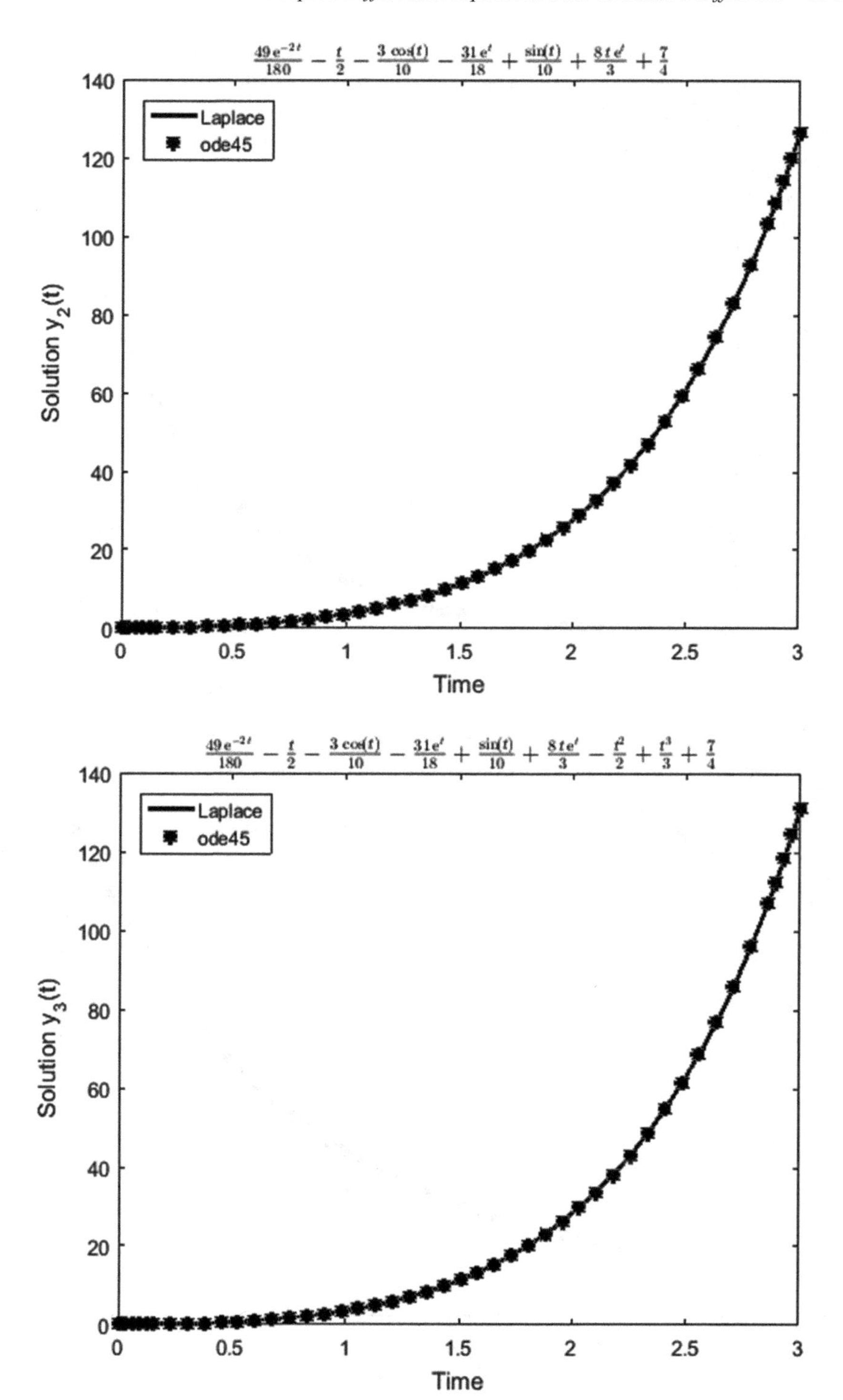

$\frac{49e^{-2t}}{180} - \frac{t}{2} - \frac{3\cos(t)}{10} - \frac{31e^t}{18} + \frac{\sin(t)}{10} + \frac{8te^t}{3} + \frac{7}{4}$
Laplace
ode45
Solution $y_2(t)$
Time
0
20
40
60
80
100
120
140
0.5
1
1.5
2
2.5
3
$\frac{49e^{-2t}}{180} - \frac{t}{2} - \frac{3\cos(t)}{10} - \frac{31e^t}{18} + \frac{\sin(t)}{10} + \frac{8te^t}{3} - \frac{t^2}{2} + \frac{t^3}{3} + \frac{7}{4}$
Laplace
ode45
Solution $y_3(t)$
Time

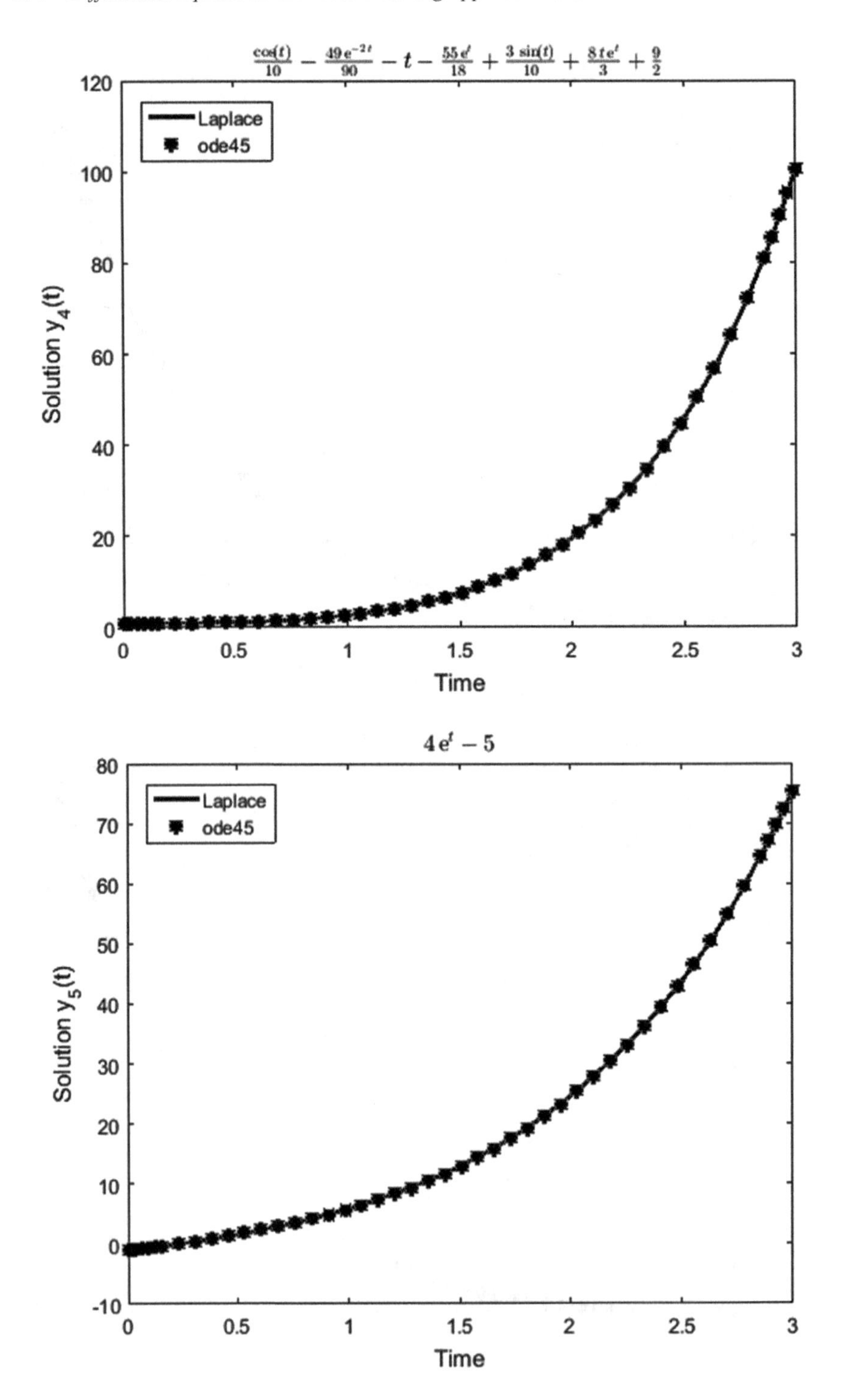

cos(t)/10 − 49e^{−2t}/90 − t − 55e^t/18 + 3 sin(t)/10 + 8te^t/3 + 9/2
Laplace
ode45
Solution y4(t)
Time
4e^t − 5
Laplace
ode45
Solution y5(t)
Time

5.6 Summary

First order coupled differential equations have been presented in this chapter. The case of a pair of homogeneous first order equations is presented first. The analysis of the properties of systems described by these equations is carried through the use of phase portraits and state of equilibrium. The solution is obtained using the concept of eigenvalues and eigenvectors with detailed attention given to the case of defective coefficient matrices and generalized eigenvectors. The solution is also obtained by converting the set of equations to a second order homogeneous differential equation with the solution set obtained using the roots of the characteristic equation, thus avoiding the need for generalized eigenvectors. The solution set is also obtained using Laplace transforms, thus providing a very general means of obtaining and comparing the solutions. Detailed examples are given, with each example annotated with the appropriate theory, along with comparison of results and verification using Runge-Kutta methods. The analysis is then extended to the case of a pair of non-homogeneous first order differential equations with the particular solutions set obtained using the method of variation of parameters. Once again, the solution set is validated through the use of Laplace transforms and verified through the use of Runge-Kutta methods.

The study is extended to sets of multiple coupled first order non-homogeneous systems. The solution is once again obtained using eigenvalues and eigenvectors with an increase in complexity as one goes from a pair of equations to three, four and more. The use of the eigenvalue based approach is limited to $n<5$ because of the computational complexity associated with defective matrices. Laplace transform based solution has been used as a means for verification and for sets with 5 or more equations, only the Laplace transform method is used to obtain the solution. In every case, the solution is further verified using the Runge-Kutta method.

The examples shown cover a wide array of possibilities with each example fully annotated with the appropriate theory, explanations, justifications and even verification carried out symbolically and displayed.

5.7 Exercises

1. For the set of values a and b given below, examine the phase plots and verify that they reflect the state of equilibrium suggested by the eigenvalues.
   a=[-3,-2,-1,0,1,2,3];
   b=[4,-4,0,-2,1,-1];
   A=[1,a;-1,b];
2. For all the cases above, obtain solutions to associated homogeneous system. Verify your results using the Laplace transforms, conversion to a second order differential equation and finally ode 45.
   $A=[A_{11},A_{12},A_{21},A_{22},x_1(0),x_2(0)]$
   A= [1, -3, -1, 1, -1, 1];
   A= [-1, -3, 1, -1, -1, 1];
   A=[1, 0, -1, -2,1,0];
   A=[1, 2, 3, 2, -1, 0];
   A=[2,-2, -1,-2,0, 1];
   A=[1 3 -1 -1, -1, 1];

A=[0,3,1,2,-1,1];

3. Particular and complete solutions. A is A=[A_{11},A_{12},A_{21},A_{22},x_1(0),x_2(0)]
   A= [1, -3, -1, 1, -1, 1];g1(t)='t*exp(2*t)';g2(t)='t';
   A= [-1, -3, 1, -1, -1, 1];g1(t)='cos(t)';g2(t)='sin(t)';
   A=[1, 0, -1, -2,1,0];g1(t)='t*exp(t)';g2(t)='exp(-2*t)';
   A=[1, 2, 3, 2, -1, 3];g1(t)='exp(-t)';g2(t)='4+t';
   A=[2,-2, -1,-2,1, 1];g1(t)='cosh(sqrt(6)*t)';g2(t)='t';
   A=[3, 0, -4,-2, 0, 1];g1(t)='t';g2(t)='t+5';
   A=[0,3,1,2,-1,1];g1(t)='t*exp(-t)';g2(t)='exp(-t)';
   A=[3,0,1,-3,-1,1];g1(t)='cosh(3*t)';g2(t)='sinh(3*t)';
4. For the following coefficient matrices, obtain the general homogeneous solution using the eigenvalues and eigenvectors. Provide the eigenvectors, eigenvalues and state whether the matrix is defective and if generalized eigenvalues were used. Provide the fundamental matrix as well. Compare the solution to one obtained from MATLAB.
   A=[1,1,-1;0,-1,0;1,0,1];
   A=[1,-1,1;0,-1,0;1,0,1];
   A=[1,1,1;-1,1,1;1,0,0];
   A=[-3,0,-1;-1,1,0;1,0,0];
   A=[-3,1,-1;-1,-1,-1;0,2,0];
   A=[1,1,0,0;1,0,1,1;1,0,0,1;0,0,1,0];
   A=[1,1,0,0;1,1,1,1;1,0,0,-1;0,1,0,1];
   A=[1,0,0,4;1,0,-1,-1;-1,0,0,1;2,0,0,3];
   A=[0,-2,0,0;2,0,0,0;4,0,0,0;0,0,1,0];
   A=[0,0,0,6;1,0,0,0;0,0,0,-2;0,0,5,0];
   A=[1,1,3,1;0,0,0,1;0,1,0,1;0,0,0,5];
   A=[1,0,0,0;1,-3,0,0;0,-4,0,0;0,0,2,0];
   A=[1,-1,0,0;-3,-1,0,0;0,-4,0,0;0,0,2,0];
5. For the coupled first order differential equations characterized by the coefficient matrix, obtain the particular solution sets and complete solutions using (1) eigenvalues and eigenvectors (2) Laplace transforms and (3) ode45 methods with the given set of forcing functions and initial values.
   A=[1,1,-1;0,-1,0;1,0,1]; x0=[0 -1 -1];g(t)={'cos(t)';'sin(t)';'exp(-t)'};
   A=[1,-1,1;0,-1,0;1,0,1];x0=[0 1 -1];g(t)={'t';'t^2';'exp(-t)'};
   A=[1,1,1;-1,1,1;1,0,0];x0=[0 -1 1];g(t)={'4';'t';'t*exp(t)'};
   A=[-3,1,-1;-1,-1,-1;0,2,0];x0=[1 -1 -1];g(t)={'exp(-2*t)';'4';'t'};
   A=[1,1,0,0;1,0,1,1;1,0,0,1;0,0,1,0];x0=[0 1 -1 -1]; g(t)={'t';'exp(-2*t)';'4';'t^2'};
   A=[1,0,0,4;1,0,-1,-1;-1,0,0,1;2,0,0,3];x0=[0 -1 1 -1]; g(t)={'5';'0';'t';'0'};
   A=[0,-2,0,0;2,0,0,0;4,0,0,0;0,0,1,0];x0=[0 -1 -1 0]; g(t)={'cos(2*t)';'sin(2*t)' ;'t';'5'};
   A=[0,0,0,6;1,0,0,0;0,0,0,-2;0,0,5,0];x0=[0 0 -1 0]; g(t)={'t';'t^2';'t';'t+2'};
   A=[1,-1,0,0;-3,-1,0,0;0,-4,0,0;0,0,2,0]; x0=[0 0 -1 0]; g(t)={'t';'exp(-2*t)';'exp(2*t)';'4'};

Appendices

Appendix A

Numerical Techniques for Solving Differential Equations

A-1 Euler's Method 385
A-2 Runge-Kutta Method 386
A-2.1 Runge-Kutta methods for coupled first order systems 391
A-2.2 Runge-Kutta methods for higher order differential equations 393

A-1 Euler's Method

Analytical solutions to differential equations might not always exist and it becomes necessary to use numerical methods to obtain solutions to differential equations. Numerical methods also offer a means to verify solutions obtained using other approaches. One of the simplest methods relies on the Taylor series expansion for a function and it can be used as a starting point for obtaining an approximate solution to a differential equation. Consider the case of a first order differential equation

$$\frac{dy}{dt} = y' = f(t,y), \quad y(0) = y_0 \ . \tag{A.1}$$

In eqn. (A.1), t is the independent variable and f(.) depends on both t and the dependent variable y. The initial condition is given as y_0. For a small change in time Δt, the Taylor series provides

$$y(t+\Delta t) = y(t) + y'(t)\Delta t + y''(t)\frac{(\Delta t)^2}{2!} + \cdots \ . \tag{A.2}$$

If the series is truncated by ignoring the terms in quadratic and higher order in Δt, eqn. (A.2) becomes

$$y(t+\Delta t) = y(t) + y'(t)\Delta t + O(\Delta t) \ . \tag{A.3}$$

The last term in eqn. (A.3) represents the error in limiting the expansion to the linear term in Δt. Ignoring the error, eqn. (A.3) becomes

$$y(t+\Delta t) = y(t) + y'(t)\Delta t = y(t) + f(t,y)\Delta t \; . \tag{A.4}$$

If the time window where the solution exists is divided into n-segments, eqn. (A.4) can be interpreted as follows

$$y_m(t) = y_{m-1}(t) + f(t, y_{m-1})\Delta t, \quad m \le n \, . \tag{A.5}$$

Thus, the solution to the differential equation can be obtained at different time instants in [0, t] in steps of Δt. It can be seen from eqn. (A.5), the accuracy of the solution will depend on the step size and the nature of y(t). If y(t) is highly non-linear, the linear approximation provided in eqn. (A.5) will be insufficient and error in the solution will increase. Equation (A.5) represents the numerical solution based on Euler's method.

The numerical approach to the solution of the differential equation can be improved if there is a better way to estimate f(t,y) for each iteration. In other words, while Euler's method uses the local value of $f(t,y_{n-1})$, an improved method would require that the estimate of $f(t,y_{n-1})$ is made by taking more samples of f(t,y) from the neighborhood. Euler's method uses uniform step sizes and if step sizes could be adjusted, the errors are likely to be less because varying step sizes will model nonlinear functions better. Another approach lies in using an improved average value of the function at each step. One such method is the Runge-Kutta method. Depending on the number of terms used for averaging, it is possible to have a 2nd order, 3rd order, 4th order method, etc. The 4th order method which is extensively used (normally called the Runge-Kutta method) is described below.

A-2 Runge-Kutta Method

The Runge-Kutta method uses a weighted average as the local estimate. Equation (A.5) is modified as

$$y_m(t) = y_{m-1}(t) + f(t,y)_{av}\,\Delta t \; . \tag{A.6}$$

The average value of the local estimate is

$$\begin{aligned}
f(t,y)_{av} &= \frac{k_1 + 2k_2 + 2k_3 + k_4}{6} \\
k_1 &= f(t_{m-1}, y_{m-1}) \\
k_2 &= f\left(t_{m-1} + \frac{\Delta t}{2}, y_{m-1} + k_1\frac{\Delta t}{2}\right) \\
k_3 &= f\left(t_{m-1} + \frac{\Delta t}{2}, y_{m-1} + k_2\frac{\Delta t}{2}\right) \\
k_4 &= f(t_{m-1} + \Delta t, y_{m-1} + k_3\Delta t)
\end{aligned} \tag{A.7}$$

The notion of the average is clear from eqn. (A.7).

In the adaptive step size approach, the step size can be automatically adjusted as the calculation proceeds based on the local truncation error (LTE), defined as the difference in the numerical value of the solution at any point and its actual value. These procedures will now be explained using an example which compares the Euler's and the Runge-Kutta methods. A first order differential equation is

$$\frac{dy}{dt} = 3 + 2t - y, \quad y(0) = 2 \; . \tag{A.8}$$

The analytical solution to the differential equation is

$$y(t) = e^{-t} + 2t + 1 \, . \tag{A.9}$$

Two step sizes (Δt=0.4 and 0.1) are used. The local truncation error (LTE) at each step is given by

$$er_k = y_k(analyt) - y_k(numer), k = 1, 2, \ldots \; . \tag{A.10}$$

The mean square error is

$$MSE = \frac{1}{N}\sum_k [er_k]^2 \; . \tag{A.11}$$

In eqn. (A.11), N is the total number of steps. A MATLAB® code that obtains the solution using both Euler's and Runge-Kutta methods and examination of the errors is given below. MATLAB automatically chooses the appropriate number of step sizes needed in ode45(.).

```
function euler_ode_demo
% dy/dt=3+2t-y, y(0)=2
% Comparison of Euler's and Runge-Kutta methods. Two steps sizes are used
% for Euler's while the step size is automatically chosen (adaptive sizing)
% by MATLAB. The local truncation errors and the mean square errors are
% compared. It is seen that while the error with Euler's declines with
% reduction in step size, Runge-Kutta method uses the lowest number of
% steps and provides the most accurate estimate of the solution
% P M Shankar, August 2016
close all
tmin=0;%initial time
tMax=2;%maximum value of time
% test the error with two step sizes for Euler
dt1=0.4;
t1=tmin:dt1:tMax; % steps, size 0.4
dt2=.2;
t2=tmin:dt2:tMax; % steps, size 0.2
y0=2; % initial condition
y1(1)=y0; % initial condition restated
y2(1)=y0; % initial condition restated
```

```
N1=length(t1);N2=length(t2);
for k=2:N1 % step size 0.4
   yy=ode_eulerfun1(t1(k-1),y1(k-1));
   y1(k)=y1(k-1)+dt1*yy;
end;
for k=2:N2 % step size 0.1
   yy=ode_eulerfun1(t2(k-1),y2(k-1));
   y2(k)=y2(k-1)+dt2*yy;
end;
% Runge-Kutta (ODE45)
tm=[0:tMax]; % MATLAB chooses the step size
[T,YDE]=ode45(@ode_eulerfun1,tm,y0);
% now verify the accuracy of the methods
% Symbolic solution ----> dy/dt=3+t-y, y(0)=2
syms x y t
y=dsolve('Dy=3+2*t-y','y(0)=2');
yys=MATLABFunction(y);

% disp(['Analytical solution of dy/dt= 3+2t-y, y(0)=2 is y(t)=' ,yyt])
% Convertion to MATLAB Function for plotting and error calculations

% create samples of analytical solution at appropriate time instants
ya1=yys(t1);% samples at step size 0.4
ya2=yys(t2); % samples at step size 0.1
yad=yys(T);% % samples at the instants generated by Runge-Kutta (ODE)
figure, plot(t1,ya1,t1,y1,'k*',t2,y2,'bd',T,YDE,'rs')
xlabel('time t'),ylabel('solution y(t)')
leg1=['Euler''s \Deltat = ',num2str(dt1)];
leg2=['Euler''s \Deltat = ',num2str(dt2)];
leg3=['Runge-Kutta'];
legend('Analytical',leg1,leg2,leg3,'location','best')
text(0.78,2.52,'y(t) = ','fontsize',12)
text(1,2.5,['$' latex(y) '$'],'interpreter','latex','fontsize',14)
N3=length(T);% number of samples from ODE
text(0.2,4.2,[leg1,', N = ',num2str(N1)])
text(0.2,4,[leg2,', N = ',num2str(N2)])
text(0.2,3.8,[leg3,', N = ',num2str(N3)])
% Error calculations
```

```
er1=y1-ya1;
er2=y2-ya2;
er3=YDE-yad;
MSE1=sum(er1.^2)/N1;
MSE2=sum(er2.^2)/N2;
MSEOD=sum(er3.^2)/N3;
tit=['Anylt Numer LTE'];
val1=[ya1;y1;er1]';
val2=[ya2;y2;er2]';
val3=[yad,YDE,er3];
figure,xlim([0,5]),ylim([0,5])
title('Error Analysis: Comparison of raw data','backgroundcolor','w')
text(.5,4.5,tit,'color','b')
text(3,4.5,tit,'color','b')
text(1,1.2,tit,'color','b')
text(.5,3.5,num2str(val1))
text(3,3,num2str(val2))
text(1,0.65,num2str(val3))
text(.5,4.8,[leg1,', N = ',num2str(N1)],'color','r','fontweight','bold')
text(3.3,4.8,[leg2,', N = ',num2str(N2)],'color','r','fontweight','bold')
text(1.1,1.5,[leg3,', N = ',num2str(N3)],'color','r','fontweight','bold')
%text(4.2,0.1,'p m shankar','color','g')
axis off

figure,plot(t1,er1,'k*',t2,er2,'bd',T,er3,'rs')
xlabel('time t')
ylabel('Local Truncation Error [y_{Analyt}-y_{numerical}]')
ylim([-.1,.1])
legend(leg1,leg2,leg3,'location','best')
title('Error Analysis')
text(0.2,.05,[leg1,', MSE = ',num2str(MSE1)])
text(0.2,.04,[leg2,', MSE = ',num2str(MSE2)])
text(0.2,.03,[leg3,', MSE = ',num2str(MSEOD)])
end

function dy= ode_eulerfun1(t,y)
%dy/dt=3+2*t-y
dy=3+2*t-y;

end
```

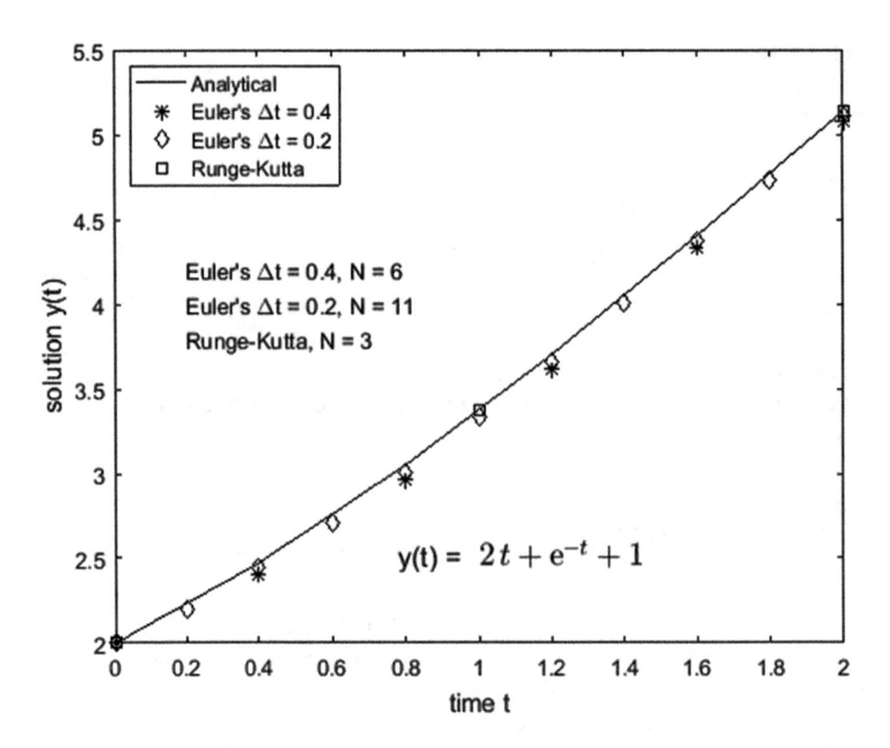

Error Analysis: Comparison of raw data

Euler's Δt = 0.4, N = 6

Anylt	Numer	LTE
2	2	0
2.4703	2.4	-0.07032
3.0493	2.96	-0.089329
3.7012	3.616	-0.085194
4.4019	4.3296	-0.072297
5.1353	5.0778	-0.057575

Euler's Δt = 0.2, N = 11

Anylt	Numer	LTE
2	2	0
2.2187	2.2	-0.018731
2.4703	2.44	-0.03032
2.7488	2.712	-0.036812
3.0493	3.0096	-0.039729
3.3679	3.3277	-0.040199
3.7012	3.6621	-0.03905
4.0466	4.0097	-0.036882
4.4019	4.3678	-0.034124
4.7653	4.7342	-0.031081
5.1353	5.1074	-0.027961

Runge-Kutta, N = 3

Anylt	Numer	LTE
2	2	0
3.3679	3.3679	4.5507e-08
5.1353	5.1353	3.3482e-08

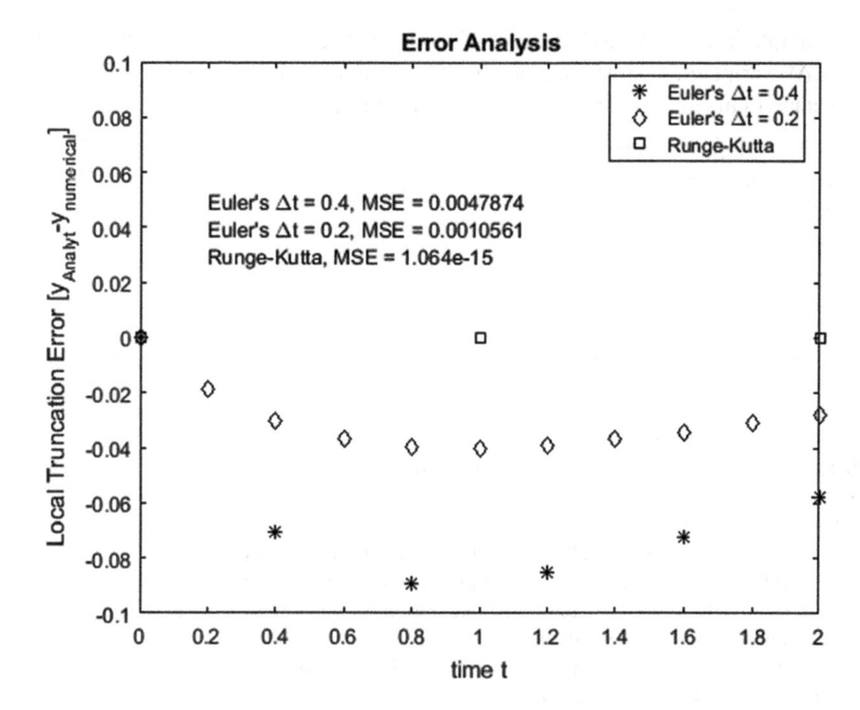

A-2.1 Runge-Kutta methods for coupled first order systems

Runge-Kutta methods can easily be applied to solve coupled first order differential equations. An example of a system with a two first order non-homogeneous equation is given. Consider a system defined by the coefficient matrix A, initial conditions $x_1(0)$ and $x_2(0)$, and forcing functions gt_1 and gt_2 as

$$A = \begin{bmatrix} 1 & 1 \\ 4 & 1 \end{bmatrix}. \tag{A.12}$$

$$\begin{bmatrix} x_1(0) \\ x_2(0) \end{bmatrix} = \begin{bmatrix} 0 \\ -1 \end{bmatrix}. \tag{A.13}$$

$$\begin{bmatrix} gt_1 \\ gt_2 \end{bmatrix} = \begin{bmatrix} e^{-t} \\ t \end{bmatrix}. \tag{A.14}$$

The set of differential equations associated with the coefficient matrix and forcing functions is

$$\begin{aligned} x_1^{'}(t) &= x_1(t) + x_2(t) + e^{-t}, \quad x_1(0) = 0 \\ x_2^{'}(t) &= 4x_1(t) + x_2(t) + t, \quad x_2(0) = -1 \end{aligned}. \tag{A.15}$$

The example in eqn. (A.15) has been solved using the Runge-Kutta method and the MATLAB script and results are given below. The numerical solution set is compared to the analytical solution set obtained using dsolve(.).

```
function example_rungekutta_coupled
% define variables
close all
syms x_1(t) x_2(t)
xt=[x_1(t);x_2(t)];
D1y=diff(x_1,t);
D2y=diff(x_2,t);
A=[1,1;4,1]; % coefficient matrix
gt=[exp(-t);t]; % forcing functions
Dy=[D1y;D2y];% initial conditions
diffX=[Dy==A*xt+gt]; % create the differential equation set
[x1,x2]=dsolve(diffX,[x_1(0)==0;x_2(0)==-1]); % solve in MATLAB
x1=simplify(x1,'steps',100);
x2=simplify(x2,'steps',100);
mf1=MATLABFunction(x1); % create inline function solution x1(t)
mf2=MATLABFunction(x2);% create inline function solution x2(t)
tspan=[0 5]; % time span for numerical integration
[tt,yy]=ode45(@fun2f,tspan,[0; -1]);
% yy is the numerical solution in two columns. First column corresponds
% to x_1(t) and the second one to x_2(t)
plot(tt,mf1(tt),'-k',tt,yy(:,1),'k*',tt,mf2(tt),'--r',tt,yy(:,2),'ro')
legend('x_1(t)-Analyt','x_1(t)-Runge-Kutta',...
    'x_2(t)-Analyt','x_2(t)-Runge-Kutta','location','best')
xlabel('time'),ylabel('solution')
title('Solutions(s): Analytical vs. Numerical','color','b')
xx1=[x_1(t)==x1]; % create equation for display
xx2=[x_2(t)==x2];% create equation for display
 % display the differential equations
text(0.5,-1e5,['$' latex(feval(symengine,'rewrite',diffX,'D')) '$'],...
    'interpreter','latex','fontsize',12,'color','b')
 % display the solutions
 text(.5,-5e5,['$' latex(xx1) '$'],'interpreter','latex','fontsize',16,'color','b')
 text(.5,-6e5,['$' latex(xx2) '$'],'interpreter','latex','fontsize',16,'color','b')

end

function dxdt=fun2f(t,x)
% create the function for numerical evaulation
dxdt=zeros(2,1);
dxdt(1,1)=x(1,1)+x(2,1)+exp(-t);
dxdt(2,1)= 4*x(1,1)+x(2,1)+t;
end
```

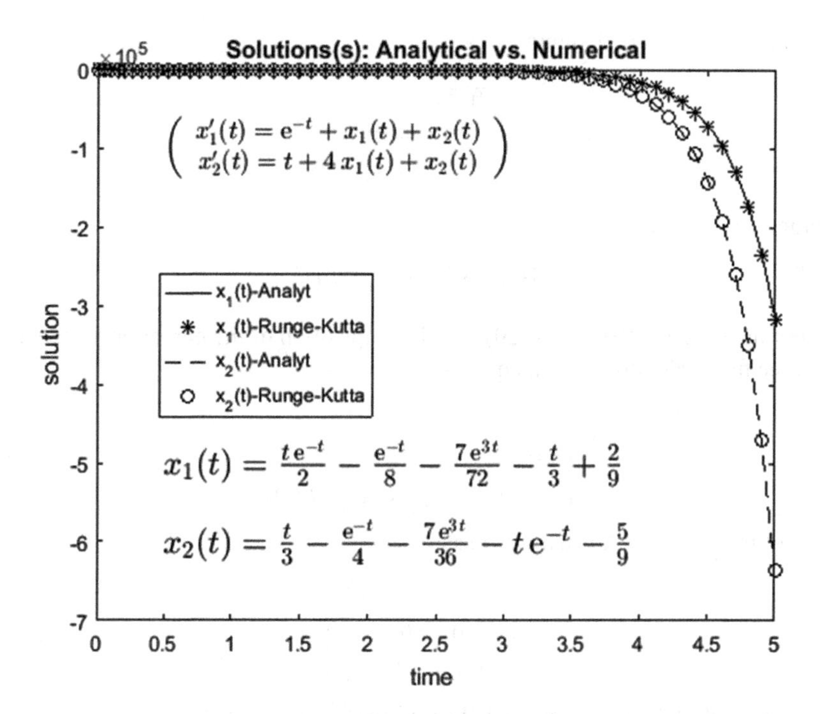

A-2.2 Runge-Kutta methods for higher order differential equations

Runge-Kutta methods are applicable to a set of first order differential equations as seen above. This means that a higher order differential equation must be decomposed to several first order differential equations before the Runge-Kutta numerical method can be applied. Consider a 3rd order differential equation

$$A_3 y''' + A_2 y'' + A_1 y' + A_0 y = f(t), \quad y''(0) = y_2, y'(0) = y_1, y(0) = y_0 \,. \tag{A.16}$$

Rewriting the differential equation as

$$y''' + \frac{A_2}{A_3} y'' + \frac{A_1}{A_3} y' + \frac{A_0}{A_3} y = \frac{f(t)}{A_3} \Rightarrow y''' + ay'' + by' + cy = h(t) \tag{A.17}$$

In eqn. (A.17),

$$\begin{aligned} a &= \frac{A_2}{A_3} \\ b &= \frac{A_1}{A_3} \\ c &= \frac{A_0}{A_3} \\ h(t) &= \frac{f(t)}{A_3} \end{aligned} \tag{A.18}$$

For the decomposition, defining

$$\begin{aligned} x_1 &= y \\ x_2 &= y' = x_1' \\ x_3 &= y'' = x_2' \end{aligned} \tag{A.19}$$

Equation (A.19) becomes

$$x_3' + ax_3 + bx_2 + cx_1 = h(t) \; . \tag{A.20}$$

Combining eqns. (A.19) and (A.20), the decomposition of the higher order equation into three first order differential equations becomes

$$\begin{aligned} x_1' &= x_2 \\ x_2' &= x_3 \\ x_3' &= -[ax_3 + bx_2 + cx_1] + h(t) \end{aligned} \; . \tag{A.21}$$

The coefficient matrix A is expressed as

$$A = \begin{bmatrix} 0 & 1 & 0 \\ 0 & 0 & 1 \\ -c & -b & -a \end{bmatrix} . \tag{A.22}$$

In matrix notation (see Appendix E), the differential equation in eqn. (A.17) becomes

$$\begin{bmatrix} x_1' \\ x_2' \\ x_3' \end{bmatrix} = \begin{bmatrix} 0 & 1 & 0 \\ 0 & 0 & 1 \\ -c & -b & -a \end{bmatrix} \begin{bmatrix} x_1 \\ x_2 \\ x_3 \end{bmatrix} + \begin{bmatrix} 0 \\ 0 \\ h(t) \end{bmatrix} . \tag{A.23}$$

The initial conditions become the vector

$$X(0) = \begin{bmatrix} y_0 \\ y_1 \\ y_2 \end{bmatrix} . \tag{A.24}$$

As a specific example, consider the following 3rd order differential equation with initial conditions

$$2y''' + 4y'' + 3y' + y = \cos(t),\; y(0) = 1, y'(0) = 0, y''(0) = \text{-}1 \tag{A.25}$$

The coefficient matrix A is

$$A = \begin{bmatrix} 0 & 1 & 0 \\ 0 & 0 & 1 \\ -\frac{1}{2} & -\frac{3}{2} & -2 \end{bmatrix} . \tag{A.26}$$

The example has been worked out in MATLAB. The solution has been obtained using the Runge-Kutta method and verified by comparing it to the solution obtained using dsolve(.) in MATLAB. Two different approaches exist for implementing the Runge-Kutta method. One uses the coefficient matrix in eqn. (A.26). In the other case, the direct conversion of higher order differential equation into appropriate first order differential equations is carried out in MATLAB. This will be shown following the discussion of the example of a second order differential equation.

A second order differential equation is

$$A_2 y''+A_1 y'+A_0 y=f(t),\quad y''(0)=y_2, y'(0)=y_1\ . \tag{A.27}$$

Proceeding similarly, the coefficient matrix becomes

$$A=\begin{bmatrix} 0 & 1 \\ -\dfrac{A_0}{A_2} & -\dfrac{A_1}{A_2}\end{bmatrix}. \tag{A.28}$$

The decomposition leads to

$$\begin{bmatrix} x_1' \\ x_2' \end{bmatrix}=\begin{bmatrix} 0 & 1 \\ -\dfrac{A_0}{A_2} & -\dfrac{A_1}{A_2}\end{bmatrix}\begin{bmatrix} x_1 \\ x_2 \end{bmatrix}+\begin{bmatrix} 0 \\ \dfrac{f(t)}{A_2}\end{bmatrix}. \tag{A.29}$$

The initial conditions become

$$X(0)=\begin{bmatrix} y_0 \\ y_1 \end{bmatrix}. \tag{A.30}$$

Consider the example

$$y''+4y'+3y=te^{-t},\ y(0)=1, y'(0)=-1\ . \tag{A.31}$$

The two examples, the 3rd order differential equation in eqn. (A.25) and the 2nd order differential equation in eqn. (A.31) are solved in MATLAB. The analytical solutions obtained using dsolve(.) are compared to the corresponding solutions obtained using the Runge-Kutta method. All the steps in the MATLAB script are commented at the appropriate lines of the script. The script also demonstrates the creation of symbolic differential equations matching the given examples so that one can use the MATLAB function odeToVectorfield(.) to create the inline function needed for the Runge-Kutta method. For the case of the 3rd or equation, the external function based approach is also used. The use of MATLABFunction(.) to convert the symbolic solution to an inline form is also shown.

```
function higherorder_ode_example
```

Part 1: 3rd order differential equation and Part 2: 2nd order equation

Part 1: example of solution of a 3rd order differential equations using Runge-Kutta method. Results are verified using dsolve(.) by plotting the results. 2y'''+4y''+3y'+y=cos(t), y(0)=1,y'(0)=0,y''(0)=-1 There are two ways of generating the function for ODE45 input. One is the traditional approach creating the function externally and the other is based on odeToVectorField(.) which can decompose a higher order differential equation into several first order ones. Both approaches are implemented.

```
close all
A=[0,1,0;0,0,1;-1/2,-3/2,-2];%The coefficient matrix A
X0=[1;0;-1]; % initial conditions
tspan=[0 2]; [T1,yy2]=ode45(@higherorderf,tspan,X0);
% note that yy(:,1) is the solution and other columns are the first and
% second derivatives
% now get the analytical solution using dsolve for verification
 syms t y
yt1=dsolve('2*D3y+4*D2y+3*Dy+y-cos(t)=0','y(0)=1','Dy(0)=0','D2y(0)=-1');
yt1=simplify(yt1,'steps',100);
% now verify the solution by plotting
yyt1=MATLABFunction(yt1);% create an inline function of the symbolic solution
plot(T1,yyt1(T1),'k',T1,yy2(:,1),'r*')
xlabel('time'),ylabel('Solution y(t)'),legend('Analyt','Numerical')
title('2y"""+4y""+3y"+y=cos(t), y(0)=1,y"(0)=0,y""(0)=-1')
text(.2,.5,'Using external function')
% it is also possible to create an inline function using the following
% approach

clear y t
syms y(t)
% define the differentials
Dy=diff(y);D2y=diff(y,2);D3y=diff(y,3);
ff1=[2*D3y+4*D2y+3*Dy+y-cos(t)==0]; % create a symbolic differential equation
V1 = odeToVectorField(ff1);% break higher order DE into 3 first order ones
FF1 = MATLABFunction(V1,'vars', {'t','Y'});% create inline Function for ODE
[T2,yx1]=ode45(FF1,tspan,X0);
figure,plot(T2,yyt1(T2),'k',T2,yx1(:,1),'r*')
xlabel('time'),ylabel('Solution y(t)'),legend('Analyt','Numerical')
title('2y"""+4y""+3y"+y=cos(t), y(0)=1,y"(0)=0,y""(0)=-1')
text(.2,.5,'Using odeToVectorField')
% part 2 example of a second order differential equation
clear all
% y"+4y'+3y=texp(-t),y(0)=1,y'(0)=-1;
y0=1;y1=-1;
syms y(t)
Dy=diff(y);D2y=diff(y,2); % define the differentials
ff2=[D2y+4*Dy+3*y-t*exp(-t)==0]; % create a symbolic differential equation
V2 = odeToVectorField(ff2);% break higher order DE into 3 first order ones
FF2 = MATLABFunction(V2,'vars', {'t','Y'});% create inline Function for ODE
tspan=[0 2];
[T3,yx2]=ode45(FF2,tspan,[y0;y1]);
% get the solution using dsolve
yt2=dsolve('D2y+4*Dy+3*y-t*exp(-t)=0','y(0)=1','Dy(0)=-1');
yt2=simplify(yt2,'steps',100);
% now verify the solution by plotting
yyt2=MATLABFunction(yt2);% create an inline function of the symbolic solution
figure,plot(T3,yyt2(T3),'k',T3,yx2(:,1),'r*')
xlabel('time'),ylabel('Solution y(t)'),legend('Analyt','Numerical')
title('y""+4y"+3y=te^{-t}, y(0)=1,y"(0)=-1')
```

```
function dy = higherorderf(t,x)
% external function for the 3rd order differential equation
dy=zeros(3,1);% this corresponds to the first order derivatives x1',x2',x3'
A=[0,1,0;0,0,1;-1/2,-3/2,-2];%The coefficient matrix A
ff=[0;0;cos(t)/2];
dy=A*x+ff;% three first order differential equations
end
```

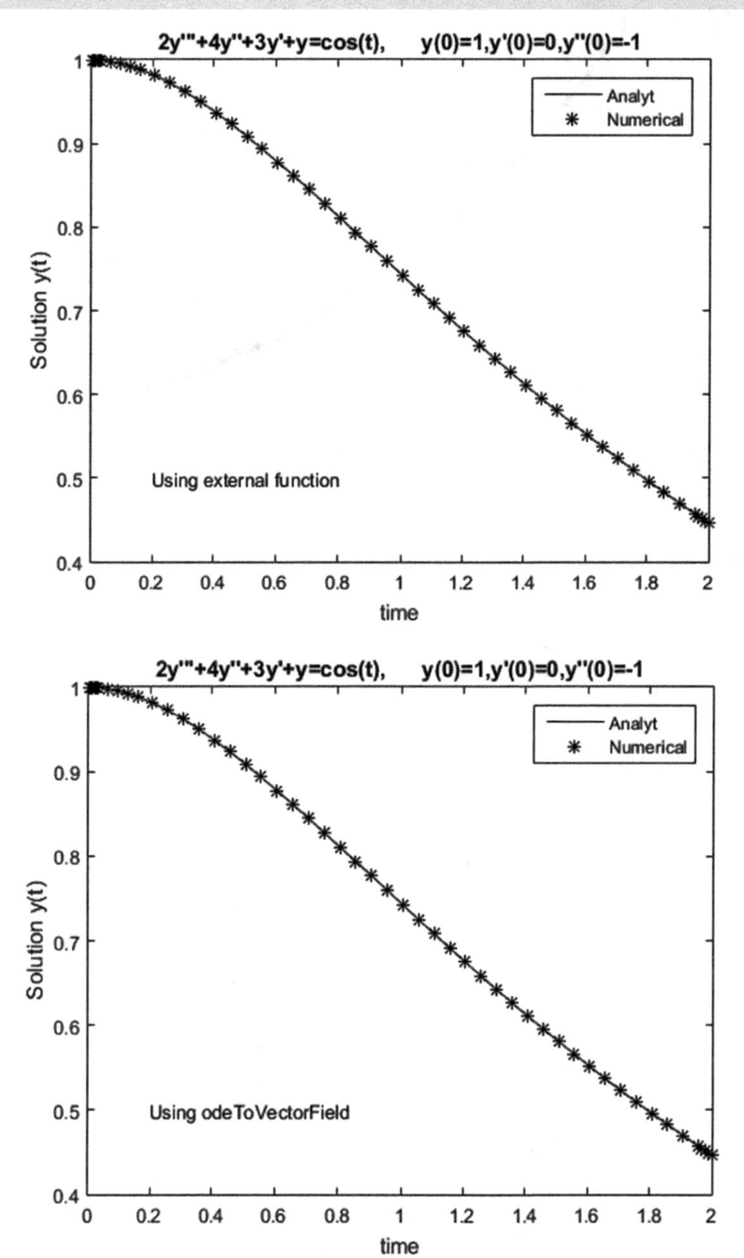

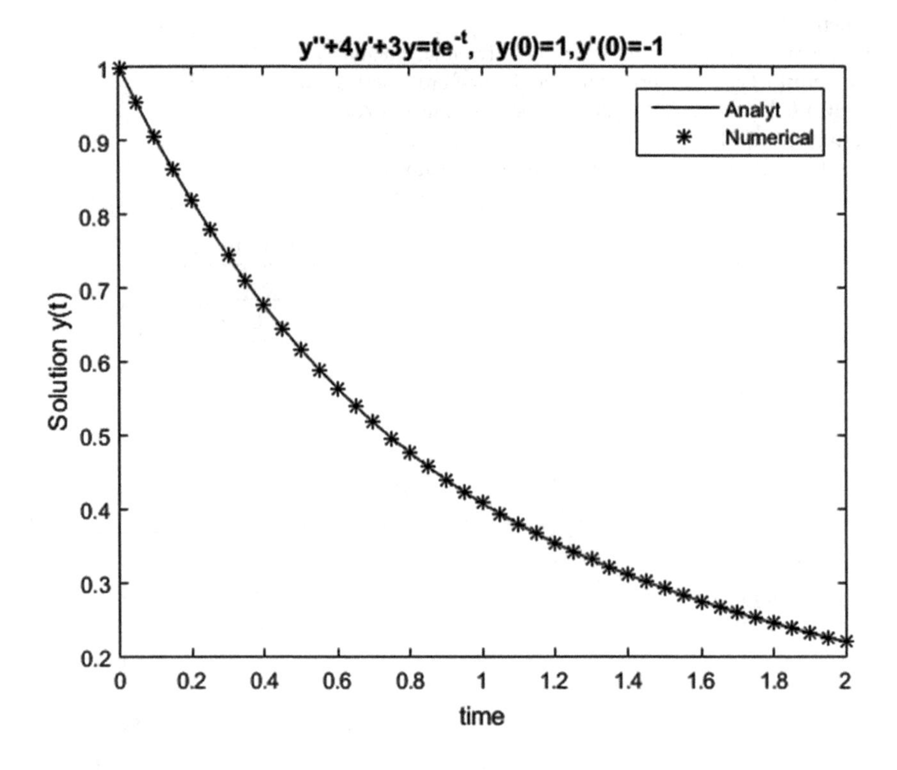
y''+4y'+3y=te^-t, y(0)=1,y'(0)=-1
Analyt
Numerical
Solution y(t)
time
1
0.9
0.8
0.7
0.6
0.5
0.4
0.3
0.2
0
0.2
0.4
0.6
0.8
1
1.2
1.4
1.6
1.8
2

Appendix B

Laplace and Inverse Laplace Transforms for Solving Differential Equations

B-1 Laplace and Inverse Laplace Transforms 399
B-1.1 Second order differential equation 401
B-1.2 Higher order differential equations 404
B-1.3 Coupled first order equations 406

B-1 Laplace and Inverse Laplace Transforms

The Laplace transform F(s) of a temporal function y(t) is the integral

$$L\left[y(t)\right]=Y(s)=\int_{0}^{\infty}y(t)e^{-st}dt\ . \tag{B.1}$$

In eqn. (B.1), $L[]$ represents the Laplace operator and Y(s) is the Laplace transform of y(t). Often, y(t) is defined to be in the 't' domain and Y(s) is defined to be in the 's' domain. Behavior of linear systems could be analyzed using Laplace transforms by taking advantage of their properties which allow a differential equation to be converted to an algebraic equation in the 's' domain. Such an algebraic equation can then be rearranged to obtain an expression for Y(s) which can be inverted to obtain the solution to the differential equation. This concept is illustrated in Figure B1 below.

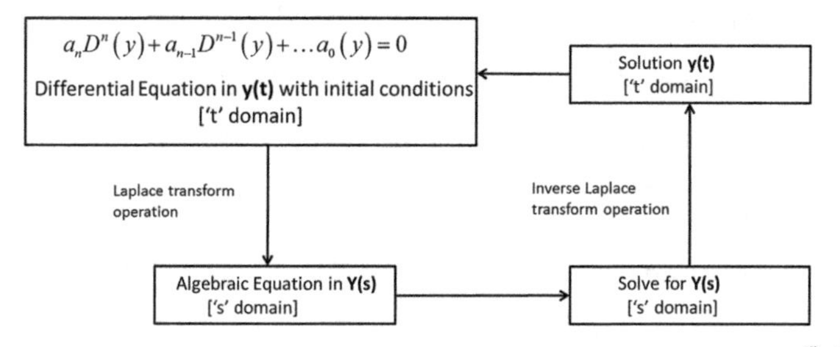

Figure B.1 Use of Laplace Transforms for solving differential equation $D^n(y)=\dfrac{d^n y(t)}{dt^n}$.

The usefulness of the Laplace transform arises from its properties, some of which are listed below.

Linearity

$$L\left[af(t)+bg(t)\right]=aL\left[f(t)\right]+bL\left[g(t)\right]=aF(s)+bG(s) \tag{B.2}$$

Scaling by t^n, n=1, 2, 3, …

$$L\left[t^n f(t)\right]=(-1)^n F^{[n]}(s) \tag{B.3}$$

$$F^{[n]}(s)=\frac{d^n}{ds^n}F(s)$$
$$f^{[n]}(t)=\frac{d^n}{dt^n}f(t) \tag{B.4}$$

Differentiation

$$L\left[f'(t)\right]=sF(s)-f(0)$$
$$L\left[f^{[n]}(t)\right]=s^nF(s)-s^{n-1}f(0)-s^{n-2}f^{[1]}(0)-\ldots-sf^{[n-2]}(0)-f^{[n-1]}(0) \tag{B.5}$$

For the special case of the derivative of second order

$$L\left[f^2(t)\right]=s^2F(s)-sf(0)-f^{[1]}(0)\ . \tag{B.6}$$

Multiplication by an exponential function

$$L\left[e^{bt}f(t)\right]=F(s-b) \tag{B.7}$$

The Table of Laplace transforms is given in Table B.1. Some of the additional properties are listed in Table B.2. These tables were generated using MATLAB®.

Table B.1 Laplace Transforms.

$y(t)$	$Y(s)=\int_0^\infty y(t)e^{-st}\,dt$
1	$\frac{1}{s}$
e^{at}	$-\frac{1}{a-s}$
t^n	$\frac{\Gamma(n+1)}{s^{n+1}}$
$\sin(at)$	$\frac{a}{a^2+s^2}$
$\cos(at)$	$\frac{s}{a^2+s^2}$
$e^{bt}\cos(at)$	$-\frac{b-s}{a^2+(b-s)^2}$
$e^{bt}\sin(at)$	$\frac{a}{a^2+(b-s)^2}$
$t^n e^{at}$	$\frac{\Gamma(n+1)}{(s-a)^{n+1}}$
$t^2\cos(at)$	$\frac{8s^3}{(a^2+s^2)^3}-\frac{6s}{(a^2+s^2)^2}$

s > 0; *a, b, c > s;* *n=0, 1, 2, 3,..*

Table B.2 Some additional properties of Laplace Transforms.

Properties of Laplace Transforms

$b\,h(t) + a\,y(t)$	$b\,H(s) + a\,Y(s)$
$y'(t)$	$s\,Y(s) - y(0)$
$y''(t)$	$s^2\,Y(s) - s\,y(0) - y'(0)$
$y'''(t)$	$s^3\,Y(s) - s\,y'(0) - s^2\,y(0) - y''(0)$

Use of the Laplace transform method requires that the differential equations are given along with the initial conditions. This means that if the differential equation is of nth order, there must be n initial values as seen from eqn. (B.5). The conceptual approach to the solution of a second order differential equation will be described first before examining differential equations of lower and higher orders and coupled first order differential equations. It is assumed that all these differential equations are linear with constant coefficients. A second order non-homogeneous differential equation with constant coefficients is

$$ay''+by'+cy=h(t);\quad y(0)=y_0,y'(0)=y_1 \tag{B.8}$$

Applying the property in eqn. (B.5), the Laplace transform of the differential equation in eqn. (B.8) becomes

$$a\left[s^2Y(s)-sy_0-y_1\right]+b\left[sY(s)-y_0\right]+cY(s)=H(s)\ . \tag{B.9}$$

The Laplace transform of y(t) is obtained by simplifying eqn. (B.9) as

$$Y(s)=\frac{(as+b)y_0+ay_1}{as^2+bs+c}+\frac{H(s)}{as^2+bs+c}=\frac{P(s)}{Q(s)} \tag{B.10}$$

The quantities P(s) and Q(s) are introduced to indicate that the transform of y(t) will be the ratio of two polynomial functions in s. The ratio P(s)/Q(s) needs to be decomposed into simpler forms of rational functions using the method of partial fractions so that the Table of Laplace transforms may be used. Laplace and inverse Laplace transforms are readily available in the Symbolic Toolbox of MATLAB.

B-1.1 *Second order differential equation*

Consider the following second order differential equation

$$y''-3y'-4y=2e^{-t},y(0)=1,y'(0)=-1 \tag{B.11}$$

Using eqn. (B.10), the Laplace transform of y(t) becomes

$$Y(s)=\frac{(s-3)-1}{s^2-3s-4}+\frac{\frac{2}{s+1}}{s^2-3s-4}=\frac{(s-4)}{s^2-3s-4}+\frac{2}{\left(s^2-3s-4\right)(s+1)} \tag{B.12}$$

Decomposing into partial fractions gives,

$$\frac{(s-4)}{s^2-3s-4}=\frac{1}{s+1} \tag{B.13}$$

$$\frac{2}{(s^2-3s-4)(s+1)}=\frac{2}{25(s-4)}-\frac{2}{5(s+1)^2}-\frac{2}{25(s+1)} \tag{B.14}$$

Taking the inverse Laplace transforms (from the Laplace transform Table) leads to

$$y(t)=y_1(t)+y_2(t) \tag{B.15}$$

$$y_1(t)=L^{-1}\left[\frac{1}{s+1}\right]=e^{-t} \tag{B.16}$$

$$y_2(t)=L^{-1}\left[\frac{2}{25(s-4)}-\frac{2}{5(s+1)^2}-\frac{2}{25(s+1)}\right]=\frac{2}{25}e^{4t}-\frac{2}{5}te^{-t}-\frac{2}{25}e^{-t} \tag{B.17}$$

Combining and simplifying, the solution becomes

$$y(t)=\frac{23}{25}e^{-t}+\frac{2}{25}e^{4t}-\frac{2}{5}te^{-t}\ . \tag{B.18}$$

It should be noted that partial fractions can also be obtained from MATLAB. For example, the decomposed components in eqn. (B.14) can be obtained as follows:

```
syms y(t) t s
f=laplace(2*exp(-t),t,s)
f1=f/(s*s-3*s-4)
p = feval(symengine,'partfrac',f1,'s')
MATLAB generates the following output

f = 2/(s + 1)

f1 =-2/((s + 1)*(- s^2 + 3*s + 4))

p = 2/(25*(s - 4)) - 2/(5*(s + 1)^2) - 2/(25*(s + 1))
```

The MATLAB code which performs the solution is given below.

```
% May 7, 2016
clear all
% solve 'D2y-3*Dy-4*y=2*exp(-t)','y(0)=1','Dy(0)=-1'
syms y(t) t s
ff=(s*s-3*s-4);% the denominator as^2+bs+c
term1=(s-4)/ff;
disp('display the Laplace transform of term 1')
disp(term1)
pp = feval(symengine,'partfrac',term1,'s');% decompose into partial fractions
pretty(pp)
y1=ilaplace(term1,s,t) % inverse Laplace transform of term 1
f=laplace(2*exp(-t),t,s) % get the Laplace transform of the forcing function
```

```
term2=f/ff;
disp('display the Laplace transform of term 2')
disp(term2)
y2=ilaplace(term2);% inverse Laplace directly without decomposition
p = feval(symengine,'partfrac',term2,'s'); % decompose into partial fractions
pretty(p) % only for display purposes
% get the solution using Laplace transform method
yLap=simplify(y1+y2,'steps',100);
disp('display the solution obtained using Laplace transforms')
disp(yLap)
% solution using dsolve(.) in MATLAB
yd=dsolve('D2y-3*Dy-4*y=2*exp(-t)','y(0)=1','Dy(0)=-1');
disp('display the solution obtained using dsolve')
disp(yd)
% verify that the solutions match: find the difference between the two
% solutions
disp('Verify that the Laplace solution is correct: Obtain the symbolic difference between
the two solutions')
disp('Difference will be 0 if the approach is correct')
disp(simplify(yLap-yd)) % verify that the solution is correct; must be equal to 0
```

```
display the Laplace transform of term 1
-(s - 4)/(- s^2 + 3*s + 4)

  1
-----
s + 1

y1 =
exp(-t)
f =
2/(s + 1)
display the Laplace transform of term 2
-2/((s + 1)*(- s^2 + 3*s + 4))

    2            2          2
---------- - ---------- - ----------
25 (s - 4)            2  25 (s + 1)
             5 (s + 1)

display the solution obtained using Laplace transforms
(exp(-t)*(2*exp(5*t) - 10*t + 23))/25

display the solution obtained using dsolve
(23*exp(-t))/25 + (2*exp(4*t))/25 - (2*t*exp(-t))/5

Verify that the Laplace solution is correct: Obtain the symbolic difference between the two
solutions
Difference will be 0 if the approach is correct
0
```

A figure automatically generated form the MATLAB based results from Chapter 3 appears below.

Complete Solution using Inverse Laplace Transform

$$\begin{array}{l} a\,y''(t)+b\,y'(t)+cy=g(t) \\ [y(0)=y_0,y'(0)=y_1] \end{array} \rightarrow Y_L(s)=\frac{(as+b)y_0+ay_1+G(s)}{(as^2+bs+c)}$$

$y''(t)-3\,y'(t)-4\,y(t)=2\mathrm{e}^{-t}$ $\quad (y_0=1 \quad y_1=-1)$

Laplace transform Y(s) of y(t)

$$Y_L(s)=-\frac{-s^2+3\,s+2}{(s+1)^2\,(s-4)}$$

Solution using inverse Laplace

$$y_L(t)=\frac{\mathrm{e}^{-t}\left(2\,\mathrm{e}^{5t}-10\,t+23\right)}{25}$$

Solution directly using dsolve(.)

$$y(t)=\frac{\mathrm{e}^{-t}\left(2\,\mathrm{e}^{5t}-10\,t+23\right)}{25}$$

B-1.2 Higher order differential equations

A higher order differential equation will now be solved using the concept of Laplace transforms. A 4th order differential equation is

$$y''''-y=\cos(t) \tag{B.19}$$

The initial conditions are

$$y'''(0)=1, y''(0)=0, y'(0)=-1, y(0)=1 \tag{B.20}$$

Taking Laplace transform of the differential equation leads to

$$s^4Y(s)-s^3y(0)-s^2y'(0)-sy''(0)-y'''(0)-Y(s)=\frac{s}{s^2+1} \tag{B.21}$$

Applying the initial conditions and collecting the terms in Y(s)

$$Y(s)\left[s^4-1\right]=\frac{s}{s^2+1}+\left(s^3-s^2+1\right) \tag{B.22}$$

The Laplace transform of y(t) is

$$Y(s)=\frac{\left(s^3-s^2+1\right)}{\left[s^4-1\right]}+\frac{s}{\left(s^2+1\right)\left[s^4-1\right]} \tag{B.23}$$

Taking the inverse Laplace transform and writing it as the sum of two terms gives

$$y(t)=y_1(t)+y_2(t)\ . \tag{B.24}$$

$$y_1(t) = L^{-1}\left(\frac{s^3 - s^2 + 1}{s^4 - 1}\right). \tag{B.25}$$

$$y_2(t) = L^{-1}\left[\frac{s}{(s^2+1)(s^4-1)}\right] \tag{B.26}$$

The solutions become

$$y_1(t) = \frac{1}{4}e^{-t} + \frac{1}{2}\cos(t) + \frac{1}{4}e^{t} - \sin(t) \tag{B.27}$$

$$y_2(t) = \frac{1}{8}\left[e^{t} + e^{-t}\right] - \frac{1}{4}\cos(t) - \frac{t}{4}\sin(t) \tag{B.28}$$

The solution becomes

$$y(t) = \frac{3}{4}\cosh(t) + \frac{1}{4}\cos(t) - \sin(t) - \frac{t}{4}\sin(t). \tag{B.29}$$

The MATLAB code that generates the solution is given below.

```
clear all;close all;clc;
syms y(t) t s
term1=(s^3-s^2+1)/(s^4-1);
y1t=ilaplace(term1,s,t);
disp('display the inverse Laplace transform of term 1')
disp(y1t)
term2=s/((s^4-1)*(s^2+1));
y2t=ilaplace(term2,s,t);
disp('display the inverse Laplace transform of term 2')
disp(y2t)
yt=simplify(y1t+y2t,'steps',100);
disp('Solution using Laplace')
disp(yt)
% solution directly using dsolve
yd=dsolve('D4y-y=cos(t)','D3y(0)=1','D2y(0)=0','Dy(0)=-1','y(0)=1');
disp('Solution using dsolve')
disp(yd)
% verify that the solutions match: find the difference between the two
% solutions
disp('Verify that the Laplace solution is correct: Obtain the symbolic difference between
the two solutions')
disp('Difference will be 0 if the approach is correct')
disp(simplify(yt-yd,'steps',100)) % must be zero

display the inverse Laplace transform of term 1
exp(-t)/4 + cos(t)/2 + exp(t)/4 - sin(t)

display the inverse Laplace transform of term 2
exp(-t)/8 - cos(t)/4 + exp(t)/8 - (t*sin(t))/4
```

```
Solution using Laplace
cos(t)/4 + (3*cosh(t))/4 - sin(t) - (t*sin(t))/4

Solution using dsolve
(3*exp(-t))/8 - cos(3*t)/16 + (5*cos(t))/16 + (3*exp(t))/8 - sin(t) - sin(t)*(t/4 + sin(2*t)/8)

Verify that the Laplace solution is correct: Obtain the symbolic difference between the two
solutions
Difference will be 0 if the approach is correct
0
```

A figure automatically generated form the MATLAB based results from Chapter 3 appears below.

Inverse Laplace based solution: Theory and Results

Differential Equation and Initial Conditions (General Case)

$$A_4\, y''''(t) + A_3\, y'''(t) + A_2\, y''(t) + A_1\, y'(t) + A_0\, y = g(t)$$

$$[y(0) = y_0, y'(0) = y_1, y''(0) = y_2, y'''(0) = y_3]$$

Laplace Transform Y(s)=A(s)+B(s)

$$A(s) = \frac{y_1\left(A_4\, s^2 + A_3\, s + A_2\right) + A_4\, y_3 + y_0\left(A_4\, s^3 + A_3\, s^2 + A_2\, s + A_1\right) + y_2\left(A_3 + A_4\, s\right)}{A_4\, s^4 + A_3\, s^3 + A_2\, s^2 + A_1\, s + A_0}$$

$$B(s) = \frac{G(s)}{A_4\, s^4 + A_3\, s^3 + A_2\, s^2 + A_1\, s + A_0}$$

Differential Equation and Initial Conditions (given)

$$y''''(t) - y(t) = \cos(t)$$

$$(\ y(0) = 1 \quad y'(0) = -1 \quad y''(0) = 0 \quad y'''(0) = 1\ )$$

Laplace Transform Y(s) of y(t)

$$Y(s) = \frac{\frac{s}{s^2+1} - s^2 + s^3 + 1}{s^4 - 1}$$

Solution y(t) using inverse Laplace

$$y(t) = \frac{\cos(t)}{4} + \frac{3\cosh(t)}{4} - \sin(t) - \frac{t\sin(t)}{4}$$

B-1.3 Coupled first order equations

As the final example, a set of coupled first order equations will now be considered. For variables x, y, and z, a set of coupled first order differential equations with initial conditions can be represented as

$$\begin{aligned} x' &= A_{11}x + A_{12}y + A_{13}z + f_x(t),\ x(0) = x_0 \\ y' &= A_{21}x + A_{22}y + A_{23}z + f_y(t),\ y(0) = y_0\ . \\ z' &= A_{31}x + A_{32}y + A_{33}z + f_z(t),\ z(0) = z_0 \end{aligned} \qquad \text{(B.30)}$$

Taking the Laplace transforms leads to

$$\begin{aligned} sX(s)-x_0 &= A_{11}X(s)+A_{12}Y(s)+A_{13}Z(s)+F_x(s) \\ sY(s)-y_0 &= A_{21}X(s)+A_{22}Y(s)+A_{23}Z(s)+F_y(s) \\ sZ(s)-z_0 &= A_{31}X(s)+A_{32}Y(s)+A_{33}Z(s)+F_z(s) \end{aligned} \quad . \tag{B.31}$$

The use of Laplace transforms in solving coupled first order equations will now be demonstrated. Consider the following set of linear coupled first order equations. Using matrix notation, eqn. (B.31) becomes

$$s\begin{bmatrix} X(s) \\ Y(s) \\ Z(s) \end{bmatrix} = \begin{bmatrix} A_{11} & A_{12} & A_{13} \\ A_{21} & A_{22} & A_{23} \\ A_{31} & A_{32} & A_{33} \end{bmatrix}\begin{bmatrix} X(s) \\ Y(s) \\ Z(s) \end{bmatrix} + \begin{bmatrix} F(s) \\ F(s) \\ F(s) \end{bmatrix} + \begin{bmatrix} x_0 \\ y_0 \\ z_0 \end{bmatrix} \tag{B.32}$$

Rewriting the left side using the notion of an identity matrix I_3 gives

$$s\begin{bmatrix} 1 & 0 & 0 \\ 0 & 1 & 0 \\ 0 & 0 & 1 \end{bmatrix}\begin{bmatrix} X(s) \\ Y(s) \\ Z(s) \end{bmatrix} = \begin{bmatrix} A_{11} & A_{12} & A_{13} \\ A_{21} & A_{22} & A_{23} \\ A_{31} & A_{32} & A_{33} \end{bmatrix}\begin{bmatrix} X(s) \\ Y(s) \\ Z(s) \end{bmatrix} + \begin{bmatrix} F_x(s) \\ F_y(s) \\ F_z(s) \end{bmatrix} + \begin{bmatrix} x_0 \\ y_0 \\ z_0 \end{bmatrix} . \tag{B.33}$$

Equation (B.33) is rewritten as

$$(sI - A)W(s) = F(s) + w_0 \ . \tag{B.34}$$

In eqn. (B.34), A is the coefficient matrix, W(s) is the 3 x 1 matrix of the Laplace transforms of x, y and z, F(s) is the [3 x 1] matrix of the Laplace transform of the forcing functions and w_0 is the [3x1] matrix with the initial conditions.

$$A = \begin{bmatrix} A_{11} & A_{12} & A_{13} \\ A_{21} & A_{22} & A_{23} \\ A_{31} & A_{32} & A_{33} \end{bmatrix} \tag{B.35}$$

$$W(s) = \begin{bmatrix} X(s) \\ Y(s) \\ Z(s) \end{bmatrix} \tag{B.36}$$

$$F(s) = \begin{bmatrix} F_x(s) \\ F_y(s) \\ F_z(s) \end{bmatrix} \tag{B.37}$$

$$w_0 = \begin{bmatrix} x_0 \\ y_0 \\ z_0 \end{bmatrix} \tag{B.38}$$

Pre-multiplying eqn. (B.34) by the inverse of the matrix $(sI - A)$, eqn. (B.34) becomes

$$W(s) = (sI - A)^{-1} F(s) + (sI - A)^{-1} w_0 \tag{B.39}$$

The solutions x(t), y(t) and z(t) are obtained by taking the inverse Laplace transform of eqn. (B.39). The use of Laplace transforms in solving coupled first order equations will now be demonstrated. Consider the following set of linear coupled first order equations

$$\begin{aligned} x' &= -x - 2y + t, \; x(0) = -1 \\ y' &= x - y + e^{-t}, \; y(0) = 1 \end{aligned} \tag{B.40}$$

The coefficient matrix is

$$A = \begin{bmatrix} -1 & -2 \\ 1 & -1 \end{bmatrix} \tag{B.41}$$

The Laplace transform set of the forcing functions F(s) is

$$F(s) = \begin{bmatrix} s^{-2} \\ (s+1)^{-1} \end{bmatrix}. \tag{B.42}$$

The initial conditions are

$$w_0 = \begin{bmatrix} -1 \\ 1 \end{bmatrix} \tag{B.43}$$

The Laplace transform of the differential equation becomes

$$\begin{bmatrix} X(s) \\ Y(s) \end{bmatrix} = \begin{bmatrix} \dfrac{(s^{-2}-1)(s+1)}{s^2+2s+3} - 2\dfrac{1+(s-1)^{-1}}{s^2+2s+3} \\ \dfrac{(s^{-2}-1)}{s^2+2s+3} + \dfrac{(s+1)\left[1+(s-1)^{-1}\right]}{s^2+2s+3} \end{bmatrix}. \tag{B.44}$$

Taking the inverse Laplace transforms leads to

$$\begin{aligned} x(t) &= \frac{t}{3} - \frac{e^t}{3} - \frac{7}{9}e^{-t}\cos\left(\sqrt{2}t\right) - \frac{8\sqrt{2}}{9}e^{-t}\sin\left(\sqrt{2}t\right) + \frac{1}{9} \\ x(t) &= \frac{t}{3} + \frac{e^t}{3} + \frac{8}{9}e^{-t}\cos\left(\sqrt{2}t\right) - \frac{7}{18}e^{-t}\sin\left(\sqrt{2}t\right) - \frac{2}{9} \end{aligned} \tag{B.45}$$

The results of the various steps and the solution set and verification of the solution set appears with the appropriate MATLAB code below.

```
clear all;close all;clc
syms x(t) y(t) z(t) t s
A=[-1,-2;1,-1];% coefficient matrix
B=sym(s*eye(2)-A); % this sI-A
disp('B=sI-A')
disp(' ')
disp(B)
f=[t;exp(-t)];% forcing functions
F=laplace(f,t,s);
disp('Display the Laplace transform of the forcing functions')
```

```
disp(F)
w0=[-1;1];
disp('Display the initial conditions')
disp(w0)
W=inv(B)*(F+w0);
disp('display the Laplace transform of x(t) and y(t) as a vector')
disp(W)
w=ilaplace(W,s,t); % solution set
disp('Display the solution set using Laplace transforms')
disp(w)
xt=w(1);
yt=w(2);
% get the solution using dsolve
[xtd,ytd]=dsolve('Dx=-x-2*y+t','Dy=x-y+exp(-t)','x(0)=-1','y(0)=1');
% verify that the solutions match: find the difference between the two
% solutions
disp('Display the solution set from dsolve')
disp(simplify(xtd,'steps',100))
disp(simplify(ytd,'steps',100))
disp('Verify that the Laplace solution is correct: Obtain the symbolic difference between
the two solutions')
disp('Difference will be [0;0] if the approach is correct')
disp(simplify([xt-xtd;yt-ytd],'steps',100)) % must give a null vector if the solutions match
```

```
B=sI-A

[ s + 1, 2]
[ -1, s + 1]

Display the Laplace transform of the forcing functions
    1/s^2
 1/(s + 1)

Display the initial conditions
   -1
    1

display the Laplace transform of x(t) and y(t) as a vector
  ((1/s^2 - 1)*(s + 1))/(s^2 + 2*s + 3) - (2*(1/(s + 1) + 1))/(s^2 + 2*s + 3)
     (1/s^2 - 1)/(s^2 + 2*s + 3) + ((1/(s + 1) + 1)*(s + 1))/(s^2 + 2*s + 3)

Display the solution set using Laplace transforms
  t/3 - exp(-t) - (exp(-t)*(cos(2^(1/2)*t) + 11*2^(1/2)*sin(2^(1/2)*t)))/9 + 1/9
       t/3 + (11*exp(-t)*(cos(2^(1/2)*t) - (2^(1/2)*sin(2^(1/2)*t))/22))/9 - 2/9
Display the solution set from dsolve
t/3 - exp(-t) - (exp(-t)*cos(2^(1/2)*t))/9 - (11*2^(1/2)*exp(-t)*sin(2^(1/2)*t))/9 + 1/9

t/3 + (11*exp(-t)*cos(2^(1/2)*t))/9 - (2^(1/2)*exp(-t)*sin(2^(1/2)*t))/18 - 2/9

Verify that the Laplace solution is correct: Obtain the symbolic difference between the two
solutions
Difference will be [0;0] if the approach is correct
 0
 0
```

Appendix C

Phase Plane Plots

C-1 Phase Planes 410

C-1 Phase Planes

While D-field plots provide insight into the behavior of a first order differential equation, an analogous display can be used to understand the stability of systems described by a pair of coupled first order differential equations or a second order differential equation. While D-field displays the directional changes in t-y(t) domain, the phase plane plots display the directional changes in the fields in the x(t)-y(t) domain where x(t) and y(t) are the solutions of a pair of coupled first order equations. For the case of a second order homogeneous system, the phase plane plots display the changes in the y(t)-y'(t) domain. These plots can point to the state of equilibrium that exists in the system described by the two first order equations or a second order differential equation. Because the analytical solutions are always available for the case of a pair of homogeneous first order linear differential equations and second order homogeneous linear systems, the generation of these plots is simple and straightforward. Consider a pair of first order homogeneous differential equations

$$\vec{x}'(t)=\begin{bmatrix} x_1'(t) \\ x_2'(t) \end{bmatrix}=\begin{bmatrix} -x_1(t)+2x_2(t) \\ 2x_1(t)+x_2(t) \end{bmatrix}. \tag{C.1}$$

The coefficient matrix associated with the system is

$$A=\begin{bmatrix} -1 & 2 \\ 2 & 1 \end{bmatrix}. \tag{C.2}$$

The first step is the creation of samples of the two differential equations. A grid of x_1-x_2 values is chosen using *meshgrid*(.) and samples of the two equations at each of these locations of the grid are calculated using the given set of differential equations. As explained earlier in connection with the D-field patterns associated with first order differential equations, *quiver*(.) allows the display of small arrows indicating the direction in which the derivatives move. The analytical solutions can then be obtained using dsolve(.) and are superimposed. Several samples of the analytical solutions are obtained by choosing initial conditions (varying the unknown constants). The direction of movement of the arrows with respect to the critical point suggests the stability of the system. The critical point is defined by

$$\vec{x}'(t)=\begin{bmatrix}0\\0\end{bmatrix}=\begin{bmatrix}-x_1(t)+2x_2(t)\\2x_1(t)+x_2(t)\end{bmatrix}. \tag{C.3}$$

The trajectories also suggest the nature of the critical point. This conclusion may be drawn from the eigenvalues of the coefficient matrix. The MATLAB® code and results are displayed below.

```
clear all;close all;
A=[-1,2;2,1];
[x1,x2]=meshgrid(-4:.5:4,-4:.5:4);
DX1=A(1,1)*x1+A(1,2)*x2;
DX2=A(2,1)*x1+A(2,2)*x2;
quiver(x1,x2,DX1,DX2,1),xlabel('x_1(t)'),ylabel('x_2(t)')
hold on
syms x y t
[xt,yt]=dsolve('Dx=-x+2*y','Dy=2*x+y'); % get the analytical solution
x1t=MATLABFunction(xt);
x2t=MATLABFunction(yt);
tt=0:.1:.5;
C=-4.5:1:4.5; % pick the unknown constants; use the same value for C1 and C2
L=length(C);
for k1=1:L
    for k2=1:L
    plot(x1t(C(k1),C(k2),tt),x2t(C(k1),C(k2),tt))
   end;
end;
xlim([-4,4]),ylim([-4,4])
lambda=eig(sym(A))

lambda =

  5^(1/2)
 -5^(1/2)
```

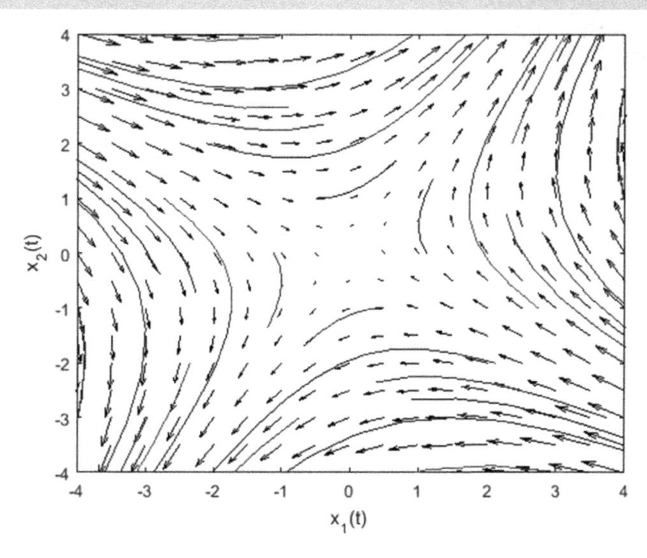

One of the eigenvalues is positive indicating that the system is unstable as seen by the trajectories moving away from the critical point [0,0]. Examining the plot also demonstrates that the critical point is a 'saddle point'.

Another example shown here corresponds to the coefficient matrix

$$A = \begin{bmatrix} -2 & -2 \\ 3 & 1 \end{bmatrix}. \tag{C.4}$$

The MATLAB code and the results are displayed below.

```
clear all;close all;
A=[-2,-2;3,1];
[x1,x2]=meshgrid(-4:.5:4,-4:.5:4);
DX1=A(1,1)*x1+A(1,2)*x2;
DX2=A(2,1)*x1+A(2,2)*x2;
quiver(x1,x2,DX1,DX2,1),xlabel('x_1(t)'),ylabel('x_2(t)')
hold on
syms x y t
[xt,yt]=dsolve('Dx=-2*x-2*y','Dy=3*x+y'); % get the analytical solution
x1t=MATLABFunction(xt);
x2t=MATLABFunction(yt);
tt=0:.1:.5;
C=-4.5:1:4.5; % pick the unknown constants; use the same value for C1 and C2
L=length(C);
for k1=1:L
  for k2=1:L
  plot(x1t(C(k1),C(k2),tt),x2t(C(k1),C(k2),tt))
  end;
end;
xlim([-4,4]),ylim([-4,4])
lambda=eig(sym(A))

lambda =

 - (15^(1/2)*1i)/2 - 1/2
   (15^(1/2)*1i)/2 - 1/2
```

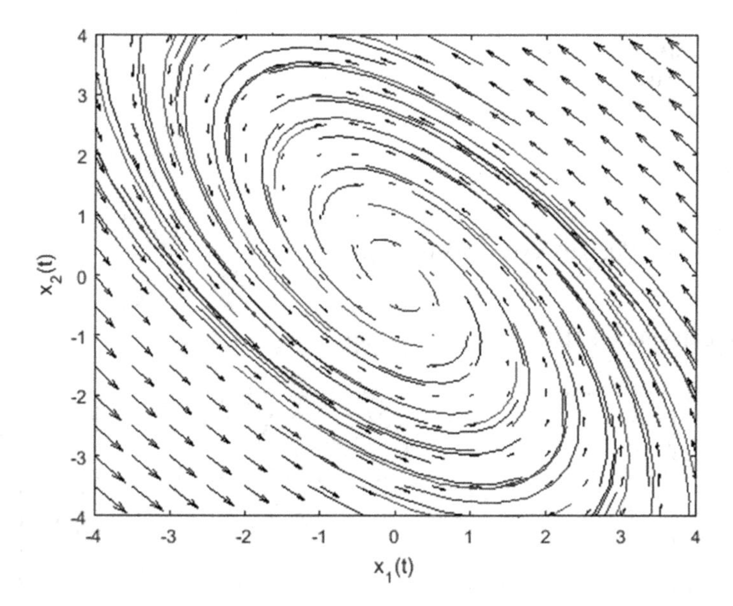

It can be seen that the system is asymptotically stable as seen by the arrows moving towards the center which is also the critical point. The stability is associated with the fact that the real part of the eigenvalue is negative.

A phase portrait associated with a linear second order homogeneous differential equation can be created similarly. The first step is to decompose the second order differential equation into two first order ones as shown in Appendix B. This allows the treatment of a second order differential equation as a pair of first order systems in variables y(t) and y'(t). The rest of the procedure is similar to the one described above in connection with the phase portraits of first order coupled systems.

The example below corresponds to the differential equation

$$y''+2y'+4y=0 \ . \tag{C.5}$$

The MATLAB code and results are displayed.

```
clear all;close all;
% y"+2y'+4y=0
a=1;b=2;c=4;
A=[0,1;-c/a,-b/a];
[x1,x2]=meshgrid(-4:.5:4,-4:.5:4);
DX1=A(1,1)*x1+A(1,2)*x2;
DX2=A(2,1)*x1+A(2,2)*x2;
quiver(x1,x2,DX1,DX2,1),xlabel('y(t)'),ylabel('dy(t)/dt')
hold on
syms y t
y=dsolve('D2y+2*Dy+4*y=0');
dy=diff(y,t);
```

```
fy=MATLABFunction(y);
fdy=MATLABFunction(dy);
tt=0:.1:.5;
C=-4:1:4;
L=length(C);
for k1=1:L
  for k2=1:L
  plot(fy(C(k1),C(k2),tt),fdy(C(k1),C(k2),tt))
 end;
end;
xlim([-4,4]),ylim([-4,4])
```

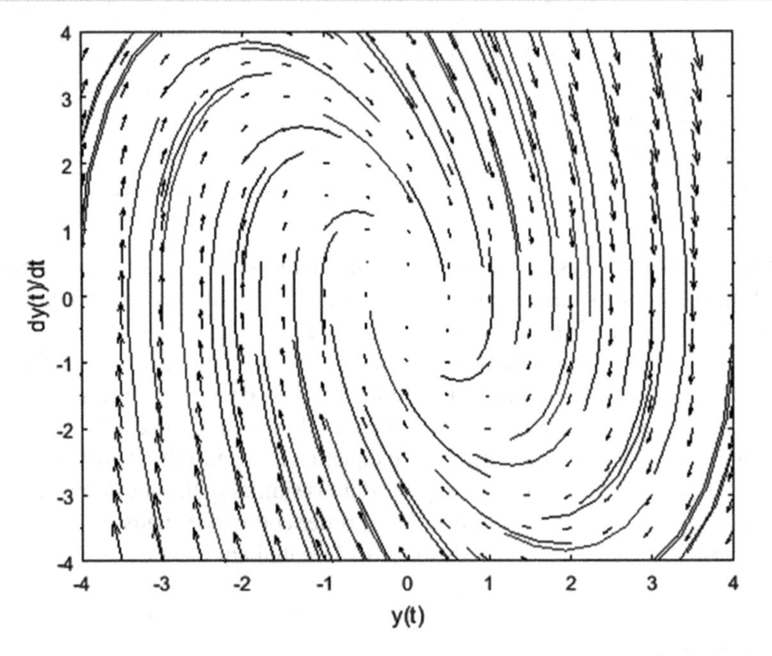

Appendix D

Elements of Linear Algebra

D-1 Concept of Matrices, Echelon and Reduced Echelon forms, and, Determinants 415
D-2 Inverse of a Matrix 423
D-3 Characteristic Equation, Eigenvalues and Eigenvectors 424
D-4 Defective Matrices and Generalized Eigenvectors 427
D-5 Use of Eigenvalues and Eigenvectors 430
D-6 Generalized Eigenvectors and the Fundamental Matrix 435
D-7 Solutions to Coupled First Order Systems Using the Concept of the Fundamental Matrix 441

D-1 Concept of Matrices, Echelon and Reduced Echelon Forms, and, Determinants

A matrix is a collection of numbers arranged in a rectangular or square form. For example, if there are m rows and n columns of numbers, the matrix has a size of [m x n] and is written as

$$A = \begin{bmatrix} a_{11} & a_{12}\cdots & a_{1n} \\ a_{21} & a_{22}\cdots & a_{2n} \\ . & . & . \\ . & . & . \\ a_{m1} & a_{m2}\cdots & a_{mn} \end{bmatrix}. \tag{D.1}$$

If the number of columns is 1, the matrix is of size [m x 1] and is also identified as a vector (column vector). For example, if one takes the first column of the matrix A in eqn. (D.1), the resulting vector is

$$\begin{bmatrix} a_{11} \\ a_{21} \\ . \\ . \\ a_{m1} \end{bmatrix} \Rightarrow R^m . \tag{D.2}$$

Thus, any column of the matrix A is [m x 1] matrix or R^m vector. If m=n, the matrix in eqn. (D.1) is a square one. If a new matrix C is obtained by scaling A by a constant c, each element of the matrix is scaled by c,

$$C = cA = \begin{bmatrix} ca_{11} & ca_{12}\cdots & ca_{1n} \\ ca_{21} & ca_{22}\cdots & ca_{2n} \\ . & . & . \\ . & . & . \\ ca_{m1} & ca_{m2}\cdots & ca_{mn} \end{bmatrix}. \tag{D.3}$$

If there are two matrices, A of size [m x n] and B of size [r x p], the product AB exists only when n=r and the resulting matrix will be of size [m x p]. Consider the simple example with A being a [3 x 3] matrix and B being a [3 x 2] matrix. The product AB will be of size [3 x 2] as shown clearly demonstrating the procedure requiring multiplication and addition of elements.

$$A = \begin{bmatrix} 3 & 4 & 5 \\ 6 & 7 & 8 \\ 2 & 9 & 10 \end{bmatrix} \tag{D.4}$$

$$B = \begin{bmatrix} a & b \\ c & d \\ e & f \end{bmatrix} \tag{D.5}$$

$$AB = \begin{bmatrix} 3*a + 4*c + 5*e & 3*b + 4*d + 5*f \\ 6*a + 7*c + 8*e & 6*b + 7*d + 8*f \\ 2*a + 9*c + 10*e & 2*b + 9*d + 10*f \end{bmatrix} \tag{D.6}$$

The multiplication principle clearly suggests that AB $\neq$ BA. In eqn. (D.6), * represents the product.

Two matrices A and B are equal only when they have the same sizes and there exists an element-by-element match between A and B.

The addition and subtraction of matrices requires that the matrices be of equal sizes and the sum or difference is obtained by taking an element-by-element subtraction or addition. Let the matrix C be a [3 x 3] matrix

$$C = \begin{bmatrix} a1 & b1 & c1 \\ d1 & e1 & f1 \\ g1 & h1 & k1 \end{bmatrix} \tag{D.7}$$

The sum or the difference of the two matrices, A and C will be

$$A \pm C = \begin{bmatrix} 3 \pm a1 & 4 \pm b1 & 5 \pm c1 \\ 6 \pm d1 & 7 \pm e1 & 8 \pm f1 \\ 2 \pm g1 & 9 \pm h1 & 10 \pm k1 \end{bmatrix}. \tag{D.8}$$

If the rows and columns of a matrix A are interchanged, the new matrix is its transpose, expressed as A^T. If a square matrix and its transpose are equal (element-by-element), the matrix said to be symmetric. There are two other types of matrices that will be seen in the study of differential equations. First one is a **null** matrix and the second one is an **identity** matrix. If all the elements of a matrix are 0's, the matrix is referred to as a null matrix,

$$Z = \begin{bmatrix} 0 & 0 & 0 \\ 0 & 0 & 0 \\ 0 & 0 & 0 \\ 0 & 0 & 0 \end{bmatrix}. \tag{D.9}$$

Therefore, if two matrices are equal, the difference of the two matrices will be a null matrix. An identity matrix on the other hand is a square matrix where the elements along the diagonals are 1's. Rest of the elements are 0's. An [n x n] identity matrix is represented as I_n,

$$I_4 = \begin{bmatrix} 1 & 0 & 0 & 0 \\ 0 & 1 & 0 & 0 \\ 0 & 0 & 1 & 0 \\ 0 & 0 & 0 & 1 \end{bmatrix}. \tag{D.10}$$

A few other characteristics and properties associated with matrices also are useful. To appreciate some of these features, consider the following matrix A

$$A = \begin{bmatrix} 2 & -3 & 2 & 5 \\ 2 & -2 & -2 & 6 \\ 2 & -3 & 2 & 8 \end{bmatrix}. \tag{D.11}$$

Matrices can be manipulated by performing row operations (no column operations!). This principle comes from the fact that a set of linear equations are solved by adding, subtracting or scaling the equations. For example eqn. (D.11) can be used to represent a set of equations whose coefficients form the matrix A as in

$$\begin{aligned} 2x_1 - 3x_2 + 2x_3 + 5x_4 &= b_1 \\ 2x_1 - 2x_2 - 2x_3 + 6x_4 &= b_2 \\ 2x_1 - 3x_2 + 2x_3 + 8x_4 &= b_3 \end{aligned}. \tag{D.12}$$

Equation (D.12) can be expressed as in the form of a matrix equation (also known as a matrix-vector equation) as

$$A\vec{x} = \vec{b}. \tag{D.13}$$

In eqn. (D.13), $\vec{x}$ is a [4 x 1] matrix or R^4 vector and $\vec{b}$ is a [3 x 1] matrix or R^3 vector,

$$\vec{x} = \begin{bmatrix} x_1 \\ x_2 \\ x_3 \\ x_4 \end{bmatrix}. \tag{D.14}$$

$$\vec{b} = \begin{bmatrix} b_1 \\ b_2 \\ b_3 \end{bmatrix} . \tag{D.15}$$

Square matrices have a few additional features and properties that are very useful in the study of differential equations. For a square matrix, the determinant of a square matrix is a single valued quantity. If P is a [2 x 2] matrix, its determinant is

$$P = \begin{bmatrix} a & b \\ c & d \end{bmatrix} \Rightarrow \det(P) = ad - bc \ . \tag{D.16}$$

But, the determinants of sizes larger than [2 x 2] require additional manipulations. Consider the matrix Q given by

$$Q = \begin{bmatrix} a & b & c \\ d & e & f \\ g & h & k \end{bmatrix} \tag{D.17}$$

Its determinant is given by

$$\det(Q) = |Q| = a*[e*k - f*h] \ - \ b[d*k - f*g\ ] + \ c[d*h \ - \ e*g] \tag{D.18}$$

In other words, the determinant of Q can be expressed in terms of determinants of matrices of smaller sizes as

$$\det(Q) = a\begin{vmatrix} e & f \\ h & k \end{vmatrix} \ - \ b\begin{vmatrix} d & f \\ g & k \end{vmatrix} + \ c\begin{vmatrix} d & e \\ g & h \end{vmatrix} . \tag{D.19}$$

If the matrix W is of size [n x n] with elements w_{ij}, i=1,2,..,n and j=1,2,..,n, the determinant of W is expressed as

$$\det(W) = \sum_{j=1}^{n} (-1)^{1+j} \ w_{1j} \det(W_{1j}) \ . \tag{D.20}$$

Note that in eqn. (D.20), W_{1j} is the matrix obtained by removing the elements of the 1st row and jth column of W. It should also be noted that a more general way of representing eqn. (D.20) is

$$\det(W) = \sum_{j=1}^{n} (-1)^{i+j} \ w_{ij} \det(W_{ij}), i = 1, 2, \cdots n \ . \tag{D.21}$$

Equation (D.21) suggests that the det(W) can be evaluated using any one of the rows. The results can be generalized to any row being replaced by a column in which case the summation is carried over the rows. Equation (D.20) is also called a **cofactor expansion** across the first row.

Another interesting and useful matrix arises from taking the product of a square matrix with itself. If A is a square matrix, then the product of k (k being a positive integer) copies of A is

$$A^k = \underbrace{A * A * \cdots A}_{k-products} \quad . \tag{D.22}$$

If A is invertible (its determinant is not zero), it is possible to define negative powers of A as

$$A^{-k} = \left(A^{-1}\right)^k = \underbrace{A^{-1} * A^{-1} * \cdots A^{-1}}_{k-products} \quad . \tag{D.23}$$

Observation of eqns. (D.22) and (D.23) results in

$$A^{-k} * A^k = A^0 = I_k \ . \tag{D.24}$$

The powers of matrix will be used later when the topic of generalized eigenvectors is discussed. The concept of the inverse of a matrix represented as A^{-1} will also be discussed later.

Going back to the basic form of a matrix, one can obtain **echelon** form (**row echelon** form) and **reduced echelon** form (**row reduced echelon** form) of the matrix. For example, consider the matrix in eqn. (D.11). Using row manipulation (replacing any row with the sum or difference of any other rows or replacing the row with all elements of that row scaled by a constant or interchanging rows), one can get the following matrix

$$A1 = \begin{bmatrix} 2 & -3 & 2 & 5 \\ 0 & 1 & -4 & 1 \\ 0 & 0 & 0 & -3 \end{bmatrix} . \tag{D.25}$$

The matrix A1 is said to be in echelon form of A in eqn. (D.11) since

1. All nonzero rows are above any rows of all zeros.
2. Each leading entry of a row is in a column to the right of the leading entry of the row above it.
3. All entries in a column below a leading entry are zeros.

Further row operations leads to the matrix A2 as

$$A2 = \begin{bmatrix} 1 & 0 & -5 & 0 \\ 0 & 1 & -4 & 0 \\ 0 & 0 & 0 & 1 \end{bmatrix} \tag{D.26}$$

The matrix M2 is referred to as the row reduced echelon form (**RREF**) of A in eqn. (D.11) since

1. The leading entry in each nonzero row is 1.
2. Each leading 1 is the only nonzero entry in its column.
3. Columns # 1, 2 and 4 are **pivot** columns and the 1's in these columns are location of the pivots.

To understand the usefulness of row reduction techniques, consider a set of linear equations,

$$\begin{aligned} a_{11}x_1 + a_{12}x_2 + \cdots + a_{1n}x_n &= b_1 \\ a_{21}x_1 + a_{22}x_2 + \cdots + a_{2n}x_n &= b_2 \\ \vdots \\ a_{m1}x_1 + a_{m2}x_2 + \cdots + a_{mn}x_n &= b_m \end{aligned} \quad . \tag{D.27}$$

The set of equations in eqn. (D.27) can be expressed in matrix form as

$$Ax \quad b \tag{D.28}$$

In eqn. (D.28), A is called the coefficient matrix of size [m x n] and $\vec{x}$ [n x 1] matrix and $\vec{b}$ [m x 1] matrix are vectors.

$$A = \begin{bmatrix} a_{11} & a_{12} \cdots & a_{1n} \\ a_{21} & a_{22} \cdots & a_{2n} \\ . & . & . \\ . & . & . \\ a_{m1} & a_{m2} \cdots & a_{mn} \end{bmatrix}. \tag{D.29}$$

$$\vec{x} = \begin{bmatrix} x_1 \\ x_2 \\ \vdots \\ x_n \end{bmatrix} \tag{D.30}$$

$$\vec{b} = \begin{bmatrix} b_1 \\ b_2 \\ \vdots \\ b_m \end{bmatrix} \tag{D.31}$$

Using these representations, the system of equations can also be expressed as

$$\begin{bmatrix} a_{11} & a_{12} \cdots & a_{1n} \\ a_{21} & a_{22} \cdots & a_{2n} \\ . & . & . \\ . & . & . \\ a_{m1} & a_{m2} \cdots & a_{mn} \end{bmatrix} \begin{bmatrix} x_1 \\ x_2 \\ . \\ x_n \end{bmatrix} = \begin{bmatrix} b_1 \\ b_2 \\ . \\ . \\ b_m \end{bmatrix}. \tag{D.32}$$

Equations (D.28) and (D.32) are the matrix equations or matrix-vector equations corresponding to the system of equations in eqn. (D.27). By concatenation of A and $\vec{b}$ leads to the augmented matrix D as

$$D = \begin{bmatrix} a_{11} & a_{12} \cdots & a_{1n} & b_1 \\ a_{21} & a_{22} \cdots & a_{2n} & b_2 \\ . & . & . & . \\ . & . & . & . \\ a_{m1} & a_{m2} \cdots & a_{mn} & b_m \end{bmatrix} \tag{D.33}$$

Note that if m < n, the number of equations is less than the number of variables (underdetermined system) and if m > n, the number of equations exceed the number of variables (over determined system). Note that a unique solution for $\vec{x}$ can exist (not assured) only if m $\geq$ n. In other words, a unique solution may or may not exist if m $\geq$ n and with m < n, no unique solutions will ever exist. It is also possible that the system of equations is inconsistent and no solution ever exists. All these can be understood by performing row operations. Note that if $\vec{b}$ is a null vector (all elements are zeros), eqn. (D.32) represents a **homogeneous** system. Otherwise, the system is a **non-homogeneous** one.

Consider a homogeneous system with a coefficient matrix A from eqn. (D.11),

$$\begin{bmatrix} 2 & -3 & 2 & 5 \\ 2 & -2 & -2 & 6 \\ 2 & -3 & 2 & 8 \end{bmatrix} \begin{bmatrix} x_1 \\ x_2 \\ x_3 \\ x_4 \end{bmatrix} = \begin{bmatrix} 0 \\ 0 \\ 0 \end{bmatrix}. \tag{D.34}$$

Creating the augmented matrix and performing row reduction operation leads to

$$\begin{bmatrix} 2 & -3 & 2 & 5 & 0 \\ 2 & -2 & -2 & 6 & 0 \\ 2 & -3 & 2 & 8 & 0 \end{bmatrix} \overset{rref}{\Rightarrow} \begin{bmatrix} 1 & 0 & -5 & 0 & 0 \\ 0 & 1 & -4 & 0 & 0 \\ 0 & 0 & 0 & 1 & 0 \end{bmatrix}. \tag{D.35}$$

It can be seen that the locations of the pivots in eqn. (D.35) are identical to those in eqn. (D.26) and there is no pivot in the last column. The RREF shows the pivot locations and those correspond to the **basic** variables (x_1, x_2 and x_4) and x_3 is the **free** variable. To interpret these results, the augmented matrix in RREF can be converted into the corresponding set of equations as

$$\begin{aligned} x_1 + 0x_2 - 5x_3 + 0x_4 &= 0 \\ 0x_1 + x_2 - 4x_3 + 0x_4 &= 0 \\ 0x_1 + 0x_2 + 0x_3 + x_4 &= 0 \end{aligned}. \tag{D.36}$$

Equation (D.36) can be rewritten in vector form as

$$\begin{bmatrix} x_1 \\ x_2 \\ x_3 \\ x_4 \end{bmatrix} = \begin{bmatrix} 5 \\ 4 \\ 1 \\ 0 \end{bmatrix} x_3 = \vec{v}_3 x_3 \tag{D.37}$$

Equation (D.37) connects the pivots with the basic variables and the notion of the free variable and basic variables are understood. The vector $\vec{v}_3$ is identified as the **null** space of the matrix A. The null space of a matrix A is the set of all solutions of the homogeneous equation $A\vec{x} = 0$.

Consider the case where homogeneous system in eqn. (D.34) is converted into a non-homogeneous with a vector $\vec{b}$,

$$\begin{bmatrix} 2 & -3 & 2 & 5 \\ 2 & -2 & -2 & 6 \\ 2 & -3 & 2 & 8 \end{bmatrix} \begin{bmatrix} x_1 \\ x_2 \\ x_3 \\ x_4 \end{bmatrix} = \begin{bmatrix} 0 \\ 5 \\ 3 \end{bmatrix}. \tag{D.38}$$

The RREF of the augmented matrix now becomes

$$\begin{bmatrix} 2 & -3 & 2 & 5 & 0 \\ 2 & -2 & -2 & 6 & 5 \\ 2 & -3 & 2 & 8 & 3 \end{bmatrix} \overset{rref}{\Rightarrow} \begin{bmatrix} 1 & 0 & -5 & 0 & 3.5 \\ 0 & 1 & -4 & 0 & 4 \\ 0 & 0 & 0 & 1 & 1 \end{bmatrix}. \tag{D.39}$$

Notice that the pivot locations remain the same as in eqn. (D.35) for the non-homogeneous case and the RREF becomes

$$\begin{aligned} x_1 + 0x_2 - 5x_3 + 0x_4 &= 3.5 \\ 0x_1 + x_2 - 4x_3 + 0x_4 &= 4 \\ 0x_1 + 0x_2 + 0x_3 + x_4 &= 1 \end{aligned} . \tag{D.40}$$

Thus, free and basic variables are identical to those associated with the homogeneous system and the solution can now be expressed as

$$\begin{bmatrix} x_1 \\ x_2 \\ x_3 \\ x_4 \end{bmatrix} = \begin{bmatrix} 5 \\ 4 \\ 1 \\ 0 \end{bmatrix} x_3 + \begin{bmatrix} 3.5 \\ 4 \\ 0 \\ 1 \end{bmatrix} = \vec{v}_3 x_3 + \vec{p} . \tag{D.41}$$

In eqn. (D.41), the vector $\vec{v}_3$ has remained the same and the solution is expressed in terms of the solution of the homogeneous part with the addition of the vector $\vec{p}$. It is clear that if the RREF of a square coefficient matrix of size [n x n] has n pivots, the corresponding homogeneous equation only has trivial solutions and there is no null space for such a matrix. If there is a pivot in the last column of the RREF of an augmented matrix, the system is inconsistent and no solutions exist.

Consider an augmented matrix D and its RREF shown below.

$$D = \begin{bmatrix} 3 & -7 & -2 & -7 \\ -3 & 5 & 1 & 5 \\ 6 & -4 & 0 & 2 \end{bmatrix} \overset{rref}{\Rightarrow} \begin{bmatrix} 1 & 0 & 0 & 3 \\ 0 & 1 & 0 & 4 \\ 0 & 0 & 1 & -6 \end{bmatrix}. \tag{D.42}$$

The augmented matrix in eqn. (D.42) corresponds to a non-homogeneous system. It can be seen that there are three pivots and three variables and therefore, the variables are all **basic**. The solution is

$$\begin{bmatrix} x_1 \\ x_2 \\ x_3 \end{bmatrix} = \begin{bmatrix} 3 \\ 4 \\ -6 \end{bmatrix}. \tag{D.43}$$

The coefficient matrix in eqn. (D.42) is a [3 x 3] square matrix. The system of equations in eqn. (D.42) has unique solutions. Such a system of equations can also be solved using other approaches using the inverse of the matrix which is discussed next.

D-2 Inverse of a matrix

Another important property of a matrix that becomes useful in solving coupled differential equations is the inverse of a matrix introduced in eqn. (D23) and eqn. (D.24). The inverse of a square matrix A exists only if

$$\det(A) \neq 0 \ . \tag{D.44}$$

The matrix A and its inverse A^{-1} are related through the property (seen in eqn. (D.23))

$$AA^{-1} = I = A^{-1}A \ . \tag{D.45}$$

A 2 x 2 matrix and its inverse are

$$A = \begin{bmatrix} a & b \\ c & d \end{bmatrix} \Rightarrow A^{-1} = \frac{1}{(ad-bc)} \begin{bmatrix} d & -b \\ -c & a \end{bmatrix} . \tag{D.46}$$

It can be seen that eqn. (D.45) is valid. Note that det(A) and det(A^{-1}) are equal. For matrices of size larger than 2 x 2, the inverse of the matrix is obtained through the following procedure. If A is an n x n matrix, create a new matrix A by (horizontal) concatenating it with an identity matrix (on the right) and calculate the RREF. The inverse of A will be the n x n elements on the right and the remaining n x n elements will constitute an identity matrix.

$$B = \left[A_{nxn} I_n\right] \ \Rightarrow \ rref(B) = \left[I_n A^{-1}\right] . \tag{D.47}$$

Consider the matrix

$$A = \begin{bmatrix} 3 & 2 \\ 5 & 6 \end{bmatrix} \tag{D.48}$$

Its determinant is 8 and hence, the matrix is invertible. An invertible matrix is also called a nonsingular matrix while a square matrix that cannot be inverted is called a singular matrix. Using eqn. (D.46), the inv(A) becomes

$$A^{-1} = \frac{1}{8} \begin{bmatrix} 6 & -2 \\ -5 & 3 \end{bmatrix} = \begin{bmatrix} \frac{3}{4} & -\frac{1}{4} \\ -\frac{5}{8} & \frac{3}{8} \end{bmatrix} . \tag{D.49}$$

Proceed along the lines of eqn. (D.47),

$$B = \begin{bmatrix} 3 & 2 & 1 & 0 \\ 5 & 6 & 0 & 1 \end{bmatrix} \Rightarrow rref(B) = \begin{bmatrix} 1 & 0 & \frac{3}{4} & -\frac{1}{4} \\ 0 & 1 & -\frac{5}{8} & \frac{3}{8} \end{bmatrix} . \tag{D.50}$$

Comparing eqns. (D.49) and (D.50), it can be seen that eqn. (D.47) offers a means to obtain the inverse of a square matrix.

Matrix inversion offers a simple means to solve a matrix vector equation such as the one in eqn. (D.28). If A is a square matrix and determinant of A exists, multiplying eqn. (D.28) from the left by A^{-1} leads to

$$A^{-1}A\vec{x} = A^{-1}\vec{b} \ . \tag{D.51}$$

Using eqn. (D.45), eqn. (D.51) becomes

$$I_n\,\vec{x} = A^{-1}\vec{b} \Rightarrow \quad \vec{x} = A^{-1}\vec{b} \ . \tag{D.52}$$

In other words, the solution to the matrix equation is obtained by taking the product of A^{-1} with $\vec{b}$.

The invertibility of matrix A also can be used in another way to obtain the solution vector of the matrix equation in eqn. (D.28) using Cramer's Rule. For the [n x n] invertible matrix A, the unique solution consists of

$$x_i = \frac{\det\left(A_i\left(b\right)\right)}{\det\left(A\right)}, i = 1, 2, \ldots, n \ . \tag{D.53}$$

In eqn. (D.53), the matrix $A_i(b)$ is the matrix obtained by replacing the ith column of A with the vector $\vec{b}$.

D-3 Characteristic Equation, Eigenvalues and Eigenvectors

Another aspect of matrices that is useful in the study of differential equations is the characteristic equation of a square matrix. Consider a square matrix A and assume that it is possible to write the matrix equation $A\vec{x}$ as a scaled version of $\vec{x}$ as in

$$A\vec{x} = \lambda\vec{x} \ . \tag{D.54}$$

In eqn. (D.54), A is a matrix of size [n x n] and $\vec{x}$ is vector of size [n x 1] and λ is a scalar quantity. The right hand side can be rewritten by multiplying $\vec{x}$ from the left by an identity matrix I_n as

$$A\vec{x} = \lambda\vec{x} = \lambda I_n\vec{x} \ . \tag{D.55}$$

Rewriting eqn. (D.55) leads to

$$\left[A - \lambda I_n\right]\vec{x} = 0 \tag{D.56}$$

A non-trivial solution exists for eqn. (D.56) only when the determinant of the matrix on the left hand side is zero,

$$\left|A - \lambda I_n\right| = 0 \ . \tag{D.57}$$

Equation (D.57) is called the characteristic equation associated with the matrix and its solutions are referred to as the **eigenvalues** of the matrix A. For example, consider a [2 x 2] matrix

$$A = \begin{bmatrix} 1 & 3 \\ 2 & 2 \end{bmatrix} . \tag{D.58}$$

Equation (D.57) becomes

$$(A-\lambda I)=\begin{bmatrix}1 & 3\\ 2 & 2\end{bmatrix}-\begin{bmatrix}\lambda & 0\\ 0 & \lambda\end{bmatrix}=\begin{bmatrix}1-\lambda & 3\\ 2 & 2-\lambda\end{bmatrix}\Rightarrow \det(A-\lambda I)=(1-\lambda)(2-\lambda)-6\cdot \tag{D.59}$$

The characteristic equation becomes

$$\lambda^2-3\lambda-4=0 \quad\Rightarrow\quad (\lambda-4)(\lambda+1)=0 \tag{D.60}$$

The solutions to eqn. (D.60) are the eigenvalues and in this case, the eigenvalues are λ_1=-1 and λ_2=4. With the availability of eigenvalues, a solution to eqn. (D.56) can be found. Equation (D.56) associated with the first eigenvalue is

$$[A-\lambda_1 I_n]\vec{v}=0\Rightarrow\begin{bmatrix}1-\lambda_1 & 3\\ 2 & 2-\lambda_1\end{bmatrix}\vec{v}=\begin{bmatrix}0\\ 0\end{bmatrix}. \tag{D.61}$$

$$\begin{bmatrix}2 & 3\\ 2 & 3\end{bmatrix}\vec{v}=\begin{bmatrix}0\\ 0\end{bmatrix} \tag{D.62}$$

Note that in eqns. (D.61) and (D.62) $\vec{v}$ replaces $\vec{x}$ and the solution vector $\vec{v}$ is identified as the **eigenvector** associated with the eigenvalue of λ_1=-1. Writing eqn. (D.62) using the concept of augmented matrices, the augmented matrix associated with the matrix equation (D.61) is

$$\begin{bmatrix}2 & 3 & 0\\ 2 & 3 & 0\end{bmatrix}\overset{rref}{\Rightarrow}\begin{bmatrix}1 & \frac{3}{2} & 0\\ 0 & 0 & 0\end{bmatrix}. \tag{D.63}$$

Note that the first row of eqn. (D.63) corresponds to the linear equation

$$v_1+\frac{3}{2}v_2=0\ . \tag{D.64}$$

This leads to

$$\begin{bmatrix}v_1\\ v_2\end{bmatrix}=\begin{bmatrix}-\frac{3}{2}\\ 1\end{bmatrix} \tag{D.65}$$

The eigenvector associated with λ_1=-1 is therefore,

$$\vec{v}=\begin{bmatrix}-\frac{3}{2}\\ 1\end{bmatrix}. \tag{D.66}$$

The eigenvector is also obtained directly using the MATLAB® command null() as shown below. The use of symbolic toolbox will yield the vector in rational form. Both results are given.

```
v=null(sym([2,3;2,3]))
v =
-3/2
  1
v=null([2,3;2,3])
v =
 -0.8321
 0.5547
```

Proceeding similarly, the eigenvector associated with λ_2=4 becomes

$$\vec{v} = \begin{bmatrix} 1 \\ 1 \end{bmatrix}. \tag{D.67}$$

Eigenvalues and eigenvectors can easily be obtained using MATLAB. The use of the symbolic toolbox will make the numerical values appear as rational numbers.

```
[V,lambda]=eig(sym([1,3;2,2]))

V =
[ -3/2, 1]
[ 1, 1]
lambda =
 [ -1, 0]
[ 0, 4]
```

If the symbolic toolbox is not used, the values appear in decimal notation as in

```
>> [V, lambda]=eig([1,3;2,2])

V =
 -0.8321 -0.7071
 0.5547 -0.7071
lambda =
  -1    0
   0    4
```

The first column of the matrix V is the eigenvector associated with the first eigenvalue of -1 and the second column of V is the eigenvector associated with the second eigenvalue of 4. Eigenvalues appear as diagonal elements of the matrix **lambda**. In general, eigenvalues and eigenvectors can be real or complex. If complex eigenvalues exist, they appear as a conjugate pair. Consider the matrix

$$A = \begin{bmatrix} 1 & -2 \\ 3 & 1 \end{bmatrix}. \tag{D.68}$$

The eigenvalues and eigenvectors are

$$\begin{aligned} \lambda &= 1 + 2.4495j; \quad \vec{v} = \begin{bmatrix} 0.8166j \\ 1 \end{bmatrix} \\ \lambda &= 1 - 2.4495j; \quad \vec{v} = \begin{bmatrix} -0.8166j \\ 1 \end{bmatrix} \end{aligned} \tag{D.69}$$

The two examples discussed so far dealt with matrices with distinct eigenvalues and distinct eigenvectors. It must be noted that when complex eigenvalues exist, they appear as a conjugate pair and are therefore always distinct.

D-4 Defective Matrices and Generalized Eigenvectors

In solving coupled first order differential equations, cases arise when the eigenvalues might be equal with the possibility that eigenvectors might be or might not be distinct. Consider the case of a matrix B,

$$B = \begin{bmatrix} 2 & 0 \\ 0 & 2 \end{bmatrix}. \tag{D.70}$$

This leads to a pair of equal eigenvalues with distinct eigenvectors,

$$\begin{aligned} \lambda_1 &= 2; \quad \vec{v} = \begin{bmatrix} 0 \\ 1 \end{bmatrix} \\ \lambda_2 &= 2; \quad \vec{v} = \begin{bmatrix} 1 \\ 0 \end{bmatrix} \end{aligned} \tag{D.71}$$

Now consider the case of a matrix A

$$A = \begin{bmatrix} -1 & 1 \\ -1 & -3 \end{bmatrix}. \tag{D.72}$$

The eigenvalues and eigenvectors are

$$\begin{aligned} \lambda_1 &= -2; \quad \vec{v} = \begin{bmatrix} -1 \\ 1 \end{bmatrix} \\ \lambda_2 &= -2; \quad \vec{v} = \begin{bmatrix} -1 \\ 1 \end{bmatrix} \end{aligned}. \tag{D.73}$$

It is clear from eqn. (D.73) that the two eigenvectors are identical. In other words, the number of distinct eigenvectors is only 1 even though the number of eigenvalues is 2. A matrix with fewer distinct eigenvectors than the number of eigenvalues is referred to as a **defective** matrix. In this case, the **algebraic** multiplicity (p) is 2 (number of equal eigenvalues) while the **geometric** multiplicity (m) is only 1 (distinct pairs of eigenvectors). Note that for symmetric matrices in eqn. (D.70), the algebraic and geometric multiplicities are equal. Therefore, symmetric matrices will not be defective. The matrix in eqn. (D.72) is not symmetric and it is defective.

Now consider the following matrix

$$A = \begin{bmatrix} 2 & 4 & 3 \\ -4 & -6 & -3 \\ 3 & 3 & 1 \end{bmatrix}. \tag{D.74}$$

The eigenvalues and eigenvectors are

$$\lambda_1 = 1; \quad \vec{v} = \begin{bmatrix} 1 \\ -1 \\ 1 \end{bmatrix}$$
$$\lambda_2 = -2; \quad \vec{v} = \begin{bmatrix} -1 \\ 1 \\ 0 \end{bmatrix} \tag{D.75}$$
$$\lambda_3 = -2; \quad \vec{v} = \begin{bmatrix} -1 \\ 1 \\ 0 \end{bmatrix}$$

In this case, there exist a pair of equal eigenvalues equal to -2 (algebraic multiplicity p=2); But, the corresponding eigenvectors are not distinct (geometric multiplicity m=1). There is an additional eigenvalue of unity and a corresponding distinct eigenvector. Consider the case of the following matrix

$$A = \begin{bmatrix} -3 & -1 & -6 \\ -2 & -1 & -4 \\ 1 & 0 & 1 \end{bmatrix}. \tag{D.76}$$

The corresponding eigenvalues and eigenvectors are

$$\lambda_1 = -1; \quad \vec{v} = \begin{bmatrix} -2 \\ -2 \\ 1 \end{bmatrix}$$
$$\lambda_2 = -1; \quad \vec{v} = \begin{bmatrix} -2 \\ -2 \\ 1 \end{bmatrix} \tag{D.77}$$
$$\lambda_3 = -1; \quad \vec{v} = \begin{bmatrix} -2 \\ -2 \\ 1 \end{bmatrix}$$

The defective matrix A has an algebraic multiplicity p=3 and geometric multiplicity m=1. These matrices in eqn. (D.74) and eqn. (D.77) are also considered as defective because m<p.

To accommodate defective matrices lacking distinct eigenvectors (we expect each eigenvalue paired with an eigenvector), the concept of generalized eigenvectors is needed. Generalized eigenvectors are solutions of

$$[A - \lambda I_n]^p \vec{v} = 0, \quad p = 2,3,\cdots. \tag{D.78}$$

Note that for p=1, eqn. (D.78) is identical to eqn. (D.61). Now, consider the matrix in eqn. (D.72) which has an algebraic multiplicity of 2 (p) and geometric multiplicity of 1. Equation (D.78) becomes

$$\left[A-\lambda I\right]^{p}\vec{v}=0,\quad p=2\ . \tag{D.79}$$

Equation (D.79) becomes

$$\left(A-\lambda I\right)^{2}\vec{v}=\left(\begin{bmatrix}-1 & 1\\ -1 & -3\end{bmatrix}-(-2)\begin{bmatrix}1 & 0\\ 0 & 1\end{bmatrix}\right)^{2}\vec{v}=\begin{bmatrix}1 & 1\\ -1 & -1\end{bmatrix}^{2}\vec{v}=\begin{bmatrix}0 & 0\\ 0 & 0\end{bmatrix}\hat{v} \tag{D.80}$$

Using the null(.) command in MATLAB, leads to the following result:

```
null([0,0;0,0])
ans =
    1  0
    0  1
```

The generalized eigenvectors are $\begin{bmatrix}1\\0\end{bmatrix}$ and $\begin{bmatrix}0\\1\end{bmatrix}$.

For the matrix in eqn. (D.74), a pair of generalized eigenvectors are needed for the eigenvalue of -2. This means that we need to solve eqn. (D.78) for (p=2)

$$\left[A-\lambda I_{n}\right]^{2}\vec{v}=0\Rightarrow\left(A-\lambda I\right)^{2}\vec{v}=\left(\begin{bmatrix}2 & 4 & 3\\ -4 & -6 & -3\\ 3 & 3 & 1\end{bmatrix}-(-2)A\begin{bmatrix}1 & 0 & 0\\ 0 & 1 & 0\\ 0 & 0 & 1\end{bmatrix}\right)^{2}\vec{v}=\begin{bmatrix}4 & 4 & 3\\ -4 & -4 & -3\\ 3 & 3 & 3\end{bmatrix}^{2}\vec{v} \tag{D.81}$$

Finding the null(.), we get the two generalized eigenvectors associated with the eigenvalue of -2 as $\begin{bmatrix}-1\\1\\0\end{bmatrix}$ and $\begin{bmatrix}-1\\0\\1\end{bmatrix}$.

For the matrix in eqn. (D.76), the algebraic multiplicity is 3 and geometric multiplicity is 1. This means that eqn. (D.78) for (p=3)

$$\left(A-\lambda I\right)^{3}\vec{v}=\left(\begin{bmatrix}-3 & -1 & -6\\ -2 & -1 & -4\\ 1 & 0 & 1\end{bmatrix}-(-1)A\begin{bmatrix}1 & 0 & 0\\ 0 & 1 & 0\\ 0 & 0 & 1\end{bmatrix}\right)^{3}\vec{v}=\begin{bmatrix}-2 & -1 & -6\\ -2 & 0 & -4\\ 1 & 0 & 2\end{bmatrix}^{3}\vec{v} \tag{D.82}$$

This leads to

$$\begin{bmatrix}0 & 0 & 0\\ 0 & 0 & 0\\ 0 & 0 & 0\end{bmatrix}\vec{v}=0 \tag{D.83}$$

The generalized eigenvectors are

$$\begin{bmatrix}1\\0\\0\end{bmatrix}, \begin{bmatrix}0\\1\\0\end{bmatrix}, \begin{bmatrix}0\\0\\1\end{bmatrix}. \tag{D.84}$$

It is clear that if the algebraic multiplicity p=n and geometric multiplicity m=1, the set of generalized vectors match the columns of the identity matrix of size n.

D-5 Use of Eigenvalues and Eigenvectors

Eigenvalues and eigenvectors find extensive use in obtaining solutions for coupled first order differential equations with constant coefficients. Consider a pair of first order coupled homogeneous equations

$$\begin{aligned} x_1' &= A_{11}x_1 + A_{12}x_2 \\ x_2' &= A_{21}x_1 + A_{22}x_2 \end{aligned} \tag{D.85}$$

The coefficient matrix

$$A = \begin{bmatrix} A_{11} & A_{12} \\ A_{21} & A_{22} \end{bmatrix}. \tag{D.86}$$

For homogeneous equations, the expected solution will be of the form

$$x(t) = \begin{bmatrix} x_1(t) \\ x_2(t) \end{bmatrix} = \vec{v}e^{\lambda t} . \tag{D.87}$$

In eqn. (D.87), $\vec{v}$ is the eigenvector. It should be noted that there exist two solutions associated with the characteristic equation

$$|A - \lambda I| = 0 . \tag{D.88}$$

If λ_1 and λ_2 are two distinct real eigenvalues with $\vec{v}_1$ and $\vec{v}_2$ being the corresponding distinct eigenvectors, the general solution will be a linear combination of the solution of the form in eqn. (D.87), one for each eigenvalue/eigenvector combination. Thus, the general solution can be expressed as

$$x(t) = \begin{bmatrix} x_1(t) \\ x_2(t) \end{bmatrix} = C_1\vec{v}_1e^{\lambda_1 t} + C_2\vec{v}_2e^{\lambda_2 t} = C_1e^{\lambda_1 t}\begin{bmatrix} v_{11} \\ v_{21} \end{bmatrix} + C_2e^{\lambda_2 t}\begin{bmatrix} v_{12} \\ v_{22} \end{bmatrix}. \tag{D.89}$$

Based on the elements of the coefficient matrix, it is possible that the eigenvalues and therefore, also the eigenvectors are complex. When a set of eigenvalues are complex, they will always form a conjugate pair. This means that a pair of eigenvalues can be expressed as

$$\lambda_{1,2} = \lambda_{\pm} = \alpha \pm j\beta . \tag{D.90}$$

The corresponding eigenvectors (just as in the case of eigenvalues, these will also form a conjugate pair) will be of the form

$$\vec{v}_{1,2} = \vec{v}_{\pm} = \vec{a} \pm j\vec{b} \ . \tag{D.91}$$

In eqn. (D.91), $\vec{a}$ and $\vec{b}$ are [2 x 1] and real,

$$\vec{a} = \begin{bmatrix} a_1 \\ a_2 \end{bmatrix}, \quad \vec{b} = \begin{bmatrix} b_1 \\ b_2 \end{bmatrix} \tag{D.92}$$

The general solution can be expressed as

$$\vec{x}(t) = \begin{bmatrix} x_1(t) \\ x_2(t) \end{bmatrix} = e^{\alpha t} \left\{ C_1 \begin{bmatrix} a_1 \cos(\beta t) - b_1 \sin(\beta t) \\ a_2 \cos(\beta t) - b_2 \sin(\beta t) \end{bmatrix} + C_2 \begin{bmatrix} a_1 \sin(\beta t) + b_1 \cos(\beta t) \\ a_2 \sin(\beta t) + b_2 \cos(\beta t) \end{bmatrix} \right\}. \tag{D.93}$$

The case of equal eigenvalues should be treated no differently if the eigenvectors are distinct and the solution will be similar to eqn. (D.89) with $\lambda_{1,2}=\alpha$.

On the other hand, if the coefficient matrix is defective with a single eigenvector, the solution requires the use of generalized eigenvectors. We will examine solutions of a set of equations when the coefficient matrix is defective following the discussion of the case of generalized vectors for larger matrices to get a general description of solutions.

A few examples to demonstrate the use of linear algebra are now provided along with appropriate MATLAB codes and verification.

First consider a coefficient matrix

$$A = \begin{bmatrix} 2 & 3 \\ 3 & -6 \end{bmatrix}. \tag{D.94}$$

The corresponding homogeneous system is

$$\vec{x}' = A\vec{x} \ . \tag{D.95}$$

The eigenvalues and eigenvectors are obtained from MATLAB as

$$\begin{gathered} \lambda_1 = -7, \quad \vec{v}_1 = \begin{bmatrix} -\dfrac{1}{3} \\ 1 \end{bmatrix}. \\ \lambda_2 = 3, \quad \vec{v}_2 = \begin{bmatrix} 3 \\ 1 \end{bmatrix} \end{gathered} \tag{D.96}$$

The solution set corresponding to the linear system described by the coefficient matrix in eqn. (D.94) is

$$\vec{x}(t) = C_1 \begin{bmatrix} -\dfrac{1}{3} \\ 1 \end{bmatrix} e^{-7t} + C_2 \begin{bmatrix} 3 \\ 1 \end{bmatrix} e^{3t} \ . \tag{D.97}$$

```
clear all;clc;
% example worked out and verification using dsolve(.)
A=[2,3;3,-6];
[V,D]=eig(sym(A))
DD=diag(D) % eigenvalues
v1=V(:,1) % eigenvector 1 corresponding to DD(1)
v2=V(:,2)% eigenvector 2 corresponding to DD(2)
clear A
syms x y t C2 C1
[xt,yt]=dsolve('Dx=2*x+3*y','Dy=3*x-6*y');
xr=symvar(xt); % find the symbolic variables
yr=symvar(yt); % % find the symbolic variables
xt=subs(xt,[xr(1),xr(2)],[C2,C1]); % replace the constants to match the text
yt=subs(yt,[yr(1),yr(2)],[C2,C1]);
X=[xt;yt];
disp(X)
```

```
V =
[ -1/3, 3]
[ 1, 1]

D =
[ -7, 0]
[ 0, 3]

DD =
 -7
 3

v1 =
 -1/3
 1

v2 =
 3
 1

3*C2*exp(3*t) - (C1*exp(-7*t))/3
 C2*exp(3*t) + C1*exp(-7*t)
```

The next example looks at the case of a symmetric coefficient matrix as

$$A = \begin{bmatrix} 2 & 0 \\ 0 & 2 \end{bmatrix}. \tag{D.98}$$

Because this is a symmetric matrix, the eigenvalues are equal with distinct eigenvectors

$$\begin{aligned} \lambda_1 &= 2, \quad \vec{v}_1 = \begin{bmatrix} 1 \\ 0 \end{bmatrix} \\ \lambda_2 &= 2, \quad \vec{v}_2 = \begin{bmatrix} 0 \\ 1 \end{bmatrix} \end{aligned}. \tag{D.99}$$

The solution set corresponding to the coefficient matrix in eqn. (D.95)

$$\vec{x}(t) = C_1 \begin{bmatrix} 1 \\ 0 \end{bmatrix} e^{2t} + C_2 \begin{bmatrix} 0 \\ 1 \end{bmatrix} e^{2t} = \begin{bmatrix} C_1 \\ C_2 \end{bmatrix} e^{2t} . \tag{D.100}$$

The MATLAB implementation is shown below.

```
clear all;clc
A=[2,0;0,2];
[V,D]=eig(sym(A))
DD=diag(D); % eigenvalues
v1=V
clear A B
syms x y t C2 C1
[xt,yt]=dsolve('Dx=2*x','Dy=2*y')
xr=symvar(xt) % find the symbolic variables
yr=symvar(yt) % % find the symbolic variables
xt=subs(xt,xr(1),C1); % replace the constants to match the text
yt=subs(yt,yr(1),C2);
 X=[xt;yt];
disp(X)
```

```
V =
[ 1, 0]
[ 0, 1]

D =
[ 2, 0]
[ 0, 2]

v1 =
[ 1, 0]
[ 0, 1]

xt =
C3*exp(2*t)
yt =
C4*exp(2*t)

xr =
[ C3, t]
yr =
[ C4, t]
C1*exp(2*t)
C2*exp(2*t)
```

Consider, the coefficient matrix

$$A = \begin{bmatrix} 4 & 3 \\ -3 & 4 \end{bmatrix} . \tag{D.101}$$

The eigenvalues and eigenvectors are complex. The eigenvalues are

$$\lambda_{\pm} = \alpha + j\beta = 4 \pm 3j \tag{D.102}$$

The eigenvectors are

$$\vec{a} \pm j\vec{b} = \begin{bmatrix} 1 \\ 0 \end{bmatrix} \pm j \begin{bmatrix} 0 \\ 1 \end{bmatrix}. \tag{D.103}$$

The solution set is

$$\vec{x}(t) = e^{\alpha t} \left\{ C_1 \begin{bmatrix} a_1 \cos(\beta t) - b_1 \sin(\beta t) \\ a_2 \cos(\beta t) - b_2 \sin(\beta t) \end{bmatrix} + C_2 \begin{bmatrix} a_1 \sin(\beta t) + b_1 \cos(\beta t) \\ a_2 \sin(\beta t) + b_2 \cos(\beta t) \end{bmatrix} \right\}. \tag{D.104}$$

Substituting for eigenvalues and eigenvectors, the solution set becomes

$$\vec{x}(t) = e^{4t} C_1 \begin{bmatrix} \cos(3t) \\ -\sin(3t) \end{bmatrix} + e^{4t} C_2 \begin{bmatrix} \sin(3t) \\ \cos(3t) \end{bmatrix}. \tag{D.105}$$

The MATLAB portion including verification using dsolve(.) appears below.

```
clear all;
A=[4,3;-3,4];
[V,D]=eig(sym(A))
DD=diag(D) % eigenvalues
v1=V(:,1) % eigenvector 1 corresponding to DD(1)
v2=V(:,2)% eigenvector 2 corresponding to DD(2)
clear A
syms x y t C2 C1
[xt,yt]=dsolve('Dx=4*x+3*y','Dy=-3*x+4*y')
xr=symvar(xt) % find the symbolic variables
yr=symvar(yt) % % find the symbolic variables
xt=subs(xt,[xr(1),xr(2)],[C2,C1]); % replace the constants to match the text
yt=subs(yt,[yr(1),yr(2)],[C2,C1]);
 X=[xt;yt];
disp(X)
```

```
V =
[ j, -j]
[ 1, 1]

D =
[ 4 – 3j, 0]
[ 0, 4 + 3j]

DD =
 4 – 3j
 4 + 3j

v1 =
 j
 1

v2 =
 -j
 1
```

```
xt =
C4*cos(3*t)*exp(4*t) + C3*sin(3*t)*exp(4*t)
yt =
C3*cos(3*t)*exp(4*t) - C4*sin(3*t)*exp(4*t)

xr =
[ C3, C4, t]
yr =
[ C3, C4, t]

C1*cos(3*t)*exp(4*t) + C2*sin(3*t)*exp(4*t)
C2*cos(3*t)*exp(4*t) - C1*sin(3*t)*exp(4*t)
```

The final example deals with a coefficient matrix A

$$A = \begin{bmatrix} 2 & 4 \\ -1 & -2 \end{bmatrix} . \tag{D.106}$$

It has two equal eigenvalues, each equal to 0. But there is only a single distinct eigenvector,

$$\vec{v} = \begin{bmatrix} 2 \\ -1 \end{bmatrix} . \tag{D.107}$$

This means that we need another vector, one from the two generalized vectors obtained by solving

$$\left(A - \lambda I_2\right)^2 \vec{u} = 0 \Rightarrow \begin{bmatrix} 2 & 4 \\ -1 & -2 \end{bmatrix} \vec{u} = 0 . \tag{D.108}$$

The generalized eigenvectors are

$$\vec{v}_g = \begin{bmatrix} 1 & 0 \\ 0 & 1 \end{bmatrix} . \tag{D.109}$$

The solution to the set of equations represented by eqn. (D.106) will be presented following a formal discussion of generalized eigenvectors and the fundamental matrix.

D-6 Generalized Eigenvectors and the Fundamental Matrix

The approach of eigenvalues and eigenvectors can be extended to coupled first order homogeneous systems of larger sizes. If the coefficient matrix is [n x n] and if the eigenvectors are real and distinct, the solution set is expressed in vector form as

$$\vec{x}(t) = \sum_{k=1}^{n} C_k \vec{v}_k e^{\lambda_k t} . \tag{D.110}$$

If n is even and all the eigenvalues are complex conjugate distinct pairs,

$$\vec{x}(t)=\sum_{k=1}^{n/2}e^{\alpha_k t}\left\{C_{1k}\begin{bmatrix}a_{1k}\cos(\beta_k t)-b_{1k}\sin(\beta_k t)\\ a_{2k}\cos(\beta_k t)-b_{2k}\sin(\beta_k t)\end{bmatrix}+C_{2k}\begin{bmatrix}a_{1k}\sin(\beta_k t)+b_{1k}\cos(\beta_k t)\\ a_{2k}\sin(\beta_k t)+b_{2k}\cos(\beta_k t)\end{bmatrix}\right\}. \tag{D.111}$$

In eqn. (D.111), $\alpha_k \pm j\beta_k$ represent the eigenvalues and a's and b's are the elements of the eigenvectors.

When more than 2 coupled variables exist, the simple approach described earlier requires generalization using the concept of the fundamental matrix, particularly when the matrix is defective. Going back to the solution of a homogeneous system with distinct eigenvectors, the solution in (D.89) can be expressed as

$$\vec{x}(t)=C_1\vec{x}_1(t)+C_2\vec{x}_2(t)=C_1\left\{e^{\lambda_1 t}\begin{bmatrix}v_{11}\\ v_{21}\end{bmatrix}\right\}+C_2\left\{e^{\lambda_2 t}\begin{bmatrix}v_{12}\\ v_{22}\end{bmatrix}\right\}. \tag{D.112}$$

In other words, if two distinct eigenvectors exist, the solution set is a linear combination of a fundamental set of vectors $\vec{X}_1(t)$ and $\vec{X}_2(t)$. The fundamental matrix X(t) associated with the coefficient matrix A is expressed as

$$X(t)=\left[\vec{X}_1(t)\;\vec{X}_2(t)\right]. \tag{D.113}$$

In eqn. (D.113), $\vec{x}_1(t)$ and $\vec{x}_2(t)$ are given by

$$\vec{X}_1(t)=e^{\lambda_1 t}\begin{bmatrix}v_{11}\\ v_{21}\end{bmatrix} \qquad \vec{X}_2(t)=e^{\lambda_2 t}\begin{bmatrix}v_{12}\\ v_{22}\end{bmatrix}. \tag{D.114}$$

This concept can be extended to a case of a coupled differential equation in n-variables and the general solution set of the differential equation will be

$$\vec{x}(t)=X(t)\vec{C}=\sum_{k=1}^{n}C_n\vec{X}_n(t)=\sum_{k=1}^{n}C_n e^{\lambda_n t}\vec{v}_n. \tag{D.115}$$

In eqn. (D.115), $\vec{x}_n$ is the eigenvector associated with the eigenvalue λ_n and $\vec{C}$ is the vector of constants. As discussed earlier, it is not necessary to have distinct eigenvalues. But, eigenvectors must be distinct. For the case of n=2,

$$\vec{x}(t)=\begin{bmatrix}x_1(t)\\ x_2(t)\end{bmatrix}. \tag{D.116}$$

For the case of a pair of complex conjugate eigenvalues, the components of the fundamental matrix $\vec{x}_1(t)$ and $\vec{x}_2(t)$ can be written using eqn. (D.104) as

$$\vec{X}_1(t)=e^{\alpha t}\begin{bmatrix}a_1\cos(\beta t)-b_1\sin(\beta t)\\ a_2\cos(\beta t)-b_2\sin(\beta t)\end{bmatrix} \qquad \vec{X}_2(t)=e^{\alpha t}\begin{bmatrix}a_1\sin(\beta t)+b_1\cos(\beta t)\\ a_2\sin(\beta t)+b_2\cos(\beta t)\end{bmatrix} \tag{D.117}$$

The fundamental matrix can be generalized to n-variables as

$$X(t)=\left[\vec{X}_1(t)\,\vec{X}_2(t)\quad\cdots\vec{X}_n(t)\right]. \tag{D.118}$$

The components of the fundamental matrix $\vec{X}_1(t), \vec{X}_2(t),\cdots,\vec{X}_n(t)$, need to be modified when the matrix defective by determining the generalized eigenvectors. If $\hat{v}_{gk}$ are the generalized vectors for k=1,2,..,n, the components of the fundamental matrix will be

$$\vec{X}_k(t)=e^{\lambda t}\left[\vec{v}_{gk}+t(A-\lambda I_n)\vec{v}_{gk}+\frac{t^2}{2!}(A-\lambda I_n)^2\vec{v}_{gk}+\cdots\frac{t^{p-1}}{(p-1)!}(A-\lambda I_n)^{p-1}\vec{v}_{gk}\right]. \tag{D.119}$$

In eqn. (D.119), p is the algebraic multiplicity and $\vec{v}_{gk}$ are the solutions (generalized eigenvector) of

$$(A-\lambda I_n)^p\vec{u}=0\ . \tag{D.120}$$

Note that when algebraic multiplicity is unity, eigenvectors are distinct. In this case, only the first term in eqn. (D.119) will be retained because

$$(A-\lambda I_n)\vec{v}=0\ . \tag{D.121}$$

Note also that n-distinct eigenvectors are obtained from eqn. (D.120) with p=1.

Several examples of coupled first order equations defined by the coefficient matrix are examined as to whether the coefficient matrix is defective. As the size of the coefficient matrix exceeds 3, the number of possibilities for the relationships among the eigenvalues and eigenvectors become much more intricate and deliberate processing is needed to identify the defective matrices and determine the algebraic and geometric multiplicities of the specific eigenvalues. These examples were created in MATLAB. Starting from the coefficient matrix, the results display eigenvalues and eigenvectors, geometric and algebraic multiplicities if the matrix is defective, identify the eigenvalues associated with the generalized eigenvectors and specific algebraic multiplicity and generalized eigenvectors. Use of these generalized vectors can be seen in Chapter 5.

```
A=[4,6,6;1,3,2;-1,-5,-2];
```

Eigenvalues and Eigenvectors

Input Matrix A

4	6	6
1	3	2
-1	-5	-2

Eigenvalues

1
2
2

Eigenvectors

-1.33	-1.5
-0.33	-0.5
1	1

Eigenvalues (3) & Eigenvectors (2): Matrix DEFECTIVE:
Generalized eigenvectors required!

Eigenvectors incl. Generalized

-1.33	3	0
-0.33	1	0
1	0	1

Algebraic multiplicity (p) of the eigenvalue λ > geometric multiplicity (m)

⇒ Generalized Eigenvectors are solutions of $[(A - \lambda I_n)^p]v = 0$

λ=2: Algebraic multiplicity =2, Geometric multiplicity =1

```
A=[0 1 0 0;0 0 1 0;0 0 0 1;-1 0 -2 0];
```

Eigenvalues and Eigenvectors

Input Matrix A

0	1	0	0
0	0	1	0
0	0	0	1
-1	0	-2	0

Eigenvalues

0-1i
0-1i
0+1i
0+1i

Eigenvectors

0-1i	0+1i
-1+0i	-1+0i
0+1i	0-1i
1+0i	1+0i

Eigenvalues (4) & Eigenvectors (2): Matrix DEFECTIVE:
Generalized eigenvectors required!

Eigenvectors incl. Generalized

-3+0i	0+2i	-3+0i	0-2i
0+2i	1+0i	0-2i	1+0i
1+0i	0+0i	1+0i	0+0i
0+0i	1+0i	0+0i	1+0i

Algebraic multiplicity (p) of the eigenvalue λ > geometric multiplicity (m)

⇒ Generalized Eigenvectors are solutions of $[(A - \lambda I_n)^p]v = 0$

Complex Equal Pairs: Algebraic multiplicity =2, Geometric multiplicity =1

```
A=[0,1,2;-5,-3,-7;1,0,0];
```

Eigenvalues and Eigenvectors

Input Matrix A			
	0	1	2
	-5	-3	-7
	1	0	0

Eigenvalues		Eigenvectors	
	-1		-1
	-1		-1
	-1		1

Eigenvalues (3) & Eigenvectors (1): Matrix DEFECTIVE:
Generalized eigenvectors required!

Eigenvectors incl. Generalized			
	1	0	0
	0	1	0
	0	0	1

Algebraic multiplicity (p) of the eigenvalue λ > geometric multiplicity (m)

⇒ Generalized Eigenvectors are solutions of $[(A - \lambda I_n)^p]v = 0$

λ=-1: Algebraic multiplicity =3, Geometric multiplicity =1

```
A=[2 1 -2 1;0 3 -5 3;0 -13 22 -12;0 -27 45 -25];
```

Eigenvalues and Eigenvectors

Input Matrix A				
	2	1	-2	1
	0	3	-5	3
	0	-13	22	-12
	0	-27	45	-25

Eigenvalues		Eigenvectors		
	-1		0	1
	-1		-0.33	0
	2		0.33	0
	2		1	0

Eigenvalues (4) & Eigenvectors (2): Matrix DEFECTIVE:
Generalized eigenvectors required!

Eigenvectors incl. Generalized				
	0	0	1	0
	2	-1	0	0
	1	0	0	0.6
	0	1	0	1

Algebraic multiplicity (p) of the eigenvalue λ > geometric multiplicity (m)

⇒ Generalized Eigenvectors are solutions of $[(A - \lambda I_n)^p]v = 0$

λ=-1: Algebraic multiplicity =2, Geometric multiplicity =1

λ=2: Algebraic multiplicity =2, Geometric multiplicity =1

```
A=[1,-1;1,3];
```

Eigenvalues and Eigenvectors

Input Matrix A

1	-1
1	3

Eigenvalues

2
2

Eigenvectors

-1
1

Eigenvalues (2) & Eigenvectors (1): Matrix DEFECTIVE:
Generalized eigenvectors required!

Eigenvectors incl. Generalized

1	0
0	1

Algebraic multiplicity (p) of the eigenvalue λ > geometric multiplicity (m)

⇒ Generalized Eigenvectors are solutions of $[(A - \lambda I_n)^p]v = 0$

λ=2: Algebraic multiplicity =2, Geometric multiplicity =1

```
A=[-2 -7 -7 -5;-3 -8 -7 -7;1 1 0 1;2 8 8 6];
```

Eigenvalues and Eigenvectors

Input Matrix A

-2	-7	-7	-5
-3	-8	-7	-7
1	1	0	1
2	8	8	6

Eigenvalues

-1-1i
-1+1i
-1+0i
-1+0i

Eigenvectors

-1.1+0.3i	-1.1-0.3i	0
-0.5+0.5i	-0.5-0.5i	-1
-0.1-0.7i	-0.1+0.7i	1
1+0i	1+0i	0

Eigenvalues (4) & Eigenvectors (3): Matrix DEFECTIVE:
Generalized eigenvectors required!

Eigenvectors incl. Generalized

-1.1+0.3i	-1.1-0.3i	0	-1.5
-0.5+0.5i	-0.5-0.5i	-1	-0.5
-0.1-0.7i	-0.1+0.7i	1	0
1+0i	1+0i	0	1

Algebraic multiplicity (p) of the eigenvalue λ > geometric multiplicity (m)

⇒ Generalized Eigenvectors are solutions of $[(A - \lambda I_n)^p]v = 0$

λ=-1: Algebraic multiplicity =2, Geometric multiplicity =1

D-7 Solutions to Coupled First Order Systems Using the Concept of the Fundamental Matrix

Let us now go back to eqn. (D.106) which represents a defective matrix. The fundamental matrix X(t) can be created from the eigenvalue of 0 and the generalized eigenvectors in terms of eqn. (D.118) and eqn. (D.119).

$$\vec{X}_1(t) = e^{\lambda t}\left\{\vec{v}_{g1} + t(A-\lambda I_2)\vec{v}_{g1}\right\} = \begin{bmatrix}1\\0\end{bmatrix} + t\begin{bmatrix}2 & 4\\-1 & -2\end{bmatrix}\begin{bmatrix}1\\0\end{bmatrix} = \begin{bmatrix}1\\0\end{bmatrix} + t\begin{bmatrix}2\\-1\end{bmatrix} = \begin{bmatrix}1+2t\\-t\end{bmatrix} \tag{D.122}$$

$$\vec{X}_2(t) = e^{\lambda t}\left\{\vec{v}_{g2} + t(A-\lambda I_2)\vec{v}_{g2}\right\} = \begin{bmatrix}0\\1\end{bmatrix} + t\begin{bmatrix}2 & 4\\-1 & -2\end{bmatrix}\begin{bmatrix}0\\1\end{bmatrix} = \begin{bmatrix}0\\1\end{bmatrix} + t\begin{bmatrix}4\\-2\end{bmatrix} = \begin{bmatrix}4t\\1-2t\end{bmatrix} \tag{D.123}$$

The fundamental matrix X becomes

$$X = \begin{bmatrix}1+2t & 4t\\-t & 1-2t\end{bmatrix}. \tag{D.124}$$

The general solution to the first order coupled system represented by the coefficient matrix in eqn. (D.106) is

$$\vec{x}(t) = c_1\vec{x}_1(t) + c_2\vec{x}_2(t) = c_1\begin{bmatrix}1+2t\\-t\end{bmatrix} + c_2\begin{bmatrix}4t\\1-2t\end{bmatrix} \tag{D.125}$$

Consider the following coefficient matrix representing a set of coupled first order differential equations,

$$A = \begin{bmatrix}1 & 0 & 0\\0 & -2 & -1\\0 & 0 & -2\end{bmatrix} \tag{D.126}$$

The eigenvalues and eigenvectors are

$$\lambda_1 = 1 \quad \vec{v}_1 = \begin{bmatrix}1\\0\\0\end{bmatrix} \tag{D.127}$$

$$\lambda_{2,3} = -2 \quad \vec{v}_{2,3} = \begin{bmatrix}0\\1\\0\end{bmatrix} \tag{D.128}$$

The coefficient matrix is therefore defective with the eigenvalue of -2 having an algebraic multiplicity (p) of 2 and geometric multiplicity (m) of 1, requiring the need for generalized eigenvectors. Generalized eigenvectors are obtained as solutions of

$$(A-\lambda I_3)^p = 0 \Rightarrow (A-(-2)I_3)^2 = 0 \tag{D.129}$$

The two generalized eigenvectors are

$$\vec{v}_{g_2} = \begin{bmatrix} 0 \\ 1 \\ 0 \end{bmatrix} \quad \vec{v}_{g_3} = \begin{bmatrix} 0 \\ 0 \\ 1 \end{bmatrix}. \tag{D.130}$$

The basic solution set can now be expressed as

$$\begin{aligned}
&\vec{X}_1(t) = e^{t}\begin{bmatrix} 1 \\ 0 \\ 0 \end{bmatrix} \\
&\vec{X}_2(t) = e^{-2t}\left\{\begin{bmatrix} 0 \\ 1 \\ 0 \end{bmatrix} + t\left(A-(-2)I_3\right)\begin{bmatrix} 0 \\ 1 \\ 0 \end{bmatrix}\right\} = e^{-2t}\begin{bmatrix} 0 \\ 1 \\ 0 \end{bmatrix} \\
&\vec{X}_3(t) = e^{-2t}\left\{\begin{bmatrix} 0 \\ 0 \\ 1 \end{bmatrix} + t\left(A-(-2)I_3\right)\begin{bmatrix} 0 \\ 0 \\ 1 \end{bmatrix}\right\} = e^{-2t}\left\{\begin{bmatrix} 0 \\ 0 \\ 1 \end{bmatrix} + \begin{bmatrix} 0 \\ -t \\ 0 \end{bmatrix}\right\} = e^{-2t}\begin{bmatrix} 0 \\ -t \\ 1 \end{bmatrix}
\end{aligned} \tag{D.131}$$

The solution to the coupled first order equations represented by the coefficient matrix is

$$\vec{x}(t) = c_1\vec{X}_1(t) + c_2\vec{X}_2(t) + c_3\vec{X}_3(t) = c_1e^{t}\begin{bmatrix} 1 \\ 0 \\ 0 \end{bmatrix} + c_2e^{-2t}\begin{bmatrix} 0 \\ 1 \\ 0 \end{bmatrix} + c_3e^{-2t}\begin{bmatrix} 0 \\ -t \\ 1 \end{bmatrix} = \begin{bmatrix} c_1e^{t} \\ c_2e^{-2t} - c_3te^{-2t} \\ c_3e^{-2t} \end{bmatrix} \tag{D.132}$$

Additional examples and applications will be seen in Chapters 3 and 5.

We can use the concept of the fundamental matrix to find the particular solution to a non-homogeneous first order system,

If X(t) is the fundamental matrix, the solution given in eqn. (D.115) implies that an analogous solution to the non-homogeneous part can be obtained by replacing the constant $\vec{C}$ by a time dependent vector p $\vec{p}(t)$ as

$$\vec{x}(t) = X(t)\vec{p}(t) \tag{D.134}$$

Differentiating eqn. (D.144) and substituting in eqn. (D.133),

$$X'(t)\vec{p}(t) + X(t)\vec{p}'(t) = AX(t)\vec{p}(t) + \vec{g}(t) \tag{D.135}$$

Note that X(t) is the fundamental matrix and therefore,

$$X'(t) = AX(t) \tag{D.136}$$

Substituting eqn. (D.136) in eqn. (D.135), we have

$$X(t)\vec{p}'(t) = \vec{g}(t) \tag{D.137}$$

Taking the inverse of the fundamental matrix, eqn. (D.137) simplifies to

$$\vec{p}'(t) = X^{-1}(t)\vec{g}(t) \tag{D.138}$$

The unknown vector satisfying eqn. (D.134) is now obtained as

$$\vec{p}(t) = \int X^{-1}(\tau)\vec{g}(\tau)d\tau \tag{D.139}$$

The particular solution to the non-homogeneous differential equation is obtained from eqn. (D.134) as

$$\vec{x}_p(t) = X(t)\int X^{-1}(\tau)\vec{g}(\tau)d\tau \tag{D.140}$$

The solution in eqn. (D.140) is identified as the particular solution based on the method of variation of parameters (similarity to the method based on Wronskian seen in Chapters 2 and 4).

Suggested Readings

A number of excellent books and monographs devoted to the topic of differential equations and linear algebra might provide supplementary information beyond what is covered in this book. Additionally, several books are also available that describe the use of MATLAB® in differential equations and linear algebra. A good source of information on the application of MATLAB is available within the help window of MATLAB itself. A partial list of some of these resources is provided below for additional reading.

Butcher, J. C. 2016. Numerical Methods for Ordinary Differential Equations, Wiley.

Chatelin, F. (Ed.). 2012. Eigenvalues of Matrices: Revised Edition. Society for Industrial and Applied Mathematics.

Courant, R. and Hilbert, D. 2008. Methods of Mathematical Physics, Volume 2: Differential Equations, Wiley.

Farlow, J., Hall, J. E., McDill, J. M. and West, B. H. 2007. Differential Equations & Linear Algebra, Pearson.

Golubitsky, M. and Dellnitz, M. 1999. Linear Algebra and Differential Equations Using MATLAB, Cengage.

Hairer, E., Lubich, C. and Roche, M. 2006. The Numerical Solution of Differential-algebraic Systems by Runge-Kutta Methods. Vol. 1409, Springer.

Hunt, B. R., Llpsman, R. L., Osborn, J. E., Outing, D. A. and Rosenberg, J. M. 2012. Differential Equations with MATLAB, Wiley.

Iseries, A. 2009. A First Course in the Numerical Analysis of Differential Equations. Cambridge University Press.

Kreyszig, E. 2010. Advanced Engineering Mathematics, Wiley.

Lang, S. 2012. Introduction to Linear Algebra, Springer.

Lopez, C. 2014. MATLAB Differential Equations, Springer.

McKibben, M. and Webster, M. O. 2014. Differential Equations with MATLAB: Exploration, Applications, and Theory, CRC Press.

Polking, J. C. and Arnold, O. 2003. Ordinary Differential Equations Using MATLAB, Pearson.

Ralston, A. 1962. Runge-Kutta Methods with Minimum Error Bounds. Mathematics of Computation 16(80): 431–437.

Robinson, J. C. 2004. Introduction to Ordinary Differential Equations. Cambridge University Press.

Stanoyevitch, A. 2004. Introduction to Numerical Ordinary and Partial Differential Equations Using MATLAB, Wiley.

Tennenbaum, M. and Pollard, H. 1985. Ordinary Differential Equations, Dover.

Xie, W-C. 2010. Differential Equations for Engineers. Cambridge University Press.

Index

A

Asymptotically stable 32, 37, 38, 40–42, 88–90, 97, 242, 243, 246–249, 251, 263, 264, 268
Autonomous 1, 2, 5, 7, 31, 32, 34–36, 43, 44, 46, 77, 79
Auxiliary equation 82

B

Basic variables 421
Boundary conditions 10, 12, 16

C

Characteristic equation 2, 81–84, 86, 88, 89, 97, 99, 119, 120, 122, 124, 177, 178, 181–183, 189–191, 193, 195–200, 233, 234, 239, 240, 252, 254, 256, 261, 263, 265, 267, 298, 381
Characteristic polynomial 3, 82
Coefficient matrix 3, 85–88, 237–239, 241, 242, 245, 246, 248, 250, 252–263, 265–267, 269–273, 275, 276, 278–280, 282, 284, 285, 287, 290, 292, 293, 295, 296, 298, 382
Complex eigenvectors 238
Complex roots 34–36, 90, 261
Constant Coefficients 2, 81, 82, 97, 99, 177, 178, 181, 182, 185, 233, 236, 239, 240, 250, 268, 274, 279, 280, 298
Coupled first order 2, 3, 84, 85, 122, 177, 236, 240, 269, 274, 275, 281, 296, 325, 331, 381, 382
Cramer's Rule 3
Critical points 31, 32, 34–36, 88, 90, 97, 240–251, 253, 257, 259, 262–264, 266, 268, 274, 276, 277, 279

D

Defective matrix 239, 293
Dependent variable 1, 5, 9, 17, 31, 77, 81, 236
Dependent vectors 280
Determinant of the matrix 275
D-fields 2, 5–9, 26, 33, 35, 44, 46, 47, 77, 79
Differential Equations 1–10, 17, 18, 26, 28, 31, 32, 34–36, 44, 46, 77–79, 81–86, 88–90, 97, 98, 119–124, 126, 177–179, 181–200, 233, 234
Directional field 6, 27, 32, 34–36
dsolve(.) 2, 10, 18–26, 77, 99, 200, 298, 331

E

Echelon form 3
Eigenvalues 3, 86–88, 236–239, 241–252, 254, 256, 258, 260, 261, 263, 265–267, 269, 271, 272, 276, 278, 280–282, 284, 285, 287, 288, 290, 291, 293, 295, 298, 331, 381, 382
Eigenvectors 3, 86–88, 236–239, 248, 250, 251, 254, 256, 258, 259–263, 265, 267, 269, 270, 272, 276, 278, 280–282, 284, 285, 287, 288, 291, 293, 295, 298, 331, 381, 382
Equilibrium points 31, 32, 46, 90, 240
Equilibrium solutions 31, 44
Euler's method 3
Exact solution 10, 12, 77, 83, 121
Explicit solution 18–26, 28, 29

F

First order 1–6, 8, 9, 17, 26, 31, 32, 35, 44, 77, 78, 81, 84, 85, 89, 122, 177, 186, 187, 195, 200, 236–240, 250, 268, 269, 274, 275, 279–281, 296, 298, 325, 331, 381, 382
Forcing functions 3, 120, 179, 195, 237, 268–270, 272, 274, 275, 279–281, 296, 382
Free variables 421

G

General solution 83, 97, 127, 178, 182, 199, 200, 238–240, 256, 260, 270, 271, 281, 289, 291, 294, 298, 331
Generalized eigenvectors 87, 239, 256, 258, 260, 265, 281, 285, 287, 288, 293, 295, 298, 331, 381
Gompertz's Law 32

H

Homogeneous 2, 3, 81–85, 88, 97, 99, 119–124, 126, 127, 177, 178, 181, 182, 185, 189, 190, 192–194, 196, 200, 233, 234, 236–240, 250, 268–281, 283, 285, 286, 288, 289, 291, 294, 298, 325, 331, 381, 382
Homogeneous linear systems 236

I

Identity matrix 87, 239, 240, 293
Implicit solution 10, 18, 26, 29, 30, 45, 46, 77
Independent solutions 238, 239
Independent variable 1, 5, 9, 17, 31, 77, 81
Integrating factor 2, 5, 9, 10, 17, 77, 78
Inverse Laplace transform 84, 98, 99, 122, 126, 187, 190, 191, 193, 195, 197, 240, 253, 255, 257, 258, 262, 263, 266, 267, 269, 270, 272, 273, 284, 286, 290, 292, 295, 331

L

Laplace transforms 2–4, 81, 84, 98, 99, 121, 122, 126, 127, 177, 179–182, 185–188, 190, 191, 193, 195–197, 199, 200, 233, 235, 236, 240, 252–258, 262, 263, 266, 267, 269, 270, 272, 273, 281, 284, 286, 290, 292, 295, 296, 298, 331 381, 382
L'Hospital Rule 32
Linear 1–6, 9, 17, 77, 81, 82, 85, 97, 119, 120, 123, 126, 127, 181, 236, 238, 274
Local Truncation Error 387

M

Matrix 3, 85–88, 186–188, 237–242, 245, 246, 248–250, 252–263, 265–276, 278–282, 284, 285, 287, 288, 290–293, 295, 296, 298, 331, 382
Matrix equation 88, 188, 237, 281
Matrix inverse 419
Matrix vector equation 281
Method of undetermined coefficients 2, 119–122, 124–126, 177, 179, 185
Method of variation of parameters 2, 3, 119, 121, 123, 124, 126, 127, 177, 179, 185, 200, 268, 381
Multiplicity 185, 293, 331
Multiplicity, Algebraic 293, 331
Multiplicity, Geometric 331

N

Newton's Law 77
Node 251
Non-homogeneous 2, 81, 82, 119, 121, 122, 126, 177, 181, 185, 236, 237, 268, 269, 271, 274–280, 325, 381
Null matrix 274
Null space 421
Numerical methods 3, 177

O

Oscillatory solutions 242

P

Partial fractions 401
Particular solution 2, 3, 10, 81, 119–127, 177, 179, 181, 182, 185, 189, 191, 192, 194, 200, 233, 234, 268, 270, 271, 273, 278–281, 283, 285, 286, 289, 291, 294, 381, 382
Phase portraits 2, 3, 89, 90, 97, 177, 178, 240–242, 245, 246, 248, 250, 253, 255, 257, 258, 262, 277, 381
Pivot columns 419
Pivots 419

R

Reduced Row echelon form 419
Roots 2, 3, 34–36, 43, 44, 46, 81–84, 86–91, 97, 99, 119, 120, 122–124, 177–180, 182–185, 188–191, 193, 195–200, 233, 234, 240, 252, 254, 256, 261, 263, 265, 267, 298, 381
Row vectors 179, 182
Runge-Kutta method 3, 26, 85, 88, 122, 182, 185, 187, 195, 196, 233, 381

S

Saddle points 251
Second order differential equations 2, 81, 177, 254
Separable functions 2, 9, 18, 26, 77
Stable, unstable, semi stable equilibrium 32
Superposition principle 83
Symmetric matrix 248
System of differential equations 258
System of equations 420

T

Taylor series 385
Transpose of a matrix 417

U

Unique solution 185

V

Vectors 26, 28, 44, 85, 122, 124, 126, 179, 180, 182, 183, 189, 191, 192, 194, 199, 234, 237–240, 274, 279, 280, 281

W

Wronskian 121, 123, 126, 185, 186, 189, 190, 192, 194, 233, 234

Z

Zero matrix 86, 186, 275